National Individual Tree Species Atlas

James R. Ellenwood, Frank J. Krist Jr., Sheryl A. Romero

United States Forest Service

Forest Health Protection, Forest Health Technology Enterprise Team

The Forest Health Technology Enterprise Team (FHTET), Fort Collins, CO, was created in 1995 by the Deputy Chief for State and Private Forestry, USDA, Forest Service, to develop and deliver technologies to protect and improve the health of American forests. http://www.fs.fed.us/foresthealth/technology

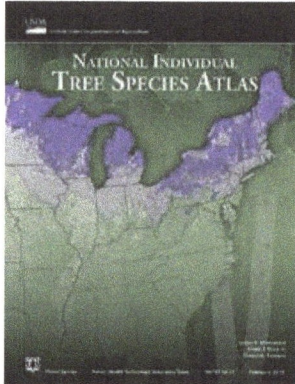

Cover design by Sheryl A. Romero, contractor for the U.S. Forest Service, Forest Health Technology Enterprise Team.

Cover Map and illustrations of quaking aspen (*Populus tremuloides*)

National Individual Tree Species Atlas. 2015, FHTET-15-01. Fort Collins, CO: U.S. Department of Agriculture, Forest Service, Forest Health Technology Enterprise Team.

An Orchard Innovations Reprint Edition
ISBN 978-1-951682-48-4

Dedication

Although it took nearly five years to produce this atlas, my love of forests and trees goes back many decades.

In 1994, as a forester on the Kaibab National Forest in Arizona, I had the good fortune to meet and observe Dr. Elbert Little in Flagstaff at a conference on piñon-juniper ecosystems. Between sessions, I introduced myself and joined him in an impassioned conversation about pine nuts—with him relating the fact that the edible piñon nut from *Pinus edulis* was far superior in flavor and nutrition to the imported, mass-marketed, store-bought, and inferior stone pine nuts (*Pinus pinea*). The conversation remained highly one-sided, and I took with me some of the passion Little had for the piñon pine. Sadly, Dr. Little passed away in 2004. He was 96.

Even before that, as a student of forestry I elected to take a course, Dendrology I, taught by Dr. Edwin Ketchledge, or "Ketch" as he was known to his students. Ketch taught the course with incredible knowledge, dedication, and enthusiasm. He was nominated and selected as a Distinguished Teaching Professor for the State University of New York. His incredible enthusiasm transcended unto his students: It has stayed with me throughout my professional career. It was his lessons that motivated me in developing this publication. I wish I could have shared this book with him, but Ketch passed away in 2010. He was 85.

This project came at an opportune moment in my career when I could address a specific need with a task that I was both passionate and enthusiastic about. Having drawn upon the energy of these two individuals, I dedicate this atlas to them. I hope for the next version of this atlas, however many years in the future it might be, that the author can ride a similar wave. --- Jim Ellenwood

Acknowledgements

For the review of species distributions—Alaska Paul Hennon, Robin Mulvey, Steve Swenson, Dustin Witwer, Ken Zogas; **Intermountain Region** Carl Jorgensen, Holly Kearns, Laura Lazarus, Daniel Ryerson; **Northeast Region** Roger Mech, Jim Steinman; **Southern Region** Carlton Cobb, Jim Meeker, Dale Starkey; **Western Region** Rod Hart, Jeff Mallory, Michael Simpson, Meghan Woods. In addition to the unnamed people they contacted for review of species distributions.

For having the vision and leadership to support this monumental effort Frank J. Sapio

For the process guidance (and keeping us based in reality) Dr. Eric L. Smith

For data development and support (the building blocks of any good geospatial product) Nathan Edberg, Aneetha Jayaraman, Arpad Lazar, and Vern Thomas

For validating the treed area layer and managing the bales of feedback Vern Thomas.

For Forest Inventory and Analysis support (without FIA, projects like this would never be possible) Elizabeth Lapoint, Andy Lister, and Richard McCullough

For doing what is necessary (it's the little things that make it just right) Mark Zweifler

For the review of the manuscript (you have made it an easy read) Mark Zweifler, Marla Downing, Cynthia Ellenwood, Andrew McMahan, and Chuck Benedict

For copy editing, photographic image support, and final preparation (it's the final touches that can make a difference) Chuck Benedict

For the final checkup (nothing goes out the door without the last thorough review) Andrew McMahan

Table of Contents

Dedication and Acknowledgements iii

List of Maps—Conifers vi

List of Maps—Hardwoods vii

Preface ix

Introduction 1

Mapped Atlas Methods. 3

   Overlays 3

   Map Preparation. 3

Modeled Atlas Methods 5

   Predictor Layers 5

      United States Geological Survey Map Zone . .6

      Climate. 6

      Terrain. 7

      Soils. 7

      Imagery 7

   Ground Data 8

   Individual Species Modeling. 8

   Validation. 9

Application of the Maps. 11

   Color Scheme. 11

   Model Fit. 12

   Leaves 12

   Other Data 12

   Specific Notes 12

      Changes to species classifications 12

      Notable differences in range 12

      Species observed 13

      Species notes. 13

   Feedback 13

Conifers. 15

Hardwoods 117

Appendices 303

   Appendix—A (Predictor Layers). 303

   Appendix—B (Species Metadata) 305

   Appendix—C (Indices) 312

   Appendix—D (Photo Credits) 316

References 319

List of Maps* – Conifers

Page	Common Name	Scientific Name	Scale
16	Pacific silver fir – Alaska	*Abies amabilis*	1:3 M
17	Pacific silver fir	*Abies amabilis*	1:4 M
18	balsam fir	*Abies balsamea*	1:10 M
19	white fir	*Abies concolor*	1:8 M
20	Fraser fir	*Abies fraseri*	1:3 M
21	grand fir	*Abies grandis*	1:6 M
22	subalpine fir – Alaska	*Abies lasiocarpa*	1:3 M
23	subalpine fir	*Abies lasiocarpa*	1:8 M
24	corkbark fir	*Abies lasiocarpa* var. *arizonica*	1:4 M
25	California red fir	*Abies magnifica*	1:4 M
26	noble fir	*Abies procera*	1:4 M
27	Shasta red fir	*Abies shastensis*	1:6 M
28	incense-cedar	*Calocedrus decurrens*	1:6 M
29	Port-Orford-cedar	*Chamaecyparis lawsoniana*	1:4 M
30	Alaska yellow-cedar – Alaska	*Chamaecyparis nootkatensis*	1:6 M
31	Alaska yellow-cedar	*Chamaecyparis nootkatensis*	1:4 M
32	Atlantic white-cedar	*Chamaecyparis thyoides*	1:8 M
33	Arizona cypress	*Cupressus arizonica*	1:3 M
34	Sargent's cypress	*Cupressus sargentii*	1:3 M
35	Ashe juniper	*Juniperus ashei*	1:6 M
36	California juniper	*Juniperus californica*	1:6 M
37	redberry juniper	*Juniperus coahuilensis*	1:6 M
38	alligator juniper	*Juniperus deppeana*	1:4 M
39	oneseed juniper	*Juniperus monosperma*	1:6 M
40	western juniper	*Juniperus occidentalis*	1:6 M
41	Utah juniper	*Juniperus osteosperma*	1:6 M
42	Pinchot juniper	*Juniperus pinchotii*	1:4 M
43	Rocky Mountain juniper	*Juniperus scopulorum*	1:10 M
44	eastern redcedar	*Juniperus virginiana*	1:12 M
45	tamarack – Alaska	*Larix laricina*	1:8 M
46	tamarack	*Larix laricina*	1:12 M
47	subalpine larch	*Larix lyallii*	1:4 M
48	western larch	*Larix occidentalis*	1:4 M
49	Brewer spruce	*Picea breweriana*	1:3 M
50	Engelmann spruce	*Picea engelmannii*	1:8 M
51	white spruce – Alaska	*Picea glauca*	1:6 M
52	white spruce	*Picea glauca*	1:12 M
53	black spruce – Alaska	*Picea mariana*	1:6 M
54	black spruce	*Picea mariana*	1:10 M
55	blue spruce	*Picea pungens*	1:6 M
56	red spruce	*Picea rubens*	1:6 M
57	Sitka spruce – Alaska	*Picea sitchensis*	1:8 M
58	Sitka spruce	*Picea sitchensis*	1:4 M
59	whitebark pine	*Pinus albicaulis*	1:6 M
60	Rocky Mountain bristlecone pine	*Pinus aristata*	1:3 M
61	knobcone pine	*Pinus attenuata*	1:6 M
62	foxtail pine	*Pinus balfouriana*	1:3 M
63	jack pine	*Pinus banksiana*	1:10 M
64	Mexican piñon pine	*Pinus cembroides*	1:4 M
65	sand pine	*Pinus clausa*	1:4 M
66	lodgepole pine – Alaska	*Pinus contorta*	1:4 M
67	lodgepole pine	*Pinus contorta*	1:8 M
68	Coulter pine	*Pinus coulteri*	1:3 M
69	border piñon	*Pinus discolor*	1:3 M
70	shortleaf pine	*Pinus echinata*	1:8 M
71	common piñon	*Pinus edulis*	1:6 M
72	slash pine	*Pinus elliottii*	1:8 M
73	Apache pine	*Pinus engelmannii*	1:3 M
74	limber pine	*Pinus flexilis*	1:8 M
75	spruce pine	*Pinus glabra*	1:6 M
76	Jeffrey pine	*Pinus jeffreyi*	1:6 M
77	sugar pine	*Pinus lambertiana*	1:6 M
78	Chihuahua pine	*Pinus leiophylla*	1:4 M
79	Great Basin bristlecone pine	*Pinus longaeva*	1:3 M
80	singleleaf piñon	*Pinus monophylla*	1:6 M
81	Arizona piñon pine	*Pinus monophylla* var. *fallax*	1:3 M
82	western white pine	*Pinus monticola*	1:8 M
83	bishop pine	*Pinus muricata*	1:4 M
84	longleaf pine	*Pinus palustris*	1:8 M
85	ponderosa pine	*Pinus ponderosa*	1:10 M
86	Table Mountain pine	*Pinus pungens*	1:4 M
87	Monterey pine	*Pinus radiata*	1:3 M
88	papershell piñon pine	*Pinus remota*	1:4 M
89	red pine	*Pinus resinosa*	1:10 M
90	pitch pine	*Pinus rigida*	1:8 M
91	California foothills pine	*Pinus sabiniana*	1:4 M
92	pond pine	*Pinus serotina*	1:6 M
93	southwestern white pine	*Pinus strobiformis*	1:4 M
94	eastern white pine	*Pinus strobus*	1:10 M
95	loblolly pine	*Pinus taeda*	1:8 M
96	Virginia pine	*Pinus virginiana*	1:8 M
97	Washoe pine	*Pinus washoensis*	1:3 M
98	bigcone Douglas-fir	*Pseudotsuga macrocarpa*	1:3 M
99	Douglas-fir	*Pseudotsuga menziesii*	1:10 M
100	redwood	*Sequoia sempervirens*	1:4 M
101	giant sequoia	*Sequoiadendron giganteum*	1:3 M
102	pondcypress	*Taxodium ascendens*	1:6 M
103	baldcypress	*Taxodium distichum*	1:10 M
104	Pacific yew	*Taxus brevifolia*	1:6 M
105	northern white-cedar	*Thuja occidentalis*	1:10 M
106	western redcedar – Alaska	*Thuja plicata*	1:3 M
107	western redcedar	*Thuja plicata*	1:6 M
108	California torrey	*Torreya californica*	1:4 M
109	eastern hemlock	*Tsuga canadensis*	1:10 M
110	Carolina hemlock	*Tsuga caroliniana*	1:3 M
111	western hemlock – Alaska	*Tsuga heterophylla*	1:6 M
112	western hemlock	*Tsuga heterophylla*	1:6 M
113	mountain hemlock – Alaska	*Tsuga mertensiana*	1:8 M
114	mountain hemlock	*Tsuga mertensiana*	1:6 M

*Map extent is within the coterminous United States unless otherwise noted after the Common name.

List of Maps* – Hardwoods

Page	Common Name	Scientific Name	Scale
118	sweet acacia	Acacia farnesiana	1:4 M
119	Florida maple	Acer barbatum	1:8 M
120	Rocky Mountain maple	Acer glabrum	1:10 M
121	bigtooth maple	Acer grandidentatum	1:8 M
122	chalk maple	Acer leucoderme	1:8 M
123	bigleaf maple	Acer macrophyllum	1:8 M
124	boxelder – West	Acer negundo	1:10 M
125	boxelder – East	Acer negundo	1:10 M
126	black maple	Acer nigrum	1:10 M
127	striped maple	Acer pensylvanicum	1:8 M
128	red maple	Acer rubrum	1:10 M
129	silver maple	Acer saccharinum	1:12 M
130	sugar maple	Acer saccharum	1:10 M
131	mountain maple	Acer spicatum	1:10 M
132	California buckeye	Aesculus californica	1:4 M
133	yellow buckeye	Aesculus flava	1:6 M
134	Ohio buckeye	Aesculus glabra	1:8 M
135	tree-of-heaven	Ailanthus altissima	1:8 M
136	Arizona alder	Alnus oblongifolia	1:3 M
137	white alder	Alnus rhombifolia	1:8 M
138	red alder – Alaska	Alnus rubra	1:4 M
139	red alder	Alnus rubra	1:8 M
140	pond-apple	Annona glabra	1:3 M
141	Pacific madrone	Arbutus menziesii	1:8 M
142	pawpaw	Asimina triloba	1:8 M
143	black-mangrove	Avicennia germinans	1:3 M
144	yellow birch	Betula alleghaniensis	1:10 M
145	sweet birch	Betula lenta	1:6 M
146	river birch	Betula nigra	1:10 M
147	paper birch – Alaska	Betula papyrifera	1:10 M
148	paper birch – West	Betula papyrifera	1:10 M
149	paper birch – East	Betula papyrifera	1:10 M
150	gray birch	Betula populifolia	1:6 M
151	American hornbeam	Carpinus caroliniana	1:10 M
152	mockernut hickory	Carya alba	1:10 M
153	water hickory	Carya aquatica	1:8 M
154	bitternut hickory	Carya cordiformis	1:10 M
155	pignut hickory	Carya glabra	1:10 M
156	pecan	Carya illinoinensis	1:10 M
157	shagbark hickory	Carya ovata	1:10 M
158	black hickory	Carya texana	1:6 M
159	American chestnut	Castanea dentata	1:8 M
160	sugarberry	Celtis laevigata	1:12 M
161	netleaf hackberry	Celtis laevigata var. reticulata	1:8 M
162	hackberry	Celtis occidentalis	1:12 M
163	eastern redbud	Cercis canadensis	1:12 M
164	curlleaf mountain-mahogany	Cercocarpus ledifolius	1:6 M
165	giant chinkapin	Chrysolepis chrysophylla var. chrysophylla	1:6 M
166	bluewood	Condalia hookeri	1:4 M
167	buttonwood-mangrove	Conocarpus erectus	1:3 M

Page	Common Name	Scientific Name	Scale
168	flowering dogwood	Cornus florida	1:10 M
169	Pacific dogwood	Cornus nuttallii	1:6 M
170	Texas persimmon	Diospyros texana	1:4 M
171	common persimmon	Diospyros virginiana	1:10 M
172	anacua knockaway	Ehretia anacua	1:3 M
173	American beech	Fagus grandifolia	1:10 M
174	Florida strangler fig	Ficus aurea	1:3 M
175	white ash	Fraxinus americana	1:10 M
176	Carolina ash	Fraxinus caroliniana	1:8 M
177	Oregon ash	Fraxinus latifolia	1:6 M
178	black ash	Fraxinus nigra	1:12 M
179	green ash	Fraxinus pennsylvanica	1:14 M
180	pumpkin ash	Fraxinus profunda	1:6 M
181	blue ash	Fraxinus quadrangulata	1:8 M
182	Texas ash	Fraxinus texensis	1:3 M
183	velvet ash	Fraxinus velutina	1:6 M
184	waterlocust	Gleditsia aquatica	1:8 M
185	honeylocust	Gleditsia triacanthos	1:10 M
186	loblolly-bay	Gordonia lasianthus	1:6 M
187	Kentucky coffeetree	Gymnocladus dioicus	1:8 M
188	American holly	Ilex opaca	1:10 M
189	Southern California black walnut	Juglans californica	1:3 M
190	butternut	Juglans cinerea	1:10 M
191	Arizona walnut	Juglans major	1:8 M
192	black walnut	Juglans nigra	1:10 M
193	white-mangrove	Laguncularia racemosa	1:3 M
194	sweetgum	Liquidambar styraciflua	1:8 M
195	yellow-poplar	Liriodendron tulipifera	1:8 M
196	tanoak	Lithocarpus densiflorus	1:4 M
197	Osage-orange	Maclura pomifera	1:10 M
198	cucumbertree	Magnolia acuminata	1:8 M
199	mountain magnolia	Magnolia fraseri	1:4 M
200	southern magnolia	Magnolia grandiflora	1:8 M
201	bigleaf magnolia	Magnolia macrophylla	1:6 M
202	sweetbay	Magnolia virginiana	1:8 M
203	Oregon crab apple	Malus fusca	1:6 M
204	red mulberry	Morus rubra	1:12 M
205	water tupelo	Nyssa aquatica	1:8 M
206	swamp tupelo	Nyssa biflora	1:10 M
207	Ogeechee tupelo	Nyssa ogeche	1:4 M
208	blackgum	Nyssa sylvatica	1:10 M
209	desert ironwood	Olneya tesota	1:4 M
210	eastern hophornbeam	Ostrya virginiana	1:12 M
211	sourwood	Oxydendrum arboreum	1:8 M
212	empress-tree	Paulownia tomentosa	1:6 M
213	redbay	Persea borbonia	1:10 M
214	water-elm	Planera aquatica	1:8 M
215	American sycamore	Platanus occidentalis	1:12 M
216	California sycamore	Platanus racemosa	1:4 M
217	Arizona sycamore	Platanus wrightii	1:4 M
218	narrowleaf cottonwood	Populus angustifolia	1:8 M

*Map extent is within the coterminous United States unless otherwise noted after the Common name.

Page	Common Name	Scientific Name	Scale
219	balsam poplar – Alaska	*Populus balsamifera*	1:8 M
220	balsam poplar	*Populus balsamifera*	1:12.5 M
221	black cottonwood – Alaska	*Populus balsamifera* ssp. *trichocarpa*	1:8 M
222	black cottonwood	*Populus balsamifera* ssp. *trichocarpa*	1:8 M
223	eastern cottonwood	*Populus deltoides*	1:12.5 M
224	plains cottonwood	*Populus deltoides* ssp. *monilifera*	1:8 M
225	Fremont cottonwood	*Populus fremontii*	1:8 M
226	bigtooth aspen	*Populus grandidentata*	1:10 M
227	swamp cottonwood	*Populus heterophylla*	1:8 M
228	quaking aspen – West	*Populus tremuloides*	1:10 M
229	quaking aspen – East	*Populus tremuloides*	1:10 M
230	quaking aspen – Alaska	*Populus tremuloides*	1:6 M
231	honey mesquite	*Prosopis glandulosa*	1:10 M
232	screwbean mesquite	*Prosopis pubescens*	1:4 M
233	velvet mesquite	*Prosopis velutina*	1:8 M
234	American plum	*Prunus americana*	1:12.5 M
235	bitter cherry	*Prunus emarginata*	1:8 M
236	pin cherry	*Prunus pensylvanica*	1:12.5 M
237	black cherry	*Prunus serotina*	1:12.5 M
238	chokecherry	*Prunus virginiana*	1:12.5 M
239	coast live oak	*Quercus agrifolia*	1:4 M
240	white oak	*Quercus alba*	1:10 M
241	Arizona white oak	*Quercus arizonica*	1:4 M
242	swamp white oak	*Quercus bicolor*	1:10 M
243	Buckley oak	*Quercus buckleyi*	1:4 M
244	canyon live oak	*Quercus chrysolepis*	1:8 M
245	scarlet oak	*Quercus coccinea*	1:8 M
246	blue oak	*Quercus douglasii*	1:4 M
247	northern pin oak	*Quercus ellipsoidalis*	1:6 M
248	Emory oak	*Quercus emoryi*	1:4 M
249	Engelmann oak	*Quercus engelmannii*	1:3 M
250	southern red oak	*Quercus falcata*	1:10 M
251	Gambel oak	*Quercus gambelii*	1:6 M
252	Oregon white oak	*Quercus garryana*	1:6 M
253	gray oak	*Quercus grisea*	1:6 M
254	silverleaf oak	*Quercus hypoleucoides*	1:4 M
255	bear oak	*Quercus ilicifolia*	1:6 M
256	shingle oak	*Quercus imbricaria*	1:8 M
257	bluejack oak	*Quercus incana*	1:8 M
258	California black oak	*Quercus kelloggii*	1:6 M
259	Lacey oak	*Quercus laceyi*	1:3 M
260	turkey oak	*Quercus laevis*	1:6 M
261	laurel oak	*Quercus laurifolia*	1:8 M
262	California white oak	*Quercus lobata*	1:6 M
263	overcup oak	*Quercus lyrata*	1:8 M
264	bur oak	*Quercus macrocarpa*	1:12 M
265	dwarf post oak	*Quercus margarettiae*	1:10 M
266	blackjack oak	*Quercus marilandica*	1:10 M
267	swamp chestnut oak	*Quercus michauxii*	1:8 M
268	dwarf live oak	*Quercus minima*	1:6 M
269	chinkapin oak	*Quercus muehlenbergii*	1:12 M
270	water oak	*Quercus nigra*	1:8 M
271	Mexican blue oak	*Quercus oblongifolia*	1:3 M
272	cherrybark oak	*Quercus pagoda*	1:8 M

Page	Common Name	Scientific Name	Scale
273	pin oak	*Quercus palustris*	1:10 M
274	willow oak	*Quercus phellos*	1:8 M
275	chestnut oak	*Quercus prinus*	1:8 M
276	northern red oak	*Quercus rubra*	1:10 M
277	netleaf oak	*Quercus rugosa*	1:4 M
278	Shumard oak	*Quercus shumardii*	1:10 M
279	Durand oak	*Quercus sinuata* var. *sinuata*	1:10 M
280	post oak	*Quercus stellata*	1:12 M
281	Texas red oak	*Quercus texana*	1:4 M
282	black oak	*Quercus velutina*	1:10 M
283	live oak	*Quercus virginiana*	1:10 M
284	interior live oak	*Quercus wislizeni*	1:6 M
285	American mangrove	*Rhizophora mangle*	1:3 M
286	New Mexico locust	*Robinia neomexicana*	1:6 M
287	black locust	*Robinia pseudoacacia*	1:12 M
288	cabbage palmetto	*Sabal palmetto*	1:4 M
289	coastal plain willow	*Salix caroliniana*	1:6 M
290	black willow	*Salix nigra*	1:12 M
291	western soapberry	*Sapindus saponaria* var. *drummondii*	1:6 M
292	sassafras	*Sassafras albidum*	1:10 M
293	American mountain-ash	*Sorbus americana*	1:10 M
294	American basswood	*Tilia americana*	1:12 M
295	Chinese tallowtree	*Triadica sebifera*	1:8 M
296	winged elm	*Ulmus alata*	1:10 M
297	American elm	*Ulmus americana*	1:12 M
298	cedar elm	*Ulmus crassifolia*	1:6 M
299	slippery elm	*Ulmus rubra*	1:12 M
300	rock elm	*Ulmus thomasii*	1:10 M
301	California laurel	*Umbellularia californica*	1:6 M

*Map extent is within the coterminous United States unless otherwise noted after the Common name.

Preface

As often happens during the course of developing something new, this *National Individual Tree Species Atlas* took shape within the context of another project. This came about while developing the *2013–2027 National Insect and Disease Forest Risk Assessment* (NIDRM, Krist et. al. 2014). My colleagues and I discovered we needed to create what would ultimately be a considerable refinement to the known extent of each major individual tree species in the United States. Specifically, not only would we need to identify precisely where each species of tree is likely to grow, but also where it *is not likely to grow*. We further discovered we were not the only people who could benefit from this new precision. Silviculturists, foresters, geneticists, researchers, botanists, wildlife habitat biologists, landscape ecologists—essentially anyone involved in natural resources management, monitoring impacts of climate change (Ellenwood et. al. 2012), or simply visiting America's forests—also would benefit. Hence, this *Atlas*.

We began by consulting what was at the time the most definitive source for the geography of trees, the *Atlas of United States Trees* series (Atlas), by U.S. Forest Service Chief Dendrologist Dr. Elbert J. Little. Published in six volumes between 1971 and 1981, this landmark compendium became our "base of operations" in 2009. During the years leading up to its publication, Little filed through untold thousands of pages of information. He understood the relationship between tree species distributions and environmental conditions. Yet, limited as he was by the technology of a paper map, he knew too much information printed on a single plate (map) would not be useful. His solution to this problem was quite clever. He separated the information that portrayed the natural range of tree species' distributions from information that represented the environmental conditions, such as rivers, lakes, and plant hardiness zones. The natural range of individual tree species became the maps bound in his Atlas and the environmental conditions were printed separately on nine, nearly transparent vellum sheets (overlays), found in an envelope at the back. By laying a vellum over a map, the reader could make a visual comparison and correlation between the natural range of a tree species and the environmental condition as defined on that overlay. Thus, in all cases, it was up to the reader to establish a relationship (if any) between an individual tree-species distribution on the map and what was depicted on a vellum overlay. By today's standards the overlays might seem primitive, but they were the best technology available at that time.

There were other shortcomings, as well. Early on, Little recognized that some of the detailed records he produced lost their precision when transferred to the coarse-scaled maps (1:10,000,000 U.S., 1:27,000,000 North America) in his Atlas. As well, he took liberal artistic license when he "connected the dots" between points around an area, within which a tree species was known to exist.

Fast forward to 2009. We could not afford to take the artistic license with the NIDRM project that Little took with his Atlas. More to the point, neither the data Little used nor the maps he produced were dependable or accurate enough for us to use in the NIDRM project. Fortunately, we had access to advanced technology with which to acquire, process, and produce more-precise data. Using data from the Forest Service's Forest Inventory and Analysis program (FIA), and from a predictor dataset consisting of climate, terrain, soils, and satellite imagery, we developed unique statistical models that predicted the spatial distribution for each of the individual tree species in the United States, as measured by FIA. Using these models, we were able to represent precisely where individual species were likely to occur. In turn we were able to use this information to make reliable assessments of risks of mortality, due to insects and diseases, faced by each tree species. As a basis for comparison, we include Little's original mapped distributions on our maps. We have no doubt that Little would have used geospatial models such as ours in the 1970s, had they been available. They simply weren't.

Some might think it odd that, at this point in history, 2014, we would produce an atlas in print form. With electronic media being firmly established in the daily lives of most citizens, a digital version might seem adequate to the task. We don't dispute that assessment. However, as we look forward to a point in time that is long after the technologies used to create and assemble both the electronic and print versions of this *Atlas* have been replaced, the print version will still be on the shelf in many libraries, accessible by anyone. What is more, this very well could be the last atlas of its kind ever printed. Time will tell.

To accommodate the need for electronic media, we have made versions of this *Atlas* available electronically. There are several digitized versions of Little's Atlas online and there are (at least) two versions of this *Atlas* available as well. One version is a static Adobe® Portable Document Format (pdf) file, nearly identical to this print version, found online (http://www.fs.fed.us/foresthealth/technology/remote_sensing.shtml). The other version is an interactive online mapping version (http://foresthealth.fs.usda.gov/host). With this version, you can view, download, incorporate other data, and conduct analysis of what you can see in this printed version of the *Atlas*, but at a much finer detail than with any other version. You will find the on-line, interactive *Atlas* more functional than an ordinary paper atlas. I invite you to access and examine both on-line versions and welcome your feedback.

Inevitably, the information printed on these pages will become obsolete, at which time this printed *Atlas* will become just a physical point in the historical record—a baseline from which anyone interested in where the trees were in the United States at the turn of the 21ˢᵗ Century can begin to look. To that end, it's how I began both this project and my lifelong career in forestry.

Suffice it to say that the maps we present here and online are as precise as we could make them. So as this publication goes to press, and with much respect and many thanks to Dr. Little and others both before and after him, we believe our *National Individual Tree Species Atlas* belongs on the same shelf with Little's *Atlas of United States Trees*.

Jim Ellenwood

National Program Manager for Remote Sensing and Image Analysis

Forest Health Technology Enterprise Team, Forest Health Protection

USDA Forest Service

March, 2014

Introduction

For over four decades the *Atlas of United States Trees* series (Little 1971, Viereck and Little 1975, Little 1976, Little 1977, Little 1978, Little 1981; henceforth referred to as the *Mapped Atlas*), has been a key reference for tree-species distributions. However, due to inherent mapping inaccuracies and the spatial coarseness of the maps, we could not use the *Mapped Atlas* to produce the *2013–2027 National Insect and Disease Forest Risk Assessment* (NIDRM, Krist et. al. 2014). Furthermore, it was clear that many other applications and analyses requiring accurate tree-species information could benefit from a more current, spatially refined dataset. And so the task was launched to create an updated Atlas, one with better spatial accuracy and finer spatial resolution. The result is this *National Individual Tree Species Atlas*, henceforth referred to as the *Modeled Atlas*.

The original *Mapped Atlas* included six volumes and featured over 700 species, many of which are mapped to just states or counties. This *Modeled Atlas* includes 264 species which are mapped to more precise locations.

The methods Little used to develop the *Mapped Atlas* (see, "*Mapped Atlas* Methods") built upon the methods used by others and were derived from recorded observations, county botanical records, and established species distributions. Objective as they might seem, often such records do not fully represent the actual distribution of a species because it is observed only sporadically throughout its actual range. Therefore, Little and his predecessors necessarily took artistic license to "connect the dots" around locations where a tree species was observed, and considered the species present within the entire delineated area. This process had inherent inconsistencies between observers, between species, and even between different areas within a given species range.

Information and technology has advanced considerably since the *Mapped Atlas* was published. The installation of a nationwide inventory of forestlands, together with the collection of climate, terrain, and remotely sensed data at the same national scale, provided an opportunity to apply a new statistical approach for estimating the potential distribution of individual tree species. Unbiased observations from the Forest Inventory and Analysis (FIA) permanent plot data provided a complete national representation from which we were able to determine not only where a specific species occurs, but also *where it does not occur*. Using these objective observations of tree species occurrence (presence or absence), we could use other data (see "Predictor Layers," page 6, and "Appendix–A") to statistically model the distribution of individual tree species. In turn, these modeled species distributions provide us with a spatial refinement that greatly improves upon the *Mapped Atlas*.

There are significant differences between Little's atlas and ours. There are several reasons for this, but putting reasons aside, the major differences between them are (1) the amount of bias in the samples (tree species observations) used to build the maps, and (2) assumptions about what exists between points of observation. For the *Mapped Atlas*, a given species-range polygon is delineated around individual observational data points derived from un-systematic methods. Statistically speaking, these observations are generally considered "biased". Thus, a polygon on a map in the *Mapped Atlas* is biased towards delineating areas where we know the species to be present and away from other places where the species might be present, but for which we don't have an observation. Also, the species-range boundary surrounding the points embodies the assumption that the species occurs everywhere inside the polygon. We know from personal observations that this assumption is not always true.

We do not make such assumptions in the *Modeled Atlas*. With the *Modeled Atlas*, the tree species observations are measured on a systematic, unbiased installation of FIA plots. Predictor variable conditions are obtained for the FIA plot locations, and relationships between predictor variables and species presence are derived via sophisticated modeling techniques described in detail, below (see "*Modeled Atlas* Methods"). These relationships are then applied to all areas to predict tree species presence for all areas "in-between" measured observations. Ultimately, what this means is the *Modeled Atlas* is much more precise than the *Mapped Atlas*.

Other reasons for differences between the *Mapped Atlas* and the *Modeled Atlas* include actual locational differences in tree species presence versus the original observation (some were made as early as 1885), mapping errors in the original dataset, survey errors, and errors developed in the modeling process. In addition, the modeling process does not distinguish between "native" species distributions and observed distributions of species planted outside of their historical range.

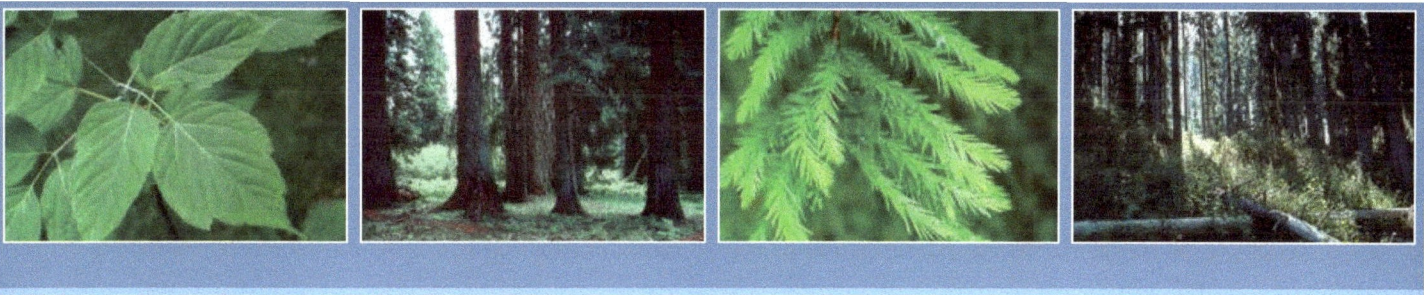

Mapped Atlas Methods

Until now the *Mapped Atlas* was the most current atlas in a long line of tree atlases of the United States dating back to 1885 (Sargent 1884). Since then, each new effort relied on a compilation of individual accounts of tree species taken from known survey records, such as county botanical records, professional observation records, and established species distributions. The *Mapped Atlas* was intended to depict the pre-Euro-American settlement extents of individual tree species and are predicated on the assumption that post-Euro-American, human-influence changes in range limits either can be determined from known activities, or are negligible in the amount of change to the species range. Tree ranges of Canada and Mexico were also included.

OVERLAYS

Little included with his *Atlas* nine transparent vellum overlays (factors or predictor variables) kept in a sleeve inside the back cover. The overlays included rivers and natural lakes, physical subdivisions or Land Surface Forms, topographic relief, plant hardiness zones, lengths of growing seasons, precipitation, climates of the United States (as a precipitation effectiveness index), maximum extent of glaciation in the Wisconsin glacial stage (Pleistocene Epoch), and major forest types. The reader placed an overlay over a map and deduced the relationship (if any) between the species on the map and the physical attribute depicted in the overlay.

MAP PREPARATION

During the years leading up to the *Mapped Atlas* publication, Little and his associates painstakingly amassed, sorted, and analyzed thousands of pages of printed information from both published and unpublished reports, and supplemented this with direct field observations in several states. They gathered additional information from state-published distribution maps; state, county, and local herbaria records; published forest survey distribution maps of commercial forests; and unpublished forest survey records by counties. They sifted through the records and transferred the data cartographically to a 1:10,000,000-scale base map of the United States, or a 1:27,000,000-scale base map of North America. During the process, Little recognized that some of the detailed records lost their precision when transferred to such a coarse-scaled map. Terrain-driven distributions were difficult to convey at such a small scale.

Compiling the data was a long process, during which Little consulted over 300 references in a multitude of formats. He observed during the process that the most reliable sources of information for species distributions were published taxonomic monographs and published accounts of single species. Unfortunately, these sources covered only a few tree species. The next-best source was state-specific publications on trees and plants. Herbariums specimens came next.

Sample criteria varied among all the sources. Conflicting information sometimes appeared. Furthermore, the taxonomic classification system was dynamic, so scientific names of species changed from time to time. Little noted that, in order to be useful, the process of aggregating all this material required considerable maintenance and regular updates. The maps were compiled through the years along with other work by Little and an assistant. Dots were connected, interpolated into polygons, and traced onto paper maps.

Elbert Little's Atlas of United States Trees, Volume 1.

Modeled Atlas Methods

We developed individual tree-species datasets for use in the 2013–2027 NIDRM assessment (Krist et. al. 2014). These datasets included estimates of tree species presence, size-weighted density (as basal area in square feet per acre), and stand density index (as number of equivalent 10-inch diameter trees per acre). The *Modeled Atlas* presents our estimates of tree species occurrence for 264 of the 387 tree species codes sampled by FIA. The remaining 123 tree species codes not included in the *Modeled Atlas* are listed in Appendix–B1. These consist of species, genera, families, and broad classification identifiers. The datasets were modeled from ground plot data measured by FIA and from predictor datasets consisting of climate, terrain, soils, and satellite imagery, described in detail below (see "Predictor Layers," page 6), and summarized in Appendix–A.

Each individual tree species was statistically modeled (see "Individual Species Modeling," page 8). Our approach used data-mining software (See5, Quinlan 2012) and an archive of geospatial information to find and model the complex relationships between independent predictor layers and the occurrence of individual tree species, as measured on over 300,000 FIA plot locations. We then applied the resultant **statistical models** to the predictor layers to generate **geospatial products** (such as Geographic Information System [GIS] raster layers and the mapped distributions you see in this *Atlas*) representing tree-species presence for the entire coterminous United States and Alaska. Post-processing techniques were performed to eliminate isolated predictions, fill-in isolated absences, limit the species distribution to treed areas, and limit the species distribution to known species occurrence from Little (1971) or sampled areas (FIA).

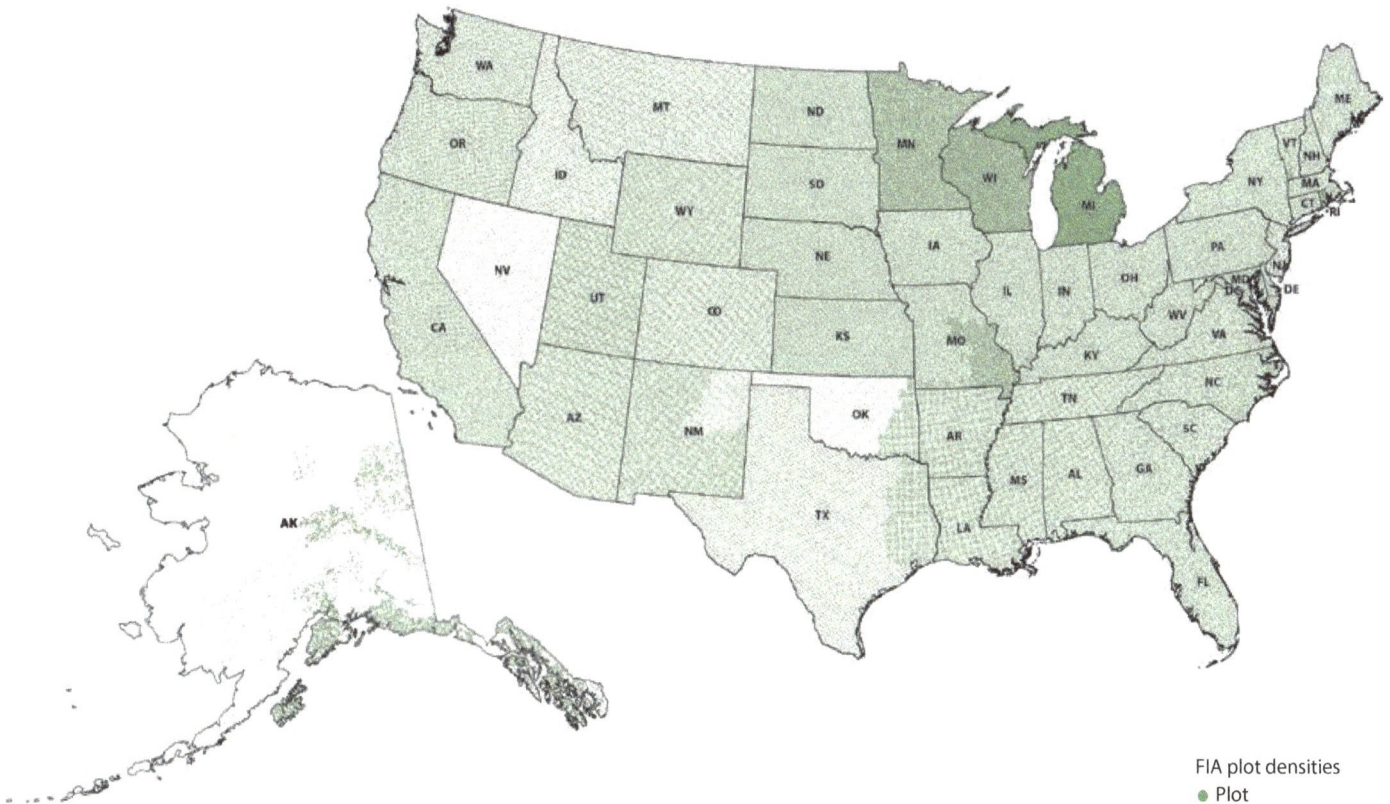

FIA plot densities
● Plot

Distribution of FIA plot data used to create tree species parameters.

PREDICTOR LAYERS

The occurrence of an individual tree species within any given area depends upon a number of factors related to the environmental conditions, cultural practices, and the biogeography of the species (Ellenwood 1984). The approach for many species distribution models has been to model representations of these factors. A number of different predictive techniques and variables have been analyzed, and few standard modeling approaches have been accepted (Austin 2007, Elith and Graham 2009).

For comparison, a predictor layer is to *the Modeled Atlas* what a vellum overlay is to the *Mapped Atlas*. Predictor layers used in the *Modeled Atlas* were derived from four national datasets and created at a 30-meter resolution (Appendix–A).

1. National Climate Data Center U.S. standard normal climate data for 7,937 climate stations in the US (NOAA-NCDC 2002).

2. National Elevation Dataset (Gesch et al. 2009).

3. USDA Natural Resource Conservation Service (NRCS) soils dataset, Soil Survey Geographic Database (SSURGO), and a regionalized dataset, STATSGO2 (Digital General Soil Map of the United States) (NRCS 2011).

4. Multi-Resolution Land Characteristics Consortium (MRLC) Landsat imagery used for the National Land Cover Database (NLCD) dataset (Homer 2004).

UNITED STATES GEOLOGICAL SURVEY (USGS) MAPZONE

The NLCD dataset is organized by USGS mapzone. The mapzones roughly follow ecological zones and their use aims to group ecologically common areas, add efficiency to the mapping process, and produce manageable file sizes. Each of the predictor datasets was organized by mapzone and the **geospatial products** were aggregated to the coterminous United States or Alaska. There are 66 USGS mapzones in the coterminous United States and 12 mapzones in Alaska. In Alaska, two zones and a portion of a third zone were not modeled, due to the limited amount of treed land (crosshatched zones in map below).

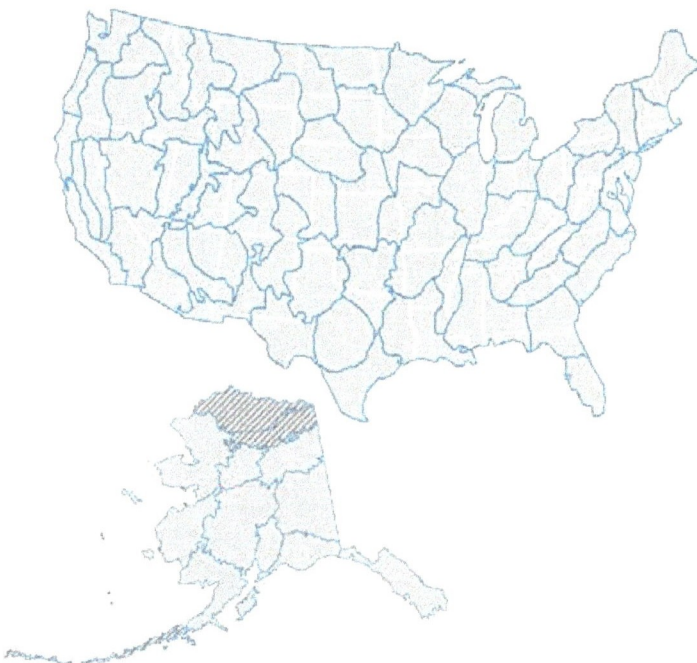

USGS mapzones.

CLIMATE

We spatially modeled monthly climate normals (three-decade averages of climatological variables) for the period of 1971 to 2000. We used a statistical technique from ANUSPLIN (Hutchinson 1991) to create 30-meter resolution data for precipitation, average mean temperature, average maximum temperature, and average minimum temperature. To produce 12 monthly layers for each of the four climate variables (48 total), we used a technique known as thin-plate spline regression, which closely follows the techniques developed by Rehfeldt (2006), on the climate variables extracted from the monthly normals from 7,939 stations in the CLIM81 30-year climate-normal dataset (NOAA-NCDC 2002).

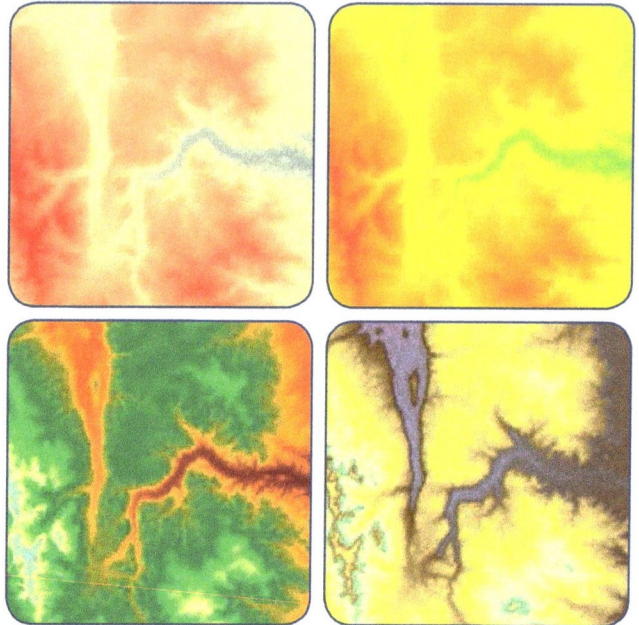

Example climate layers [from upper-left, clockwise]: total average temperature degree-days below 0 °C, total minimum temperature degree-days below 0 °C, seasonal moisture index, mean annual temperature.

We separated the monthly station normals into two sets: one for the coterminous United States and the other for Alaska. For the coterminous United States, regression splines were built for precipitation from 7,467 stations. For the three temperature variables (monthly average mean, monthly average maximum, and monthly average minimum), regression splines were built using 5,332 stations. For Alaska, regression splines for precipitation were built from 124 stations, and regression splines for the three temperature variables were built using 119 stations. For the Southeast Alaska panhandle, the narrow nature of the landform in relation to the climate data stations proved to be problematic when creating splined surfaces that reflect expected climate surfaces. So, Rehfeldt's (2006) spline models for western North America were utilized for this area.

The modeled regression splines were applied to the 30-meter National Elevation Dataset (NED) to produce 12 monthly rasters for each of the four climate variables, for a total of 48 climate layers. From the 48 climate layers, 18 generalized climate variables were derived utilizing techniques from Rehfeldt (2006). Due to model over-fitting, 13 of the 18 generalized variables were removed from the **statistical models**, leaving a subtotal of five climate variables. Two additional climatic predictor variables were derived by creating a principal components analysis of the 12 monthly precipitation variables, with the first two principal components being utilized for the modeling. The resultant seven climate variables are

1. the maximum temperature in the warmest month (tenths of degrees C),

2. growing season precipitation (during the frost-free months, inclusive, mm),

3. degree-days below 0 °C,

4. annual moisture index (the ratio of variable degree-days greater than 5 °C to mean annual precipitation),

5. seasonal moisture index (the ratio of degree-days greater than 5 °C accumulating within the frost-free period to growing season precipitation),

6. the first principal component of the average monthly precipitation, and

7. the second principal component of the average monthly precipitation.

Climate is a broad characterization of conditions over a long period of time and over a large area. Therefore, it is difficult to represent with any degree of certainty the local conditions at the time of plant germination and establishment. Climate at the fine scale is considered micro-climate and can be better represented with terrain variables. Terrain can be considered an extension of the climate layers because, depending upon the steepness and direction of terrain or the spatial relationship with surrounding terrain, the effect of terrain on localized weather can be substantial.

TERRAIN

We utilized elevation data and slope position data from the NLCD dataset. Other terrain variables were derived from the elevation data using both standard spatial routines and some customized routines. Standard routines were utilized for curvature and slope-position index. Aspect was represented via two variables using slope-weighted sine and cosine of azimuth (for north-south slope and east-west slope, respectively) to allow for the linear modeling of an otherwise non-linear

variable. Direct and indirect solar radiation variables were derived from latitude, elevation, slope, and aspect using techniques developed by Kumar and corrected by Zimmerman (Kumar et. al. 1997, Zimmerman et. al. 2007). Topographic scale calculates topographic features (ridge, slope, toe slope, etc.) at various spatial scales and hierarchically integrates these features into a single grid. The resulting grid displays the most extreme deviations from a homogenous surface from the various spatial scales. For the *Modeled Atlas*, topographic scale was computed for three spatial windows of 150-, 300-, and 450-meter radii and combined into a single predictor variable. In addition, the elevation, slope-weighted bi-directional aspect, slope-position index, curvature, direct and indirect shortwave radiation brings to eight the total number of terrain predictor layers used in the selected models.

SOILS

We extracted soils data from the SSURGO and STATSGO dataset and utilized them in the initial analysis. Two layers were created, one for drainage index (Schaetzl et al. 2009), and another for forest productivity index (Schaetzl et al. 2012). These two variables proved to be significant for many species (see Appendix–B2 and B3); however, they were problematic in developing the **geospatial products**, so they were not utilized in the **statistical models.**

IMAGERY

We utilized Landsat imagery from the NLCD 2002 project which consists of six-band imagery from the visible, near-infrared, and shortwave infrared spectrums. The NLCD project prioritized Landsat 7 cloud-free imagery collected between 1999 and 2003. Imagery gaps were filled with imagery from Landsat 5. For the coterminous United States, three-season imagery was collected between 1999 and 2004. Two-season imagery was utilized for most of Alaska, but in two of the Alaskan mapzones single-season imagery was utilized. All of the Landsat images in Alaska were collected between 1994 and 2006. The tree-species models were built with Landsat bands 1, 4, 5, and 7, and tasseled-cap transformed greenness

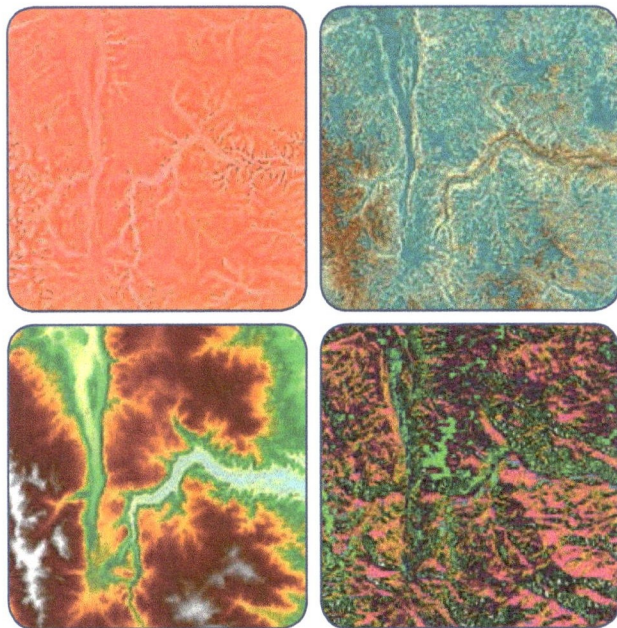

Example terrain and soil layers [from upper-left, clockwise]: topographic scale, soil drainage index (modeled), Landform, Digital Elevation Model.

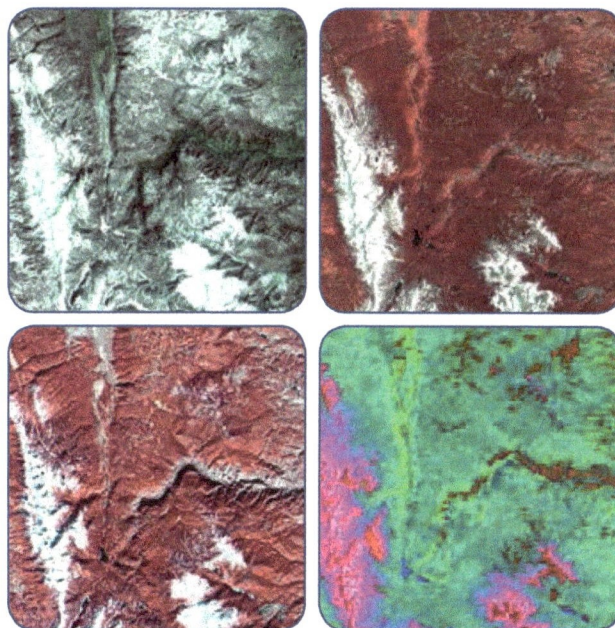

Example imagery layers [from upper-left, clockwise]: Spring Landsat (bands 4,3,2), Leaf-on Landsat (bands 4,3,2), Tasseled-cap Transformed Leaf-on Landsat (Bands W,G,B), Leaf-off Landsat (bands 4,3,2).

bands (Appendix–A). Texture was computed for an alternative vegetative index (bands 5 and 2) using a seven-pixel kernel. The combination of the four bands, one tasseled cap band, and one texture band for each of the three seasons brings the total number of imagery predictor layers to 18.

GROUND DATA

We utilized Forest Inventory and Analysis (FIA) ground sampled plot data (Woudenberg et al. 2010) to determine the occurrence of individual tree species at FIA subplot locations. Tabular plot and tree data records were extracted from the FIA database (extracted 2/2011). Plot data were limited to state inventory cycles that were aligned most closely with imagery dates, and generally ranged from 1999 to 2005, with some plots in the western United States sampled as recently as 2009. Approximately 80% of the FIA plots were sampled within five years of the corresponding Landsat imagery collection date. Tree data records were limited to live trees of 1-inch diameter or greater (measured at breast height, 4.5 feet above ground [DBH]). For Alaska, annualized FIA inventories were limited to south-central and southeast-panhandle areas, from 2004 to 2009.

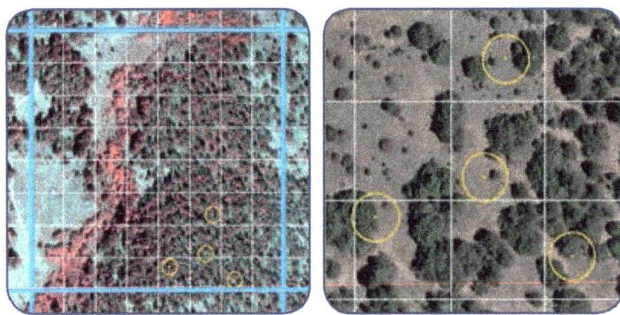

*A perturbed FIA subplot cluster (**set of 4 circles**) is super-imposed upon a high resolution image with a 240-meter grid (**blue square**) [Left] and 30-meter grid [Right] to illustrate the spatial relationships among the ground data, the predictor layer ground resolution [30m], and the Mapped Atlas product [240m].*

We improved spatial coverage of tree occurrence data by supplementing the FIA datasets described above with additional inventories. These included

- historical FIA plots from FIA-Alaska for interior Alaska collected from 1968 to 1991 in limited areas,
- intensified grid plots for U.S. Forest Service Pacific Northwest (WA and OR) and Pacific Southwest (CA) Regions, and the Bureau of Land Management (BLM) Oregon State Office, and
- the Great Plains Initiative (GPI) non-forestry inventory (Lister et al. 2012).

For the Pacific Northwest and Pacific Southwest Regions and BLM plots, locations coincidental to FIA locations were eliminated. For the Great Plains states, the Great Plains Initiative (GPI) dataset utilized single plots, while the other non-FIA datasets used a 5-point, cluster-plot design installed on an intensified grid.

FIA plots are composed of four subplots. Subplot coordinates were computed from actual plot coordinates, based upon the FIA plot design and the magnetic declination at the time the subplots were installed. Each subplot was used as a sample of the predictor layers. For each tree species modeled, all subplots within each mapzone were used to predict species presence if any subplot within the mapzone contained the species. Accordingly, the number of samples used to

model each species varied based upon the occurrence of a species across mapzones. Note that the pool of samples used to model each species contains both *presence samples* (plots containing the species) and *absence samples*.

INDIVIDUAL SPECIES MODELING

We linked subplot coordinates to predictor layers to create sample data files for statistical modeling. **Statistical models** were developed for each individual species by USGS mapzone and are referred to as "zonal-models." In addition, "global models" were built for each individual species using all mapzones. For each species/mapzone combination, **geospatial products** were built from the **statistical models**.

The statistical modeling involved a progression of trial and error processes. Early iterations in the modeling process utilized all 89 predictor variables, and many of the **statistical models** created **geospatial products** that were less than adequate. To help reduce the number of predictor variables, a single model iteration was created with less-redundant predictor layers to determine which variables were significant in modeling individual species distributions (Appendix–A, column e). This was utilized to select a consistent, narrower, predictor-layer dataset and reduce overall processing time. Predictor variables which produced significant over-fitting for a majority of the species were dropped from subsequent models. The final iteration produced zonal-models which utilized 33 predictor variables (Appendix A – column i). Zonal-models were also developed for the treed area and utilized 35 predictor variables (Appendix A – column j). The **geospatial products** created from the treed area zonal models were used as a tree "mask" to limit all individual tree-species distributions.

We used See5 (version 2.06 from Quinlan, 2012), a classification and regression tree modeling method, to independently model occurrence of live tree basal area for trees over 1-inch DBH. The See5 models were built utilizing "boosting" and "costing" functions. For poor models, these functions were adjusted to improve model outputs.

For a zonal-model that did not produce a valid **statistical model** or a valid **geospatial product**, the global-model was utilized instead for that mapzone. Two "flavors" of global models were built: one with USGS mapzone as a predictor variable and one without (Appendix A – column k and column g, respectively). If the global model with the mapzone variable did not produce a valid **statistical model** or a valid **geospatial product**, then the global model without the mapzone variable was utilized for that species mapzone.

The number of samples is critical if data mining software is to be successful. Morgan et. al. (2003) demonstrated sample sizes of 10,000 or more were adequate for data mining models; however, 3000 seems to adequately represent many applications. To give the user an idea of the adequacy of the data, Table 1 summarizes, for all 387 FIA-species models, the number of species by class of sample size used to model species occurrence. What Table 1 shows is that for a majority of the species, there may be a less-than-adequate number of samples (fewer than 3000). Generally speaking, the more samples the better the model. Infrequent species have fewer plots and would be expected to produce poorer models (Appendix B1).

The selected species-mapzone model built from the **statistical models** were converted to a raster surface using the RSAC Cubist/See5 toolset (Ruefenacht et al. 2008). All individual tree-species presence maps were generated at a 30-meter resolution. The raster surfaces were filtered to fill in isolated holes and buffered by two pixels (60 meters) to be more inclusive.

Table 1 - Summary of plots per species.

Summary of Sample Intensity per Species - All Species Codes						
Number of samples	Number of species within subplot sample group	Number of species within subplot sample group (presence only)	Number of Atlas species within plot sample group	Number of species within subplot sample group for Alaska	Number of species within subplot sample group (presence only) for Alaska	Number of Atlas species within plot sample group for Alaska
3000+	359	113	59	18	4	
300+	28	102	104	1	9	8
30+		80	68		2	5
3+		66	33		3	3
<3		26			1	
Total	**387**	**387**	**264**	**19**	**19**	**16**

Total subplots includes both total (presence + absence) subplots as well as presence-only subplots used by the statistical models. In addition, the numbers of presence-only plots used for the sensitivity validation are summarized for the 264 Modeled Atlas **geospatial products**.

The 30-meter outputs were resampled to 240-meters and buffered by four pixels to facilitate their observation at a national scale. Contiguous pixels were clumped and compared to known inventory plots and to the *Mapped Atlas*. Clumped areas outside a frequency-based distance from known locations were eliminated. The **geospatial product** depicts the remaining buffered areas.

VALIDATION

Validation is defined as the process of ensuring that a product conforms to a user's specific need or requirement. The initial objective of this project was to identify the distributions of individual or groups of tree species in order to map their potential risk of mortality from insect and diseases.

We utilized over 1.2 million subplots to model individual species and produce the *Modeled Atlas*. A comparable dataset would be needed in order to conduct a scientifically valid accuracy assessment; ultimately, the cost of producing such a dataset is prohibitive. Instead, we used the plot version of the ground dataset as a validation. This gives the reader an idea of how well the *Modeled Atlas* represents what is on-the-ground.

The validation involves the use of the original FIA plot data (with the subplot components), rather than the disaggregated FIA subplot data.

Whereas the **statistical models** we developed used the individual subplots as a plot-independent sample, the validation data utilized the subplots as a *unit of the plot*. That is, if a species was modeled as being present on any of the subplots composing a plot, and the species is actually present (in the FIA sample) on any of the subplots composing the plot, then the plot was considered in agreement. Likewise, if both the model and the inventory indicate a species absence, then the model is considered in agreement. In addition, two types of disagreement can occur. The species may be modeled as being present on a plot and the inventory may indicate its absence; conversely, the species may be modeled as being absent but the inventory may indicate its presence.

A statistical classification function, termed "sensitivity," was used to determine the percentage of agreement between the modeled extent and the FIA field plots. A high percentage would be considered good. Sensitivity was calculated for both the statistical model (global model only) and the **geospatial product** and is summarized in Appendices B2 and B3 in columns "Global" and "Validation," respectively. The statistical model sensitivity can give a sense of how strong the data were in producing the species-occurrence relationship, while the modeled geospatial extent product sensitivity reflects the soundness of the **geospatial product**.

Application of the Maps

COLOR SCHEME

What the *Modeled Atlas* does well is enable the reader to quickly differentiate the previously *mapped* species extent (*Mapped Atlas*) from the recently *modeled* species extent (*Modeled Atlas*). We use contrasting colors to highlight agreement and differences. A color contrast is used in the underlying base map to differentiate treed areas (darker) from non-treed areas (lighter) (item 1). Each map features the *Modeled Atlas* extent (item 2) for the given species and the *Mapped Atlas* extent (item 3) (if previously mapped).

The color range varies with relief to illustrate the underlying terrain. A blended color indicates where the two extents overlap (item 4). For areas where the *Modeled Atlas* extent exceeds the *Mapped Atlas* extent, approximate locations of plots having the species are shown (item 5). The plot locations are based upon perturbed FIA plot coordinates for the coterminous United States and LANDFIRE cover plots for Interior Alaska (LANDFIRE 2011).

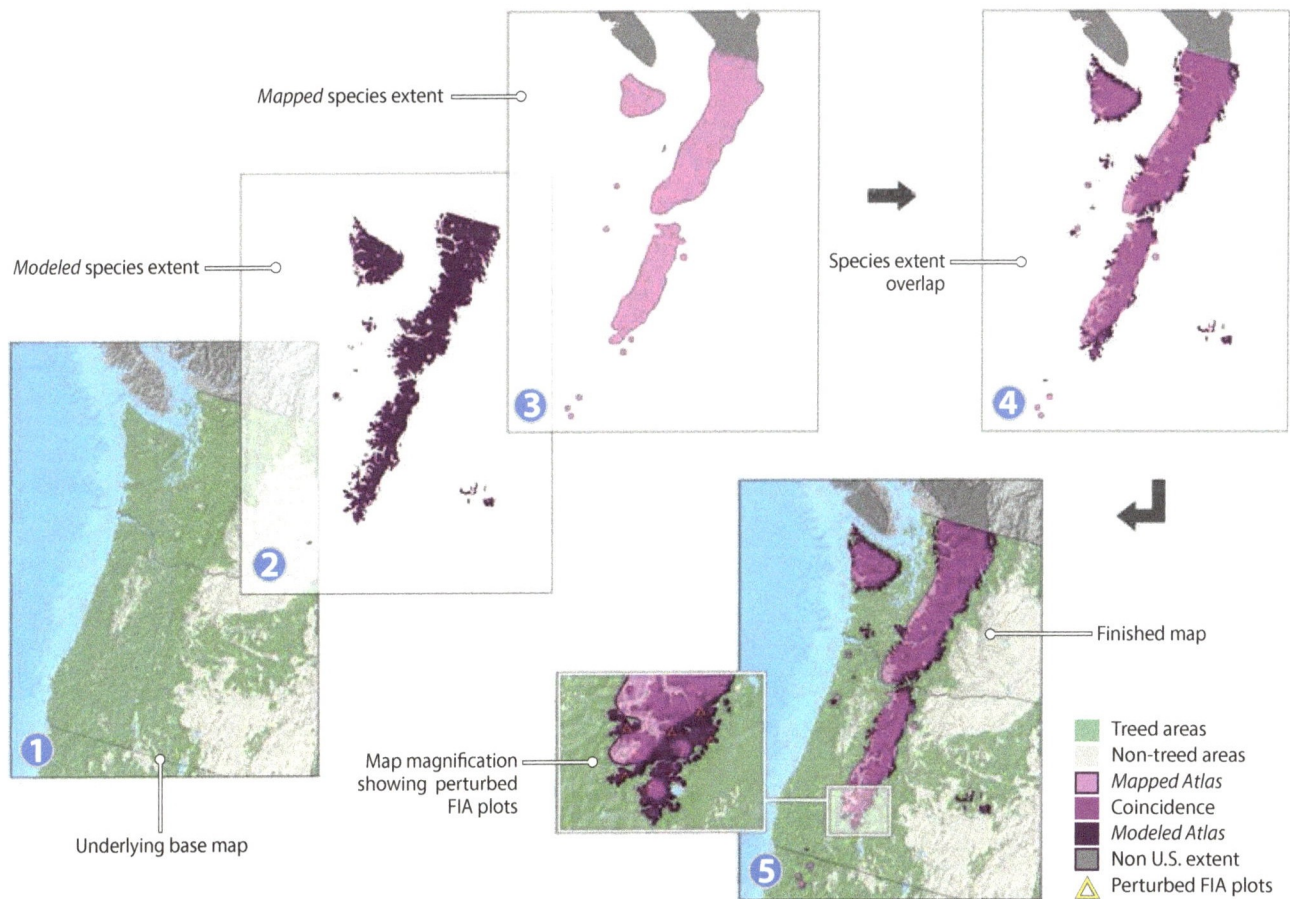

Color scheme layers used to create the Modeled Atlas.

Mapped species extent — ③

Modeled species extent — ②

Species extent overlap — ④

① Underlying base map

Map magnification showing perturbed FIA plots — ⑤

Finished map

Treed areas
Non-treed areas
Mapped Atlas
Coincidence
Modeled Atlas
Non U.S. extent
⚠ Perturbed FIA plots

National Individual Tree Species Atlas

MODEL FIT

An inset on each map allows the user to assess the relative fitness of that individual species for a given mapzone (item 10, pages 15 and 117). The inset represents the individual model correlations for each USGS mapzone (item 11, pages 15 and 117), which is an indication of how well the model fit the predictor layers, and may be an indication of how well the model predicts the actual occurrence of the individual species. Darker colors represent stronger models and lighter colors represent weaker models. The lightest gray indicates mapzones where the species did not occur. A validation summary is included in Appendices B2 and B3.

LEAVES

A representative photo of the natural foliage is included (item 16, pages 15 and 117). The photos are not intended for tree identification, but to represent what the user would see from the ground. Next to the leaf photo is a color chip (item 17, pages 15 and 117). It represents the natural-color (red, green, blue) average values of the summer Landsat image for plots wherein the species occurs. It is what the species would look like if you observed it from space, and represents not only the foliage, but the geology, soils, terrain, associated species, and morphological characteristics of that species. Although insignificant from a modeling standpoint, the color chip offers a curiosity and unique perspective on the data used to produce the *Modeled Atlas*.

OTHER DATA

The species common name, scientific name, botanical authority, family, and FIA species code are included for reference for each map (item 12-15, pages 15 and 117). Map scale is not included on the maps because visual scale is indicated by the state lines. However, map scale is stated in the section, "List of Maps" (page vi–viii). Additional species information about plot frequency, overall model sensitivity, validation sensitivity, and the most significant predictor variables is included in Appendix B2 for Alaska and Appendix B3 for the coterminous United States. The most significant variables are included for a glimpse behind the underlying models.

SPECIFIC NOTES

Digital representations of the *Mapped Atlas* (USGS 1999) were incorporated in the *Modeled Atlas* maps for comparative purposes for each species included in the *Mapped Atlas*. The *Modeled Atlas* map extents were limited to mapzones with occurrence on FIA plots and limited to mapzones that coincided with the *Mapped Atlas*. Tree species, such as Colorado blue spruce grown or planted well outside their natural range for commercial or landscaping purposes, were not included in the *Modeled Atlas* map extents. Additionally, map extents were limited to the United States and species extents in Canada and Mexico were not included in the *Modeled Atlas* maps. For species not originally in the *Mapped Atlas*, county data was utilized from the USDA NRCS Plants database (NRCS 2014). An additional map for cherrybark oak (*Quercus pagoda* Raf.) was digitized from a map online at http:\\www.efloras.org.

Differences between the *Mapped Atlas* extents and the *Modeled Atlas* extents can be due to a number of reasons. For non-overlapping areas where the *Mapped Atlas* does not show the presence of a given species (and the *Modeled Atlas* does), range extension may be suggested if the species reproduces aggressively. Other reasons could be the extensive planting of a commercially desirable species or an infrequent species that was previously not mapped very well. In cases where the *Modeled Atlas* does not show the presence of a given species (and the *Mapped Atlas* does), inventory field crews may have incorrectly identified a species, in which case the models would not accurately represent that species. It is also possible that the plot samples with the predictor variables may not have produced an adequate "signal" to accurately represent the species extent. Finally, it is also conceivable that the species no longer exists in the area delineated by the *Mapped Atlas*.

There are several possibilities why a species is omitted by the *Modeled Atlas* in areas where it is known to exist. Perhaps the species occurs infrequently and does not model well. The species may have been reduced by an invasive species, such as the near extinction of the American Chestnut (*Castanea dentata* (Marsh.) Borkh.) due to chestnut blight. Or perhaps the coarse mapping of the *Mapped Atlas* has been subject to error, something that is very apparent in the West, especially in the Basin and Range Province. And probably the most significant is widespread farming practices, which have altered many natural ranges of individual species.

CHANGES TO SPECIES CLASSIFICATIONS

Appendix B4 describes noted changes from the *Mapped Atlas* in particular to species classifications. For these descriptions FIA was utilized as the nomenclature authority, because it was the origin of the ground data. It should be noted that the official nomenclature from the International Code of Nomenclature for algae, fungi, and plants (ICN) can be more current than the FIA records, because they are the authority of botanical names.

NOTABLE DIFFERENCES IN RANGE

Little recognized that three species —black locust (*Robinia pseudoacacia* L.) (Page 287), northern catalpa (*Catalpa speciosa* (Warder) Warder ex Engelm.), and Osage-orange (*Maclura pomifera* (Raf.) C.K. Schneid.) (Page 197) — had been widely planted and had become so thoroughly naturalized that the extents of the original native species were difficult to determine. Of the three species, northern catalpa was dropped from the *Modeled Atlas* because the models were not very reliable. For the two other species, the range differences are apparent when comparing how they are depicted in the *Mapped Atlas* with how they are depicted in the *Modeled Atlas*. Other species have become conspicuous because they have been extensively planted beyond their natural ranges. This can be said for loblolly pine (*Pinus taeda* L.), slash pine (*Pinus elliottii* Engelm.), red pine (*Pinus resinosa* Aiton), Fraser fir (*Abies fraseri* (Pursh) Poir.), pecan (*Carya illinoinensis* (Wangenh.) K. Koch), honeylocust (*Gleditsia triacanthos* Marsh.), Kentucky coffee tree (*Gymnocladus dioicus* (L.) K. Koch), and white spruce (*Picea glauca* (Moench) Voss). Some species have extended their ranges naturally through aggressive reproduction in disturbed areas. A good example of this is red maple (*Acer rubrum* L.). In the East, many western species, such as blue spruce (*Picea pungens* Engelm.), Douglas-fir (*Pseudotsuga menziesii* (Mirb.) Franco), white fir (*Abies concolor* (Gord. & Glend.) Lindl. ex Hildebr.), and ponderosa pine (*Pinus ponderosa* C. Lawson), have been planted. Many species have been planted as shelterbelts in the Great Plains; however, due to the narrow nature of shelterbelts, and with the exception of jack pine (*Pinus banksiana* Lamb.) in Nebraska, most did not model well.

Taxonomic differences are noted between white fir (*A. concolor* (Gord. & Glend.) Lindl. ex Hildebr.) and grand fir (*Abies grandis* (Dougl. ex D. Don) Lindl.) in Central Oregon. Hybridization is problematic for several closely related species with overlapping ranges, and makes field identification difficult for, and subject to inconsistencies among, field crews.

Some species—American chestnut (*C. dentata* (Marsh.) Borkh.) and butternut (*Juglans cinerea* L.), to name just two—have been decimated by diseases and their original ranges substantially reduced. Other species, such as longleaf pine (*Pinus palustris* Mill.), could have been reduced through management practices which tended to favor slash pine over longleaf pine. The extent of western white pine (*Pinus monticola* Douglas ex D. Don) has been altered by a combination of management and disease (Harvey 2008).

SPECIES OBSERVED

The selected inventory cycle (complete set of annual inventory panels) included 387 different species codes, of which 264 were utilized in the creation of this *Modeled Atlas*. Of the 123 FIA species codes not included in the Atlas (Appendix B1), 37 species codes were not individual species but codes for trees where the precise species could not be determined at the time of inventory. These include (1) broad-based codes used for unknown dead conifer, unknown dead hardwood, unidentified hardwoods, other tropical species, or scrub oak; (2) most species of palm trees, (representing a family of species) which had been lumped into a single FIA species code; and (3) 31 classifications at the genus level, used when either no species code was available or the species was not determined. Models were built for the remaining 86 species codes; however, the resultant **geospatial products** were deemed inadequate. Sixty one species had too few samples (less than 30 subplots with presence); ten were non-native species; 13 species had poor models (even with more than 30 subplots with presence); and three species were combined into one species as they represented three different varieties of the same species (American basswood *Tilia americana* L.). All non-native species except tree-of-heaven (*Ailanthus altissima* (Mill.) Swingle), Chinese tallow tree (*Triadica sebifera* (L.) Small), and empress-tree (*Paulownia tomentosa* (Thunb.) Siebold & Zucc. ex Steud.) were excluded from the *Modeled Atlas*.

SPECIES NOTES

Even with extensive training and highly regimented quality assurance and control procedures in place, species can still be misidentified. Procedurally, we intended to accept the field calls as completely valid; however, a number of exceptions were applied.

Page 32—Atlantic white-cedar (*Chamaecyparis thyoides* (L.) Britton, Sterns & Poggenb.) has been recorded by FIA in the states of Kentucky, Ohio, West Virginia, Tennessee, Northern Georgia, Northern Alabama, Pennsylvania, Northern New York, and Northern Vermont. This puts it far outside what is recorded in the *Mapped Atlas*. These areas were removed from the *Modeled Atlas*.

Page 79—Great Basin bristlecone pine (*Pinus longaeva* D.K. Bailey) and Rocky Mountain bristlecone pine (*Pinus aristata* Engelm.) were once considered the same species, but were regionally separated as two distinct species during the past few decades. A single Rocky Mountain bristlecone pine was coded by FIA in an area that is considered Great Basin bristlecone pine. We corrected this for the analysis.

Page 96—Virginia pine (*Pinus virginiana* Mill.) was miscoded by FIA for 172 field plots in Texas, Louisiana, Mississippi, and Southern Alabama, which are far outside the distribution in the *Mapped Atlas*. Subsequently, field crews corrected these records during re-measurement; however, as of February, 2014, some plots had yet to be re-measured and records exist of Virginia pine in these areas. We anticipate that these records will be modified at the next field visit. Updated **statistical models** and **geospatial products** were re-created for this *Modeled Atlas*.

Page 269—Chinkapin oak (*Quercus muehlenbergii* Engelm.) is an eastern species and giant chinkapin (page 165) is a western species. A number of FIA records in the western region were miscoded as chinkapin oak and the statistical model built for chinkapin oak included these miscoded records. The Chinkapin oak **geospatial products** for the western mapzones were ultimately dropped from the *Modeled Atlas* and the published map includes only the eastern extent. (Note: In subsequent updates to the FIA database, most of the miscoded records were corrected.)

Page 290—Species in the willow genus are difficult for field crews to distinguish and many of the species are considered shrubs. Black willow (*Salix nigra* Marsh.) is the largest and the most common of the willow tree species. In North Carolina, the field crews code all willows using a generic code. For this reason, black willow was not modeled in the coastal zone associated with North Carolina. To mitigate this discrepancy, we utilized the generic **geospatial product** instead for black willow in the coastal zone of North Carolina.

Page 294—For American basswood (*Tilia americana* L.), three separate species codes, including three recognized varieties, American basswood (*Tilia americana* L. var. *americana*), Carolina basswood (*Tilia americana* L. var. *caroliniana* (Mill.) Castigl.) and white basswood (*Tilia americana* L. var. *heterophylla* (Vent.) Loudon), were utilized. The **geospatial products** for each of the varieties substantially overlapped with one another which suggested an inconsistency in the field calls. For this reason, all varieties were lumped together to create a single **geospatial product** for American basswood in the *Modeled Atlas*.

Artifacts in the models may be associated with field-crew calls rather than the actual species distribution. This may occur where closely related species are identified as one species in certain counties and a different species in another. We suggest that field crews should be able to recognize these minor anomalies in the *Atlas* and see them as an opportunity to gather additional expert opinions on future field calls.

FEEDBACK

As Little did in his *Atlas of United States Trees*, we welcome your feedback. Please let us know of any errors, omissions or anomalies you find in this publication, so that we can update the electronic versions, and include them in any revised print editions. Visit the website at http://www.fs.fed.us/foresthealth/technology/remote_sensing.shtml

Conifers

HOW TO READ THE MAPS

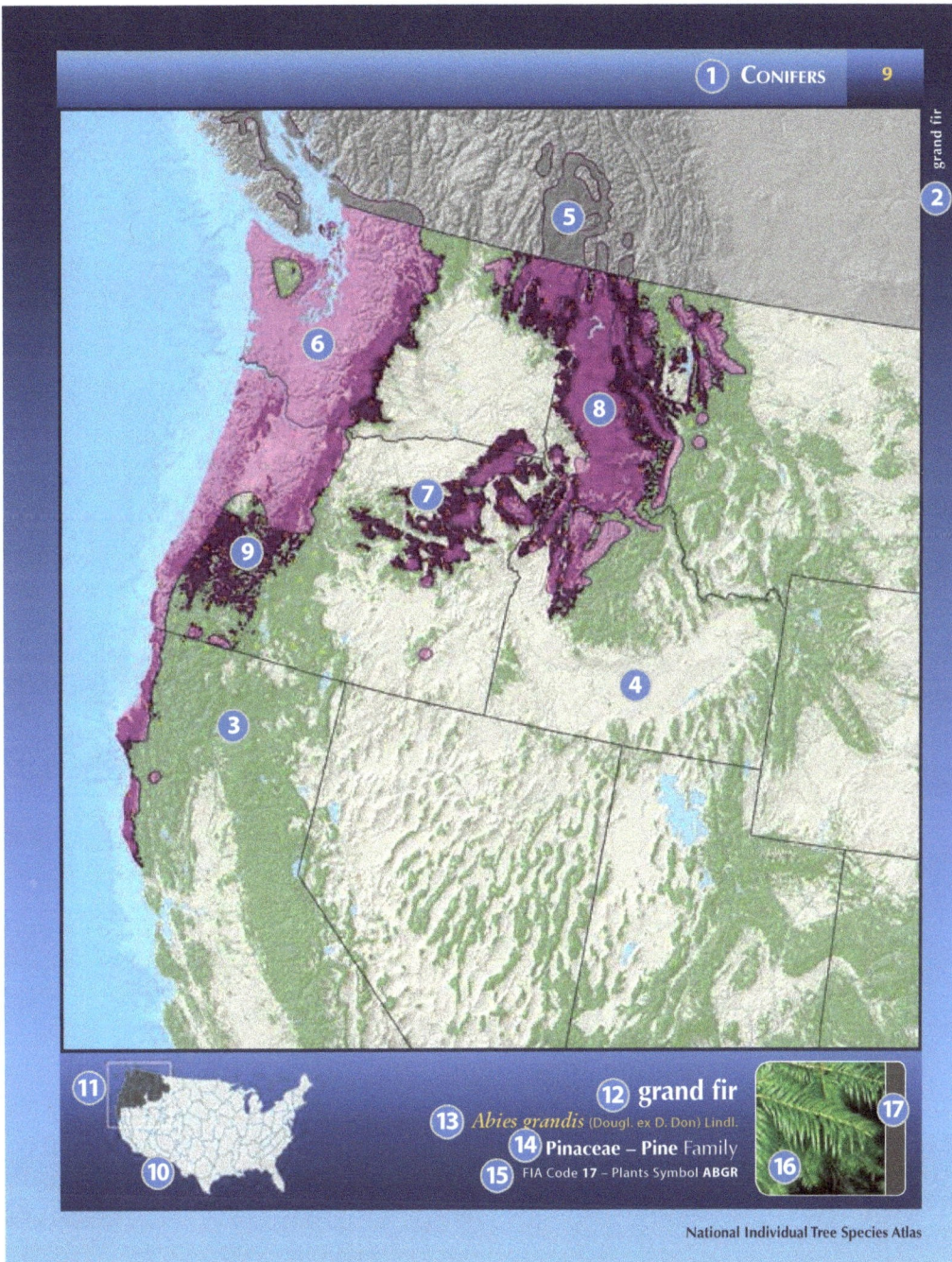

1. Division
2. Common name quick reference
3. ⬛ Treed areas
4. ⬛ Non-treed areas
5. ⬛ *Mapped* non U.S. extent
6. ⬛ *Mapped Atlas*
7. ⬛ *Modeled Atlas*
8. ⬛ Coincidence
9. ⚠ Perturbed FIA plots
10. USGS map zones
11. Map extent with model correlations (x100)

⬛ 81 -100 Strongest fit
⬛ 61 - 80
⬛ 41 - 60
⬛ 21 - 40
⬛ 0 - 20
⬛ Not modeled
▨ No data available

12. Common name
13. Scientific name and authority
14. Family name
15. FIA code and Plants symbol
16. Leaf
17. Color chip

Map labels

1 CONIFERS 9

grand fir

12 **grand fir**
13 *Abies grandis* (Dougl. ex D. Don) Lindl.
14 **Pinaceae – Pine** Family
15 FIA Code **17** – Plants Symbol **ABGR**

National Individual Tree Species Atlas

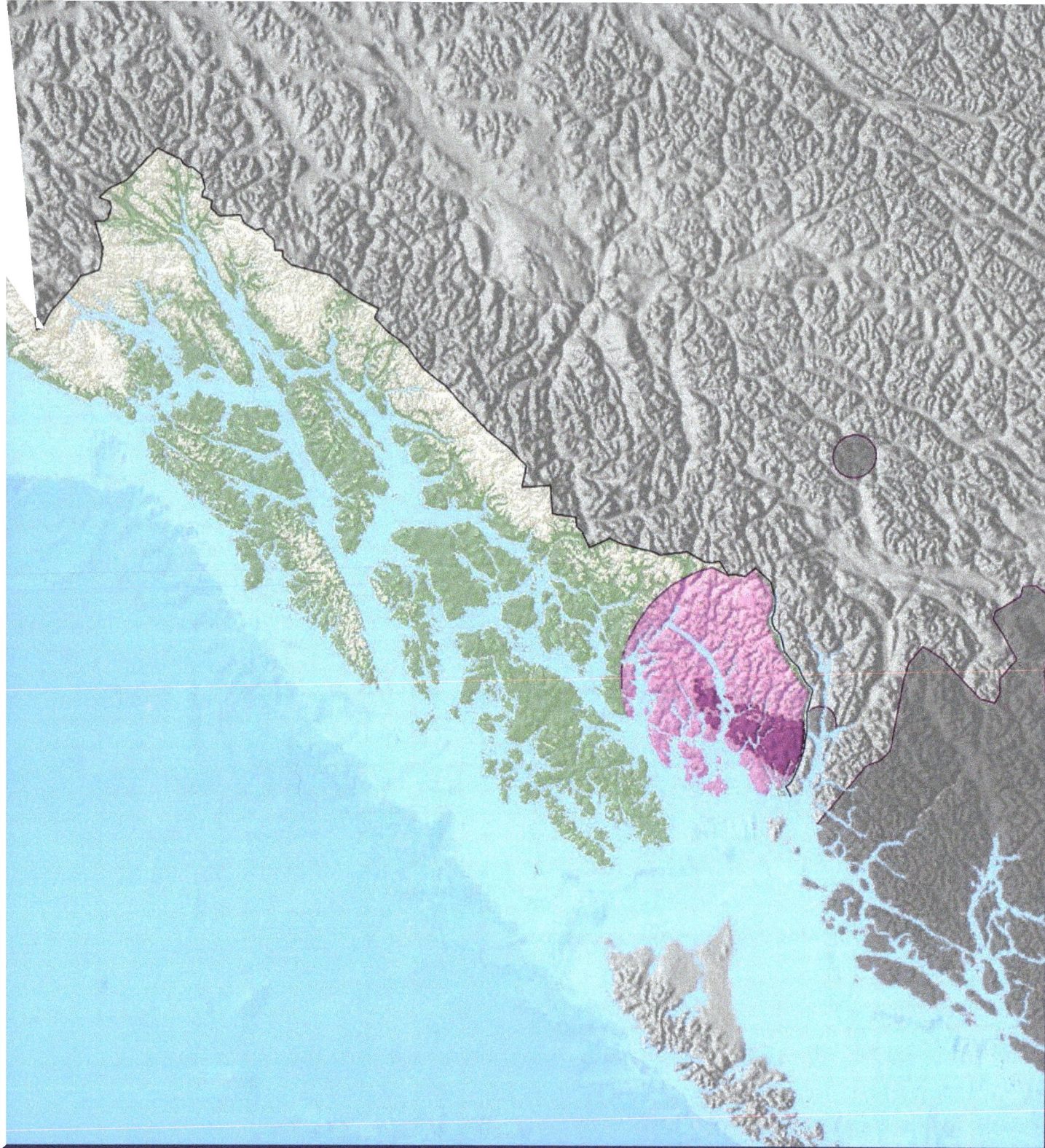

Pacific silver fir

Abies amabilis (Dougl. ex Loud.) Dougl. ex Forbes

Pinaceae – Pine Family

FIA Code **11** – Plants Symbol **ABAM**

National Individual Tree Species Atlas

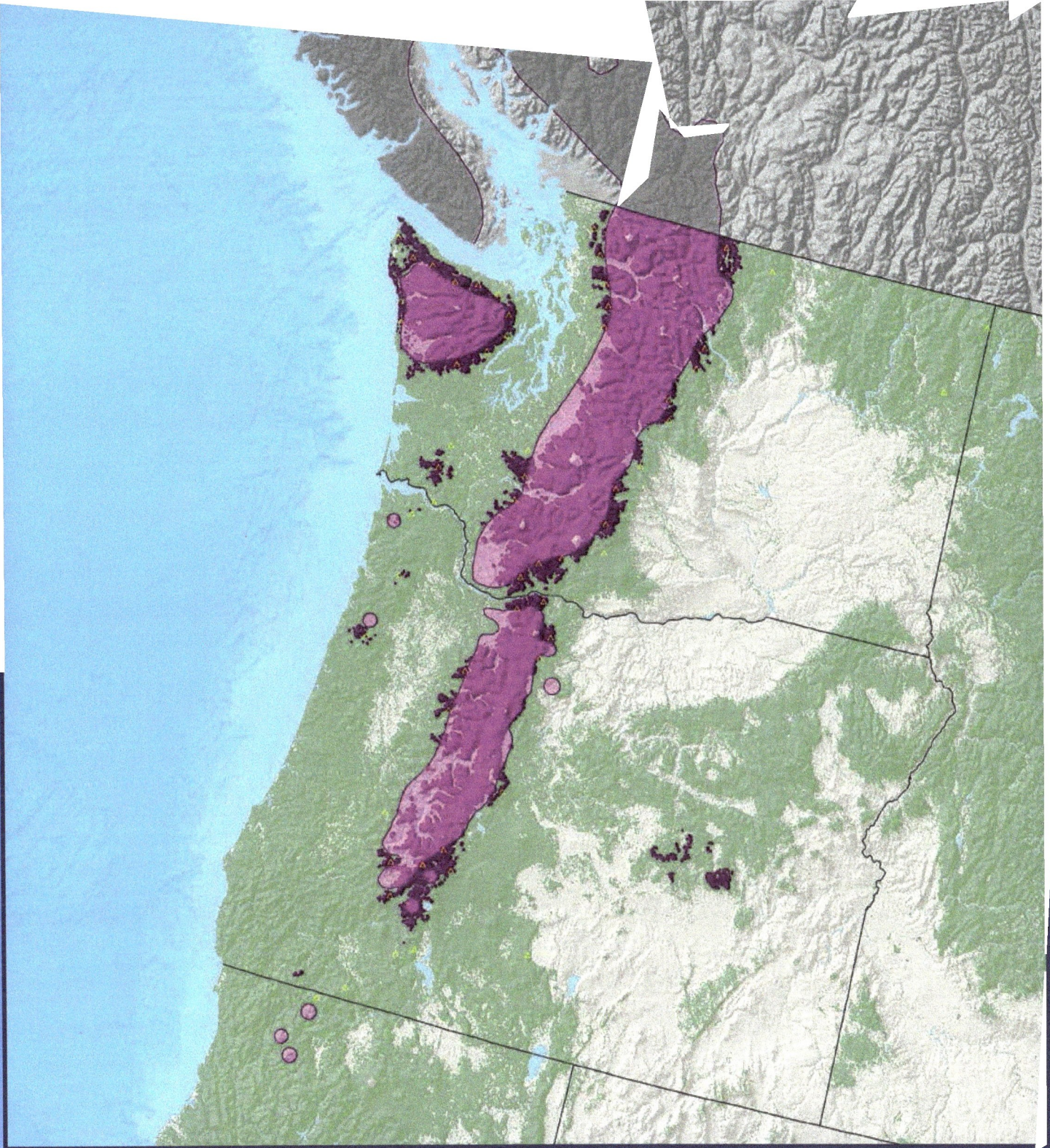

Pacific silver fir

Abies amabilis (Dougl. ex Loud.) Dougl. ex Forbes

Pinaceae – Pine Family

FIA Code **11** – Plants Symbol **ABAM**

National Individual Tree Species Atlas

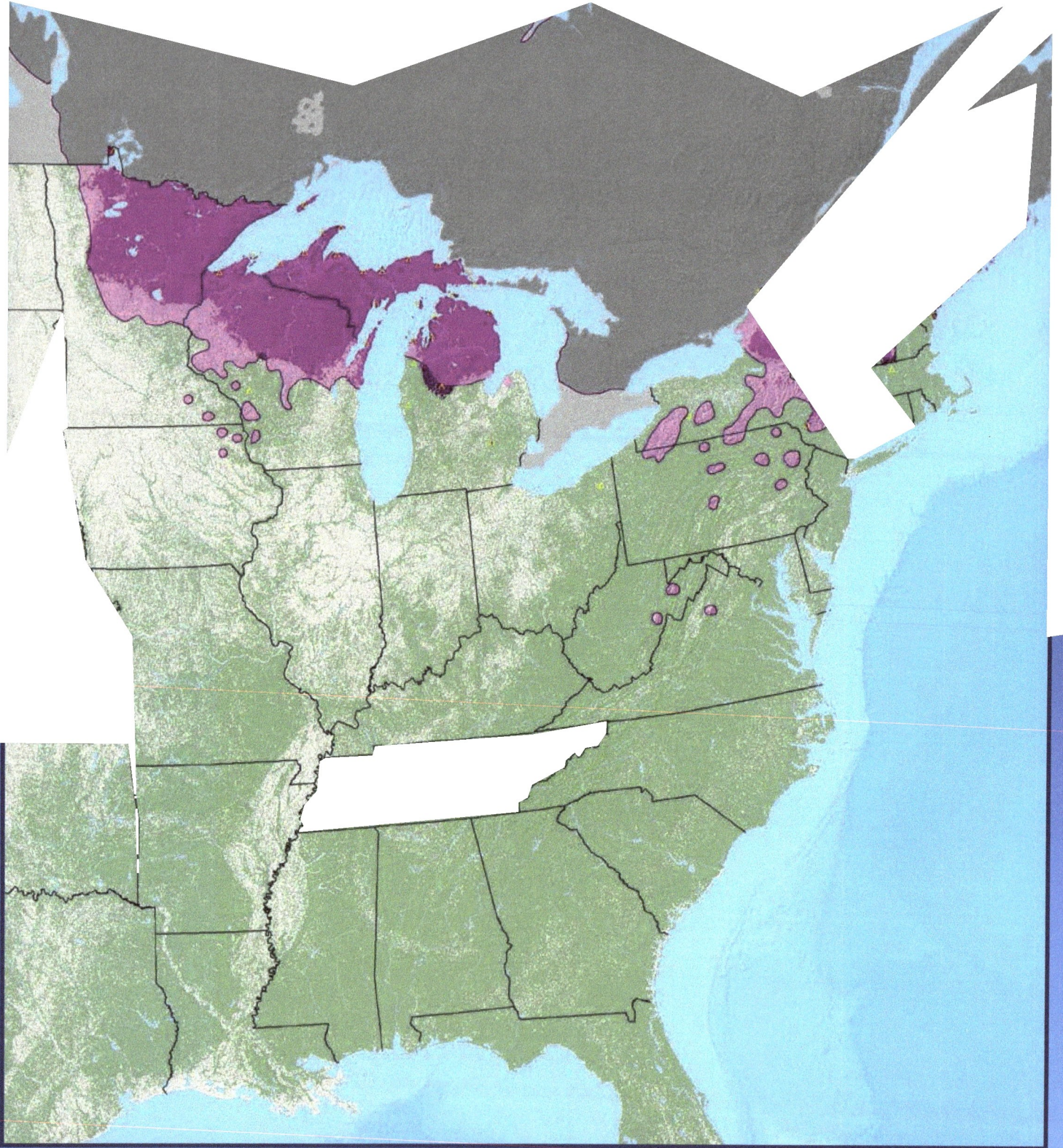

balsam fir

Abies balsamea (L.) Mill.

Pinaceae – Pine Family

FIA Code **12** – Plants Symbol **ABBA**

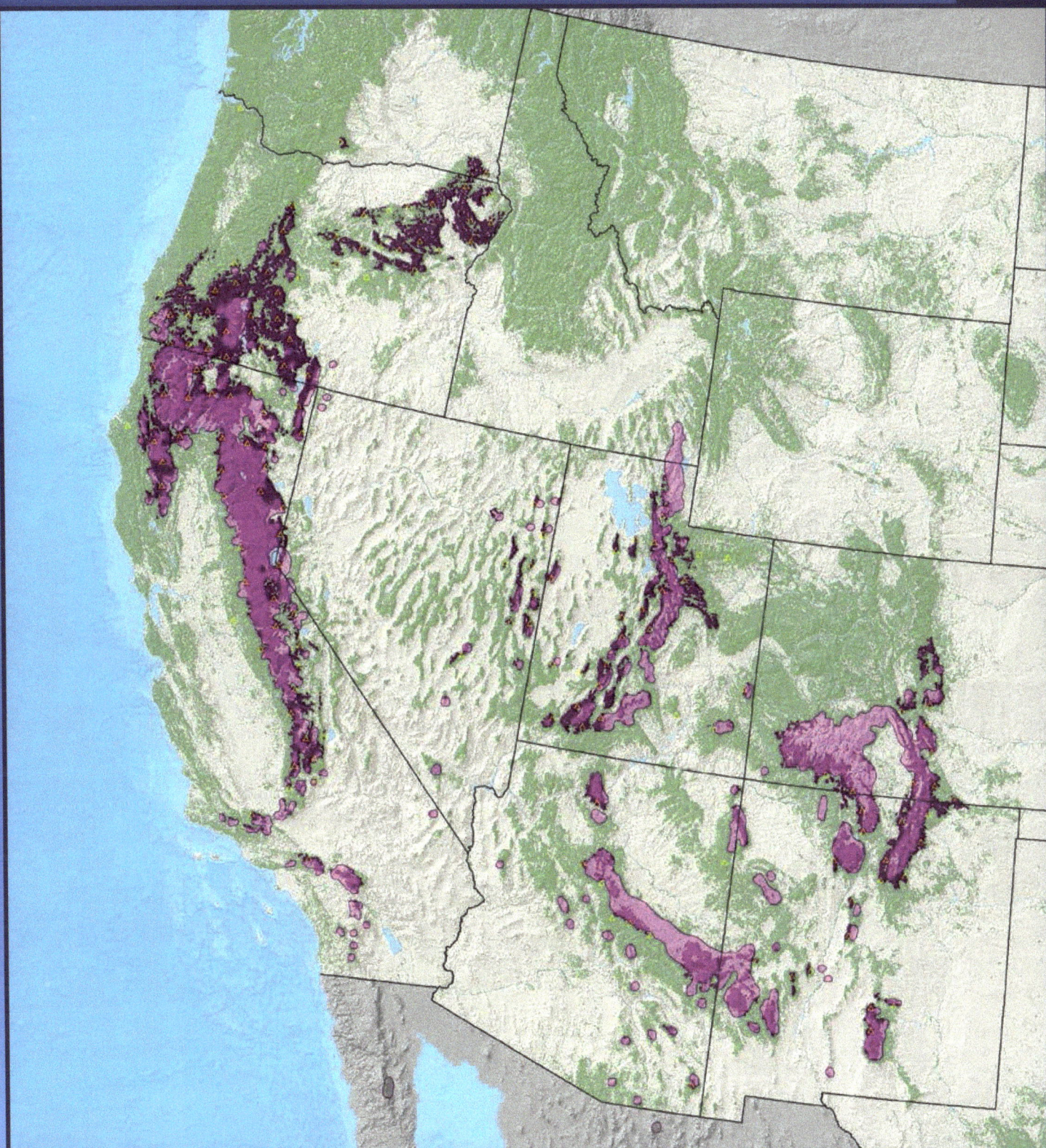

white fir

white fir

Abies concolor (Gord. & Glend.) Lindl. ex Hildebr.

Pinaceae – Pine Family

FIA Code **15** – Plants Symbol **ABCO**

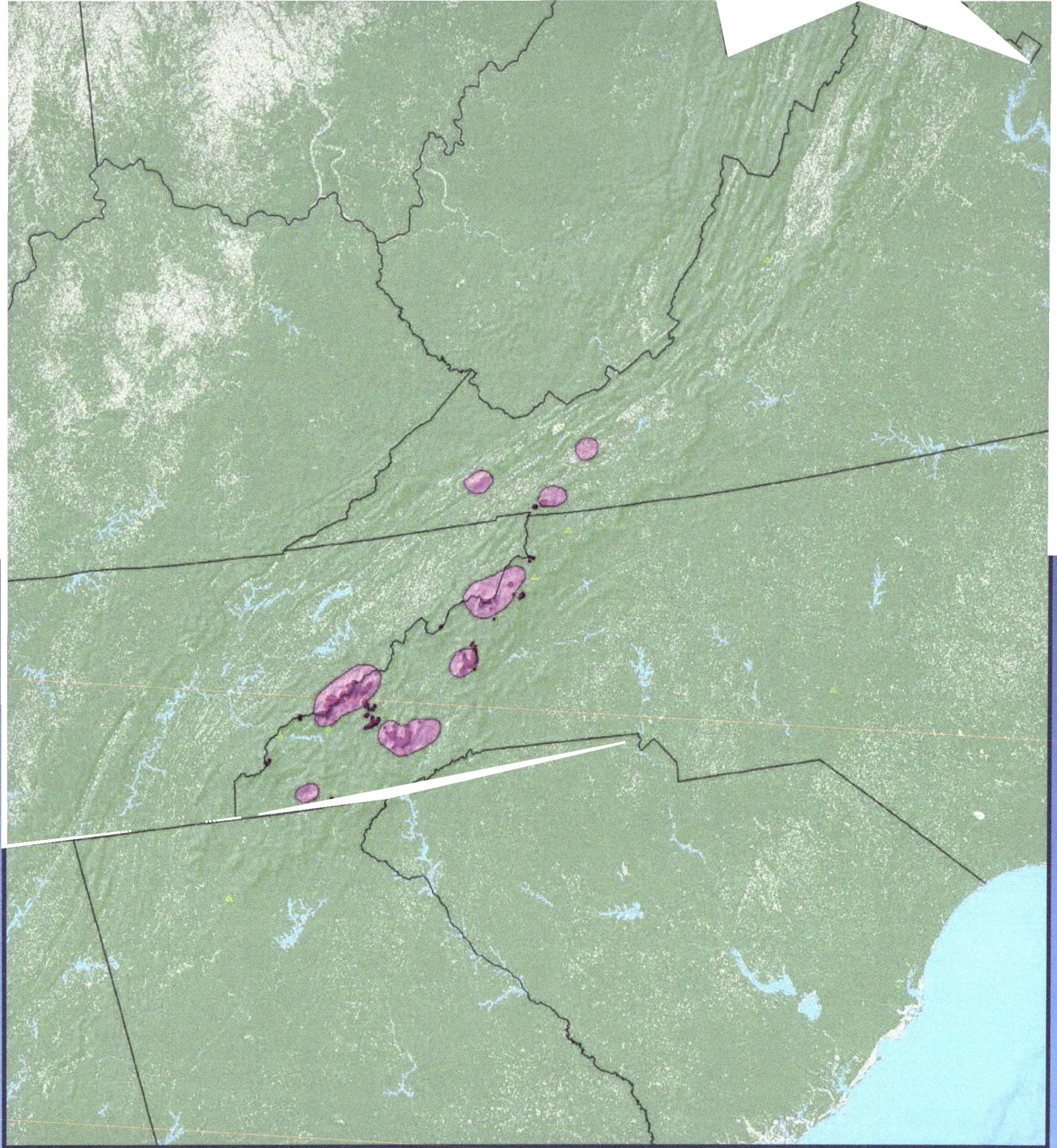

Fraser fir

Abies fraseri (Pursh) Poir.

Pinaceae – Pine Family

FIA Code **16** – Plants Symbol **ABFR**

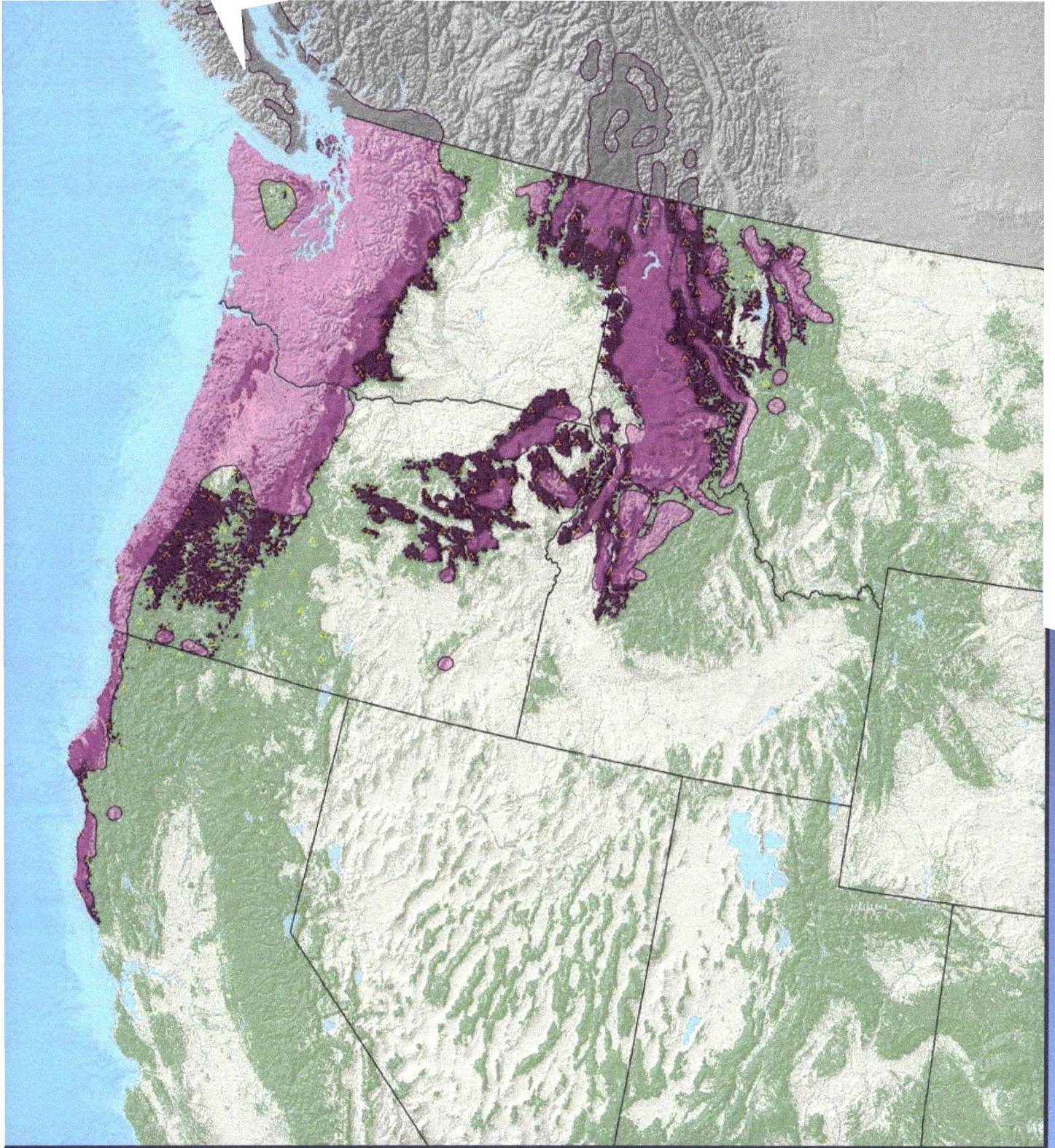

grand fir

Abies grandis (Dougl. ex D. Don) Lindl.

Pinaceae – Pine Family

FIA Code **17** – Plants Symbol **ABGR**

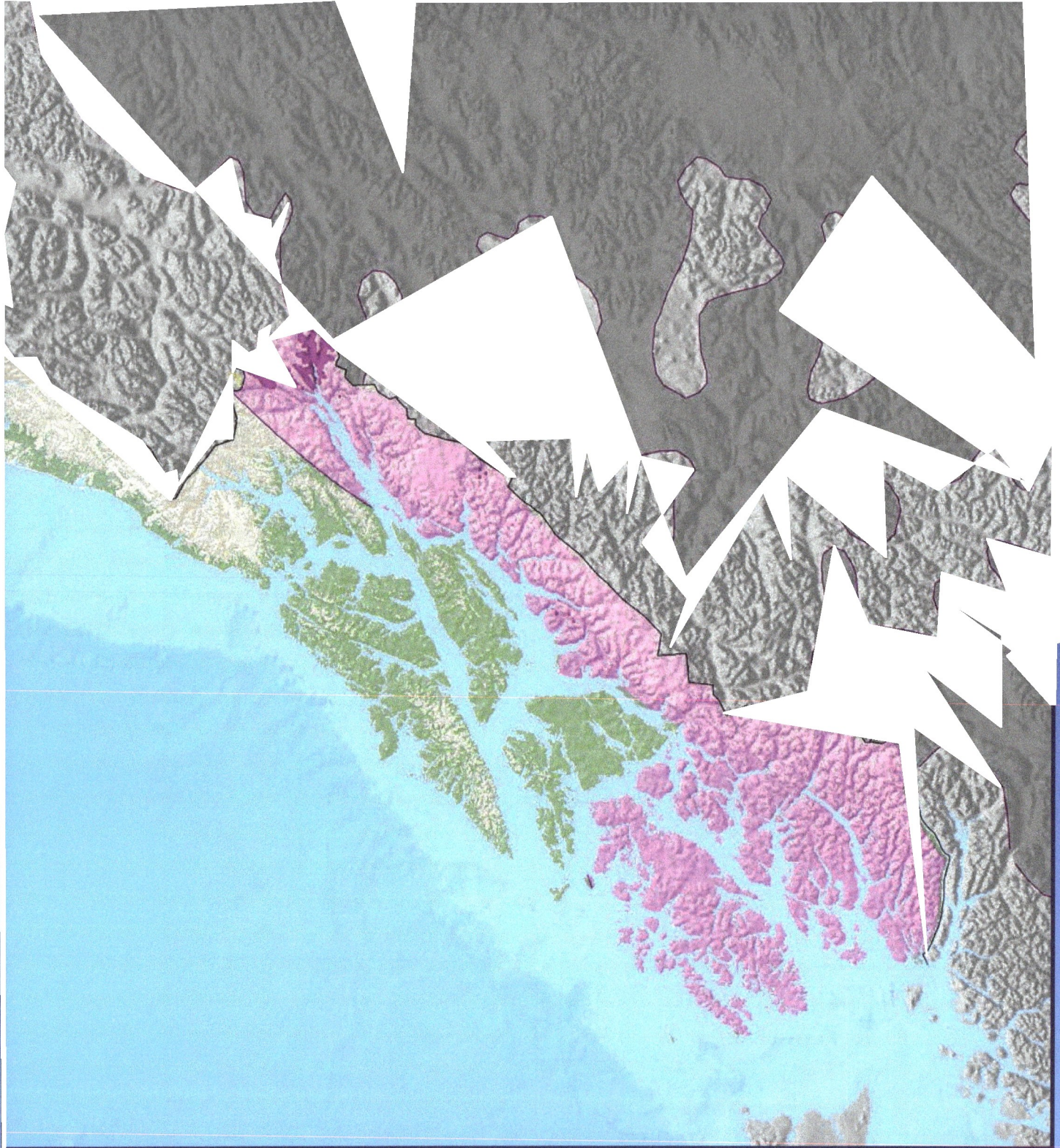

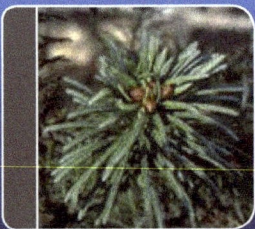

subalpine fir
Abies lasiocarpa (Hook.) Nutt.
Pinaceae – Pine Family
FIA Code **19** – Plants Symbol **ABLA**

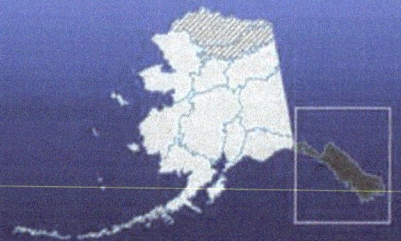

National Individual Tree Species Atlas

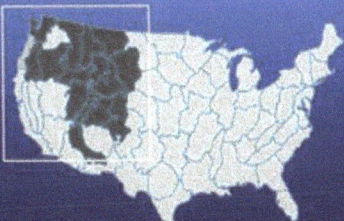

subalpine fir

Abies lasiocarpa (Hook.) Nutt.

Pinaceae – Pine Family

FIA Code **19** – Plants Symbol **ABLA**

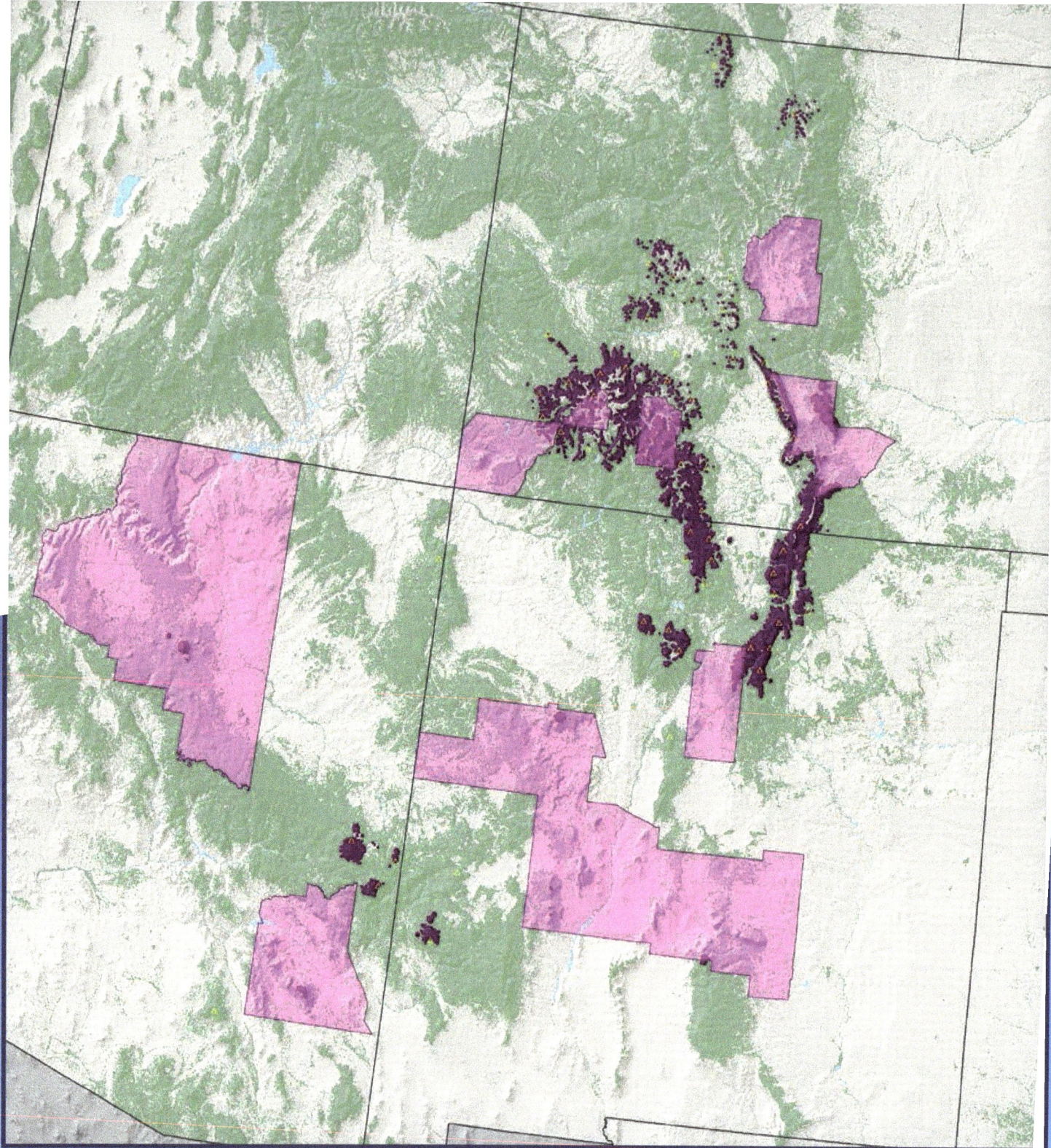

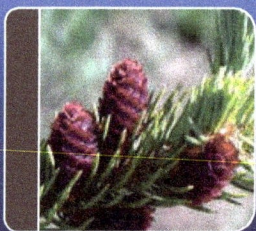

corkbark fir

Abies lasiocarpa (Hook.) Nutt. var. *arizonica* (Merriam) Lemmon

Pinaceae – Pine Family

FIA Code **18** – Plants Symbol **ABLAA**

National Individual Tree Species Atlas

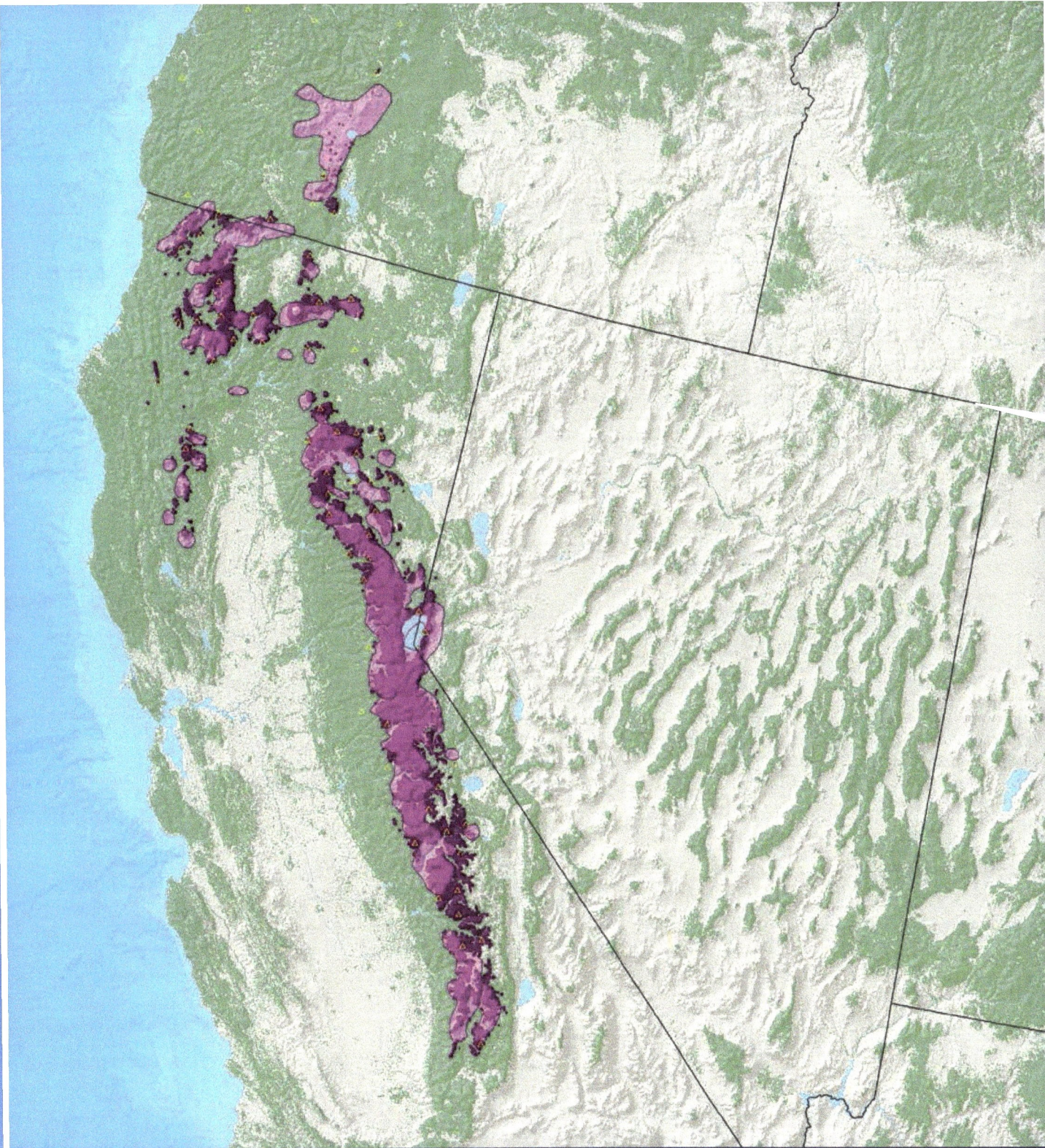

California red fir
Abies magnifica A. Murr.
Pinaceae – Pine Family
FIA Code **20** – Plants Symbol **ABMA**

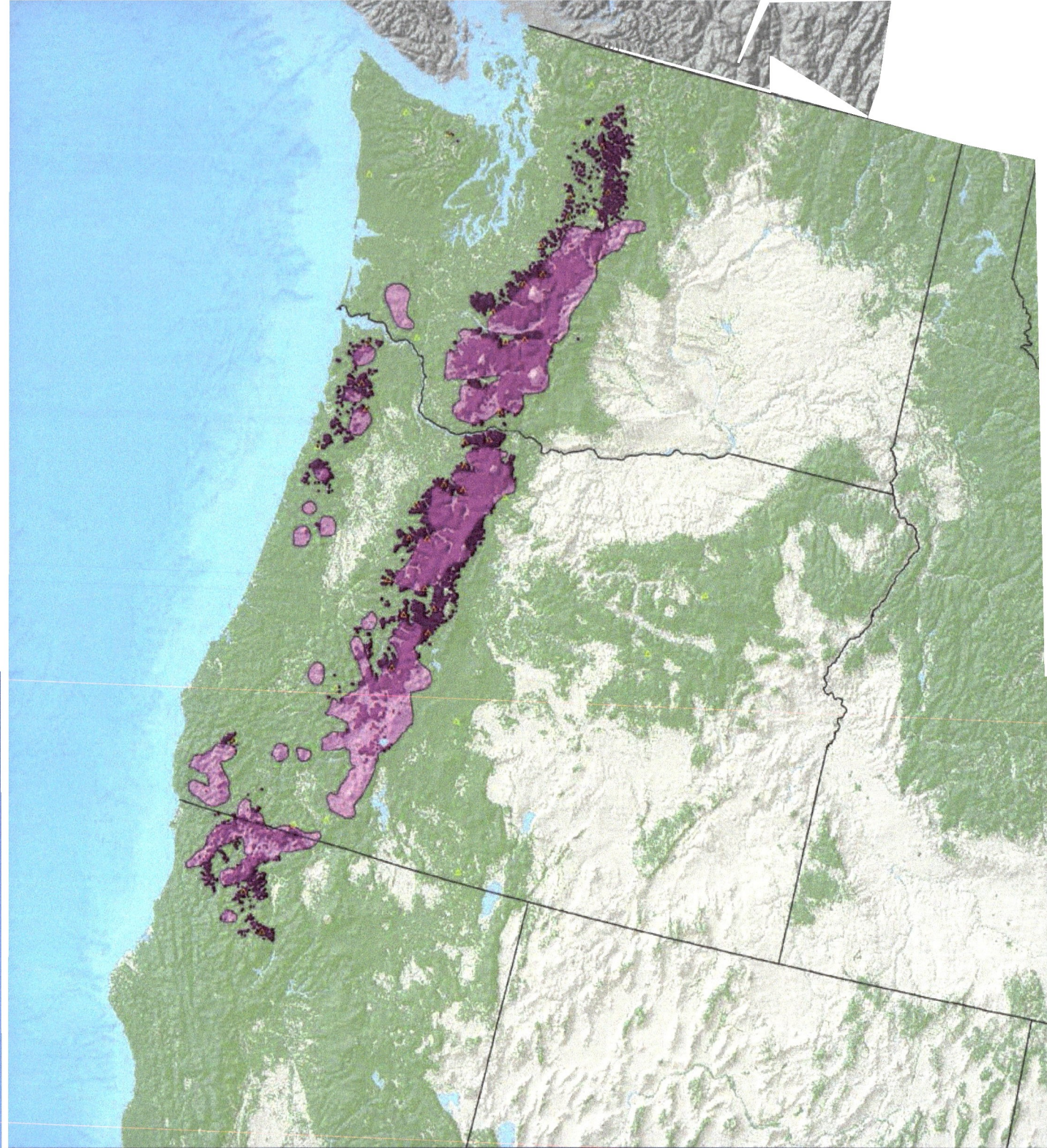

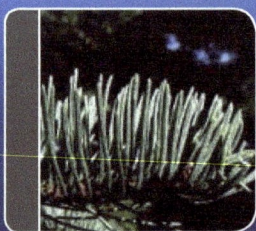

noble fir

Abies procera Rehd.

Pinaceae – Pine Family

FIA Code **22** – Plants Symbol **ABPR**

National Individual Tree Species Atlas

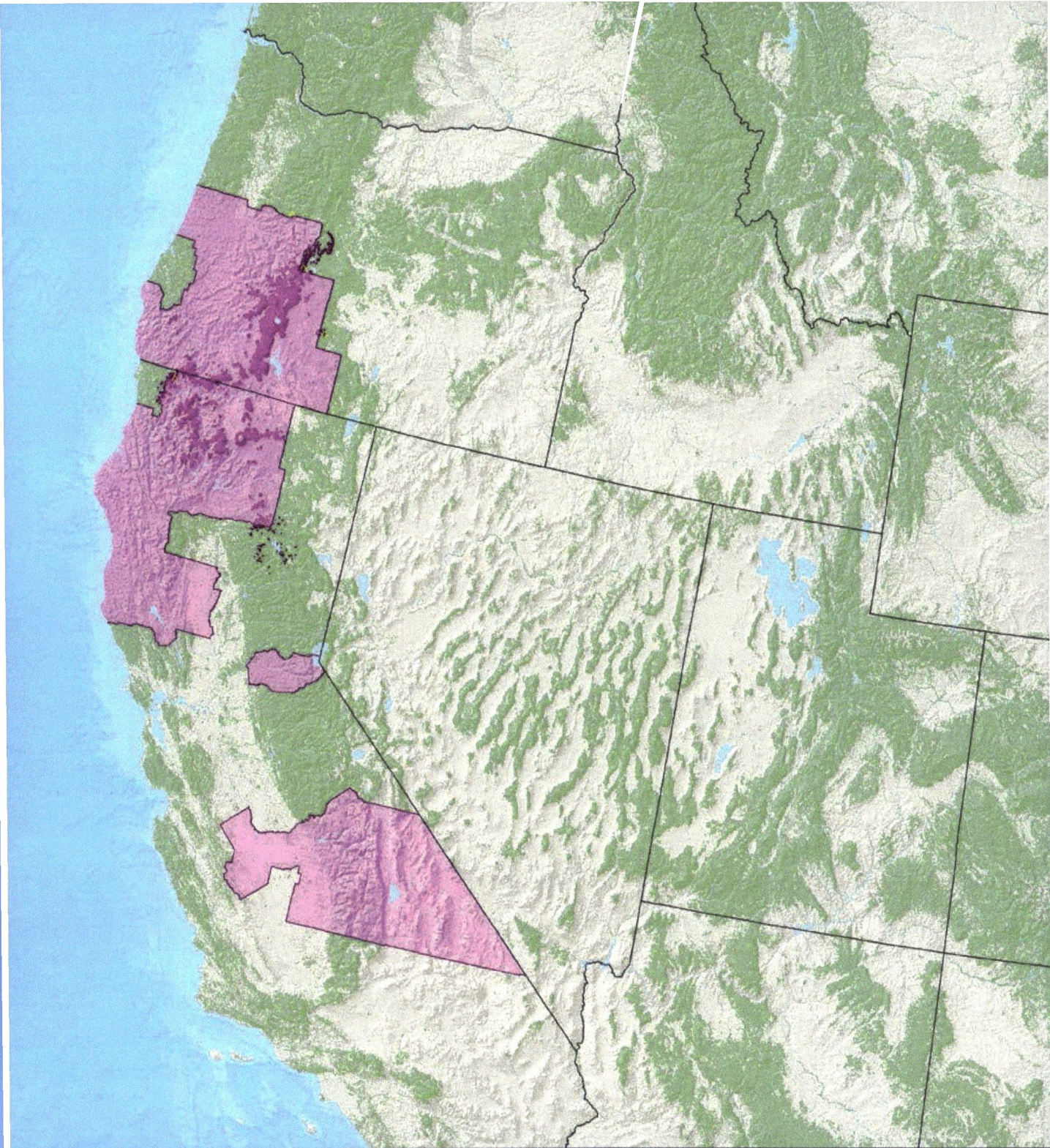

Shasta red fir

Abies shastensis Lemmon

Pinaceae – Pine Family

FIA Code **21** – Plants Symbol **ABSH**

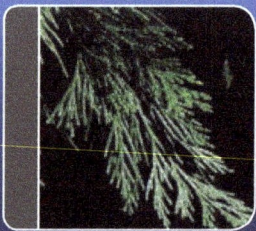

incense-cedar

Calocedrus decurrens (Torr.) Florin

Cupressaceae – Cypress Family

FIA Code **81** – Plants Symbol **CADE27**

National Individual Tree Species Atlas

Port-Orford-cedar

Chamaecyparis lawsoniana (A. Murray) Parl.

Cupressaceae – Cypress Family

FIA Code **41** – Plants Symbol **CHLA**

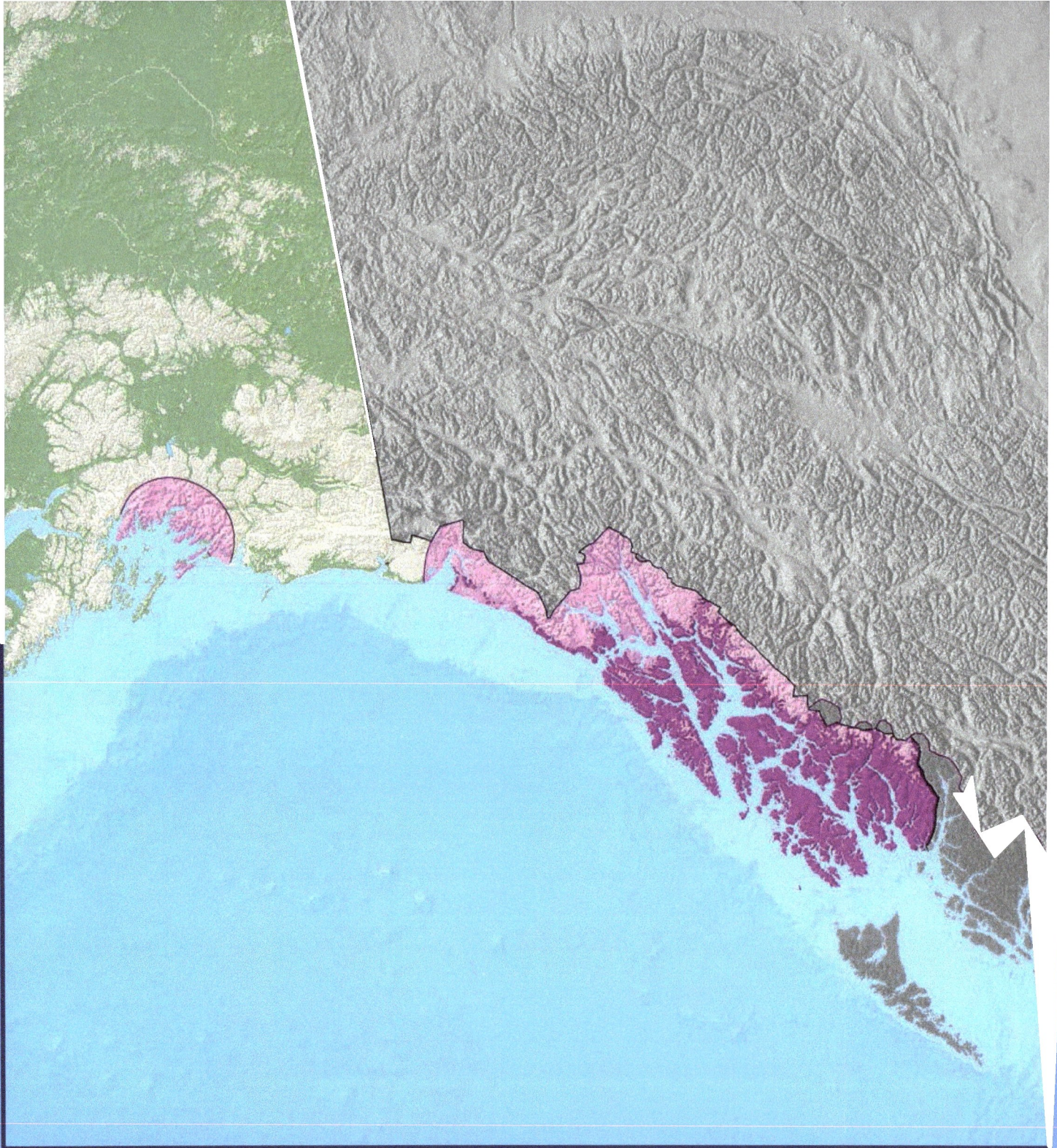

Alaska yellow-cedar

Callitropsis nootkatensis (D. Don) Oerst. ex D.P. Little

Cupressaceae – Cypress Family

FIA Code **42** – Plants Symbol **CHNO**

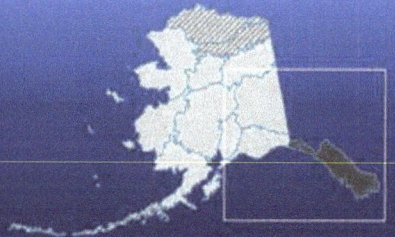

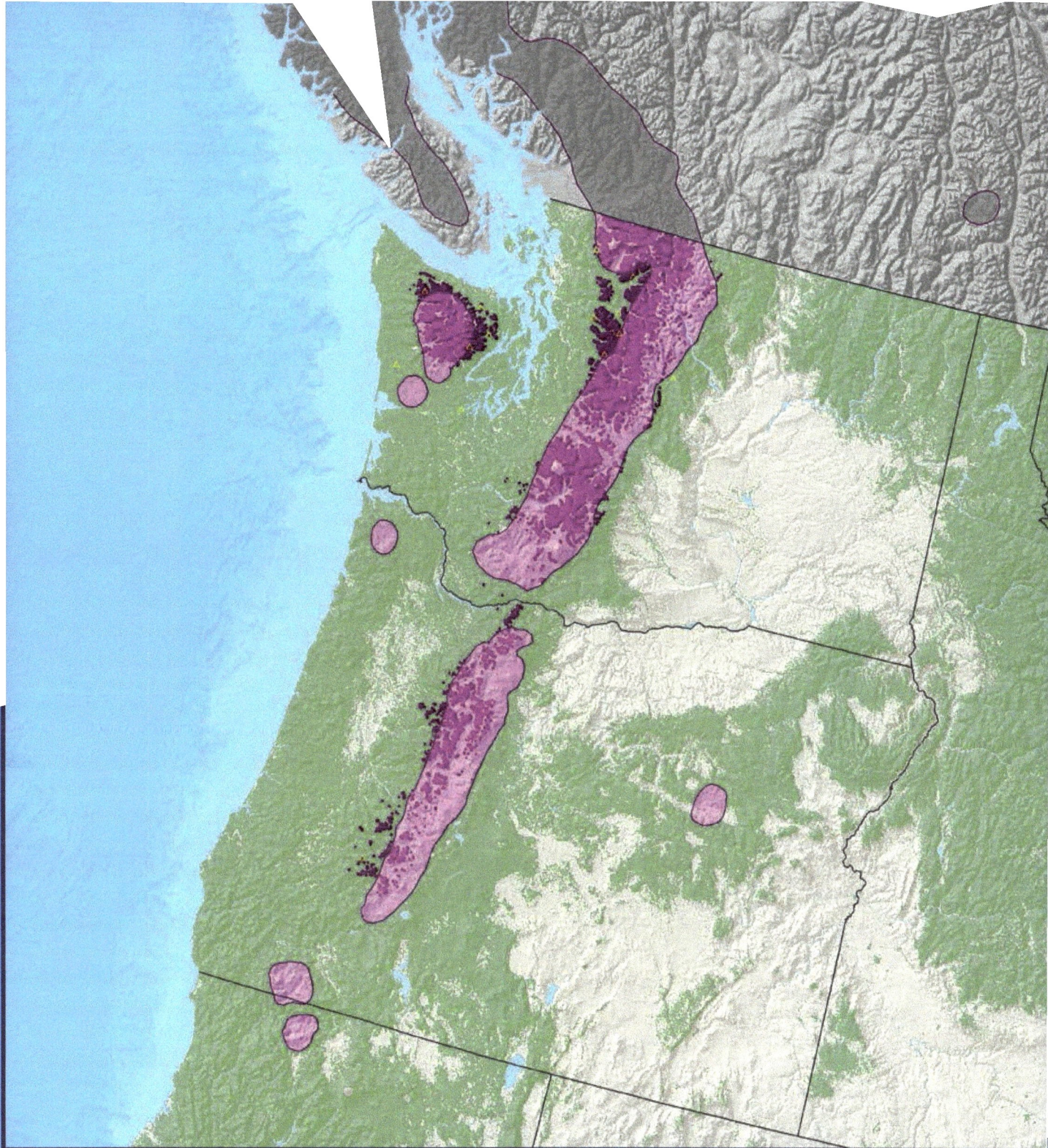

Alaska yellow-cedar
Callitropsis nootkatensis (D. Don) Oerst. ex D.P. Little
Cupressaceae – Cypress Family
FIA Code **42** – Plants Symbol **CHNO**

National Individual Tree Species Atlas

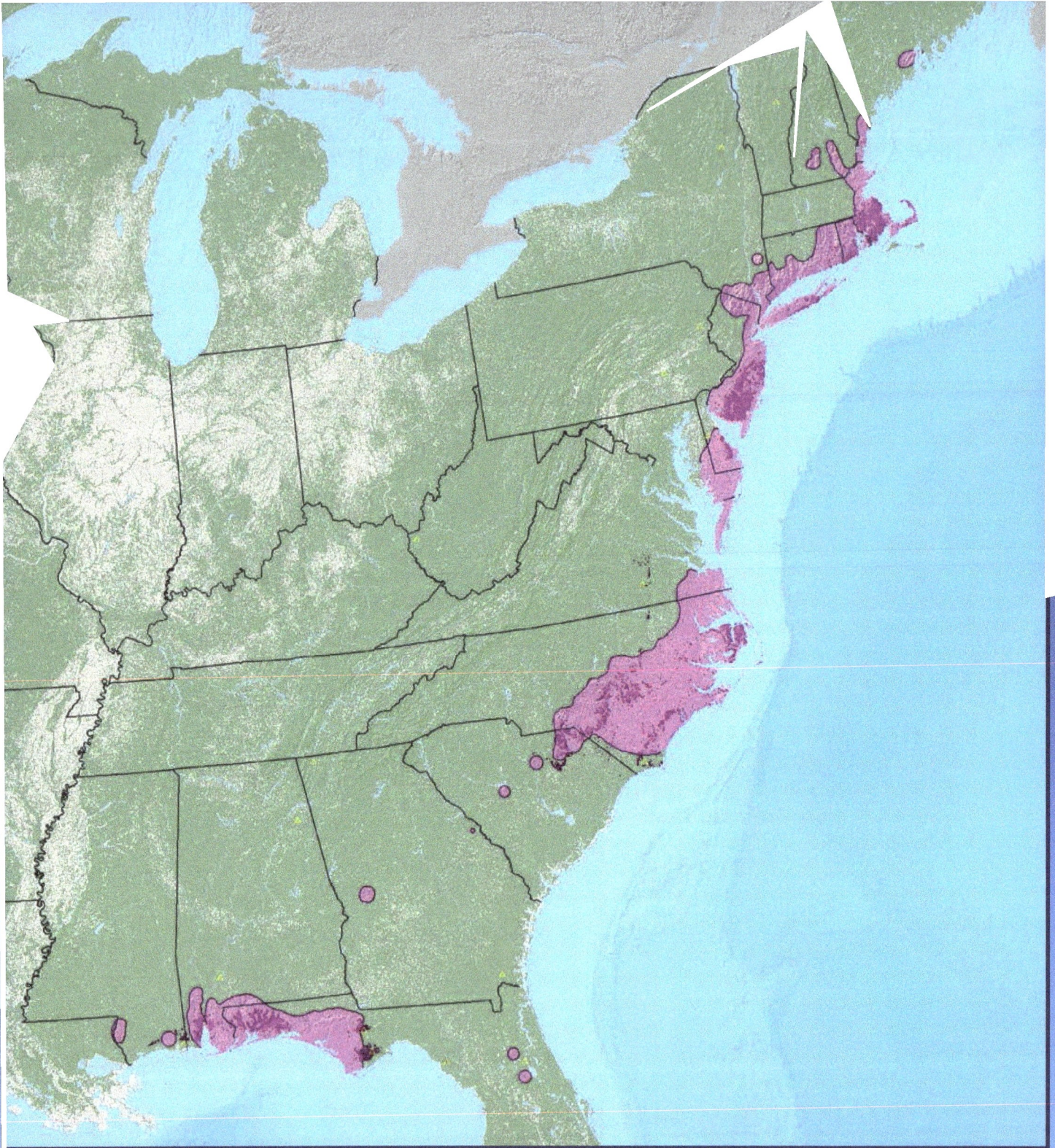

Atlantic white-cedar

Chamaecyparis thyoides (L.) Britton, Sterns & Poggenb.

Cupressaceae – Cypress Family

FIA Code **43** – Plants Symbol **CHTH2**

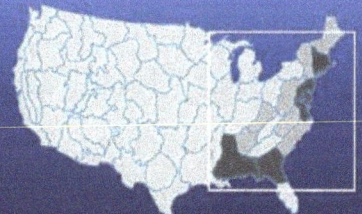

National Individual Tree Species Atlas

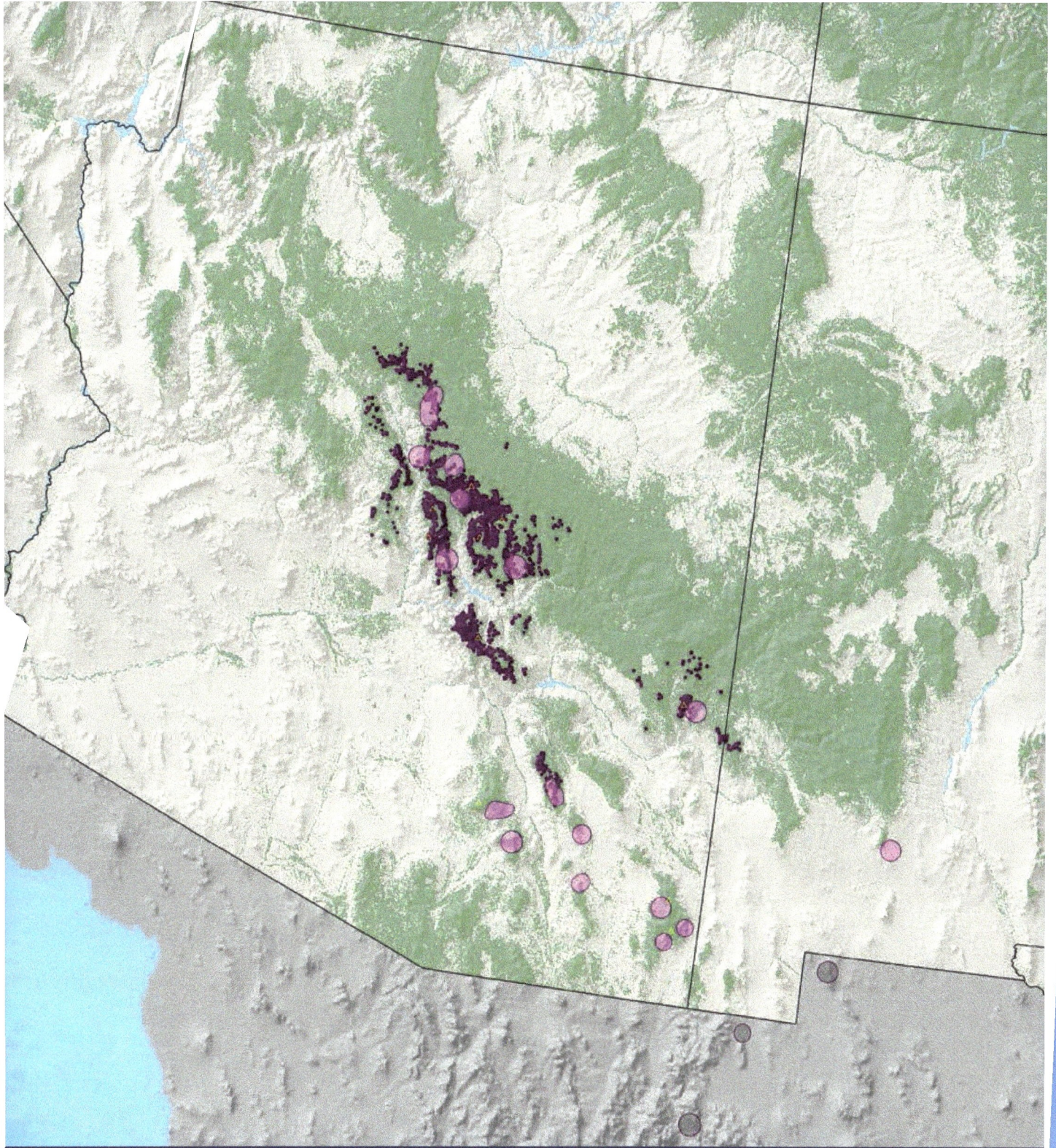

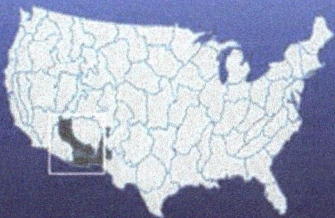

Arizona cypress
Cupressus arizonica Greene
Cupressaceae – Cypress Family
FIA Code **51** – Plants Symbol **CUAR**

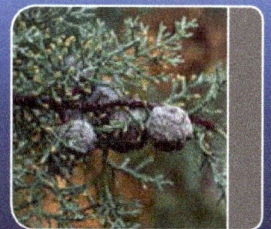

National Individual Tree Species Atlas

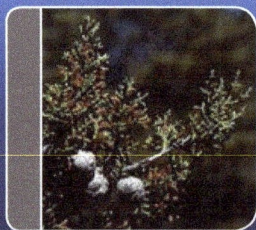

Sargent's cypress
Cupressus sargentii Jepson
Cupressaceae – Cypress Family
FIA Code **55** – Plants Symbol **CUSA3**

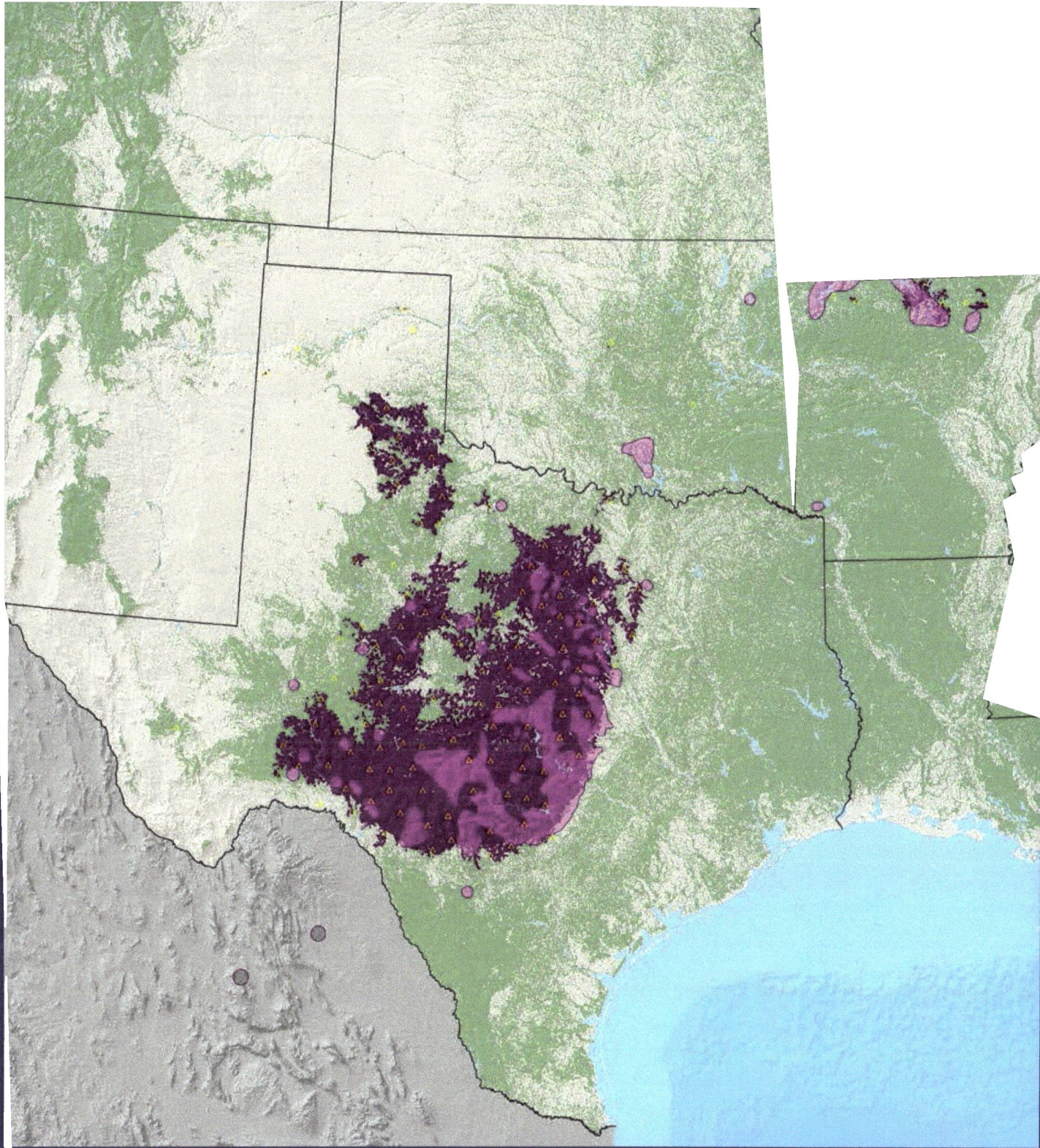

Ashe juniper
Juniperus ashei J. Buchholz
Cupressaceae – Cypress Family
FIA Code **61** – Plants Symbol **JUAS**

National Individual Tree Species Atlas

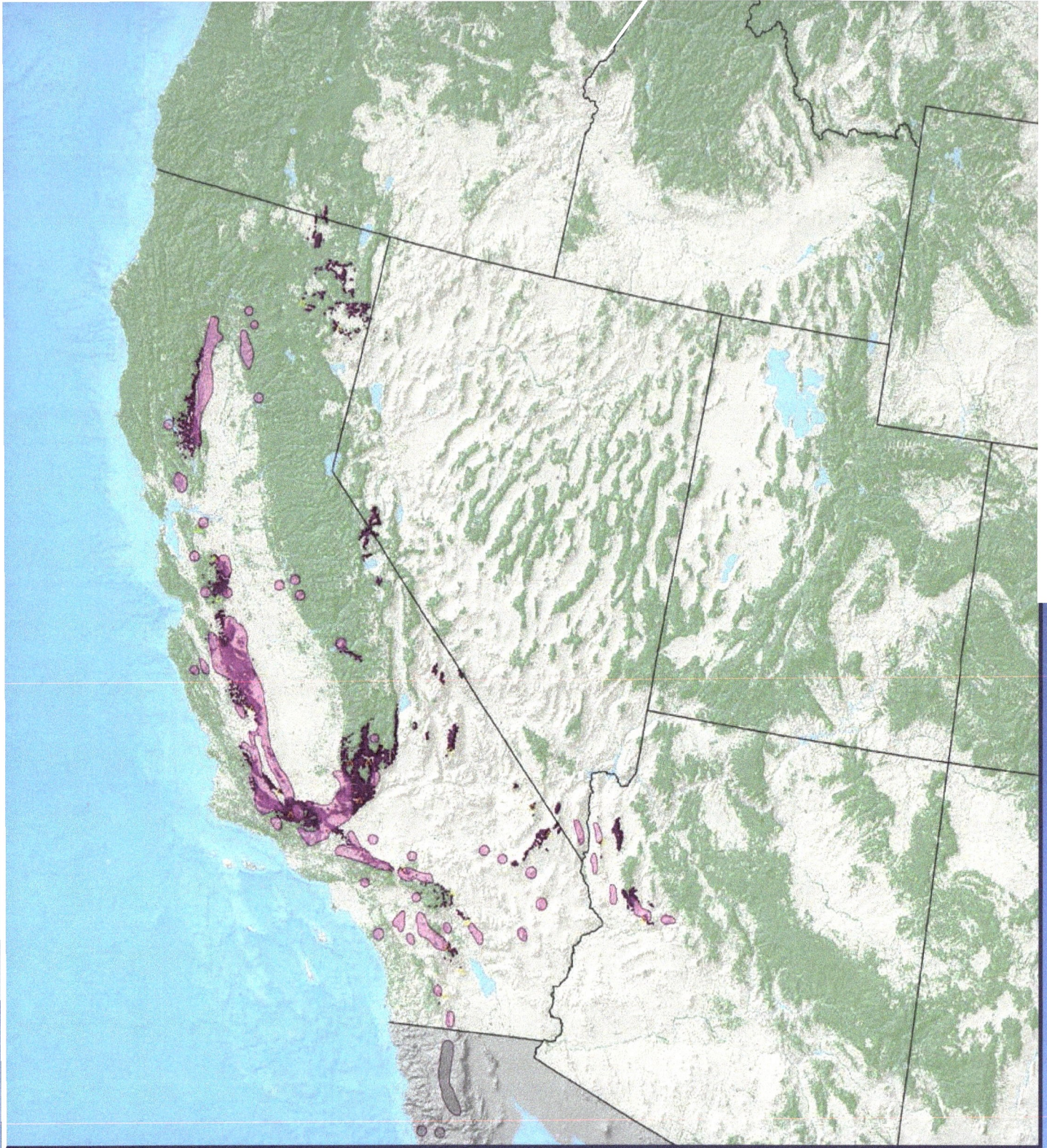

California juniper

Juniperus californica Carrière

Cupressaceae – Cypress Family

FIA Code **62** – Plants Symbol **JUCA7**

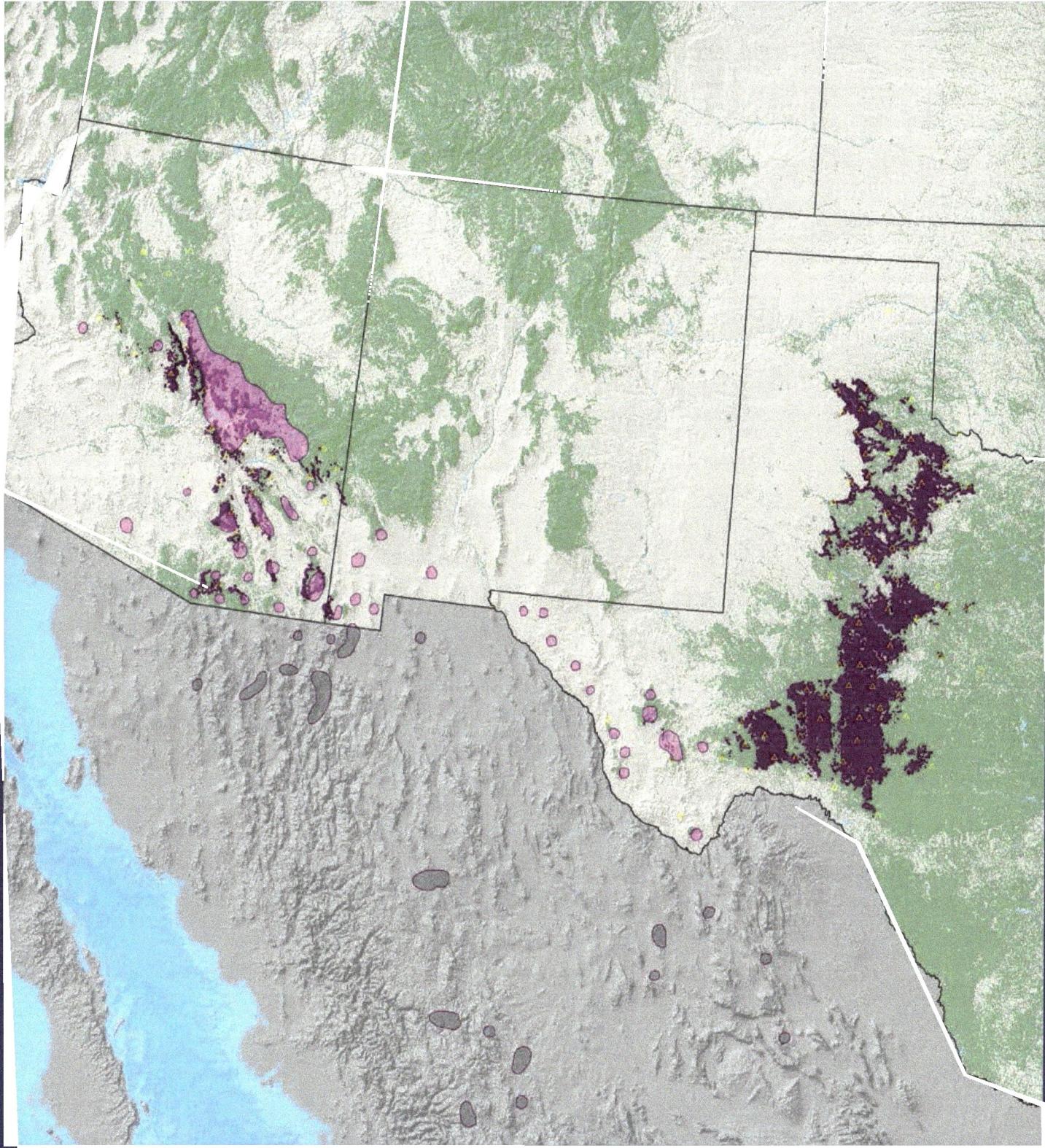

redberry juniper

Juniperus coahuilensis (Martiñez) Gaussen ex R.P. Adams

Cupressaceae – Cypress Family

FIA Code **59** – Plants Symbol **JUCO11**

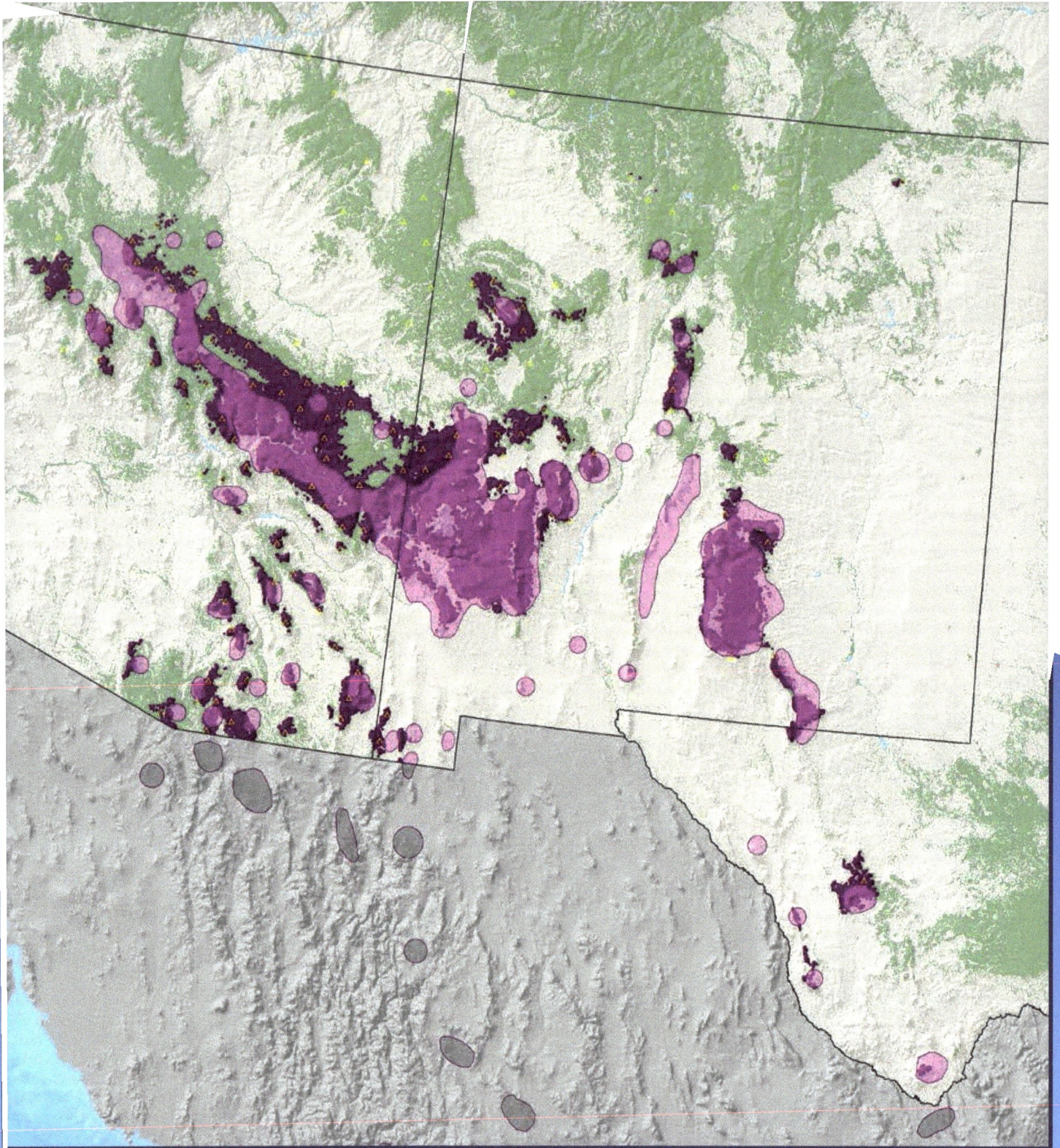

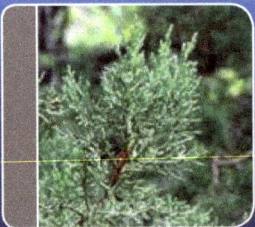

alligator juniper

Juniperus deppeana Steud.

Cupressaceae – Cypress Family

FIA Code **63** – Plants Symbol **JUDE2**

National Individual Tree Species Atlas

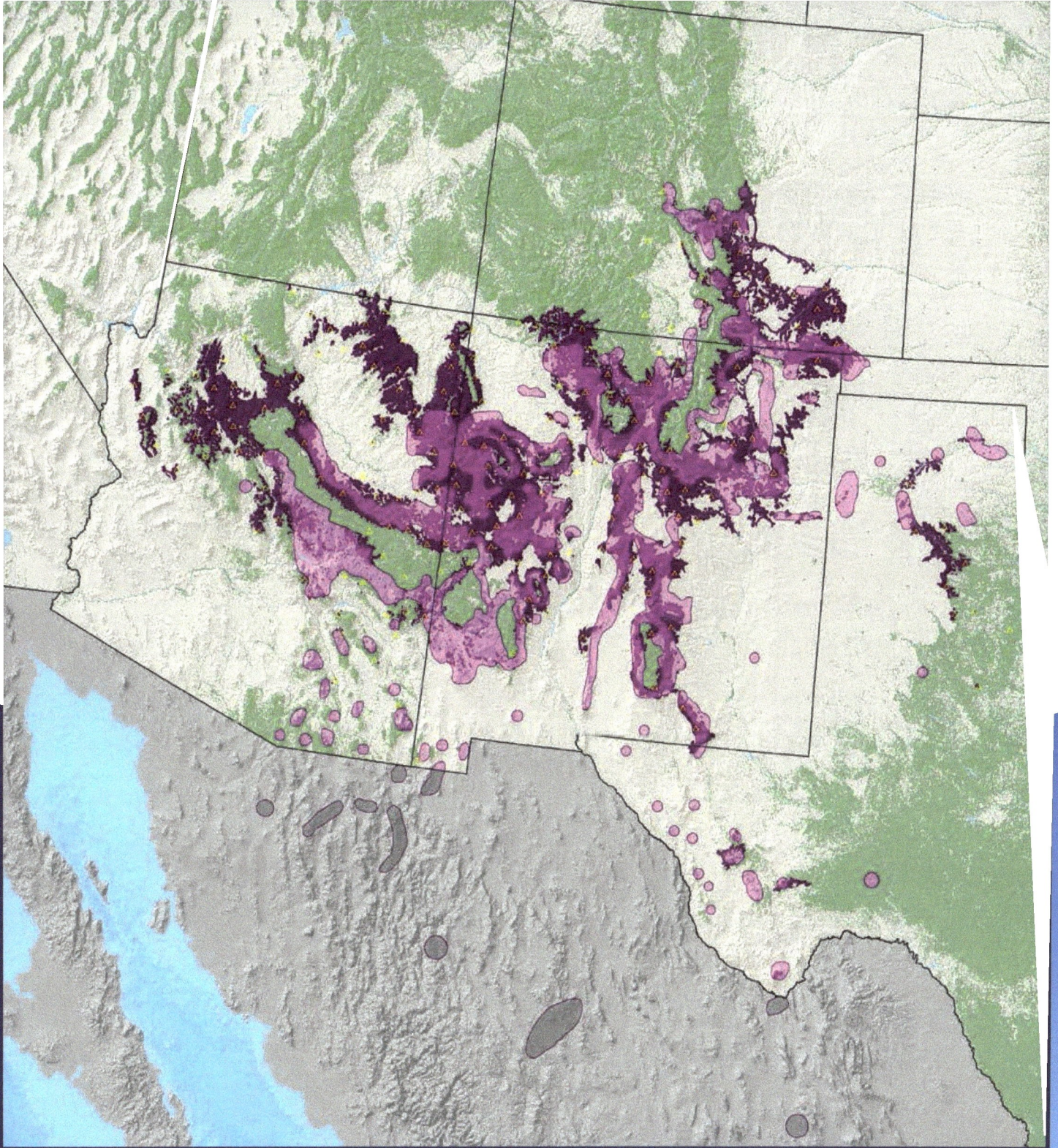

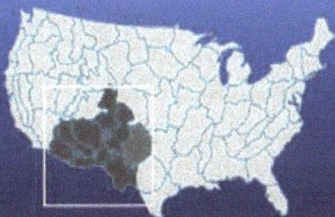

oneseed juniper
Juniperus monosperma (Engelm.) Sarg.
Cupressaceae – Cypress Family
FIA Code **69** – Plants Symbol **JUMO**

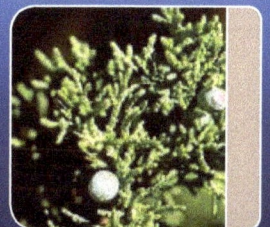

National Individual Tree Species Atlas

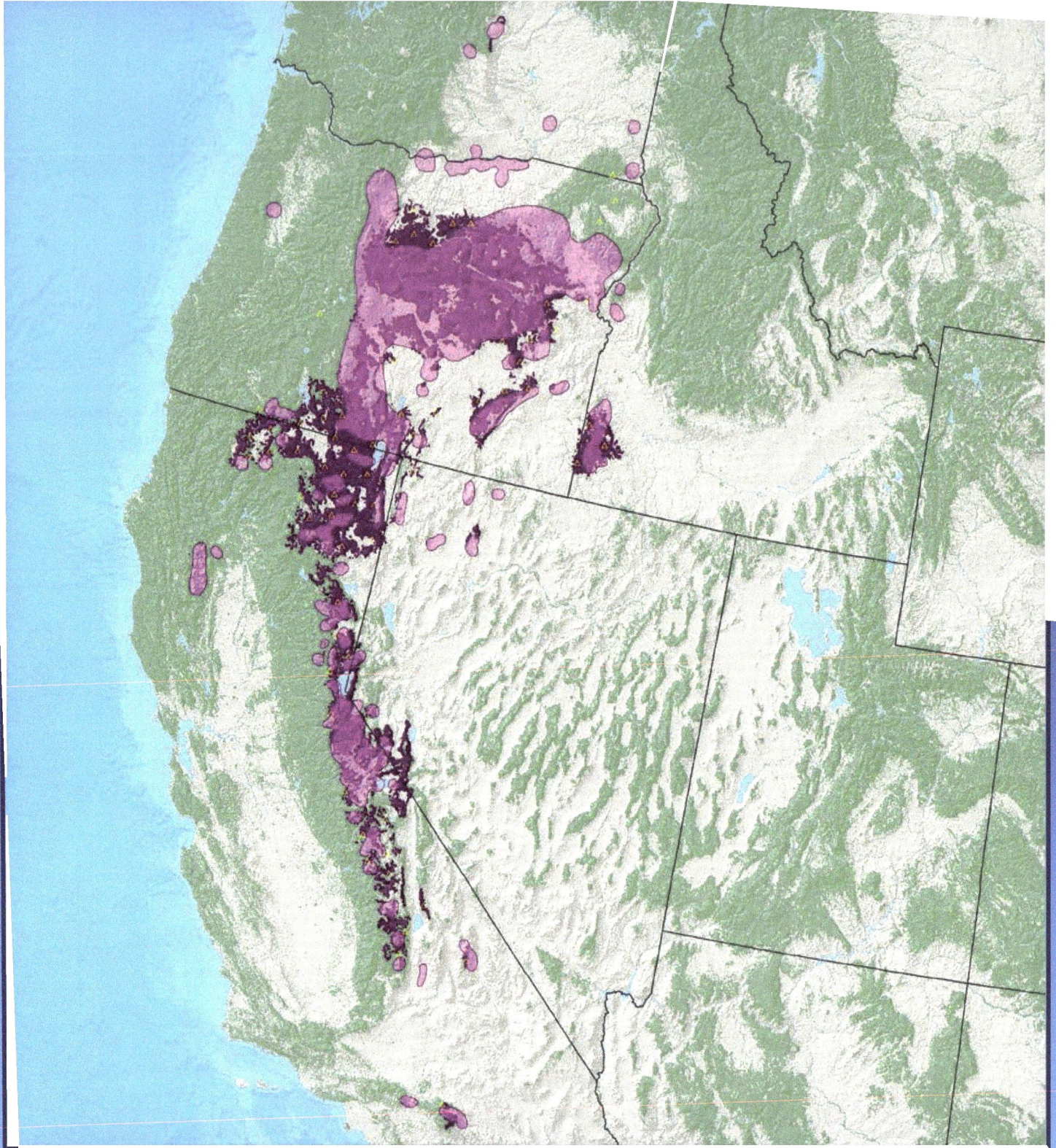

western juniper

Juniperus occidentalis Hook.

Cupressaceae – Cypress Family

FIA Code **64** – Plants Symbol **JUOC**

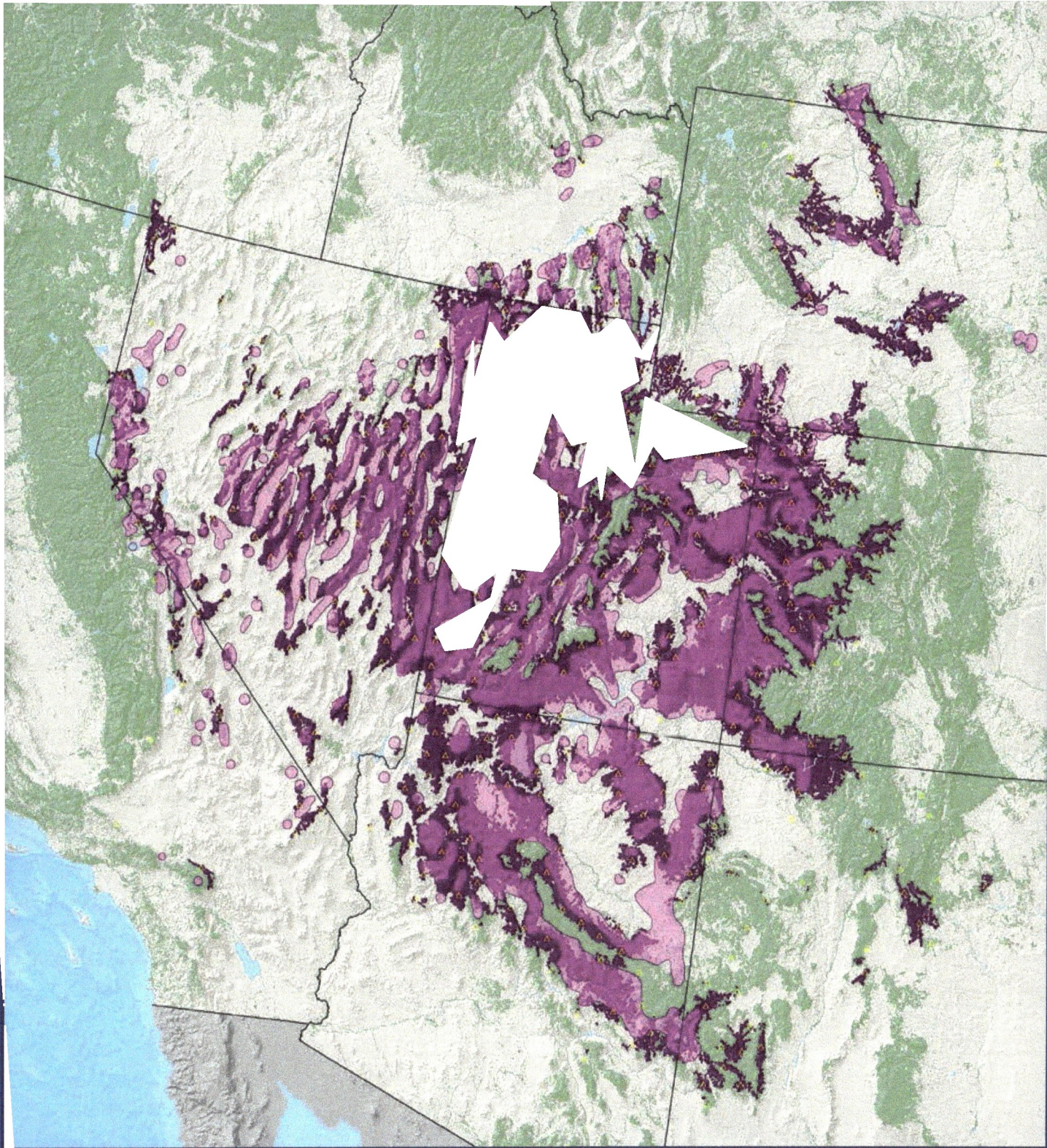

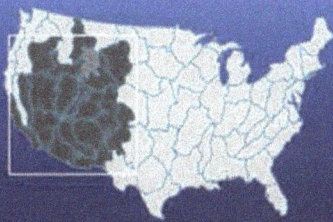

Utah juniper

Juniperus osteosperma (Torr.) Little

Cupressaceae – Cypress Family

FIA Code **65** – Plants Symbol **JUOS**

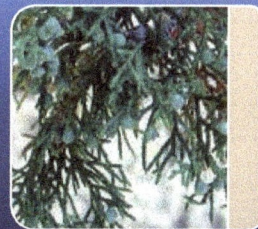

National Individual Tree Species Atlas

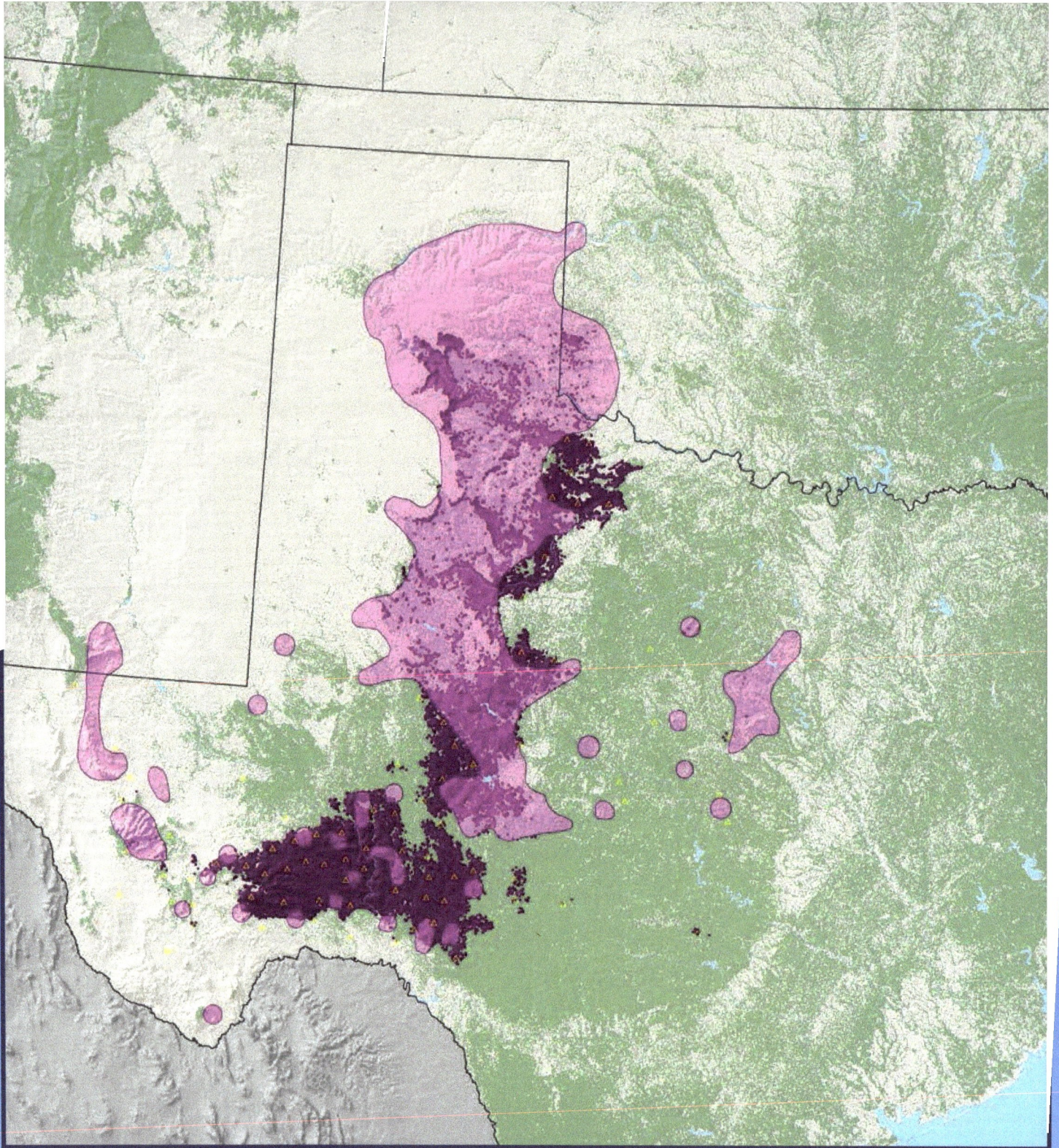

Pinchot juniper
Juniperus pinchotii Sudworth
Cuppressaceae – Cypress Family
FIA Code **58** – Plants Symbol **JUPI**

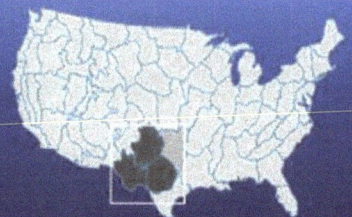

National Individual Tree Species Atlas

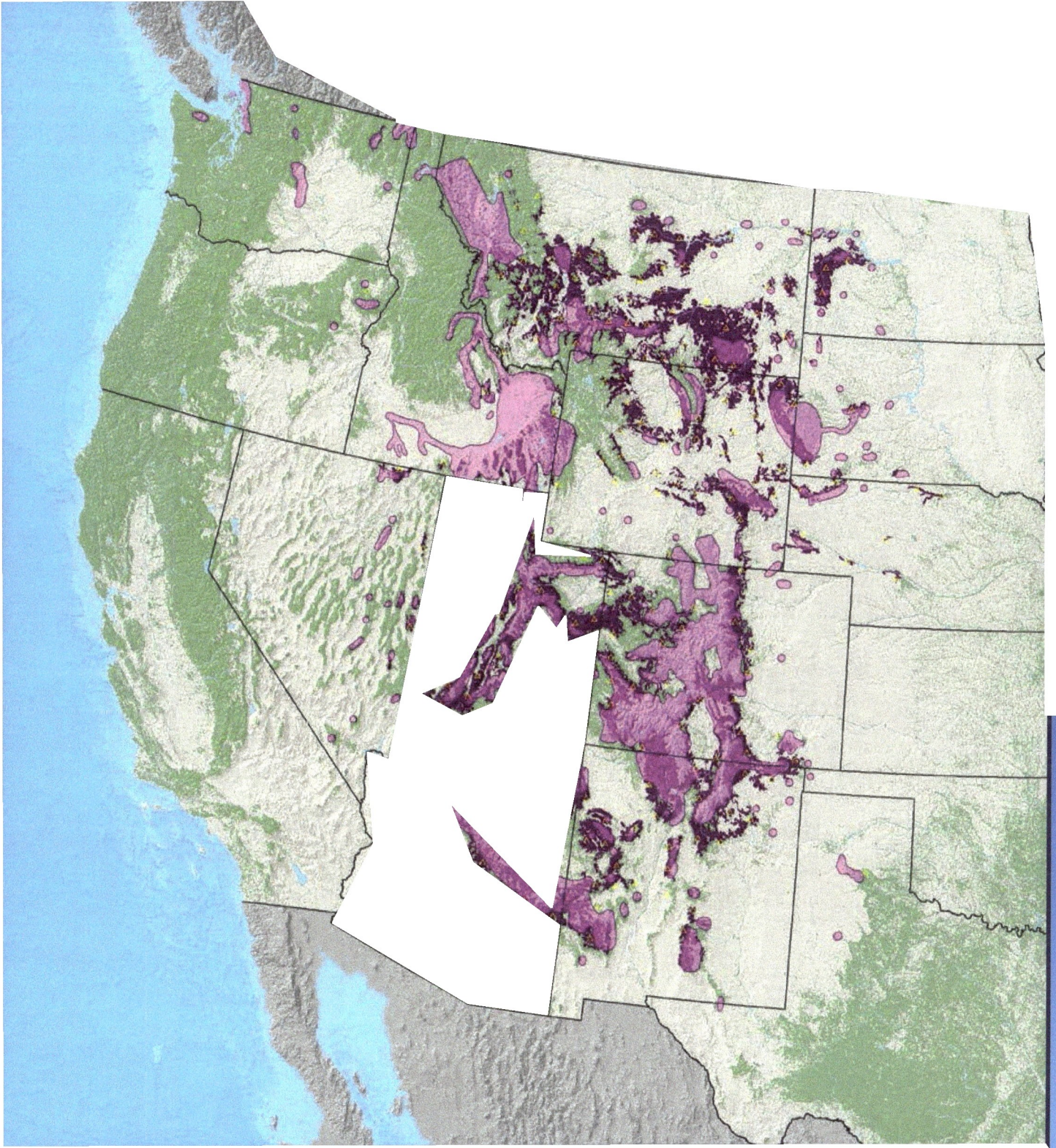

Rocky Mountain juniper
Juniperus scopulorum Sarg.
Cupressaceae – Cypress Family
FIA Code **66** – Plants Symbol **JUSC2**

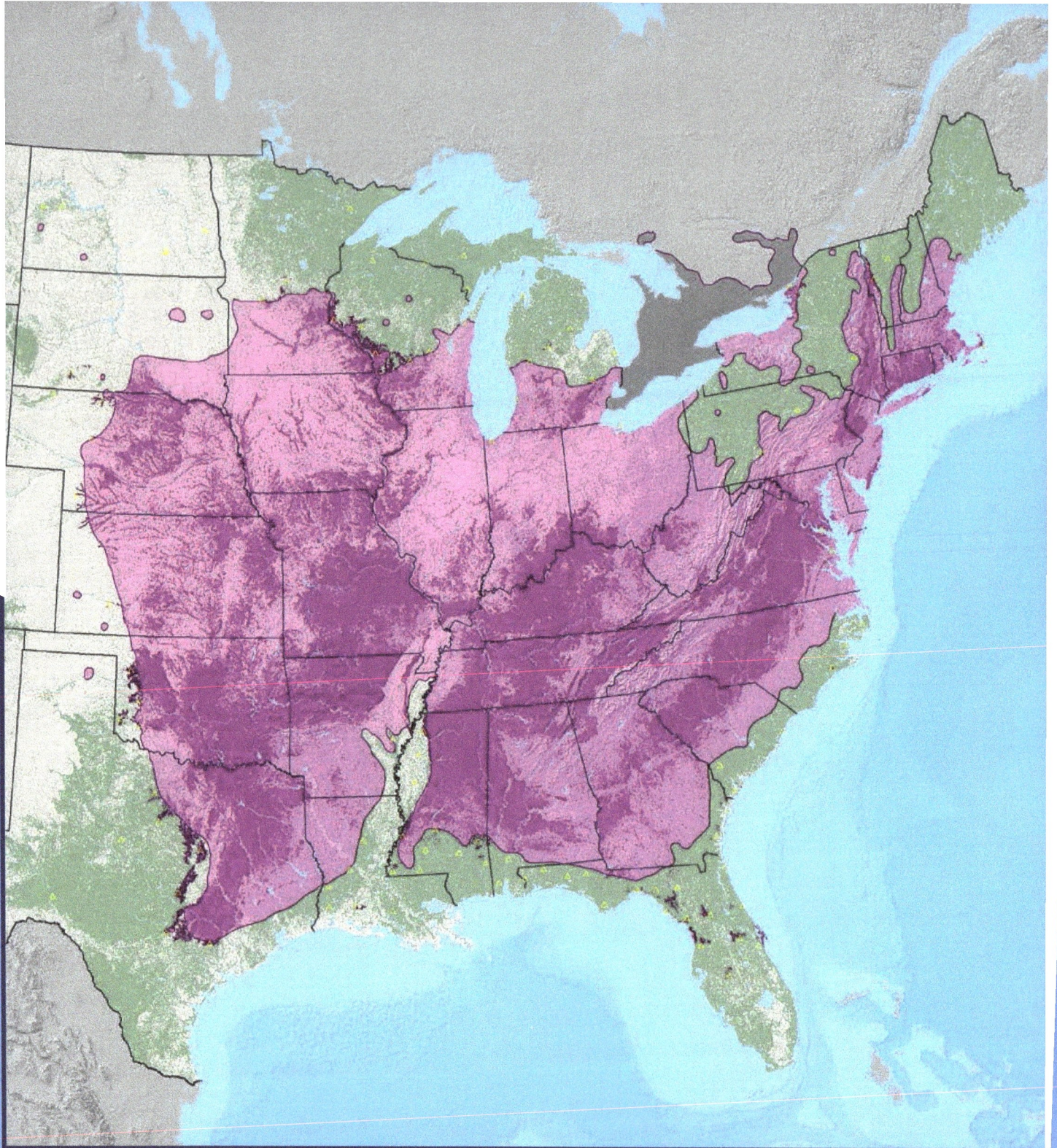

eastern redcedar

Juniperus virginiana L.

Cupressaceae – Cypress Family

FIA Code **68** – Plants Symbol **JUVI**

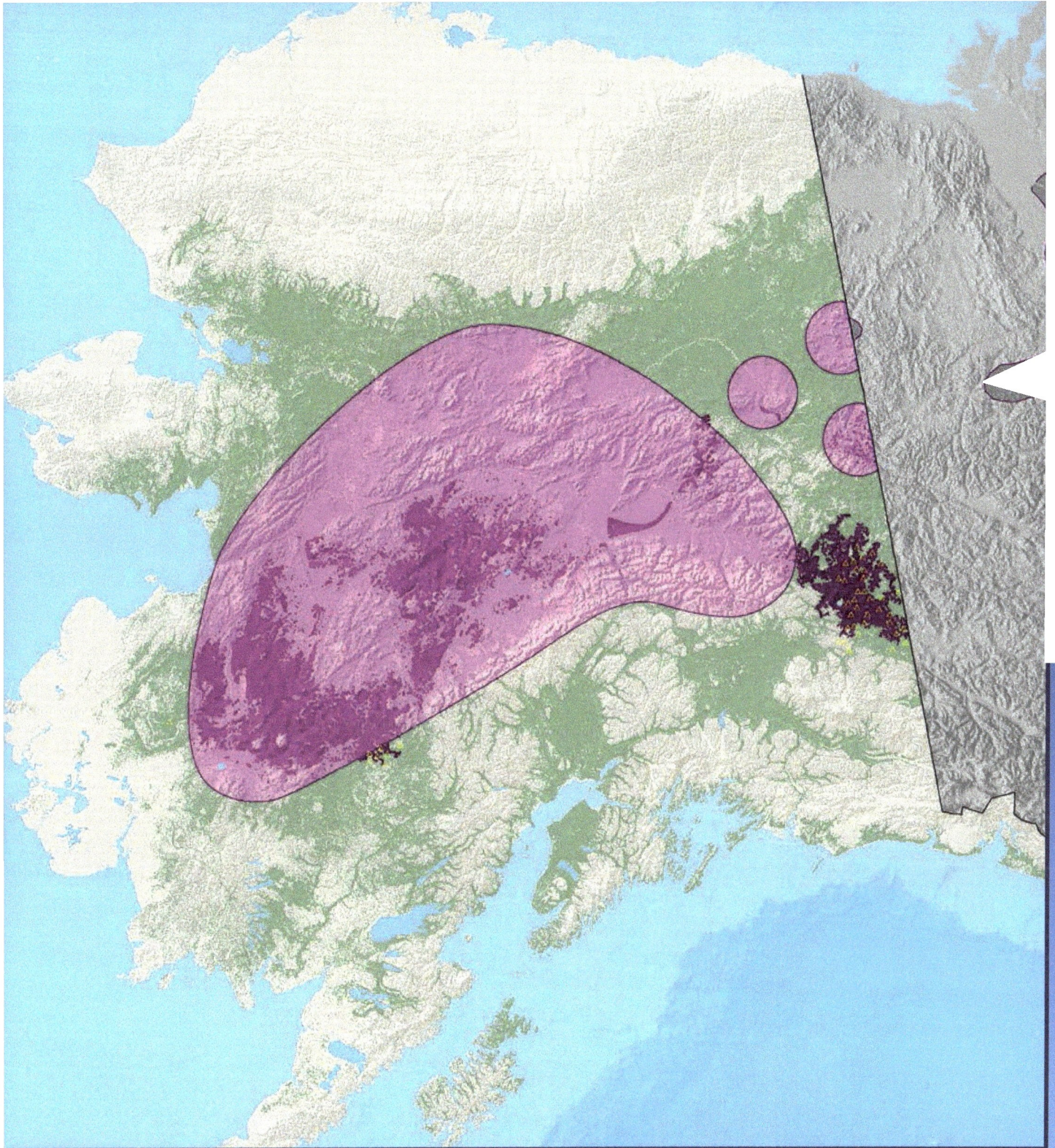

tamarack

Larix laricina (Du Roi) K. Koch

Pinaceae – Pine Family

FIA Code **71** – Plants Symbol **LALA**

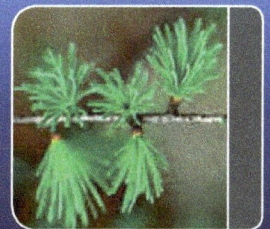

National Individual Tree Species Atlas

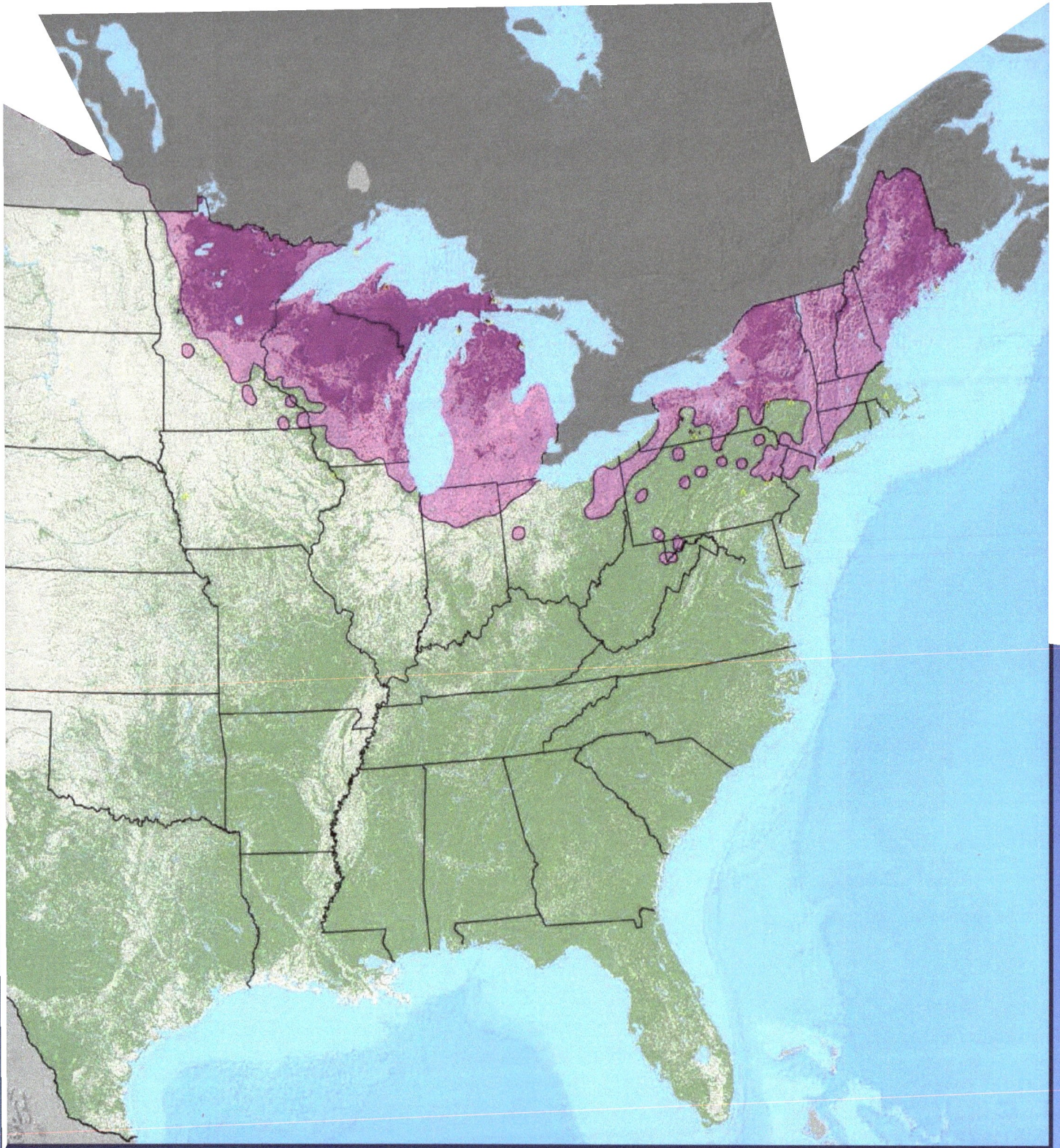

tamarack

Larix laricina (Du Roi) K. Koch

Pinaceae – Pine Family

FIA Code **71** – Plants Symbol **LALA**

National Individual Tree Species Atlas

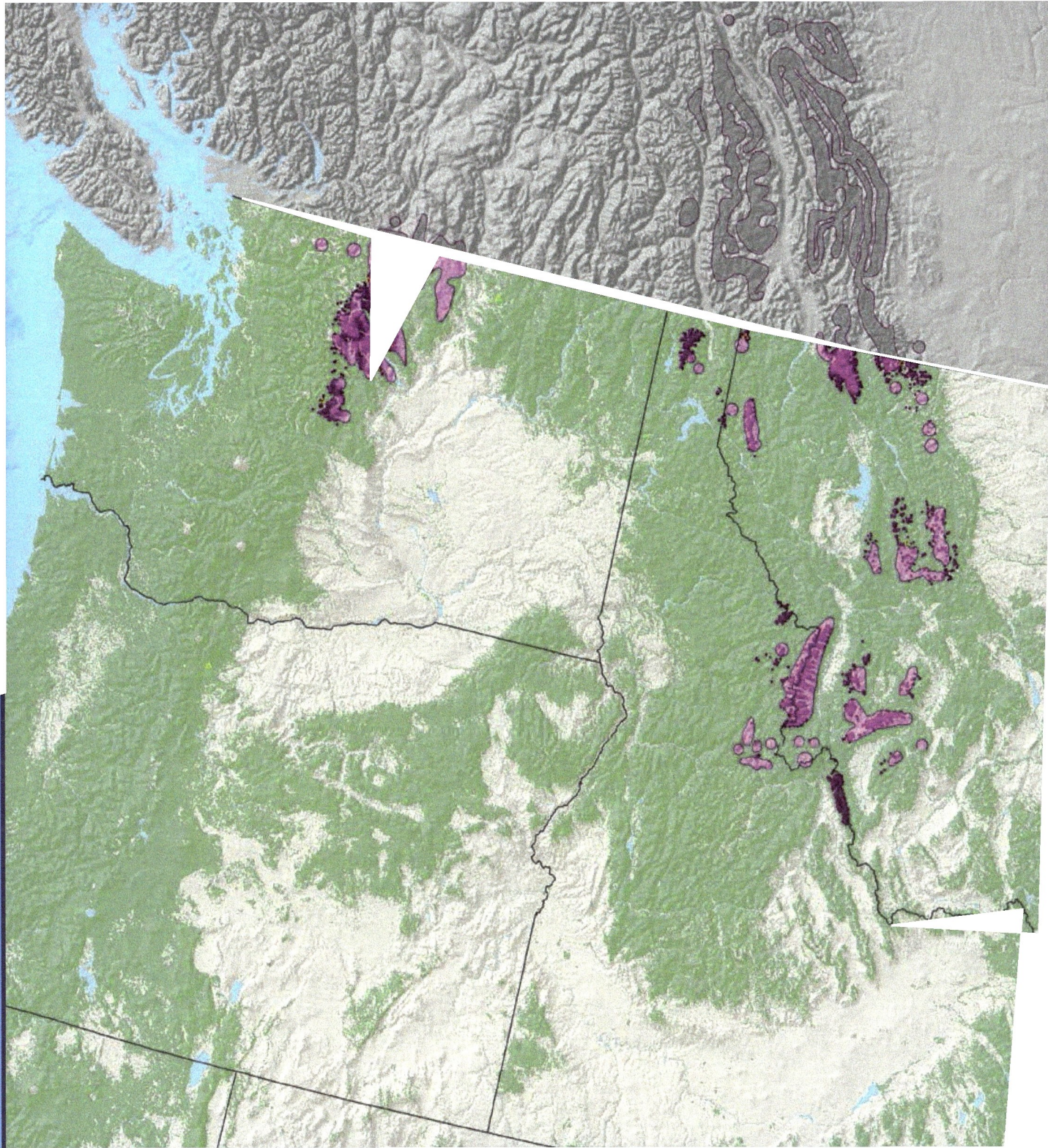

subalpine larch
Larix lyallii Parl.
Pinaceae – Pine Family
FIA Code **72** – Plants Symbol **LALY**

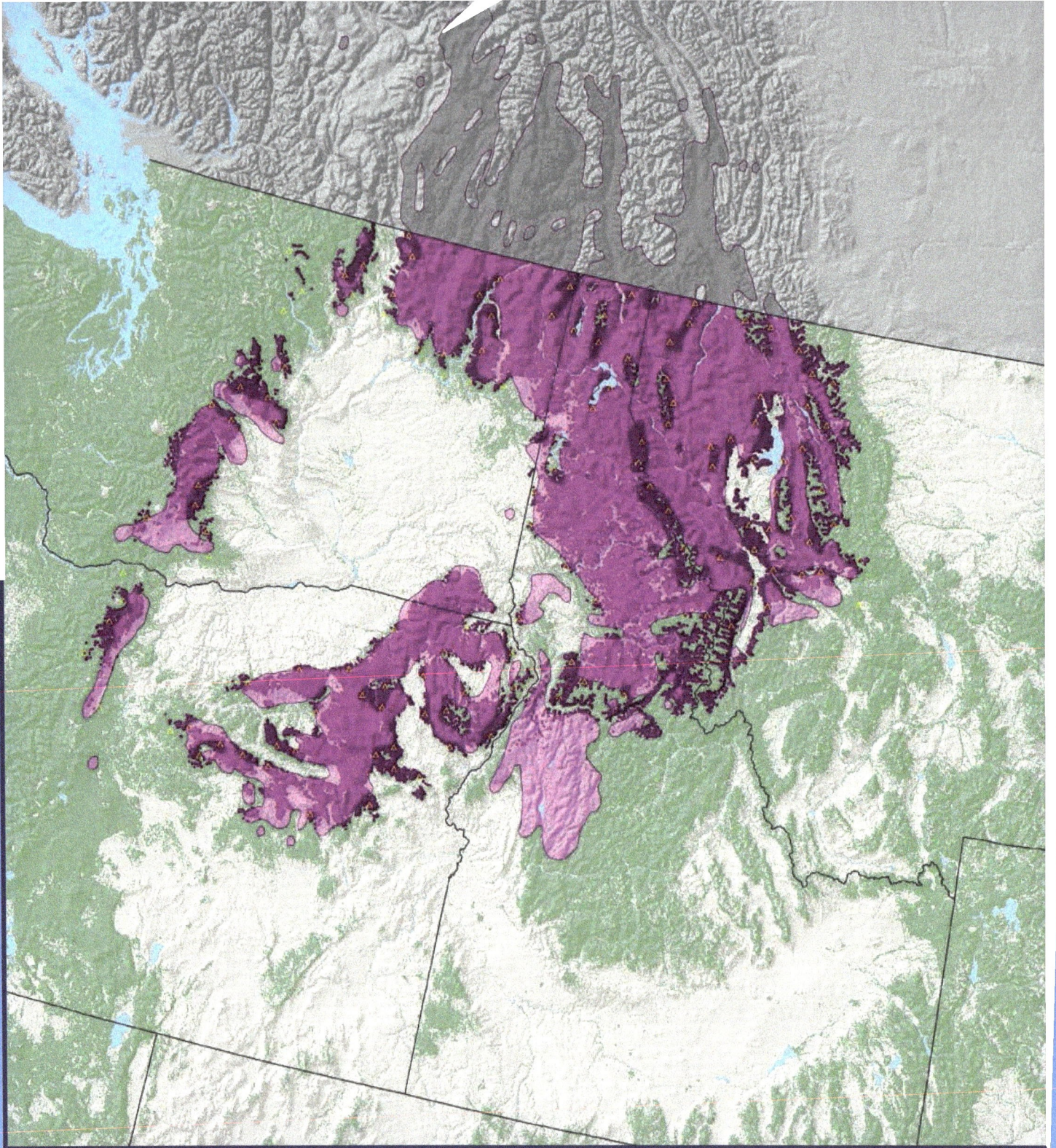

western larch

Larix occidentalis Nutt.

Pinaceae – Pine Family

FIA Code **73** – Plants Symbol **LAOC**

National Individual Tree Species Atlas

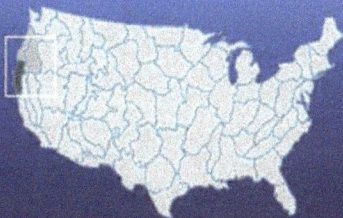

Brewer spruce

Picea breweriana S. Watson

Pinaceae – Pine Family

FIA Code **92** – Plants Symbol **PIBR**

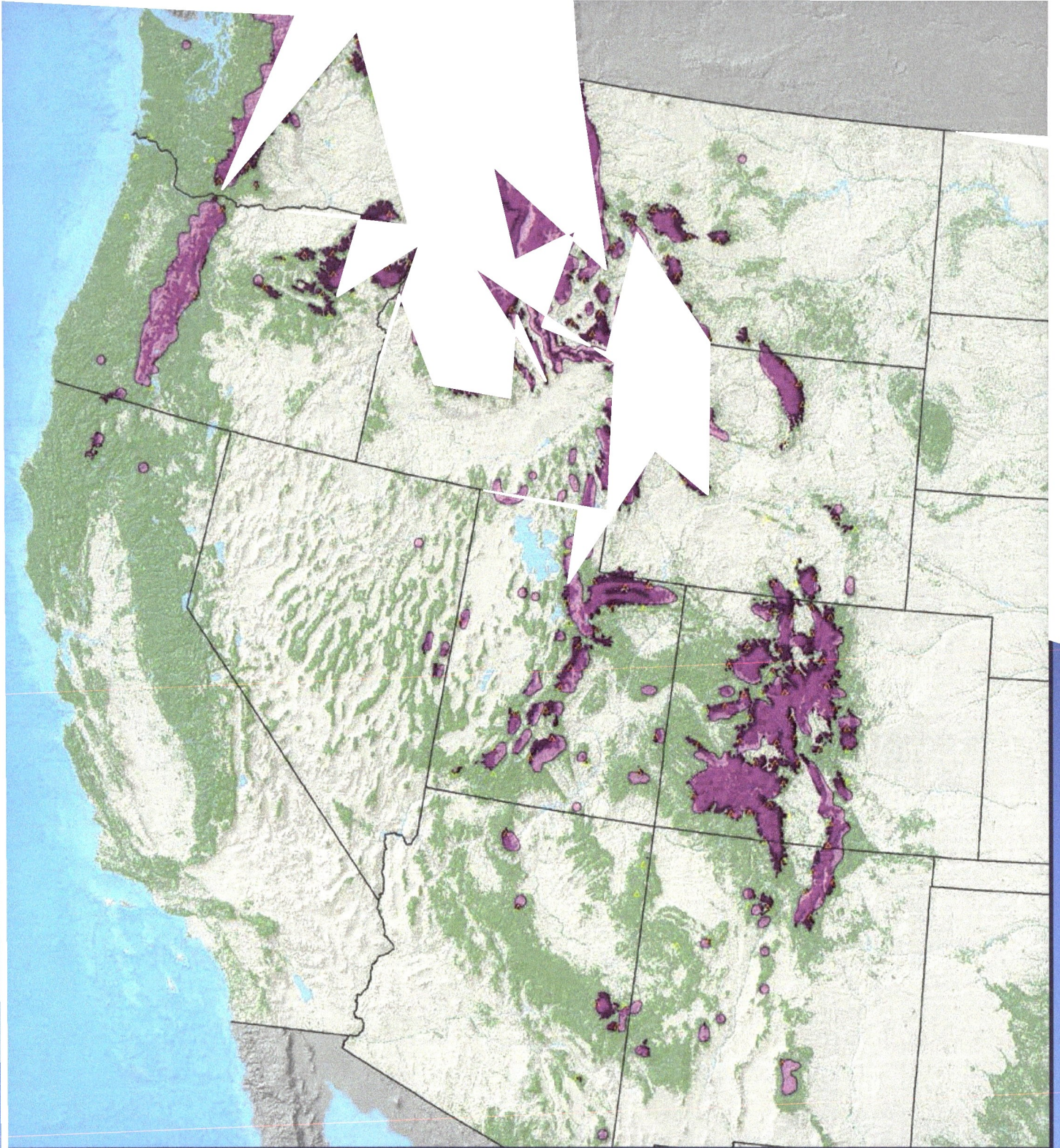

Engelmann spruce

Picea engelmannii Parry ex Engelm.

Pinaceae – Pine Family

FIA Code **93** – Plants Symbol **PIEN**

National Individual Tree Species Atlas

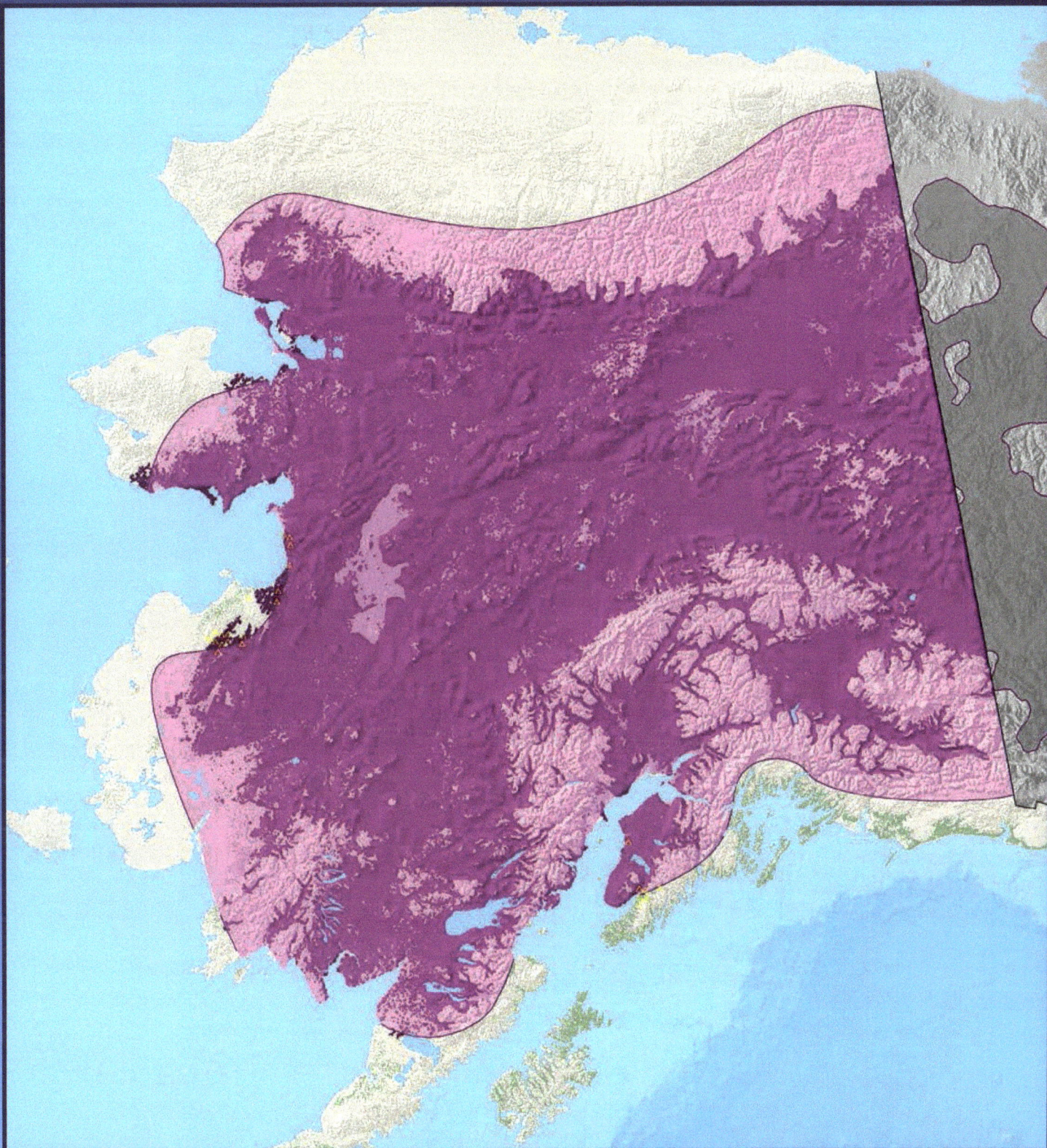

white spruce

white spruce
Picea glauca (Moench) Voss
Pinaceae – Pine Family
FIA Code **94** – Plants Symbol **PIGL**

white spruce
Picea glauca (Moench) Voss

Pinaceae – Pine Family

FIA Code **94** – Plants Symbol **PIGL**

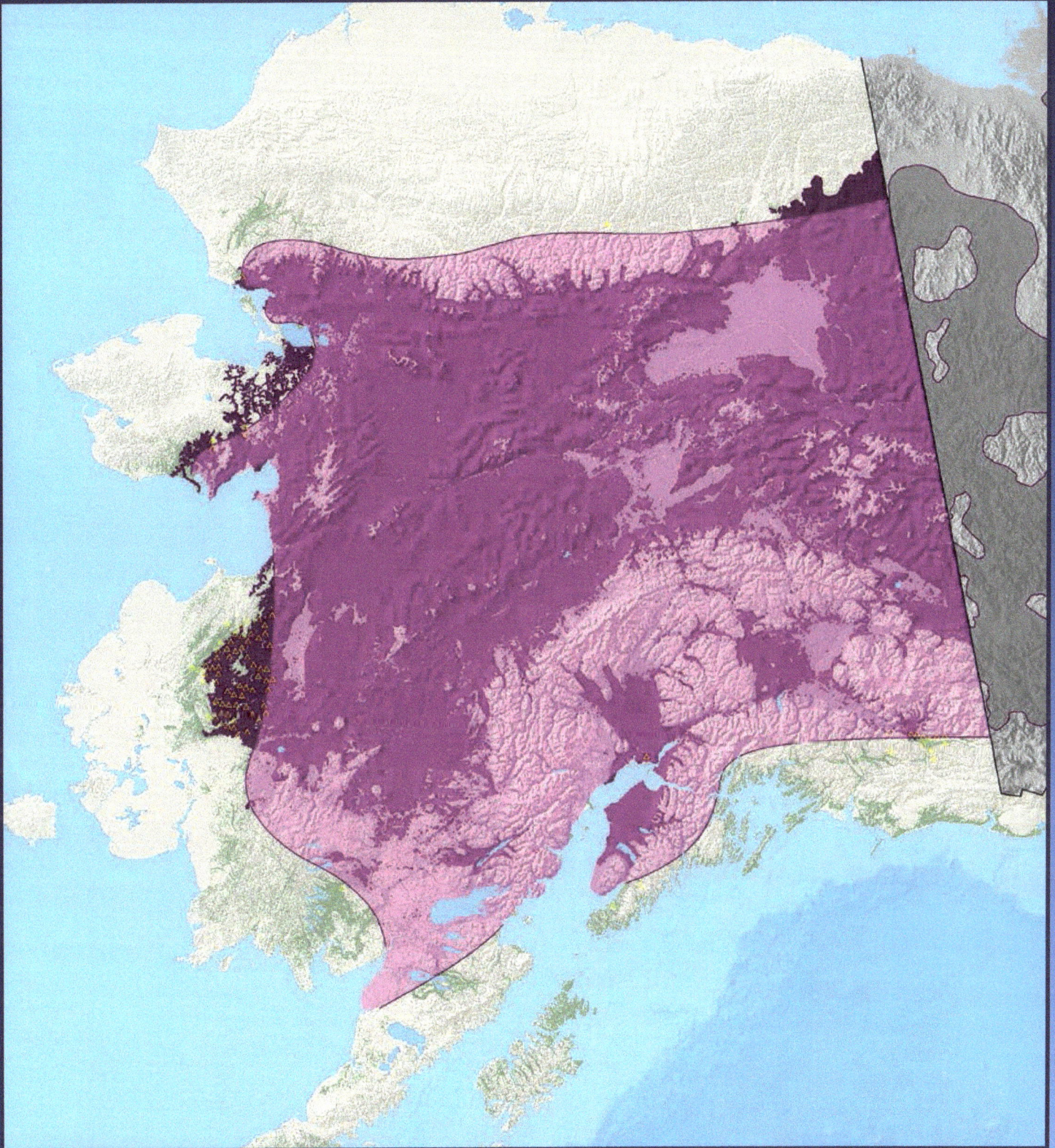

black spruce

black spruce

Picea mariana (Mill.) Britton, Sterns & Poggenb.

Pinaceae – Pine Family

FIA Code **95** – Plants Symbol **PIMA**

black spruce

Picea mariana (Mill.) Britton, Sterns & Poggenb.

Pinaceae – Pine Family

FIA Code **95** – Plants Symbol **PIMA**

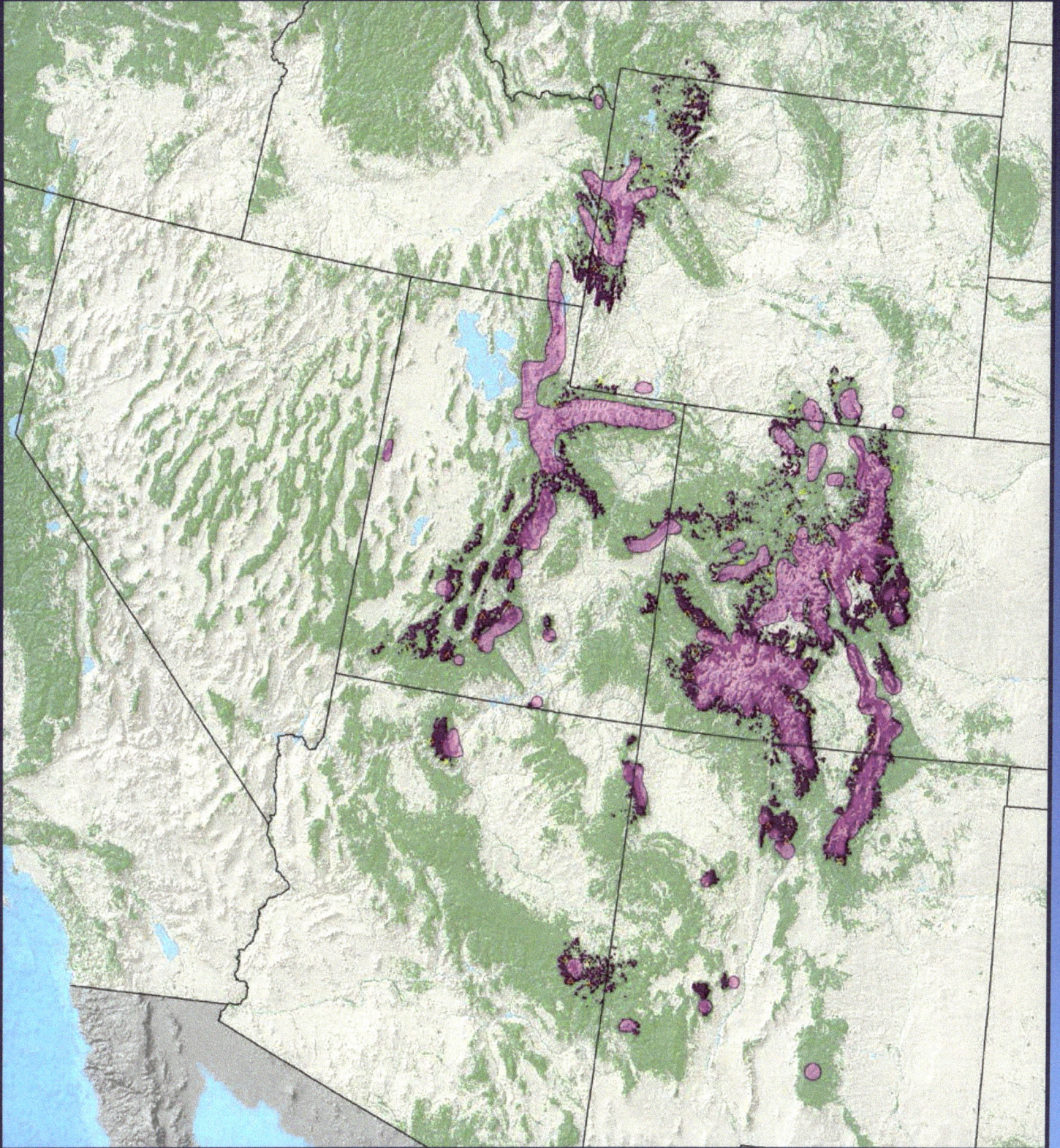

blue spruce

Picea pungens Engelm.

Pinaceae – Pine Family

FIA Code **96** – Plants Symbol **PIPU**

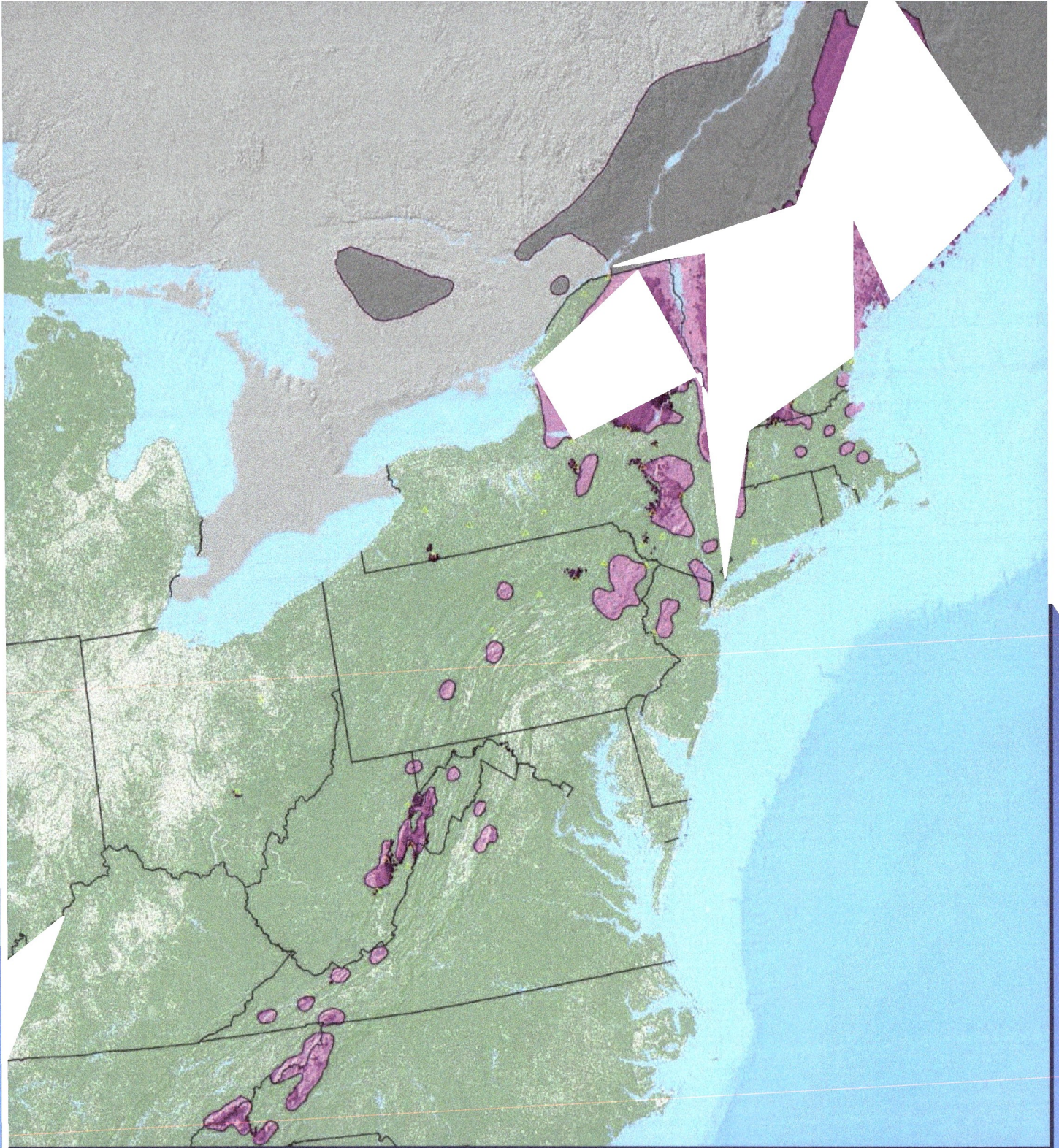

red spruce
Picea rubens Sarg.
Pinaceae – Pine Family
FIA Code **97** – Plants Symbol **PIRU**

National Individual Tree Species Atlas

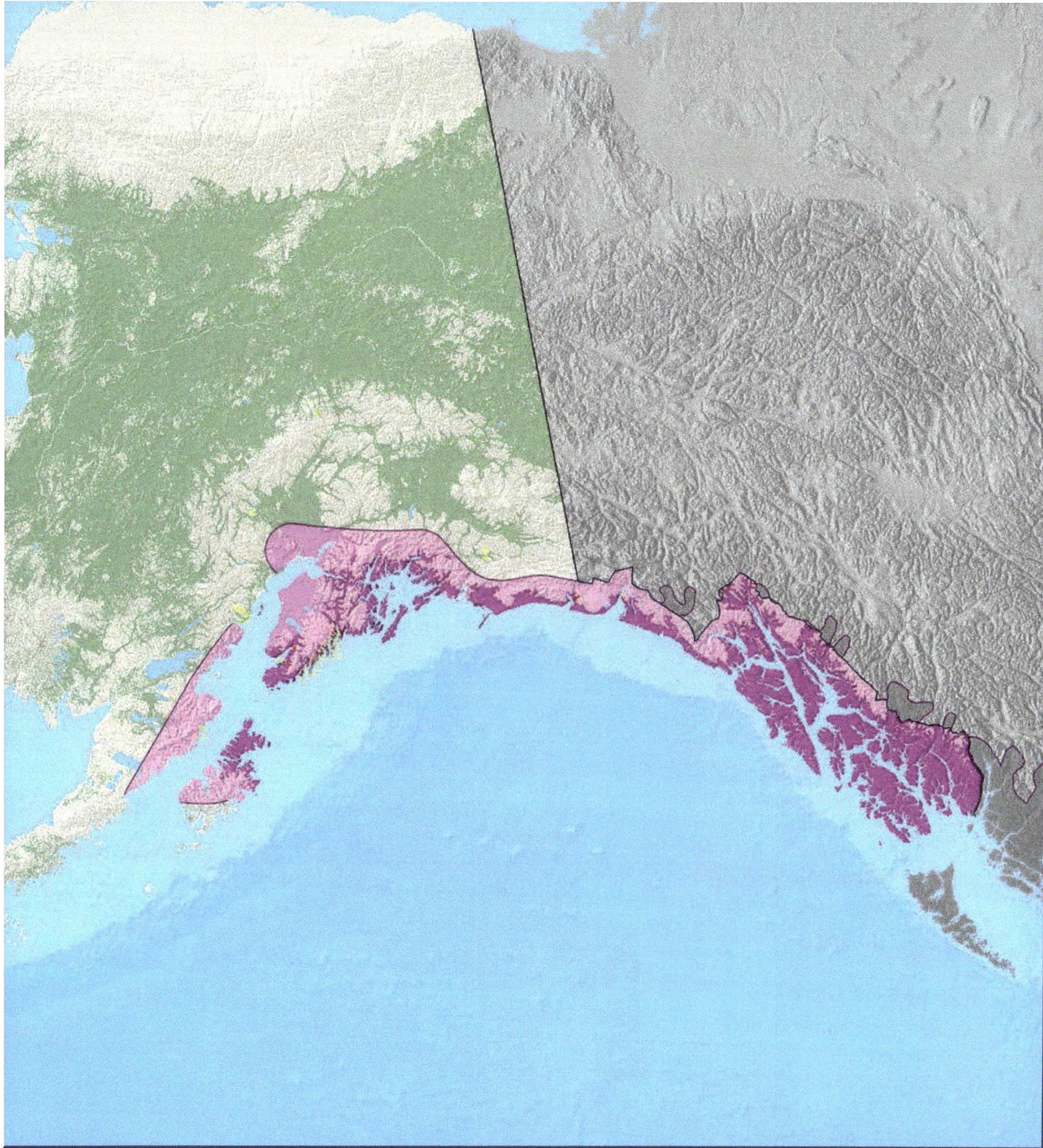

Sitka spruce

Picea sitchensis (Bong.) Carrière

Pinaceae – Pine Family

FIA Code **98** – Plants Symbol **PISI**

National Individual Tree Species Atlas

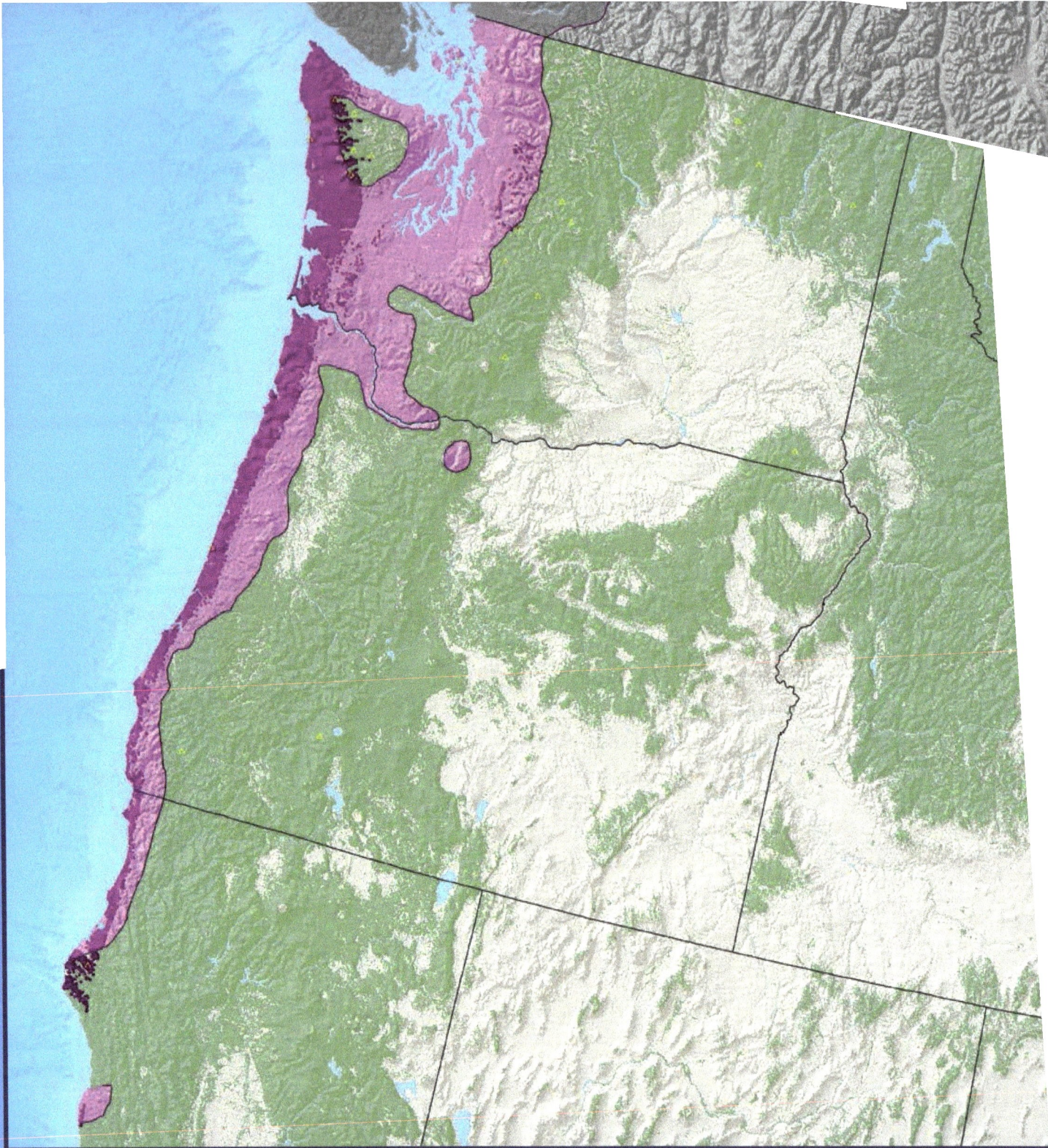

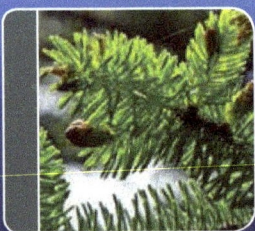

Sitka spruce

Picea sitchensis (Bong.) Carrière

Pinaceae – Pine Family

FIA Code **98** – Plants Symbol **PISI**

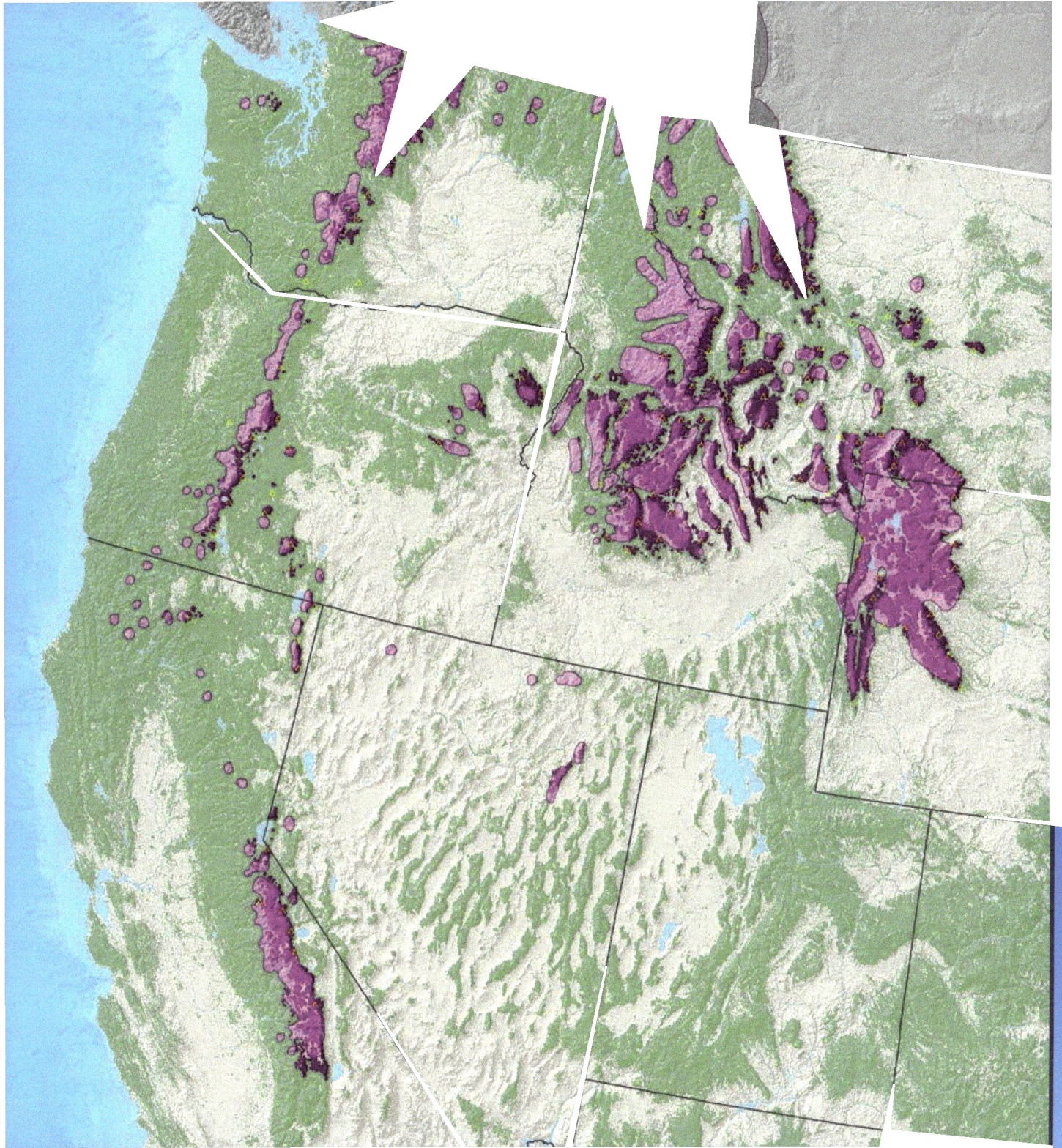

whitebark pine

Pinus albicaulis Engelm.

Pinaceae – Pine Family

FIA Code **101** – Plants Symbol **PIAL**

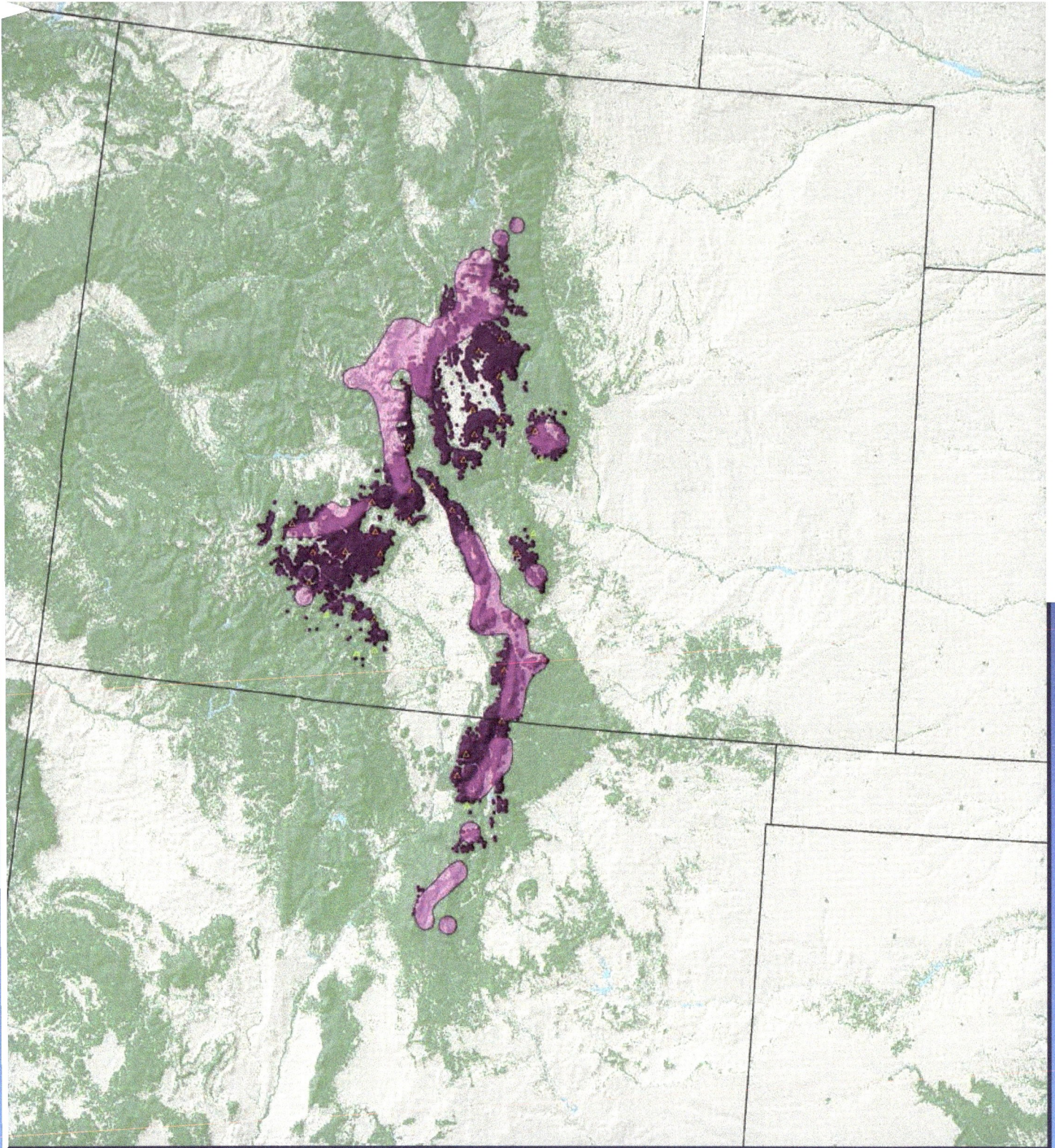

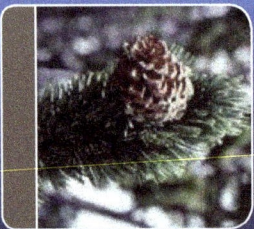

Rocky Mountain bristlecone pine

Pinus aristata Engelm.

Pinaceae – Pine Family

FIA Code **102** – Plants Symbol **PIAR**

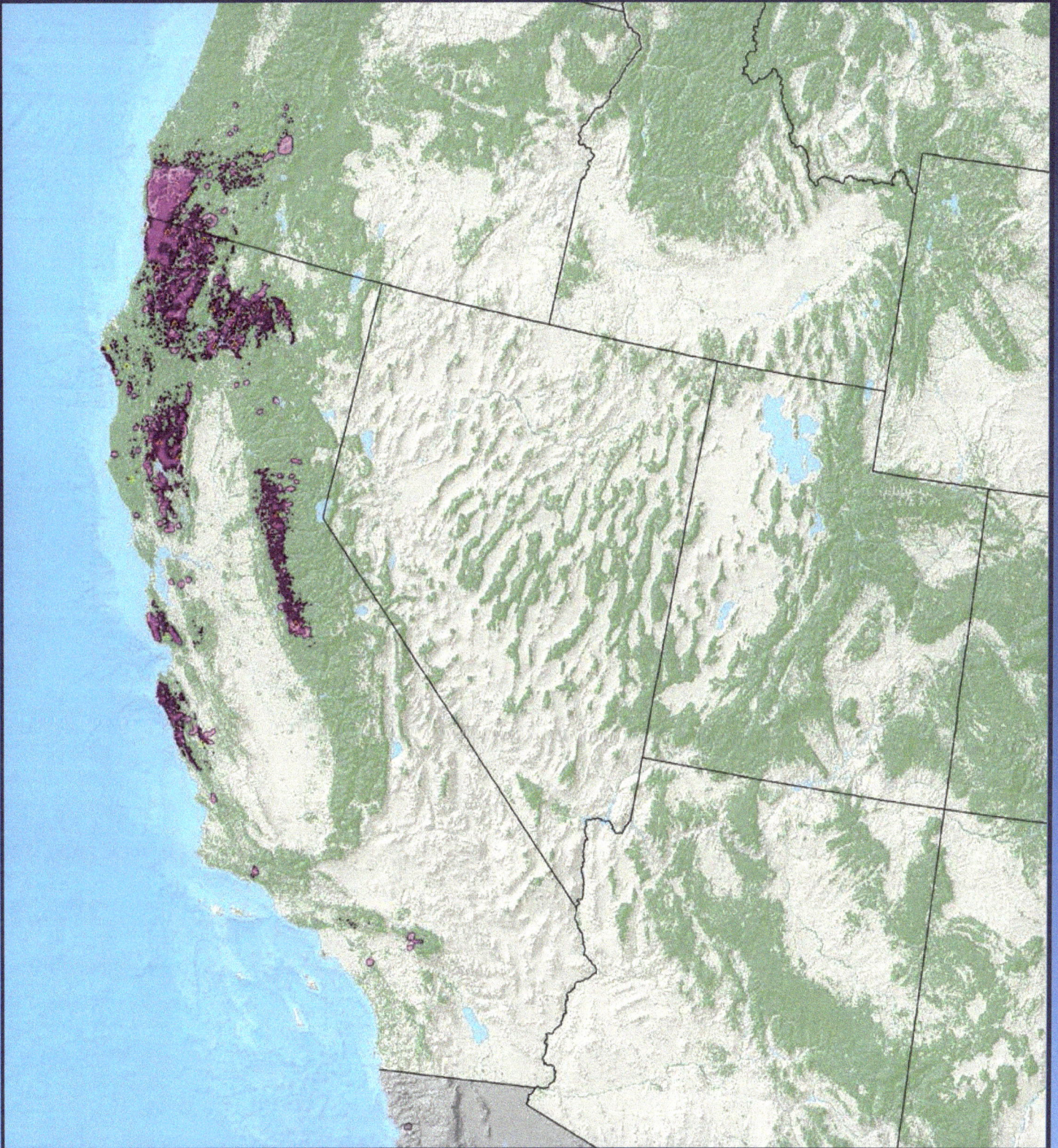

knobcone pine

Pinus attenuata Lemmon

Pinaceae – Pine Family

FIA Code **103** – Plants Symbol **PIAT**

National Individual Tree Species Atlas

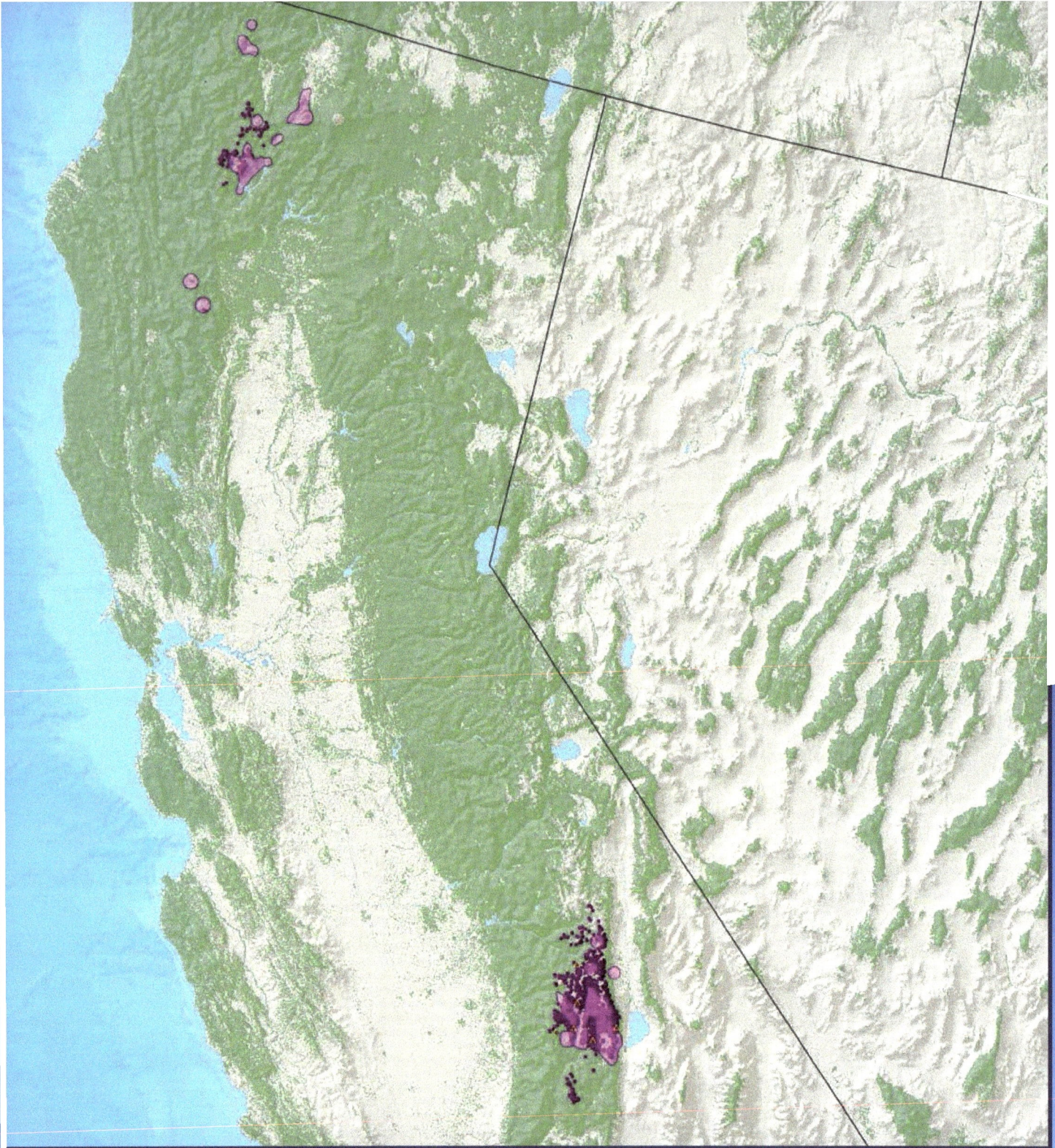

foxtail pine
Pinus balfouriana Balf.
Pinaceae – Pine Family
FIA Code **104** – Plants Symbol **PIBA**

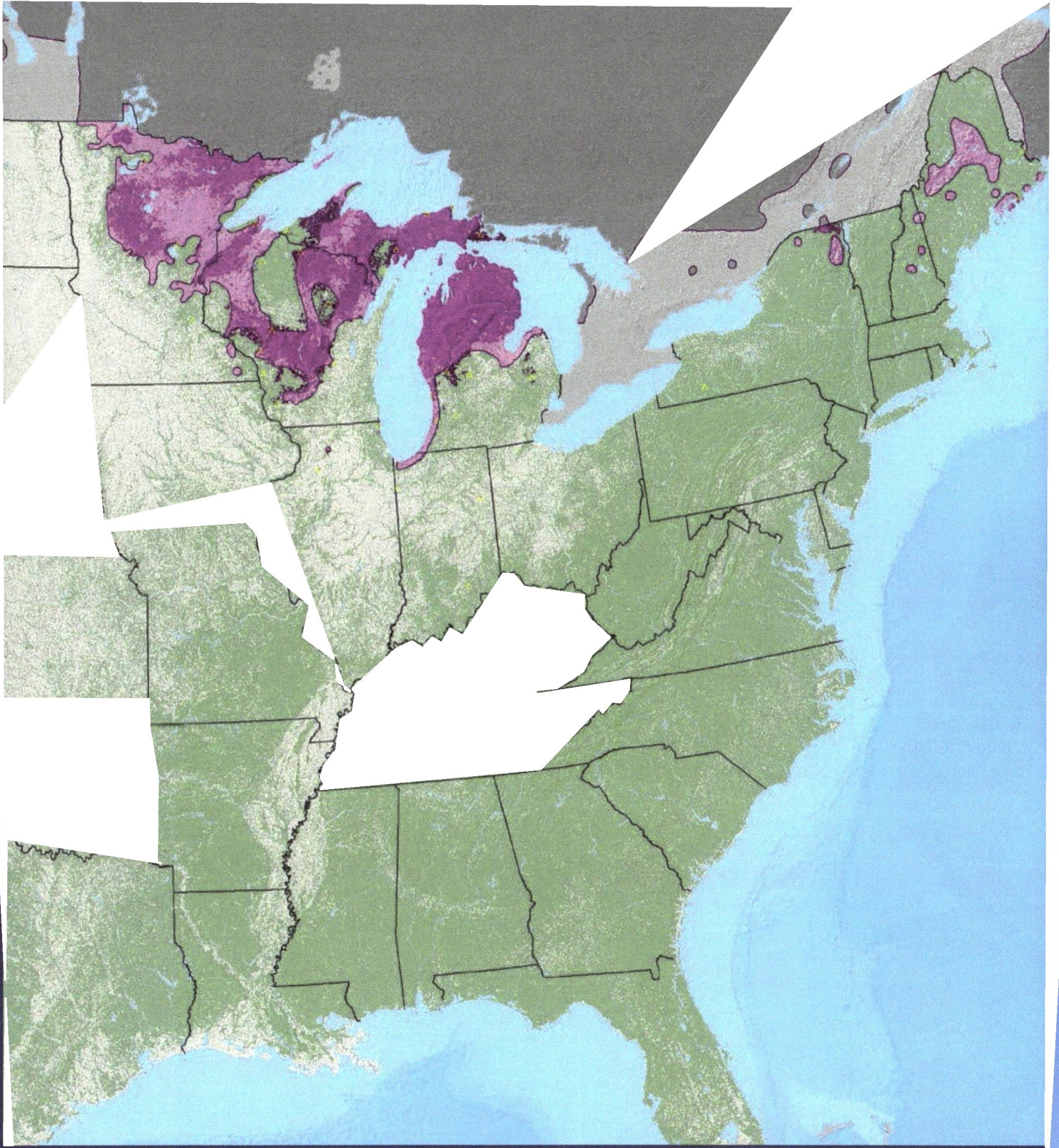

jack pine
Pinus banksiana Lamb.
Pinaceae – Pine Family
FIA Code **105** – Plants Symbol **PIBA2**

National Individual Tree Species Atlas

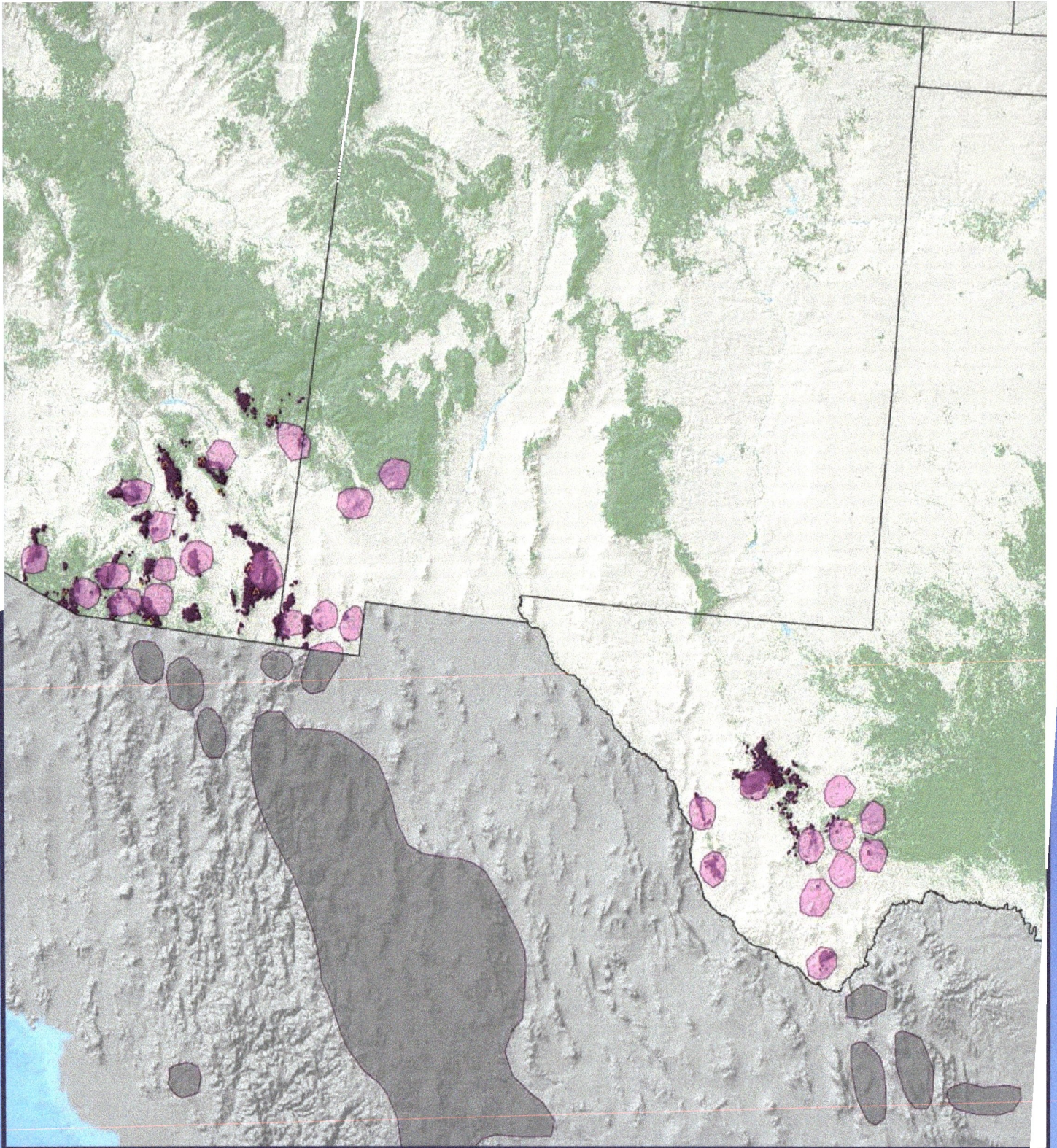

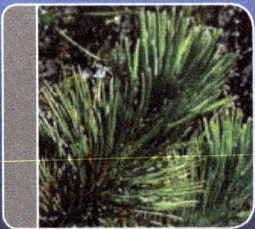

Mexican piñon pine

Pinus cembroides Zucc.

Pinaceae – Pine Family

FIA Code **140** – Plants Symbol **PICE**

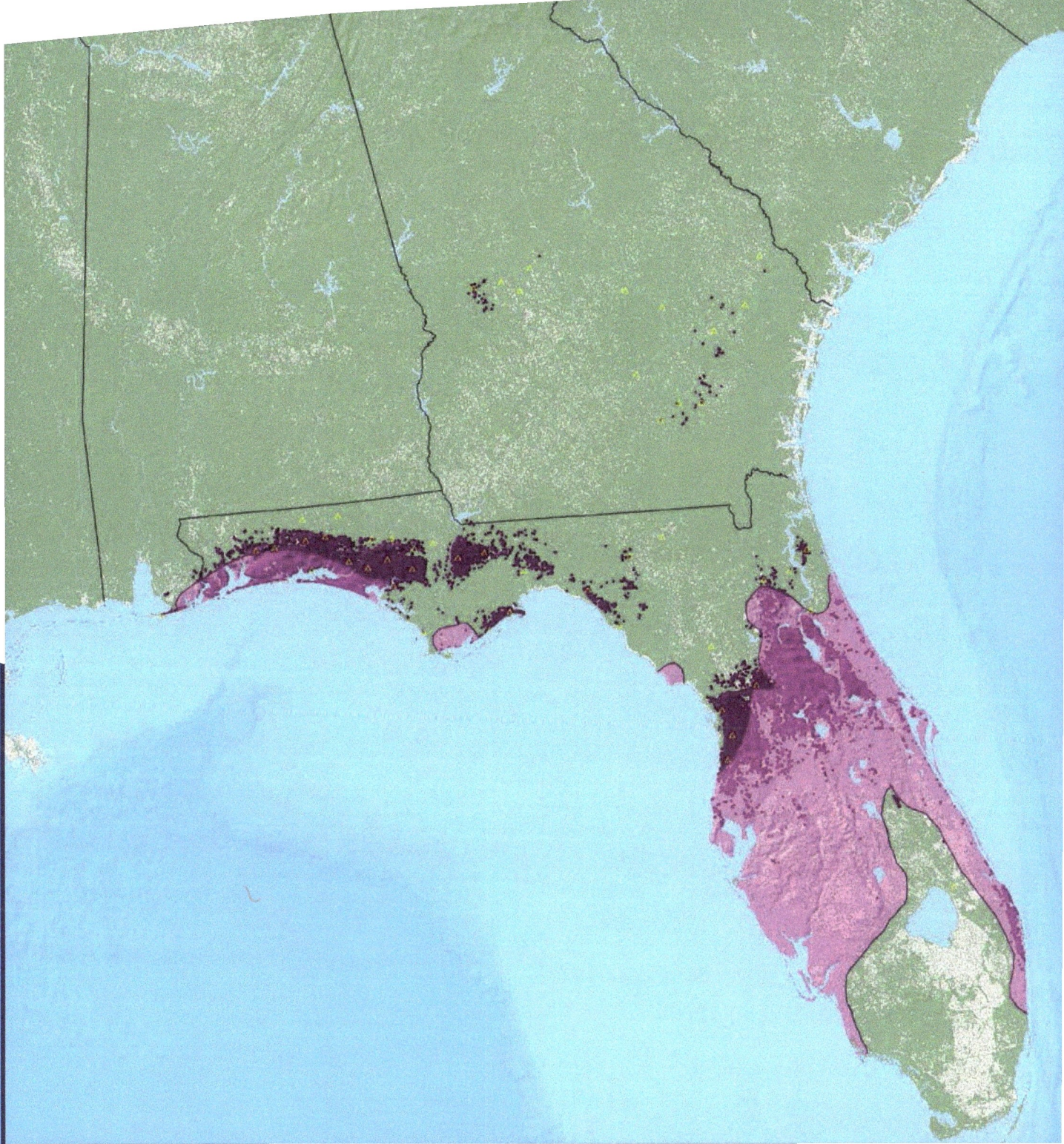

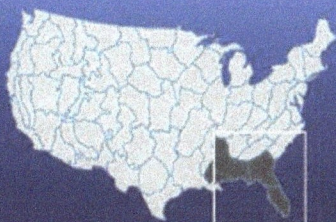

sand pine

Pinus clausa (Chapman ex Engelm.) Vasey ex Sarg.

Pinaceae – Pine Family

FIA Code **107** – Plants Symbol **PICL**

National Individual Tree Species Atlas

lodgepole pine

Pinus contorta Douglas ex Louden

Pinaceae – Pine Family

FIA Code **108** – Plants Symbol **PICO**

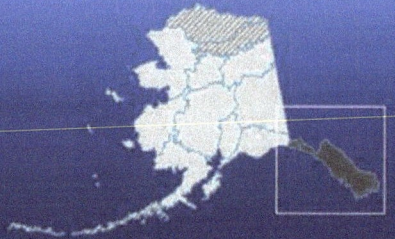

lodgepole pine

Pinus contorta Douglas ex Louden

Pinaceae – Pine Family

FIA Code **108** – Plants Symbol **PICO**

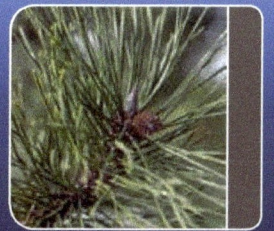

National Individual Tree Species Atlas

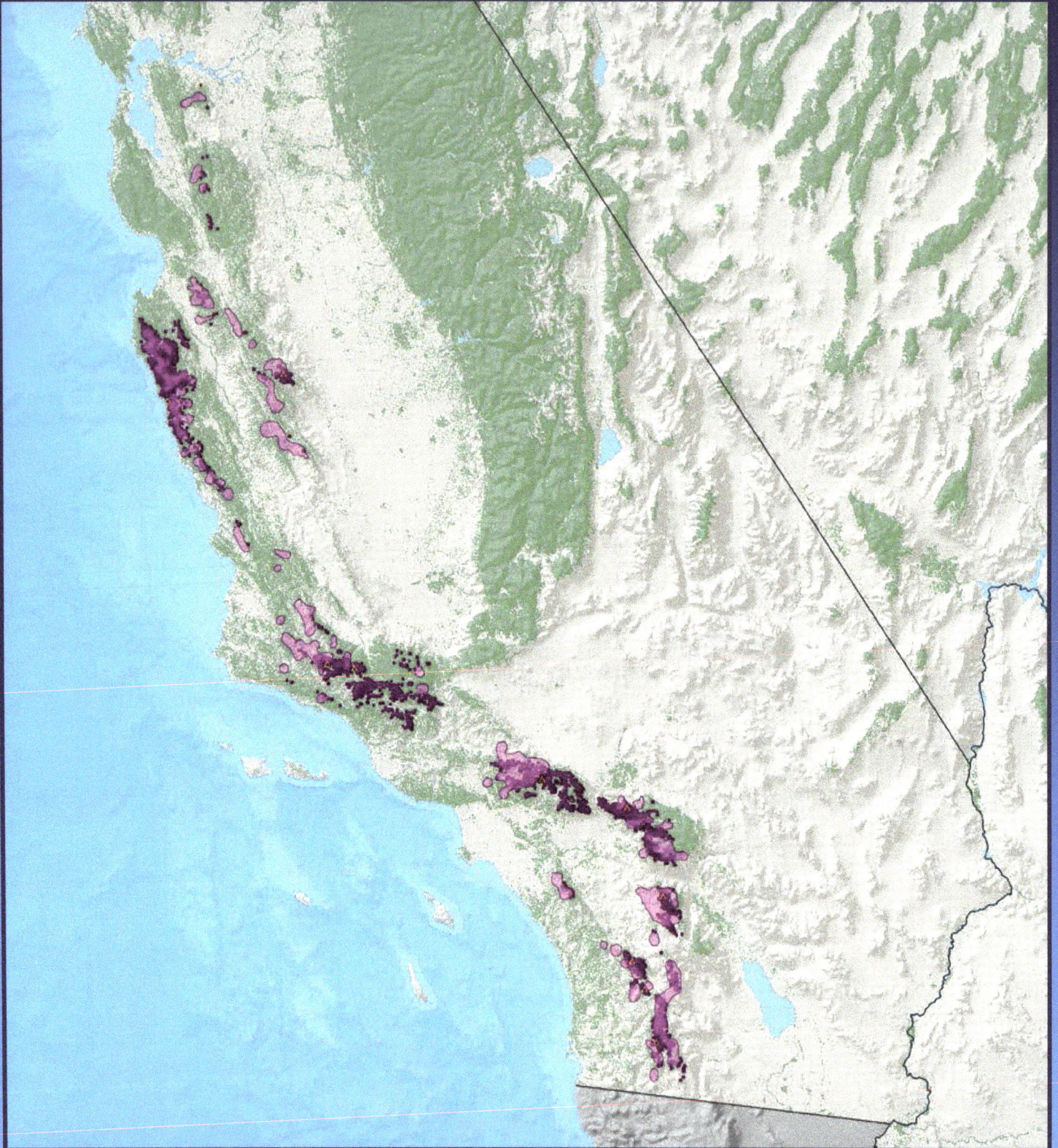

Coulter pine
Pinus coulteri D. Don
Pinaceae – Pine Family
FIA Code **109** – Plants Symbol **PICO3**

Coulter pine

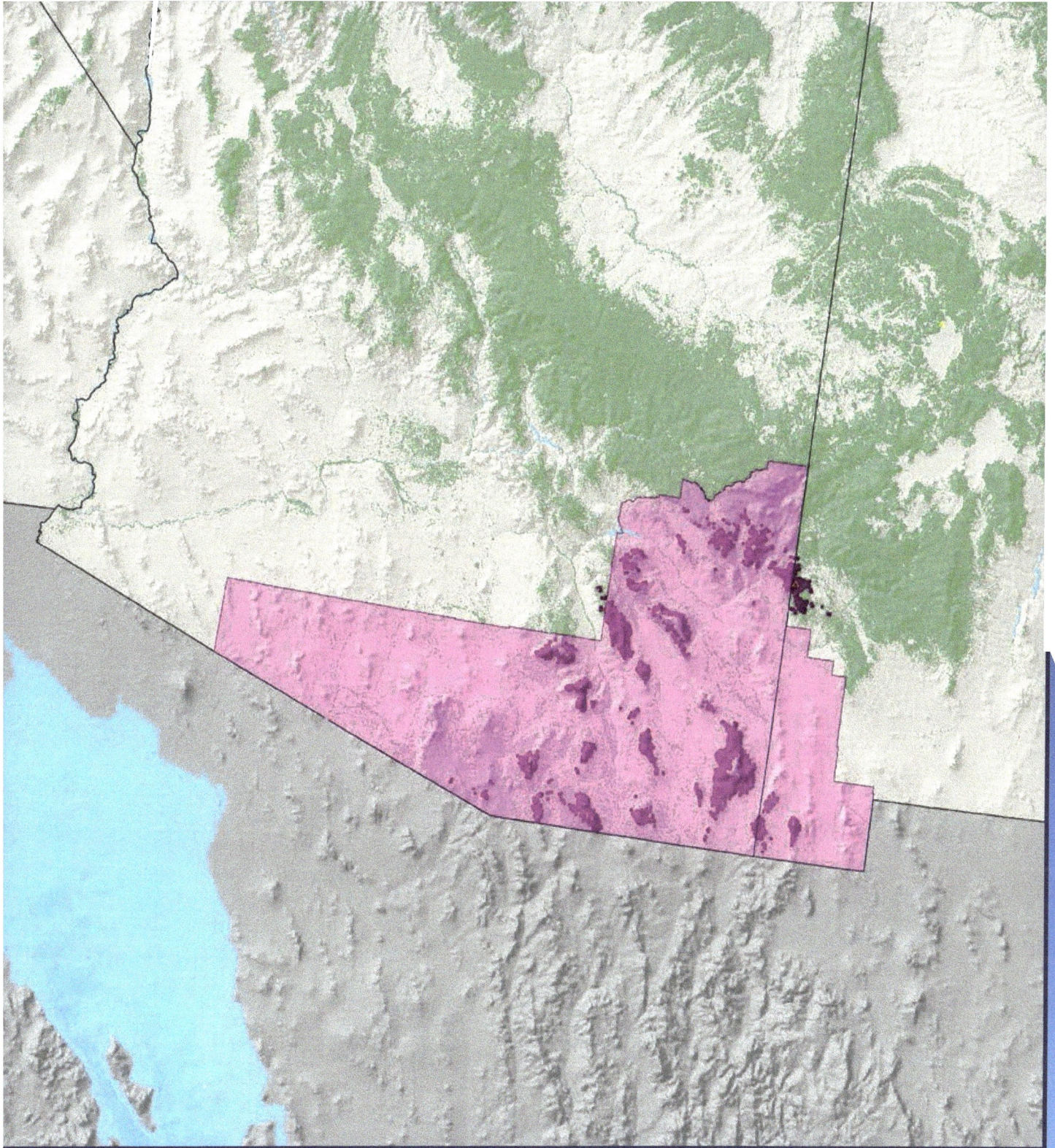

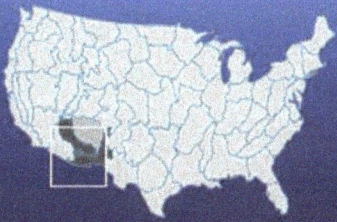

border piñon

Pinus discolor D.K. Bailey & Hawksworth

Pinaceae – Pine Family

FIA Code **134** – Plants Symbol **PIDI3**

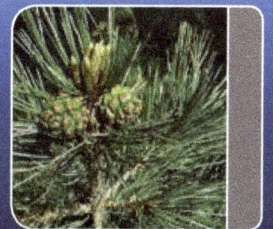

National Individual Tree Species Atlas

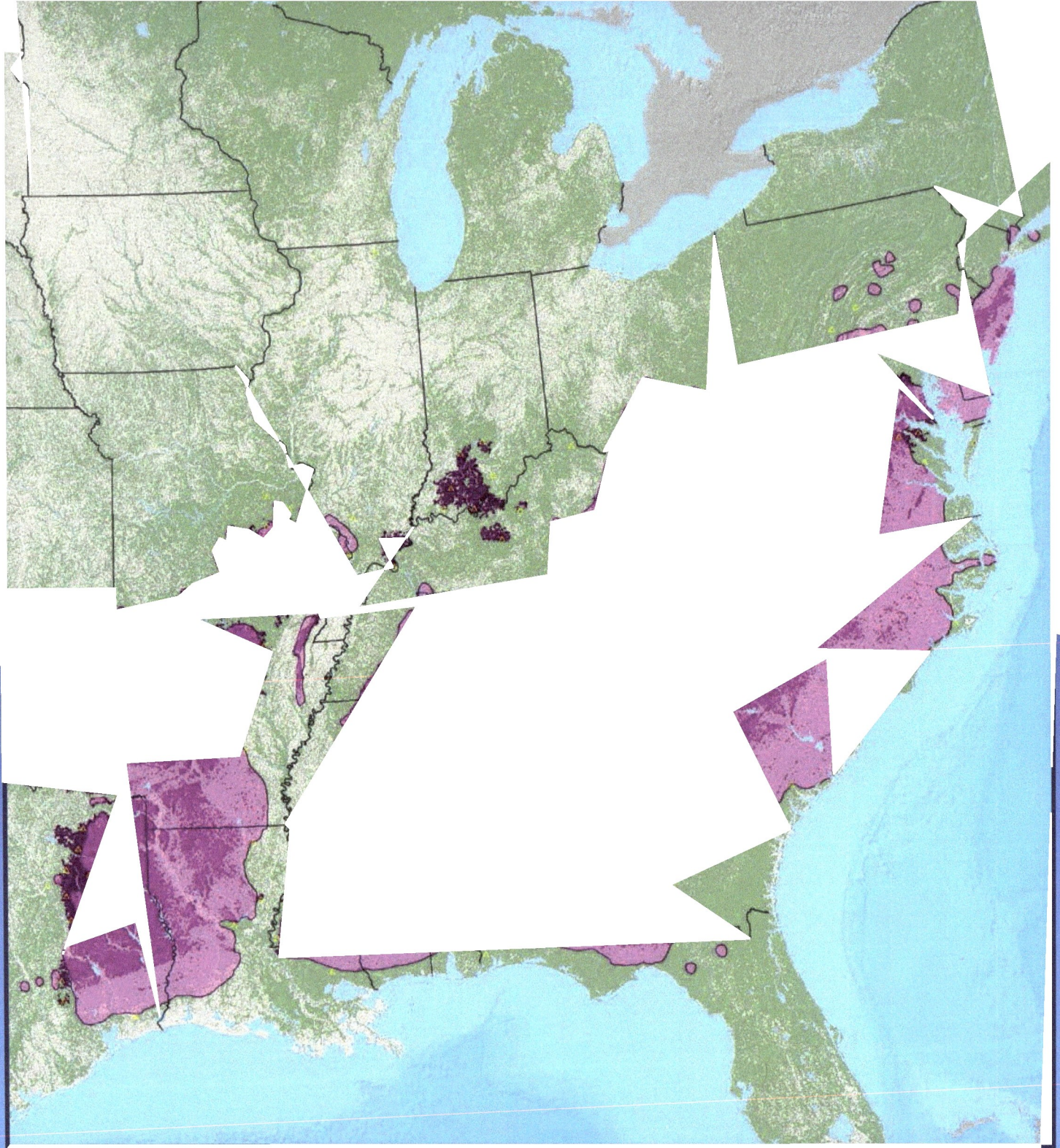

shortleaf pine

Pinus echinata Mill.

Pinaceae – Pine Family

FIA Code **110** – Plants Symbol **PIEC2**

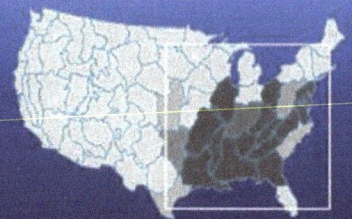

National Individual Tree Species Atlas

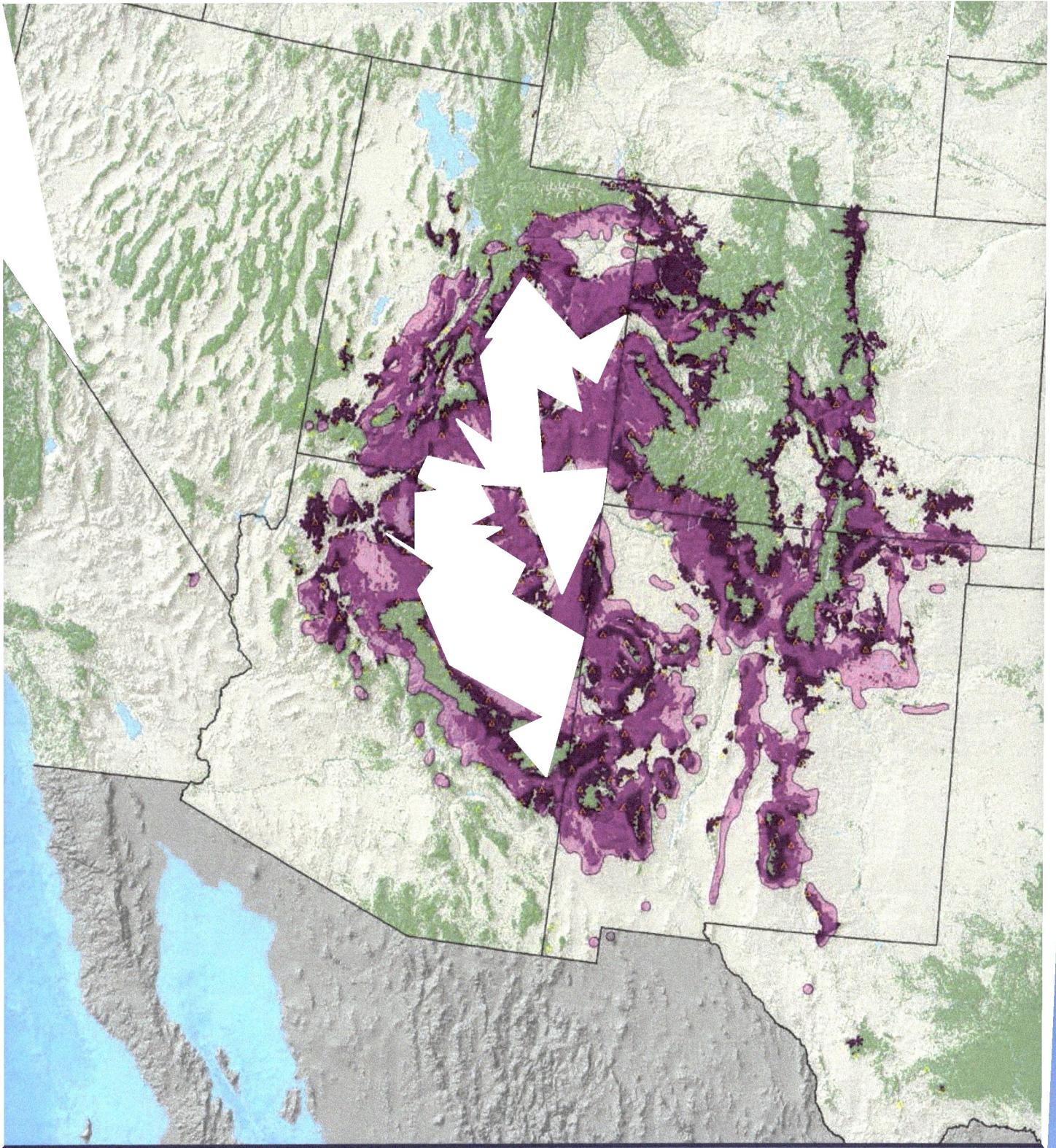

common piñon
Pinus edulis Engelm.
Pinaceae – Pine Family
FIA Code **106** – Plants Symbol **PIED**

National Individual Tree Species Atlas

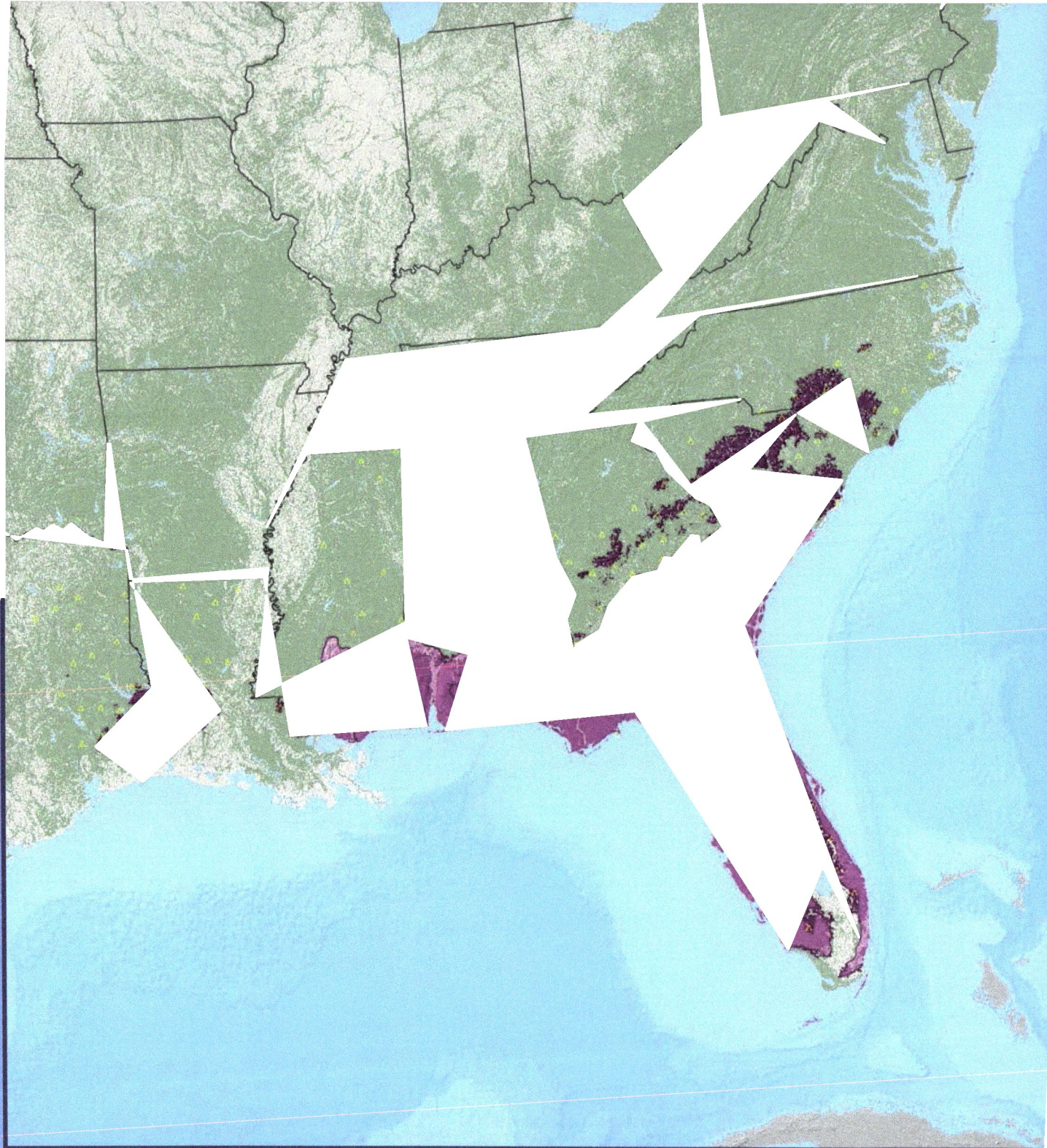

slash pine
Pinus elliottii Engelm.
Pinaceae – Pine Family
FIA Code **111** – Plants Symbol **PIEL**

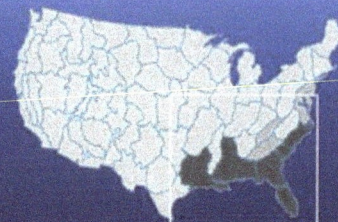

National Individual Tree Species Atlas

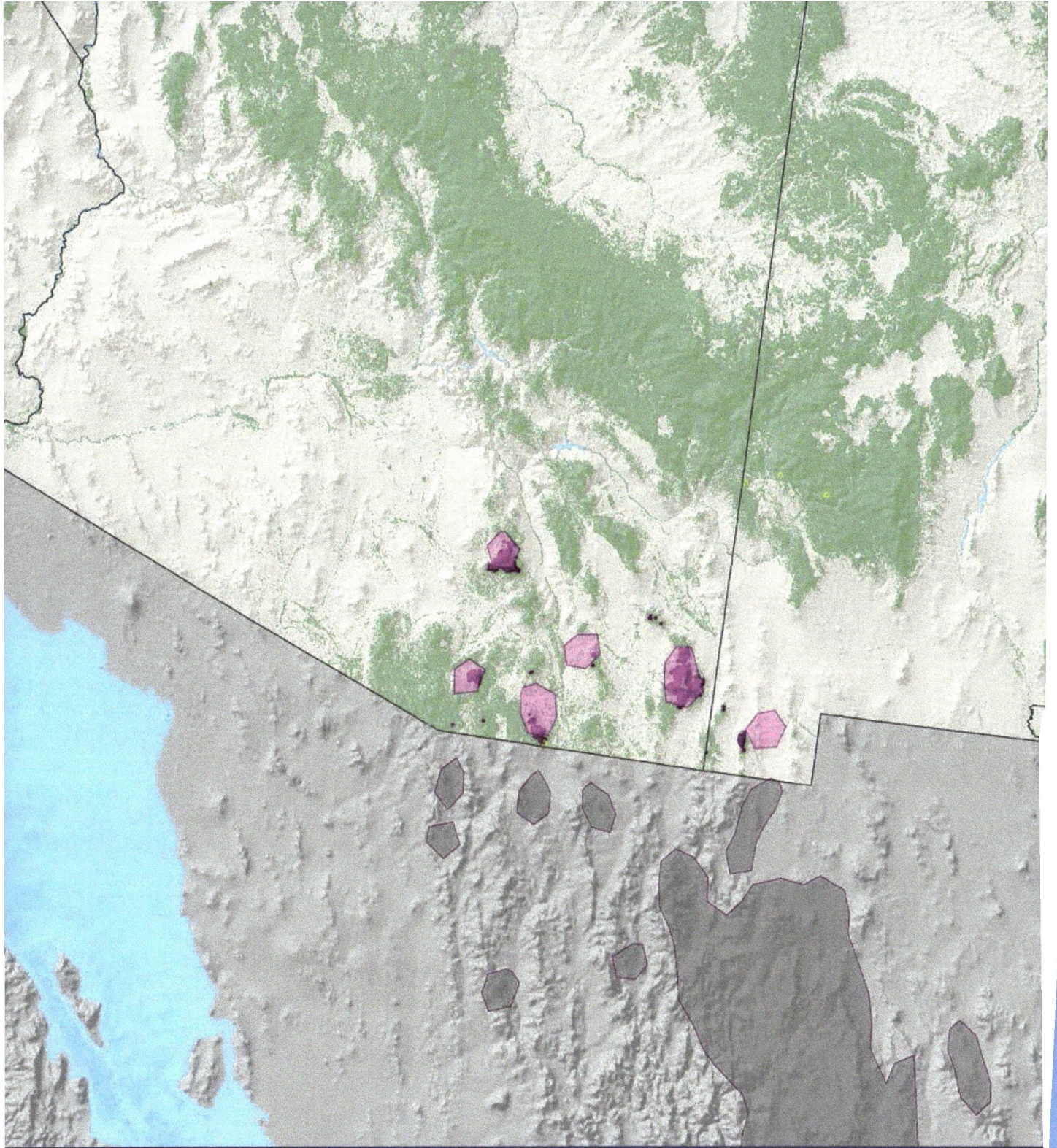

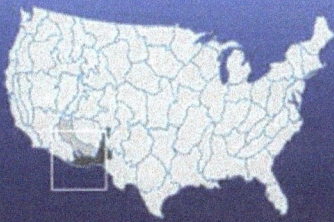

Apache pine

Pinus engelmannii Carrière

Pinaceae – Pine Family

FIA Code **112** – Plants Symbol **PIEN2**

National Individual Tree Species Atlas

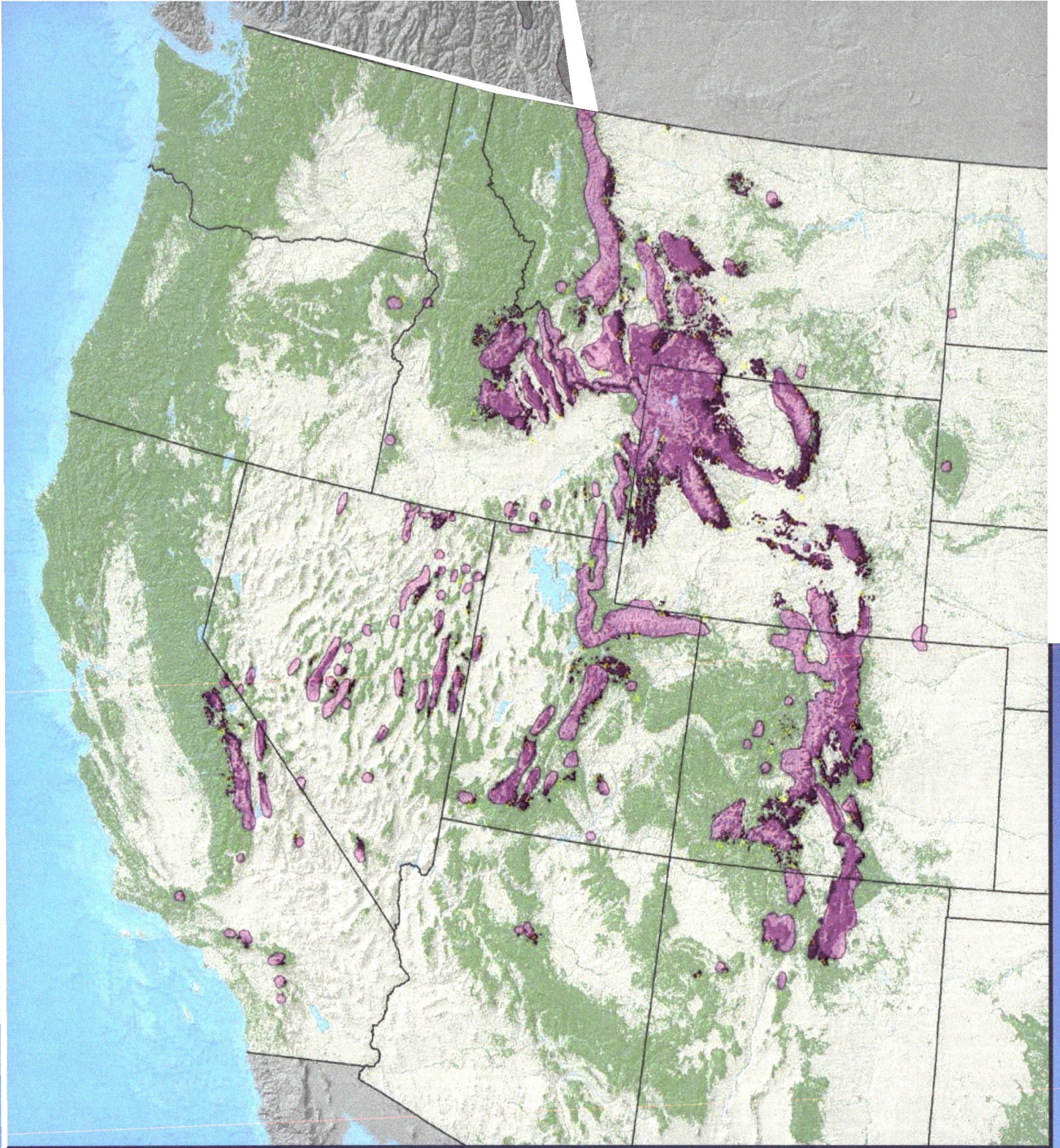

limber pine
Pinus flexilis James
Pinaceae – Pine Family
FIA Code **113** – Plants Symbol **PIFL2**

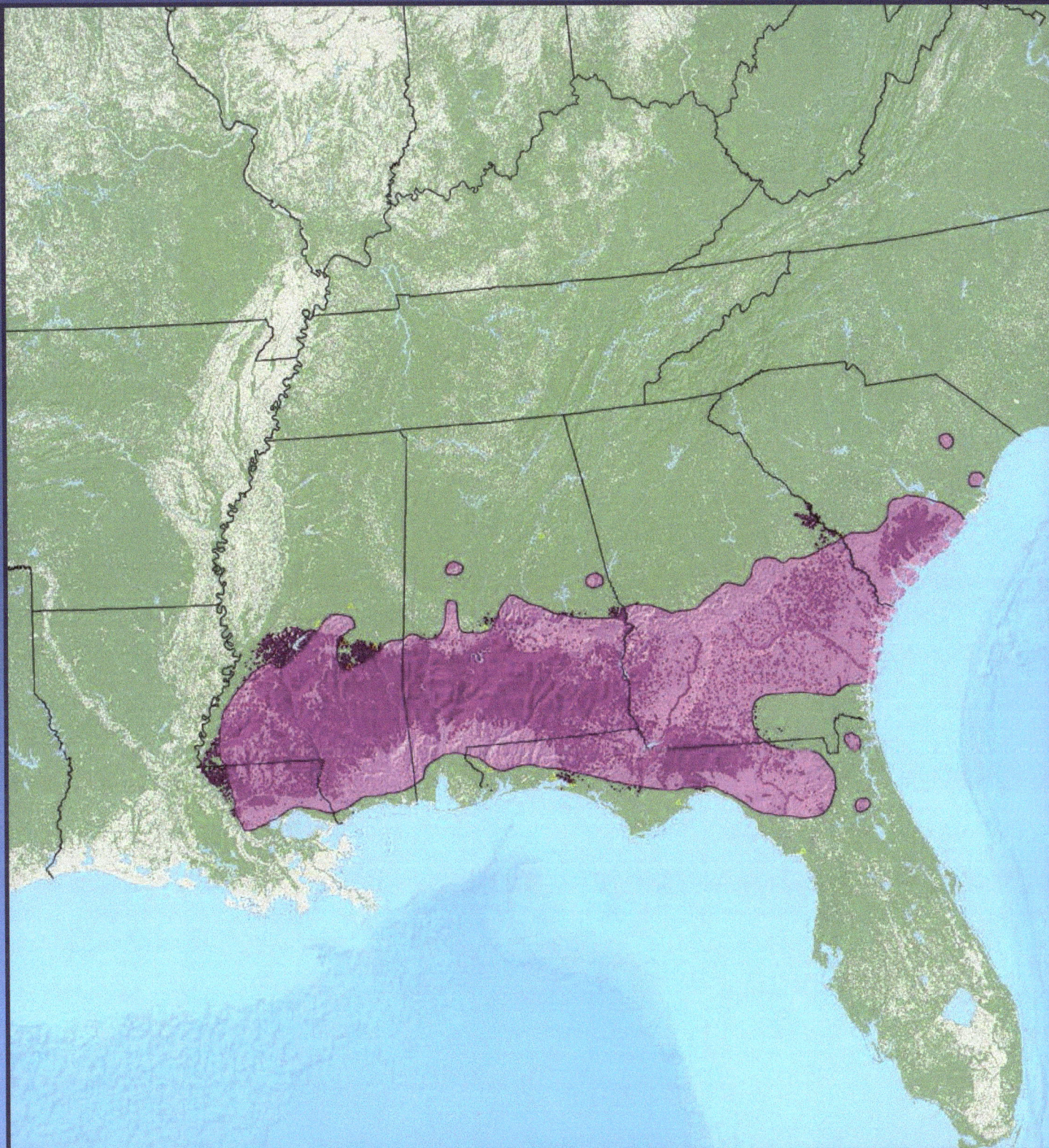

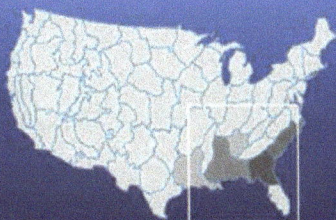

spruce pine

spruce pine

Pinus glabra Walter

Pinaceae – Pine Family

FIA Code **115** – Plants Symbol **PIGL2**

National Individual Tree Species Atlas

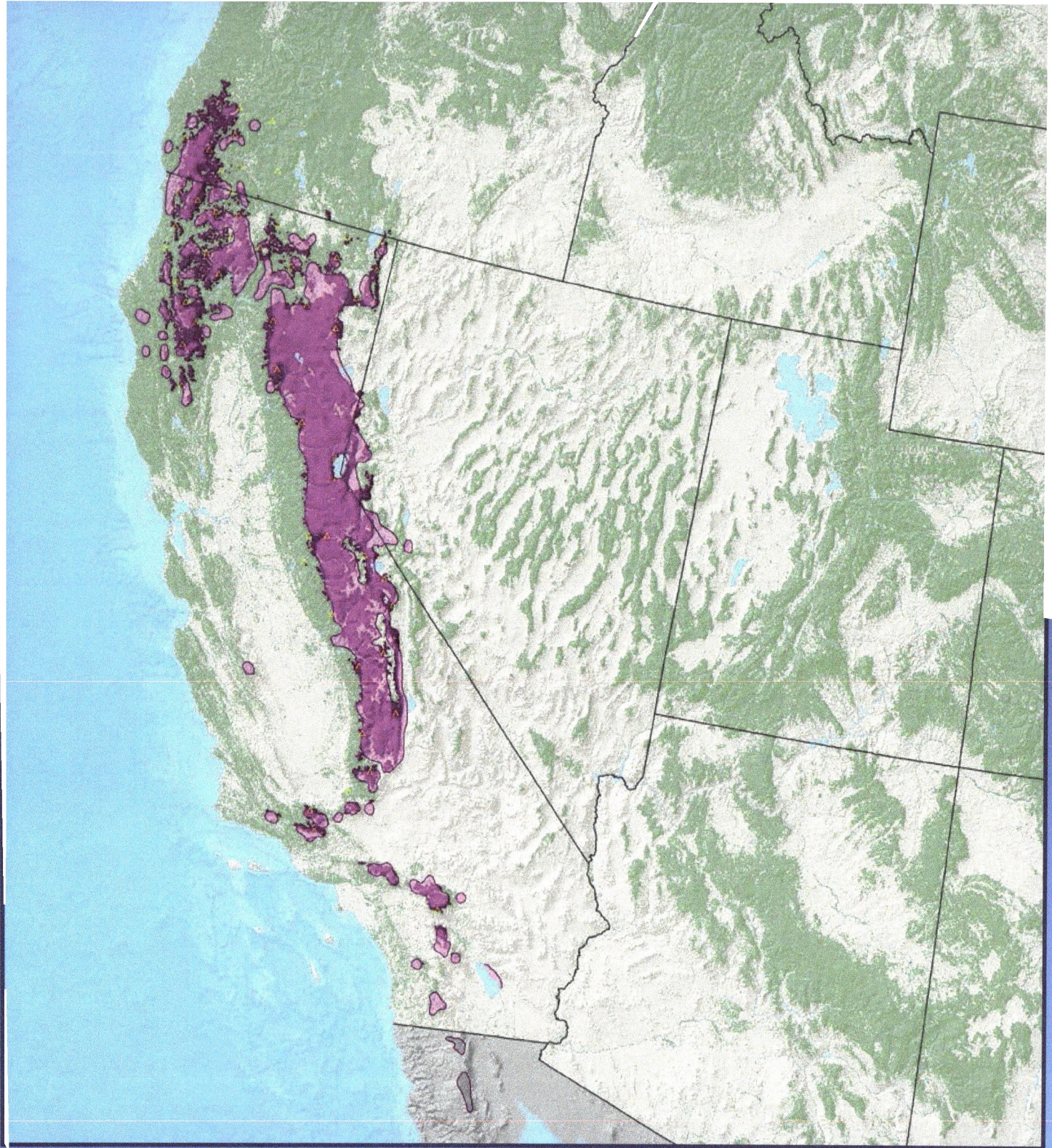

Jeffrey pine

Pinus jeffreyi Balf.

Pinaceae – Pine Family

FIA Code **116** – Plants Symbol **PIJE**

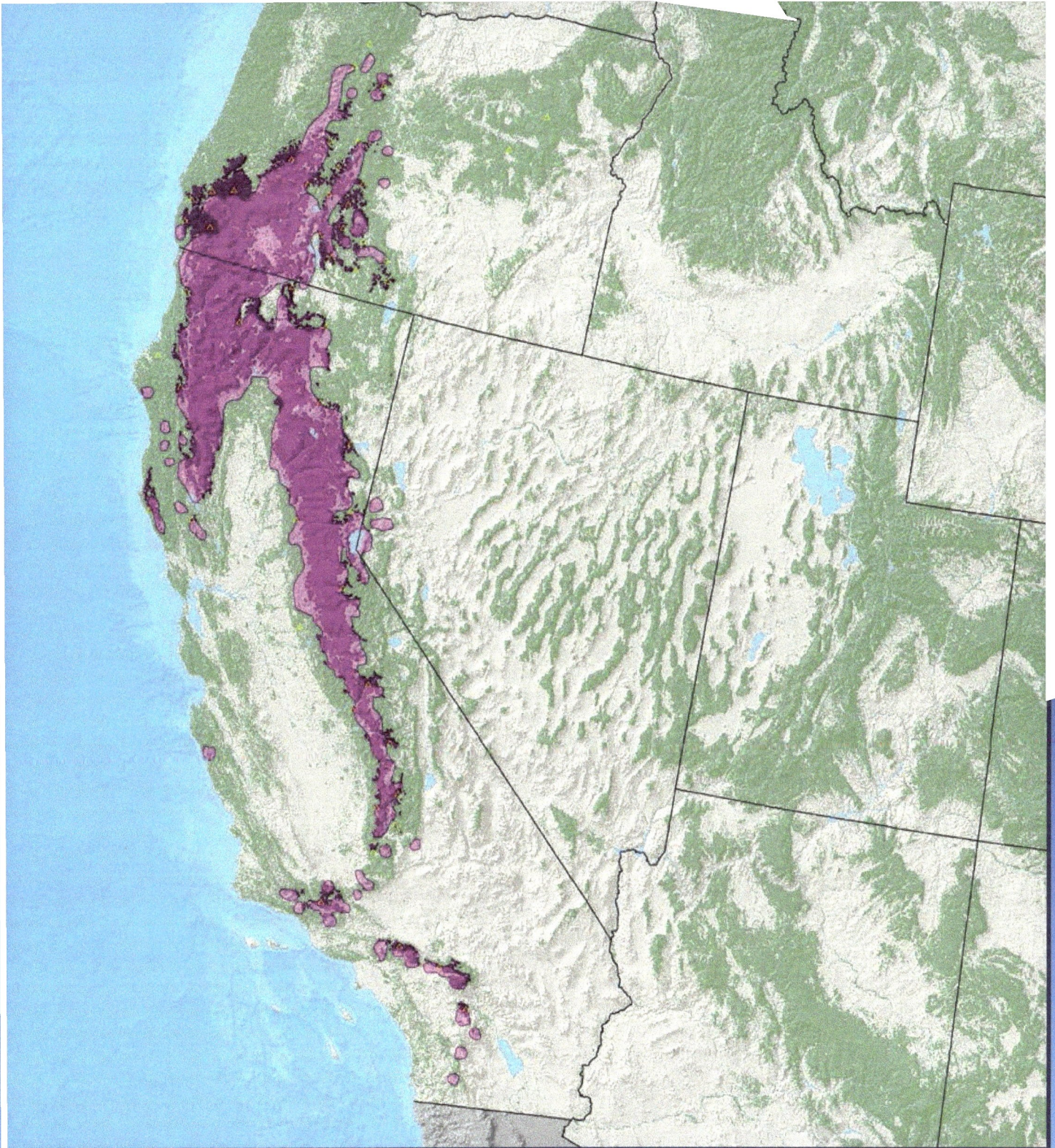

sugar pine

Pinus lambertiana Douglas

Pinaceae – Pine Family

FIA Code **117** – Plants Symbol **PILA**

National Individual Tree Species Atlas

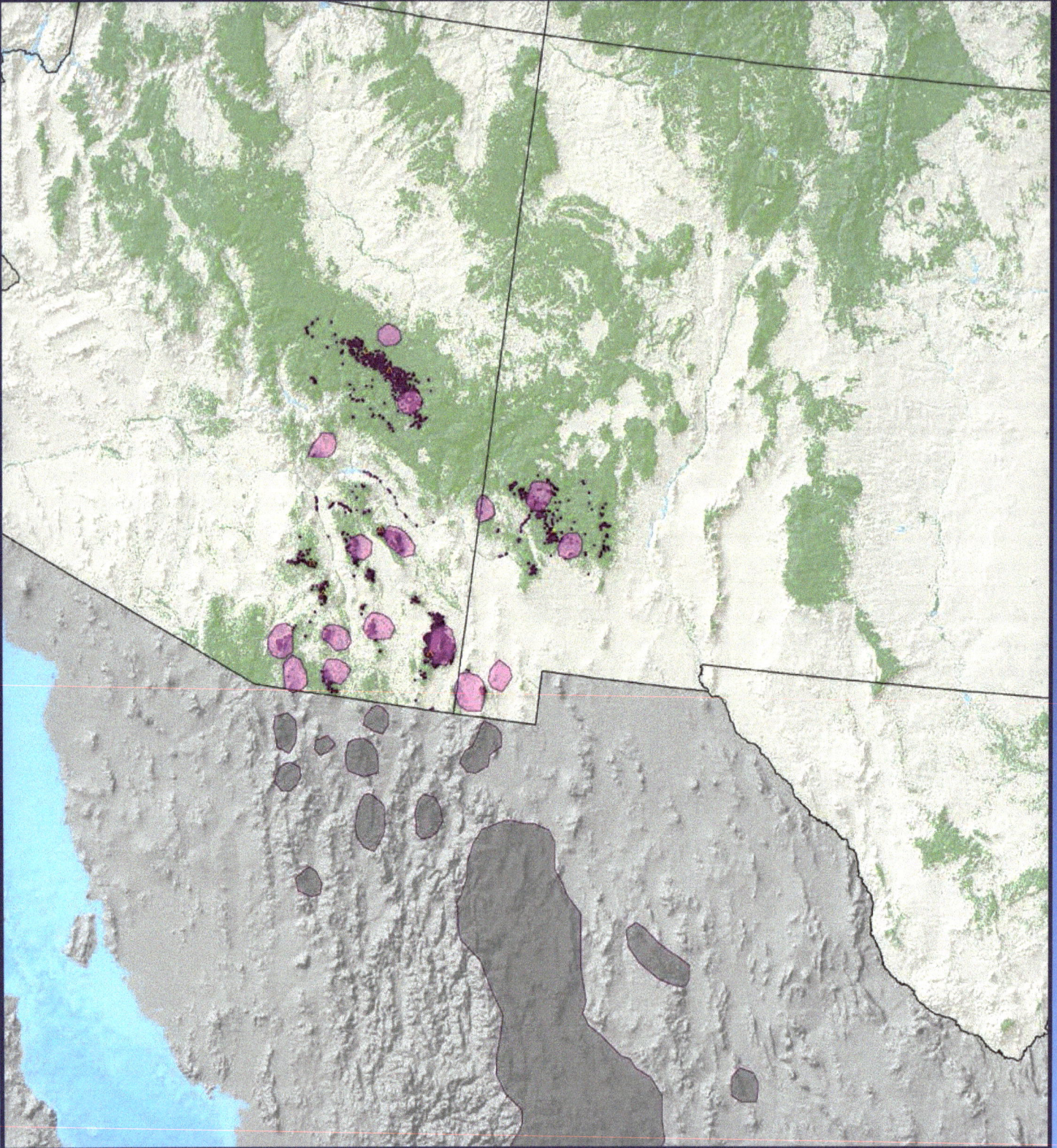

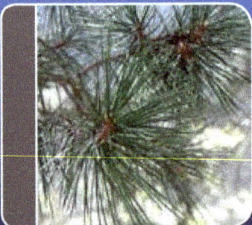

Chihuahua pine

Chihuahua pine
Pinus leiophylla Schiede & Deppe
Pinaceae – Pine Family
FIA Code **118** – Plants Symbol **PILE**

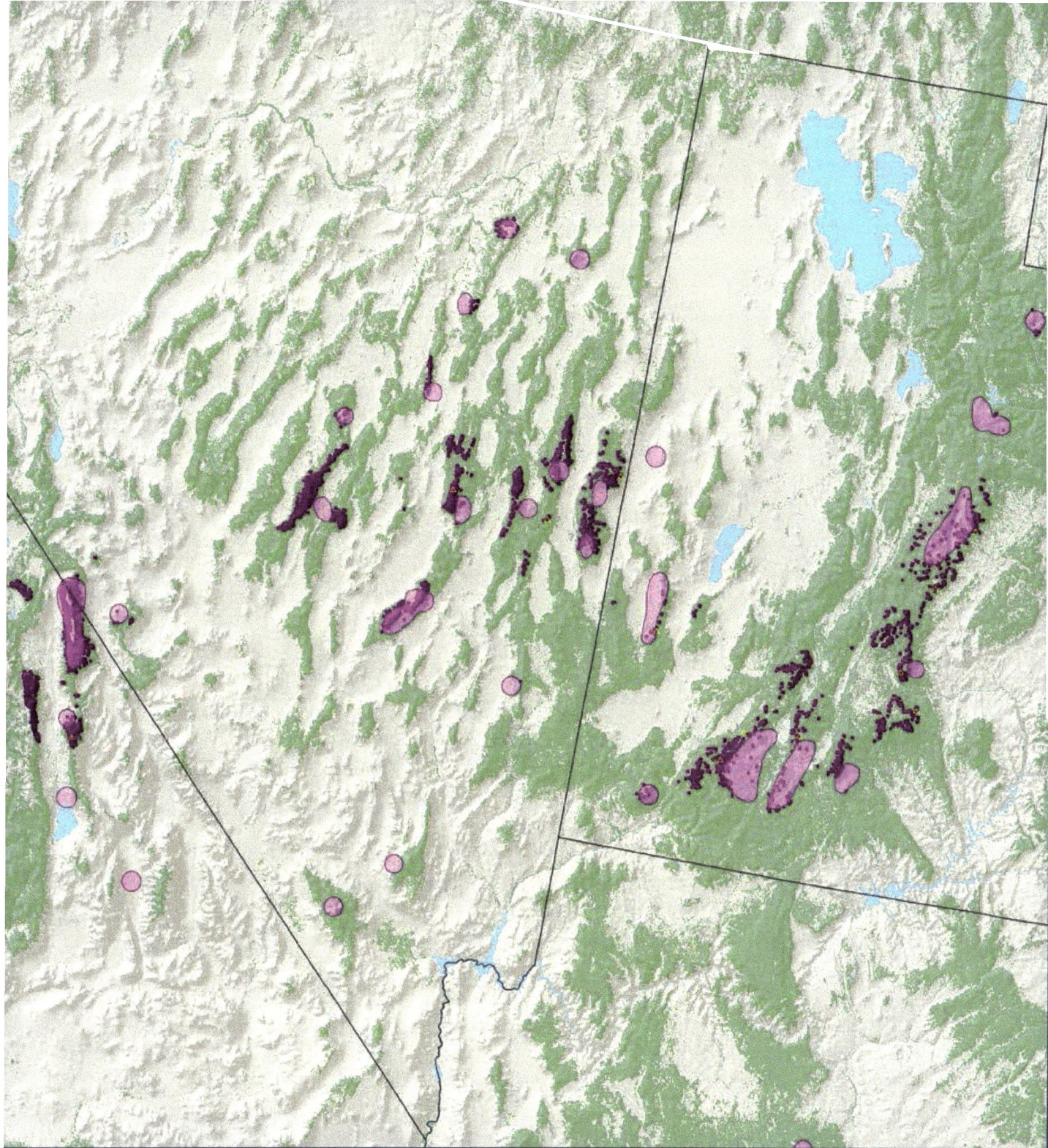

Great Basin bristlecone pine

Pinus longaeva D.K. Bailey

Pinaceae – Pine Family

FIA Code **142** – Plants Symbol **PILO**

singleleaf piñon

Pinus monophylla Torr. & Frém.

Pinaceae – Pine Family

FIA Code **133** – Plants Symbol **PIMO**

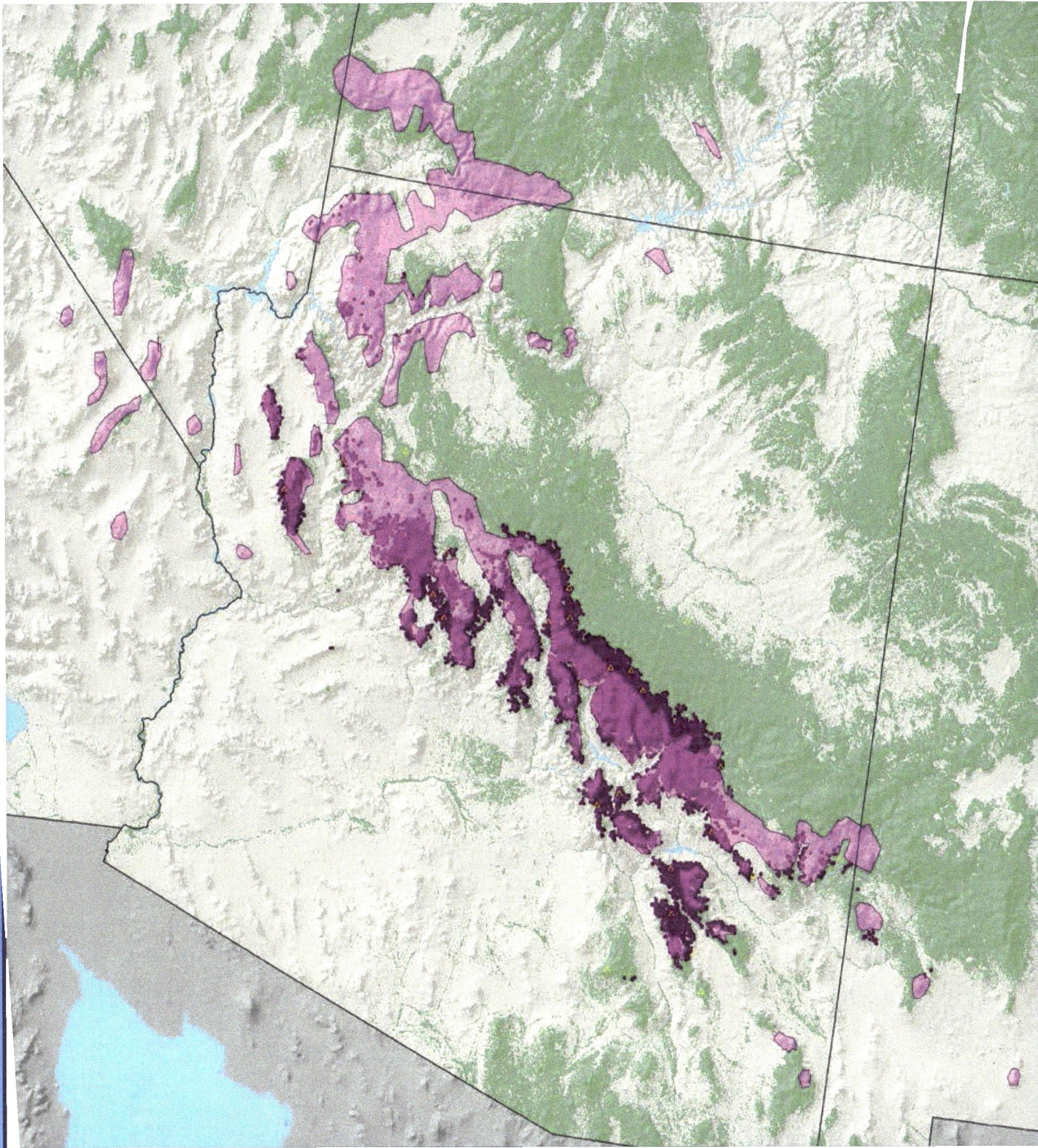

Arizona piñon pine

Pinus monophylla Torr. & Frém. var. *fallax* (Little) Silba

Pinaceae – Pine Family

FIA Code **143** – Plants Symbol **PIMOF**

western white pine

Pinus monticola Douglas ex D. Don

Pinaceae – Pine Family

FIA Code **119** – Plants Symbol **PIMO3**

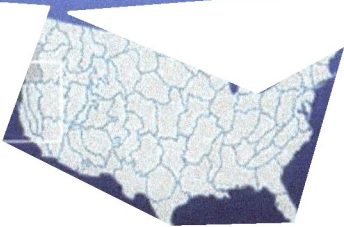

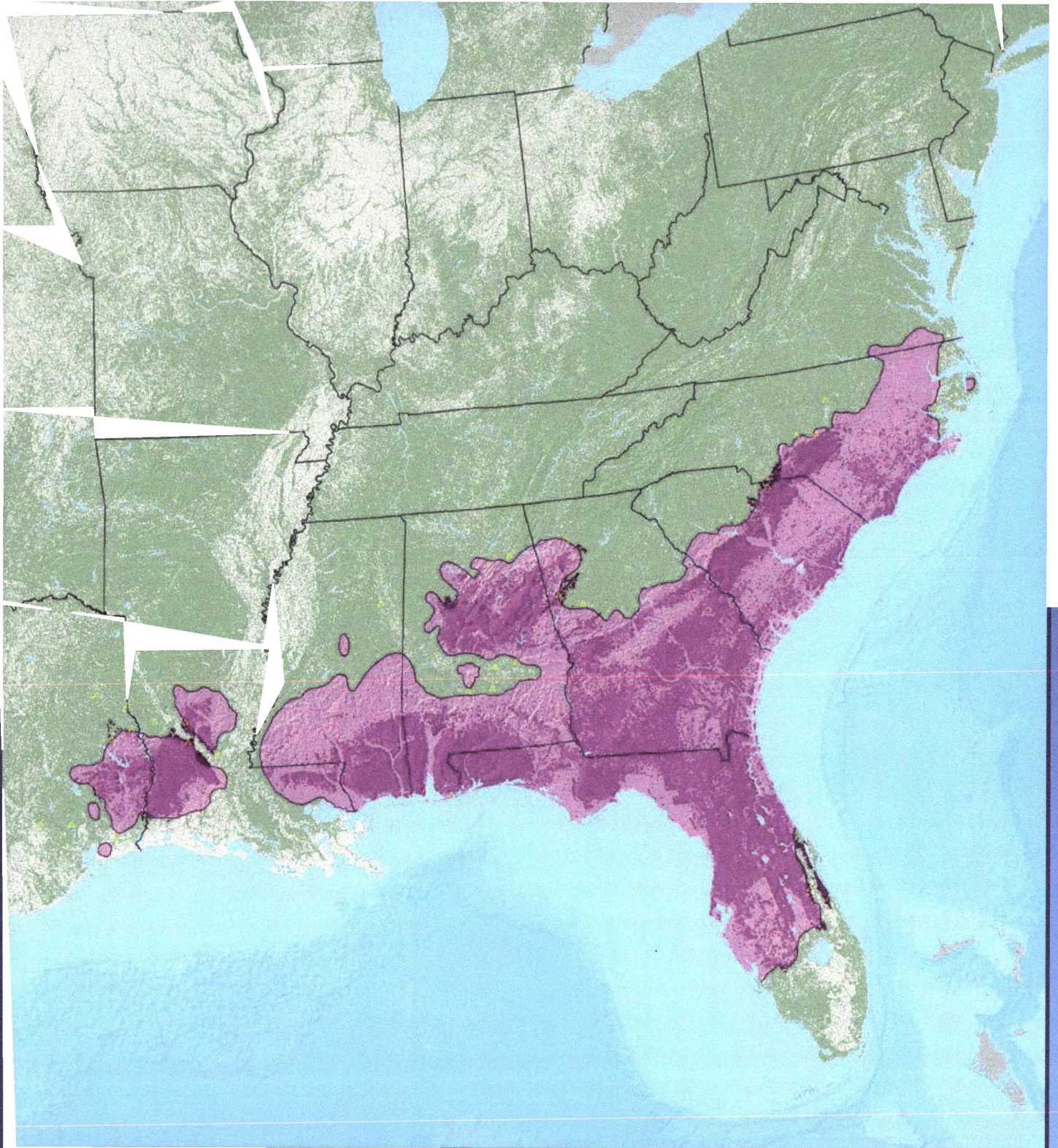

longleaf pine

Pinus palustris Mill.

Pinaceae – Pine Family

FIA Code **121** – Plants Symbol **PIPA2**

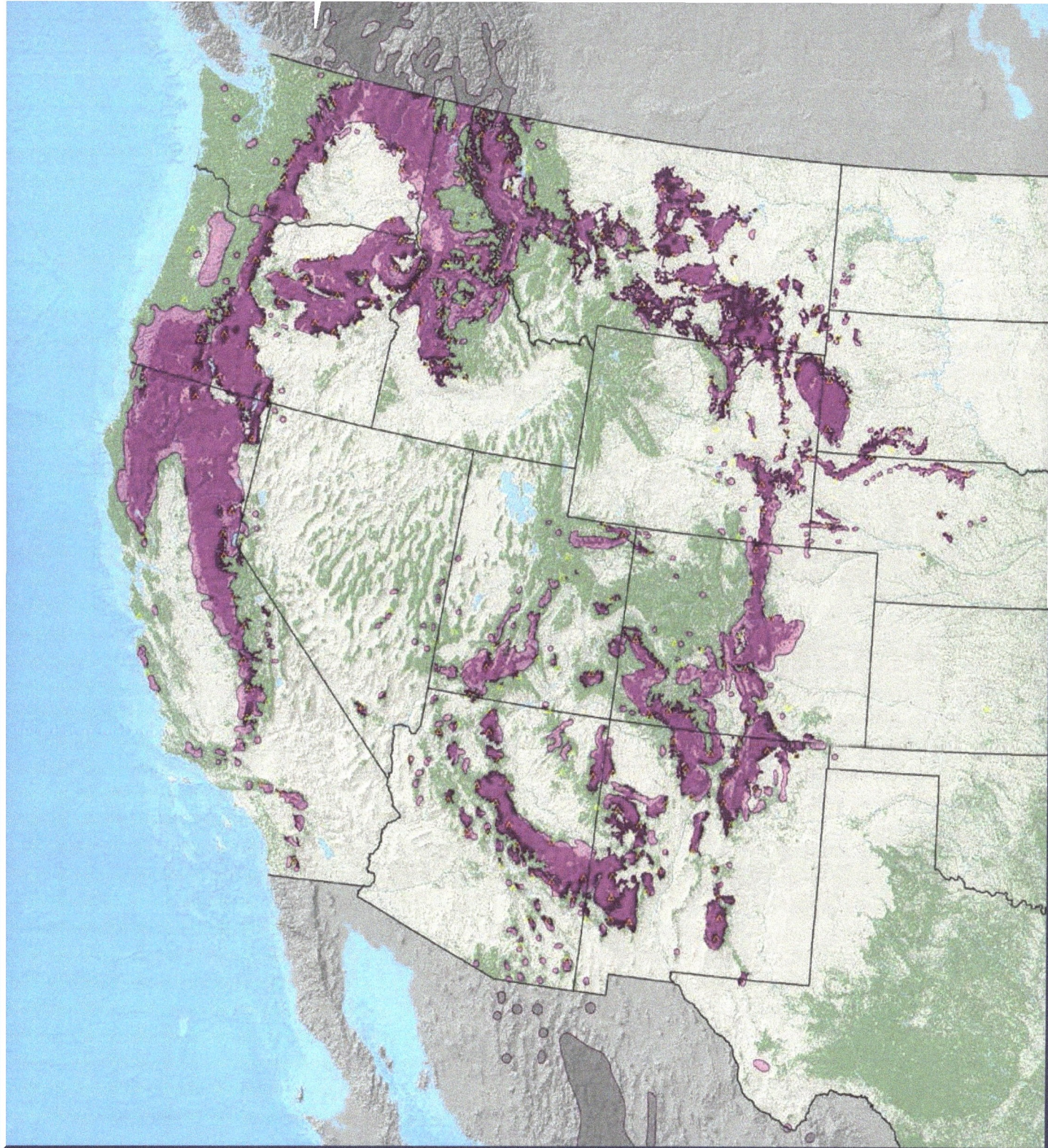

ponderosa pine
Pinus ponderosa C. Lawson
Pinaceae – Pine Family
FIA Code **122** – Plants Symbol **PIPO**

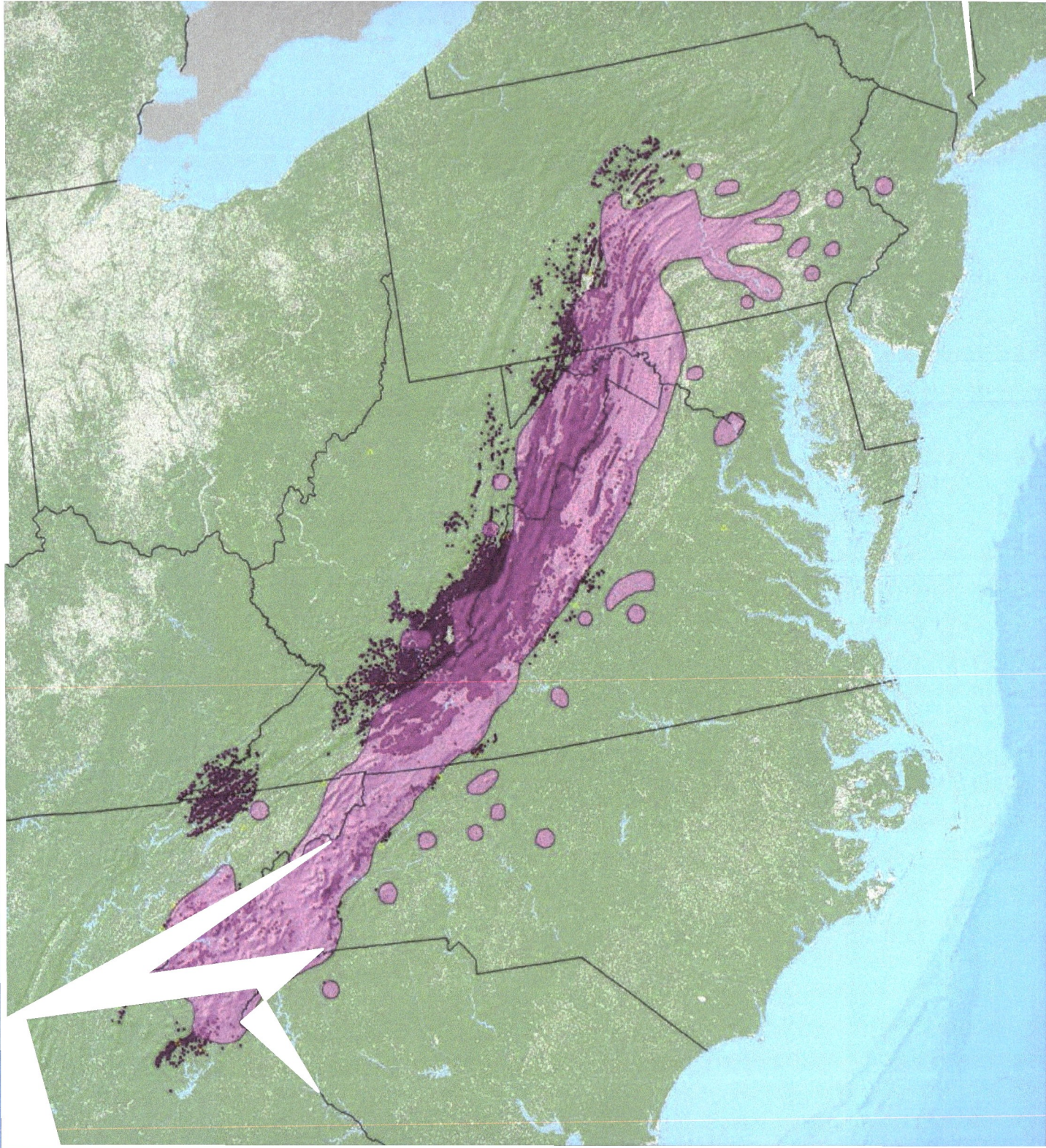

Table Mountain pine

Pinus pungens Lamb.

Pinaceae – Pine Family

FIA Code **123** – Plants Symbol **PIPU5**

Monterey pine
Pinus radiata D. Don
Pinaceae – Pine Family
FIA Code **124** – Plants Symbol **PIRA2**

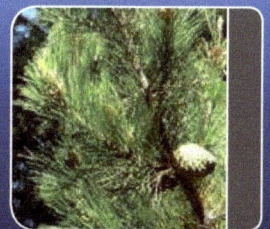

National Individual Tree Species Atlas

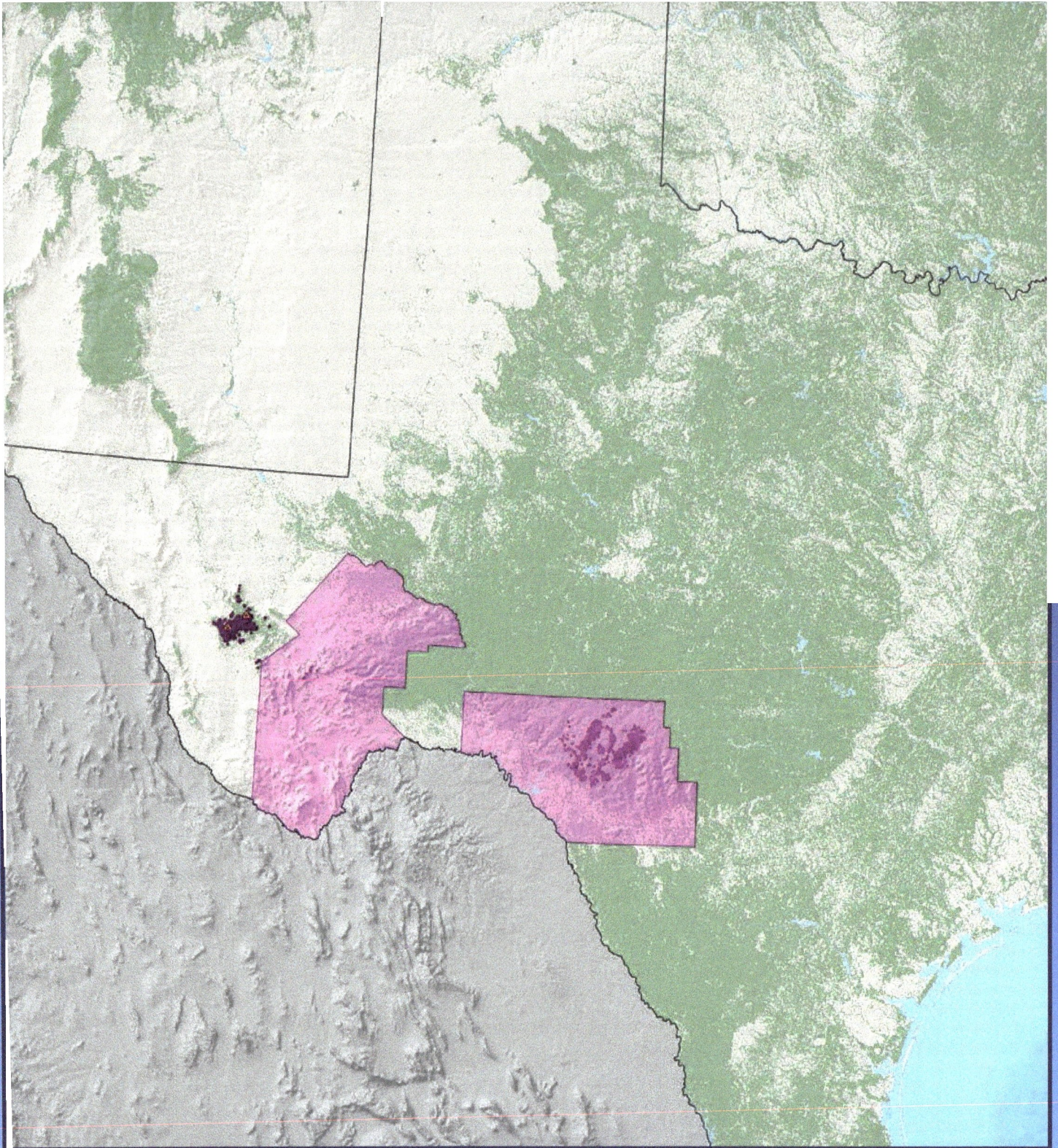

papershell piñon pine

Pinus remota (Little) D.K. Bailey & Hawksw.

Pinaceae – Pine Family

FIA Code **141** – Plants Symbol **PIRE5**

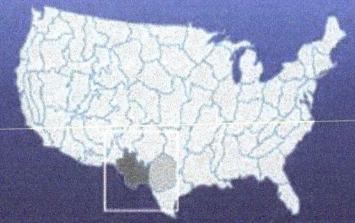

National Individual Tree Species Atlas

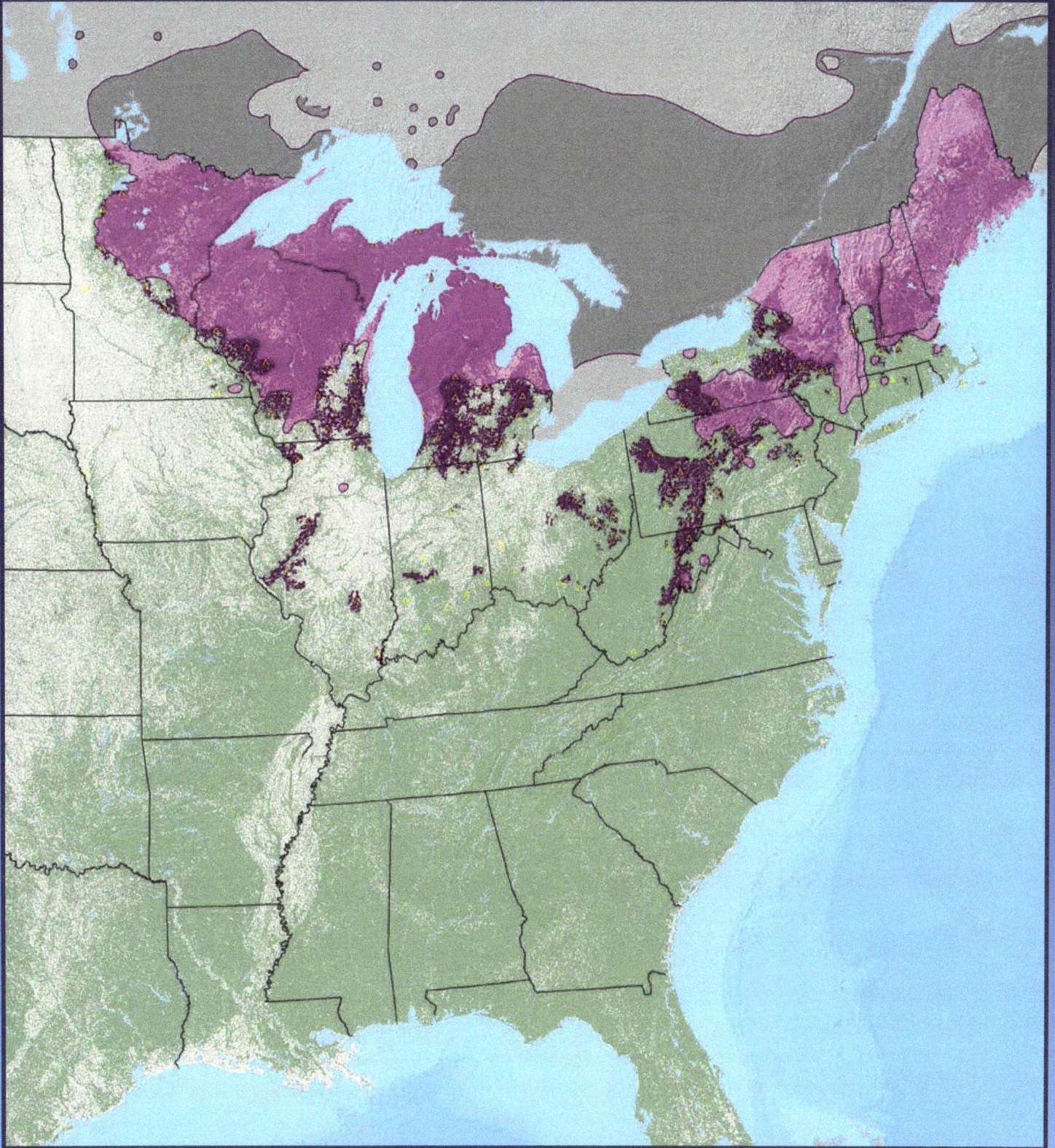

red pine

red pine
Pinus resinosa Aiton
Pinaceae – Pine Family
FIA Code **125** – Plants Symbol **PIRE**

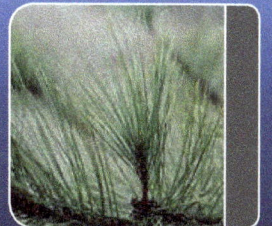

National Individual Tree Species Atlas

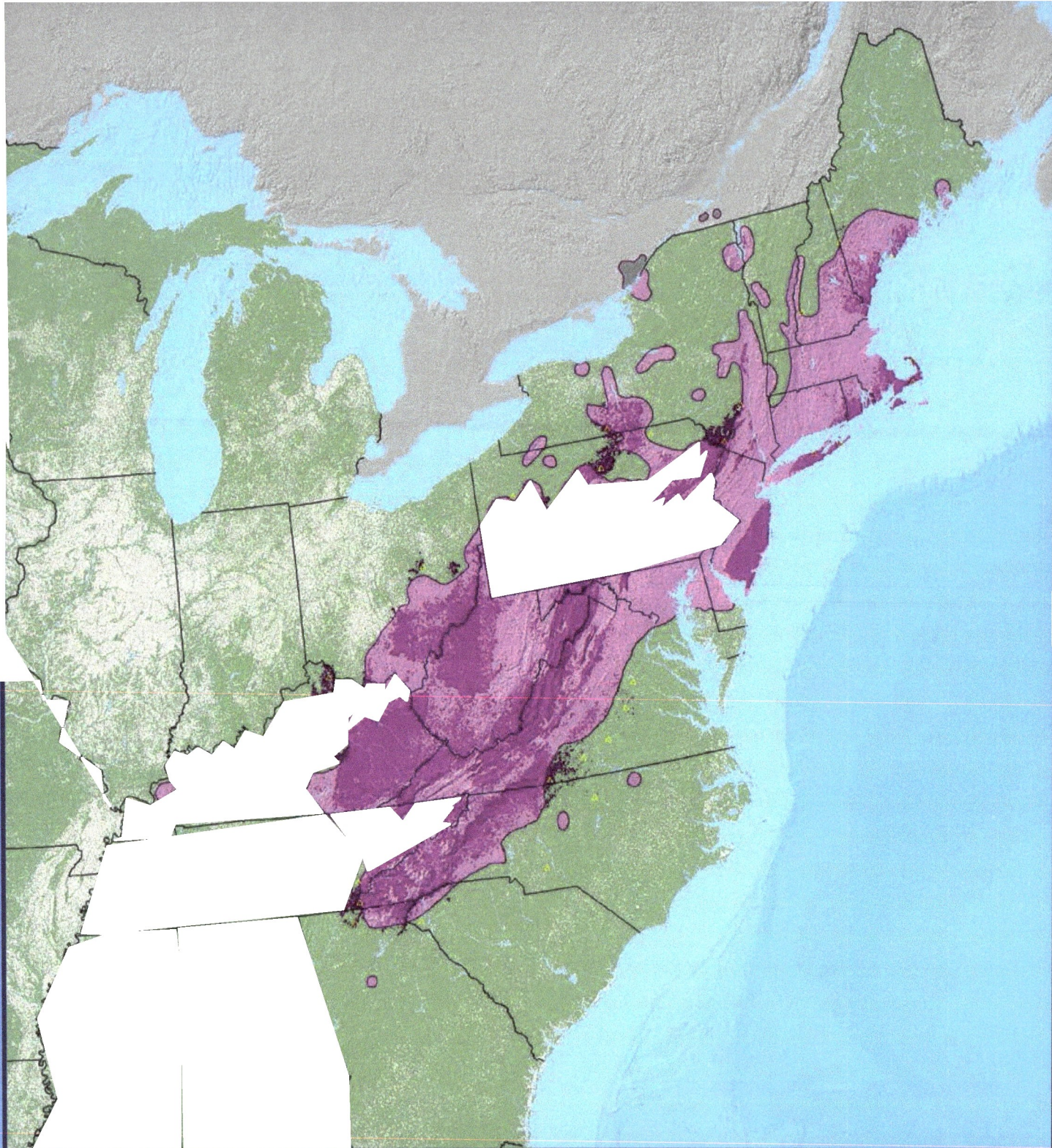

pitch pine

Pinus rigida Mill.

Pinaceae – Pine Family

FIA Code **126** – Plants Symbol **PIRI**

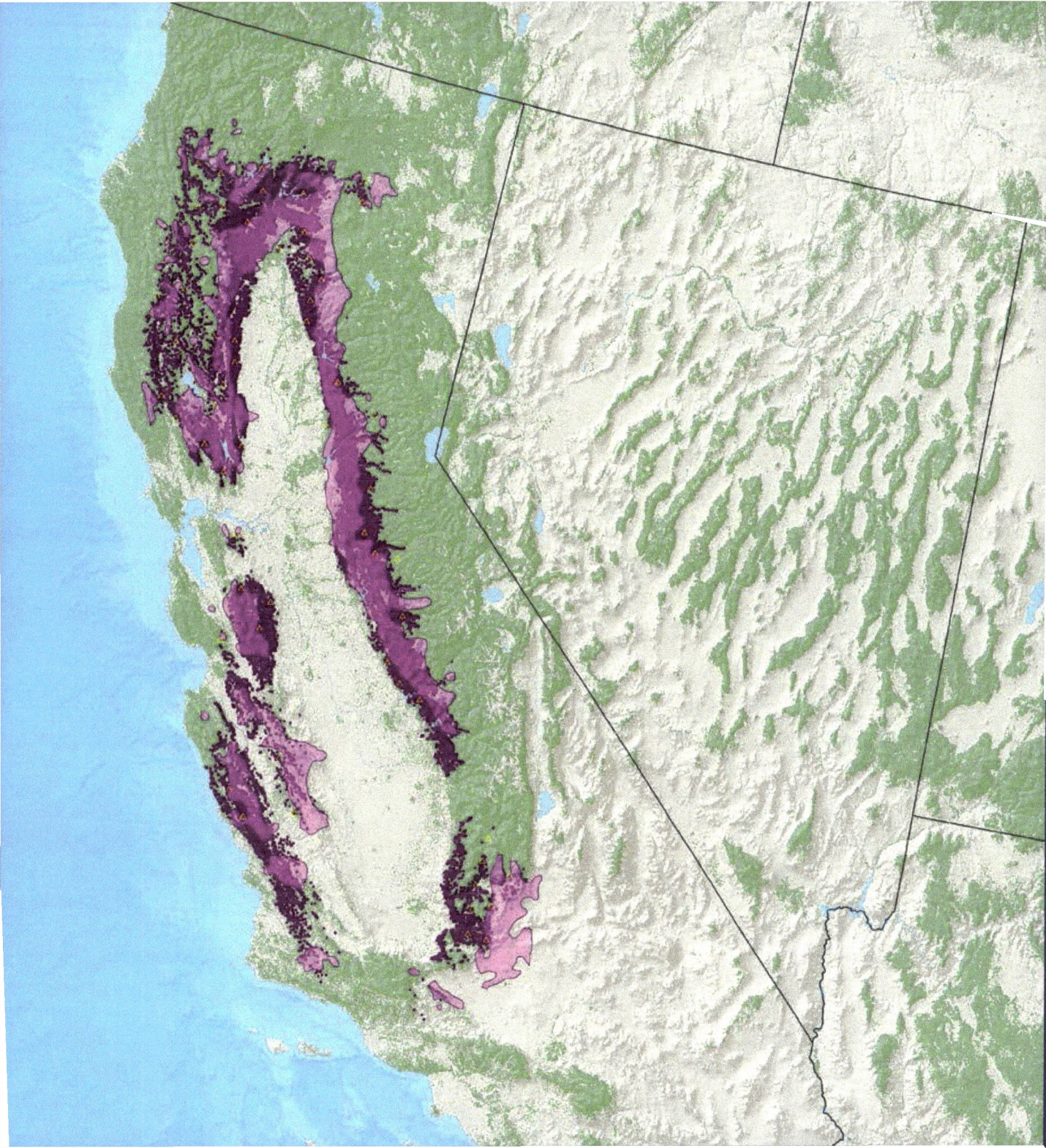

California foothills pine

Pinus sabiniana Douglas ex Douglas

Pinaceae – Pine Family

FIA Code **127** – Plants Symbol **PISA2**

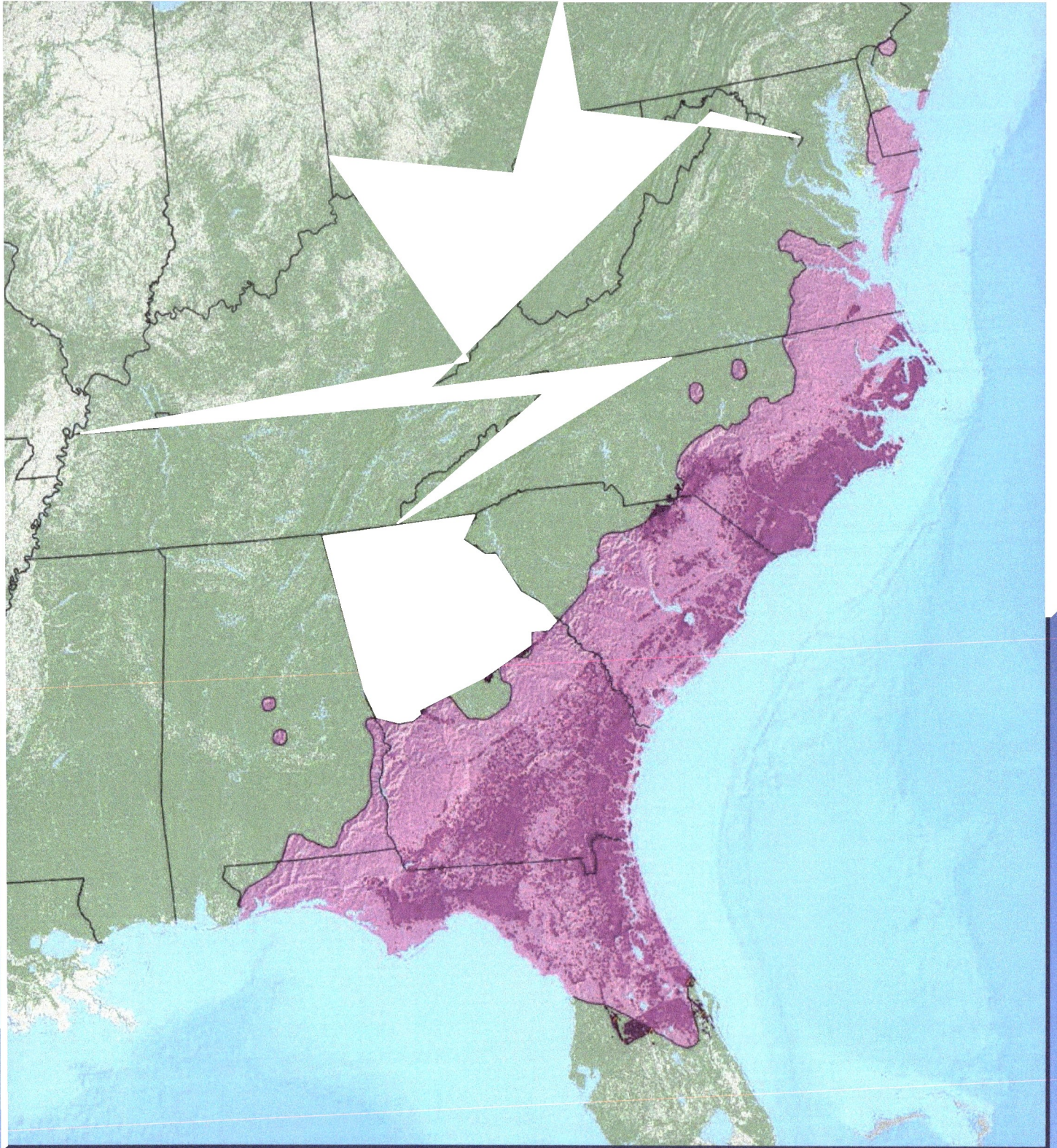

pond pine

Pinus serotina Michx.

Pinaceae – Pine Family

FIA Code **128** – Plants Symbol **PISE**

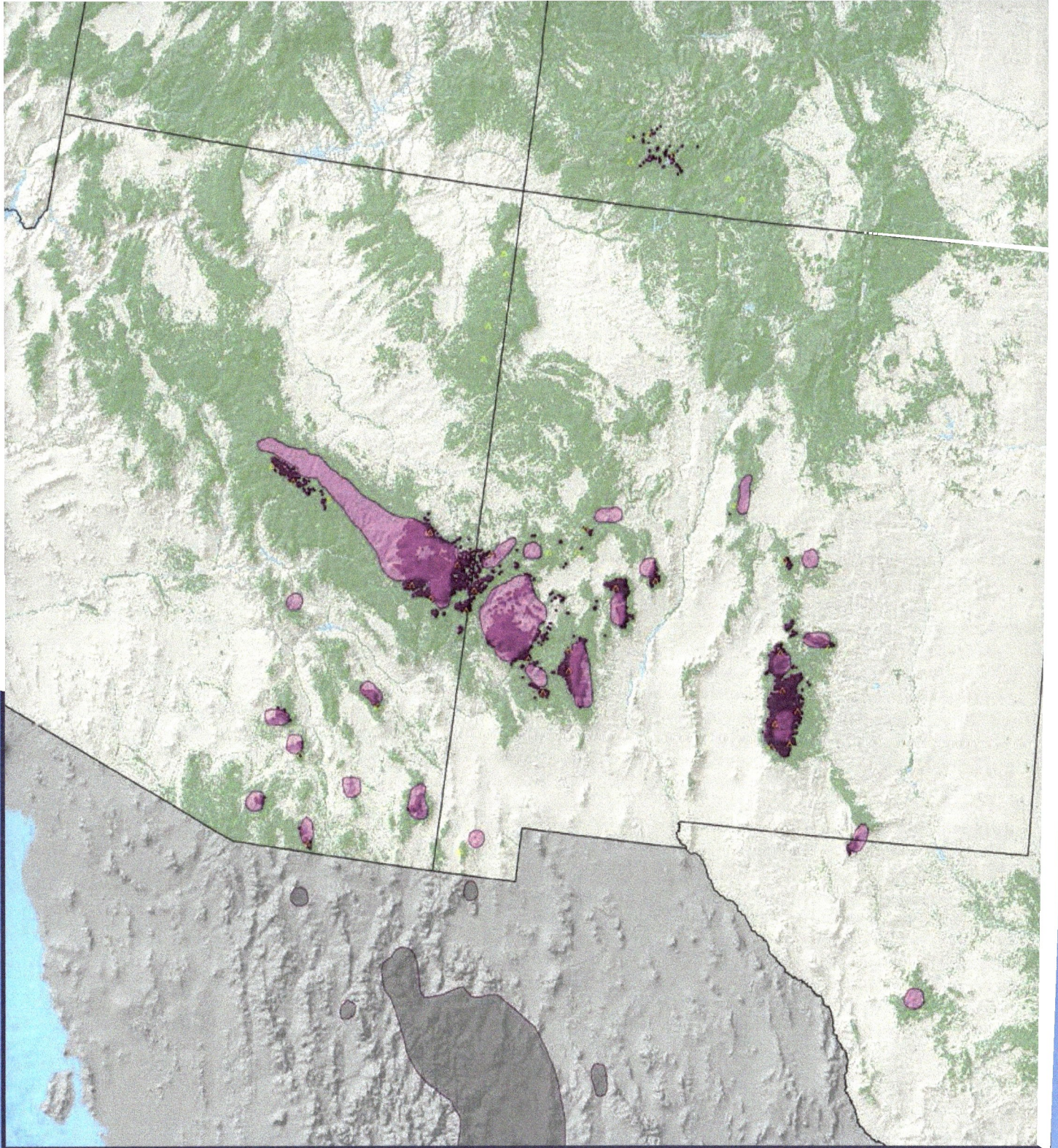

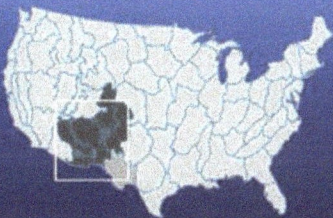

southwestern white pine

Pinus strobiformis Engelm.

Pinaceae – Pine Family

FIA Code **114** – Plants Symbol **PIST3**

National Individual Tree Species Atlas

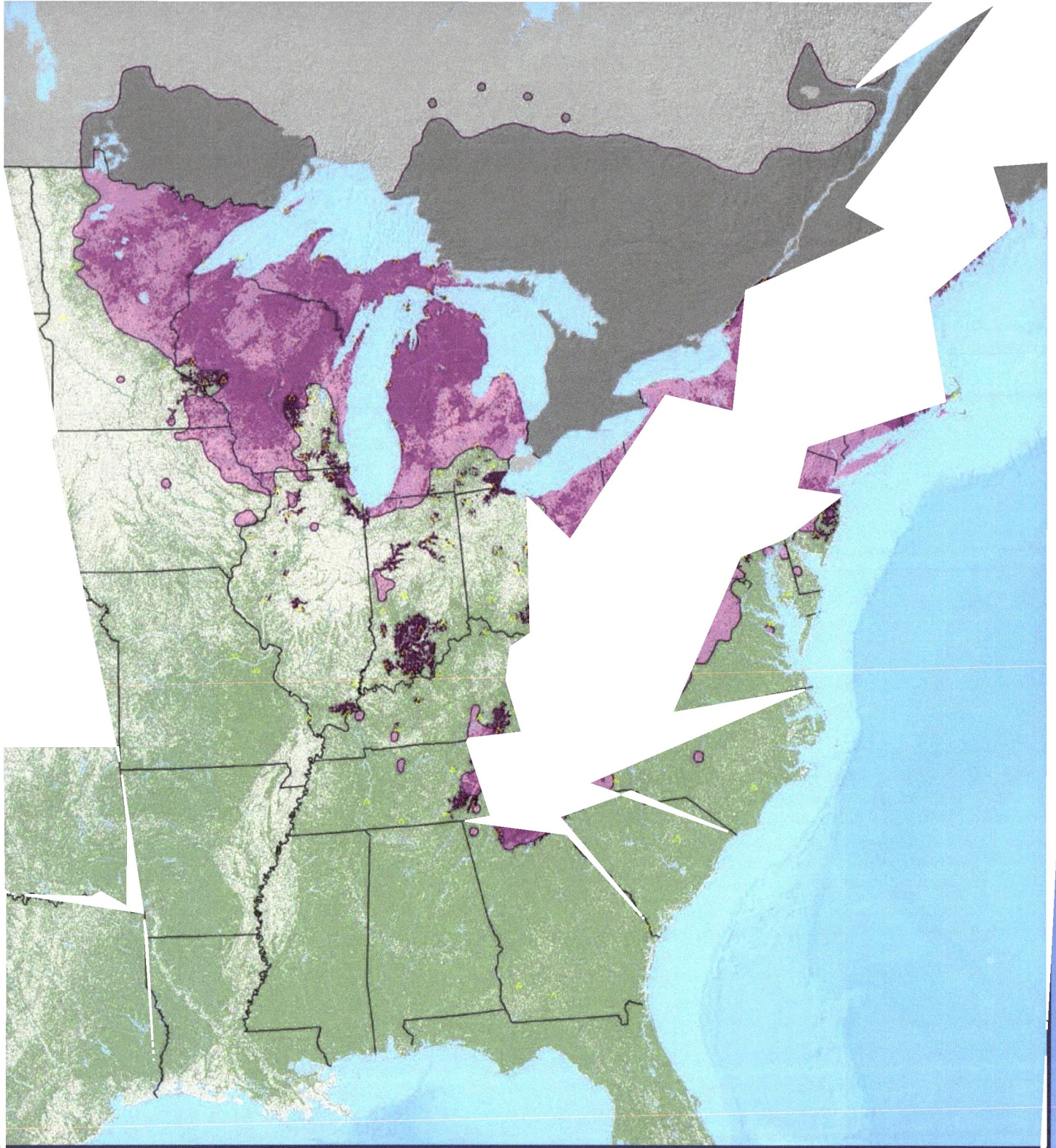

eastern white pine

Pinus strobus L.

Pinaceae – Pine Family

FIA Code **129** – Plants Symbol **PIST**

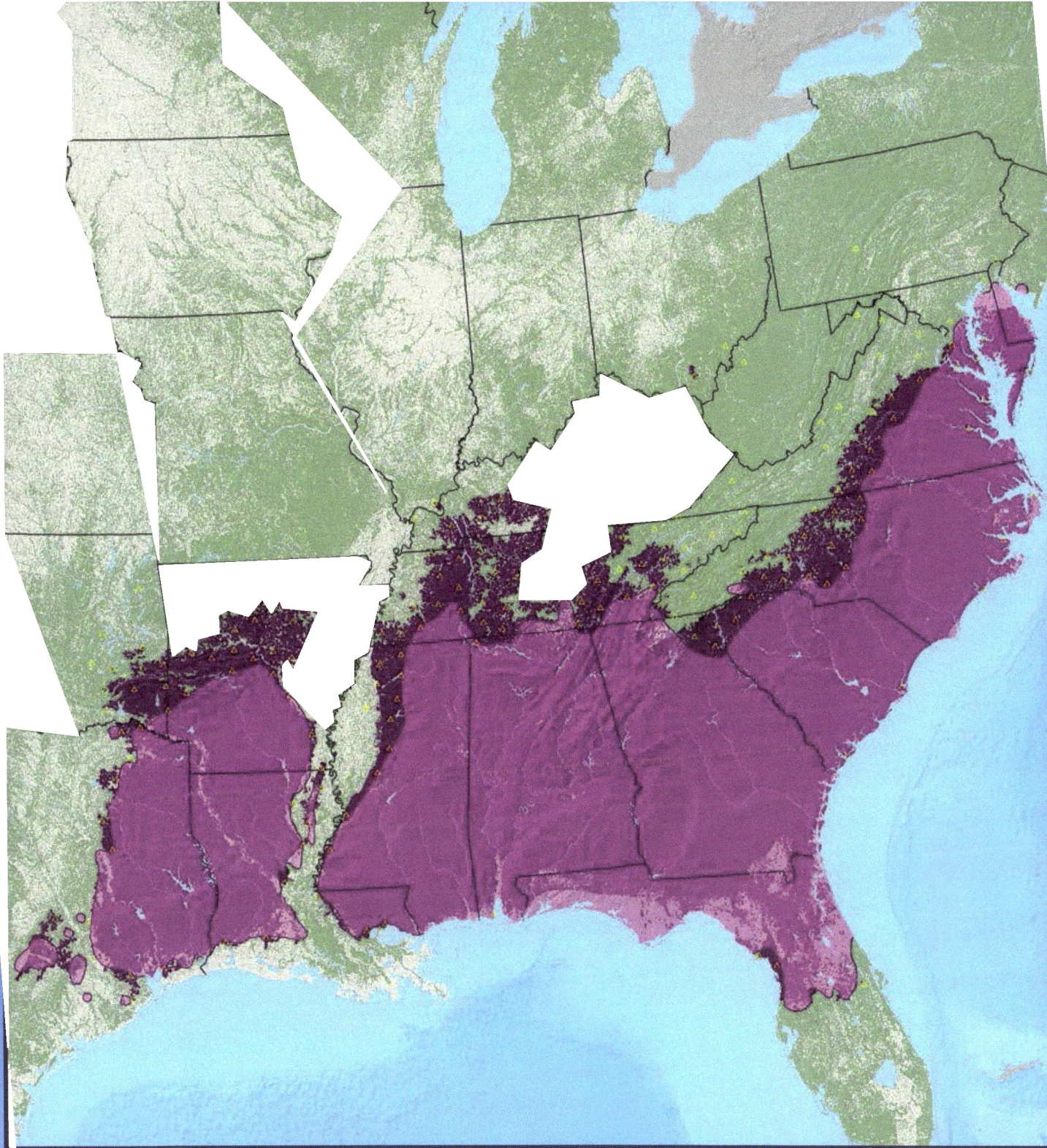

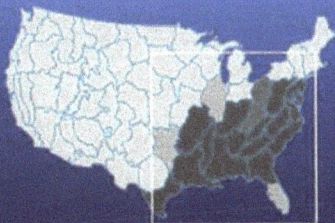

loblolly pine
Pinus taeda L.
Pinaceae – Pine Family
FIA Code **131** – Plants Symbol **PITA**

National Individual Tree Species Atlas

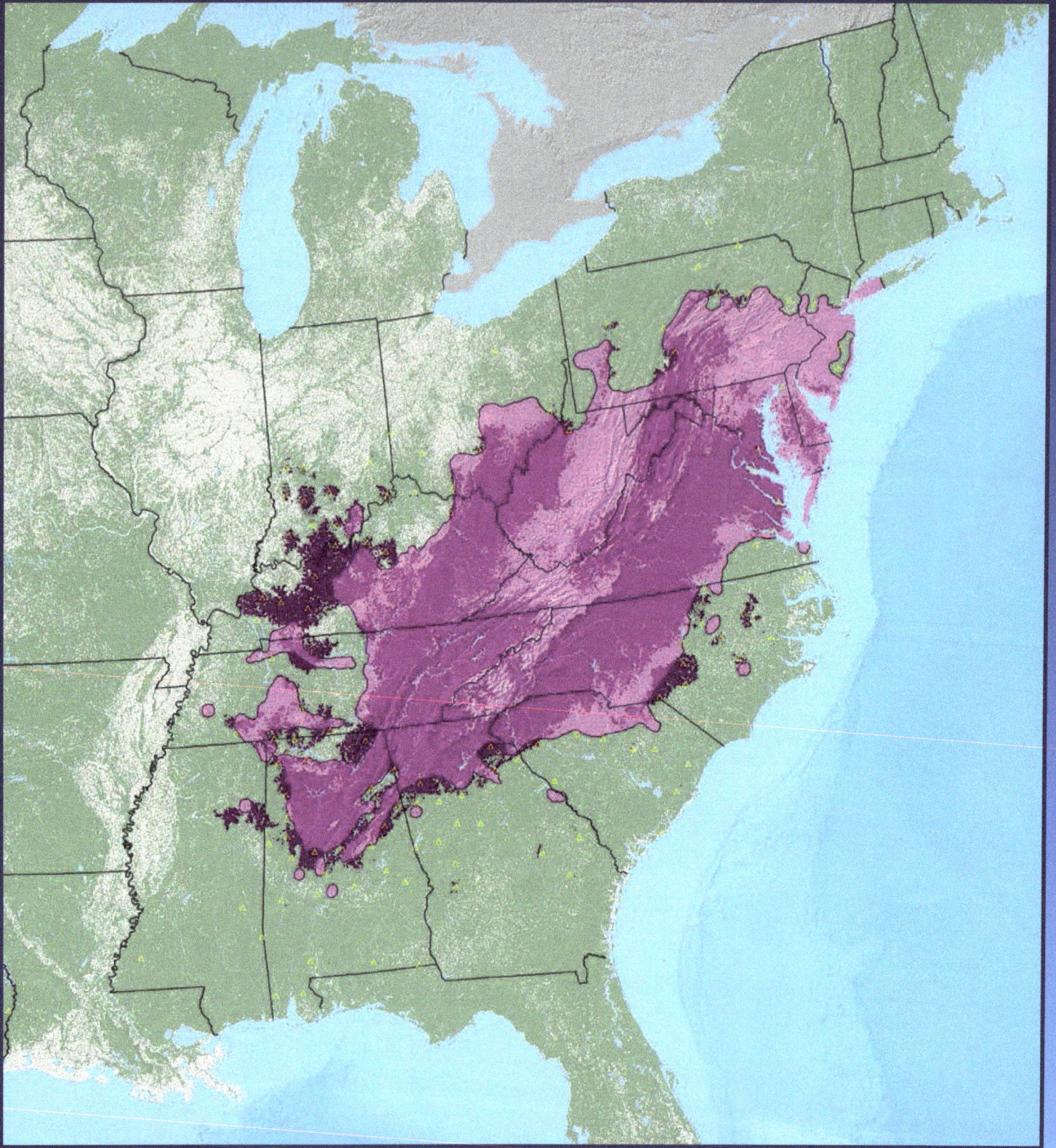

Virginia pine

Virginia pine
Pinus virginiana Mill.
Pinaceae – Pine Family
FIA Code **132** – Plants Symbol **PIVI2**

Washoe pine

Pinus washoensis H. Mason & Stockwell

Pinaceae – Pine Family

FIA Code **137** – Plants Symbol **PIWA**

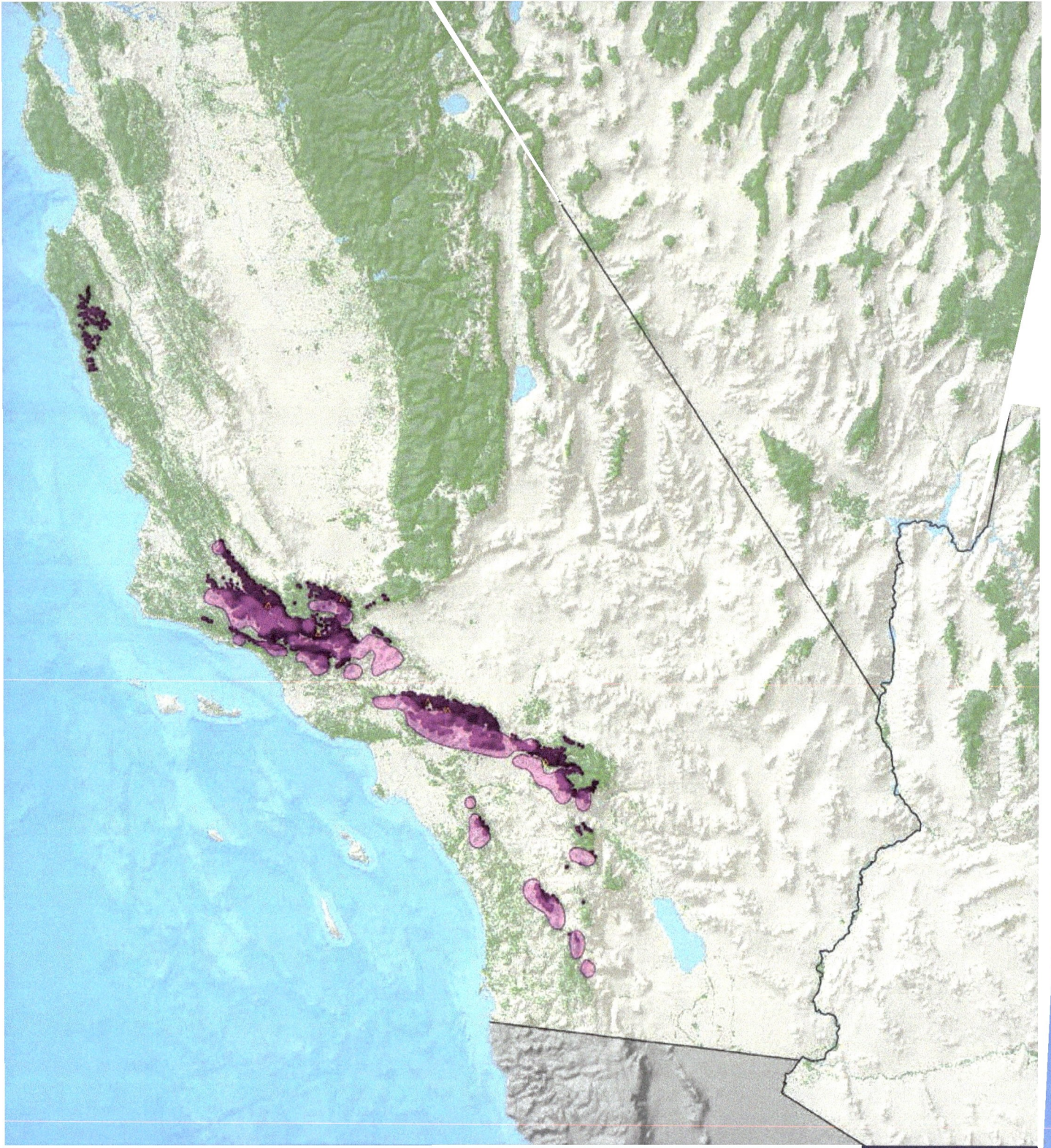

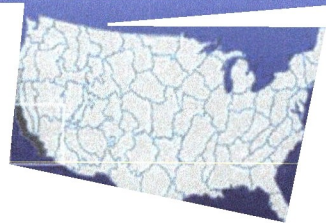

National Individual Tree Species Atlas

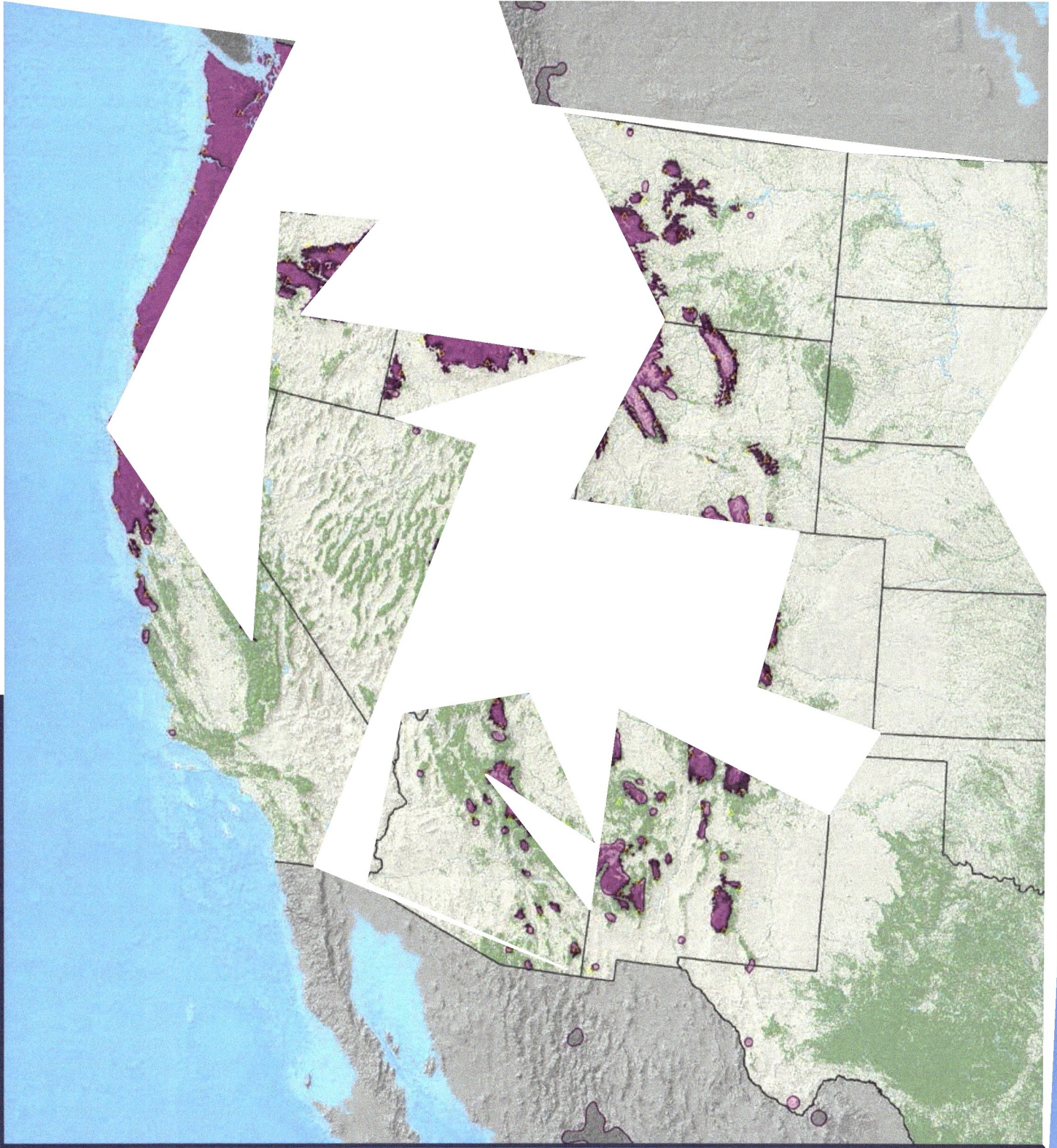

Douglas-fir

Pseudotsuga menziesii (Mirb.) Franco

Pinaceae – Pine Family

FIA Code **202** – Plants Symbol **PSME**

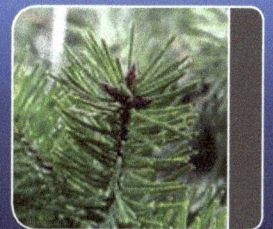

National Individual Tree Species Atlas

redwood
Sequoia sempervirens (Lamb. ex D. Don) Endl.
Cupressaceae – Cypress Family
FIA Code **211** – Plants Symbol **SESE3**

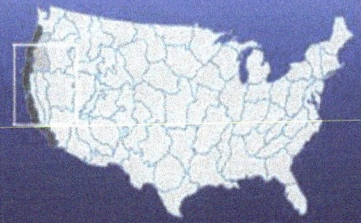

National Individual Tree Species Atlas

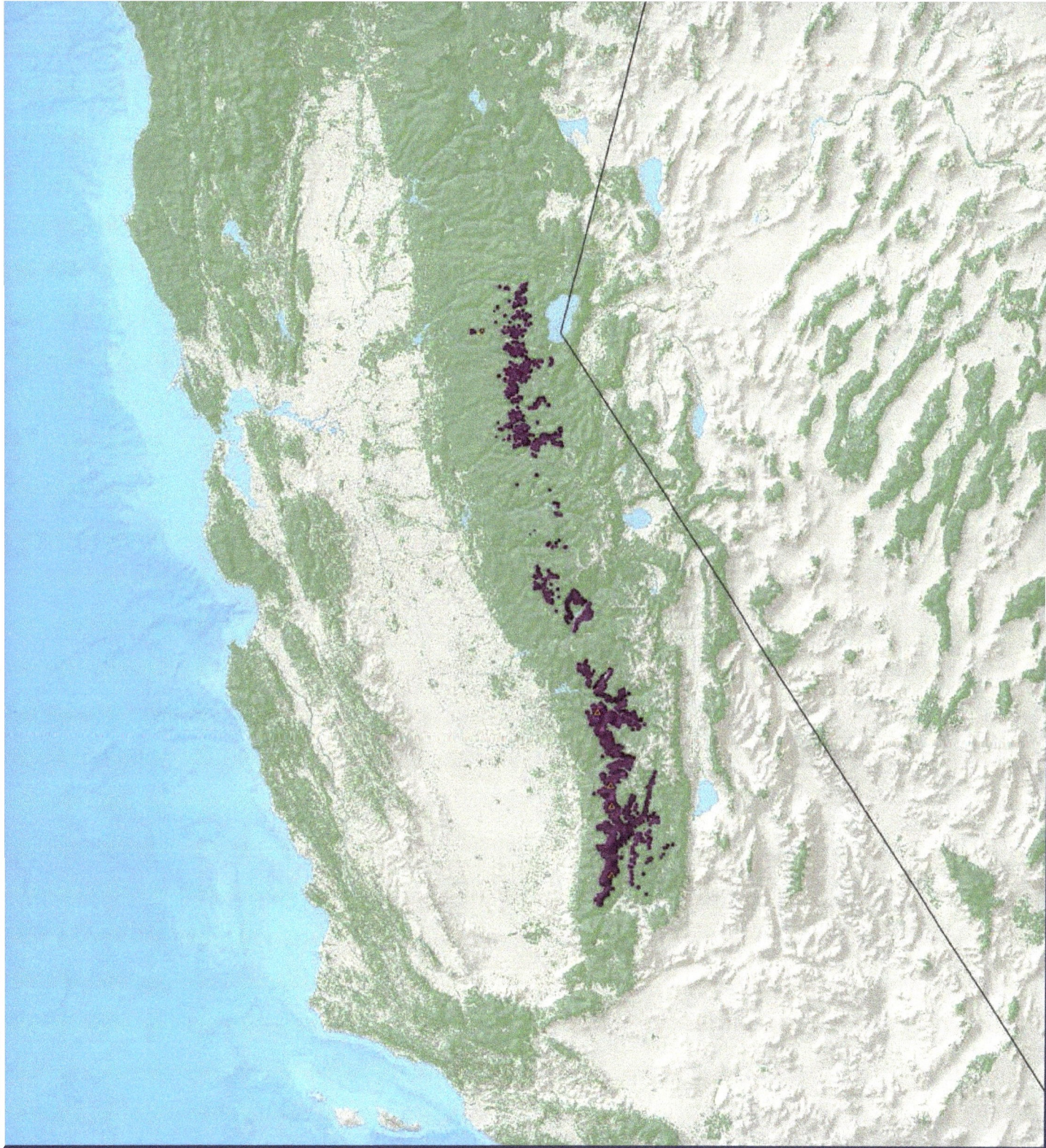

giant sequoia

Sequoiadendron giganteum (Lindl.) J. Buchholz

Cupressaceae – Cypress Family

FIA Code **212** – Plants Symbol **SEGI2**

National Individual Tree Species Atlas

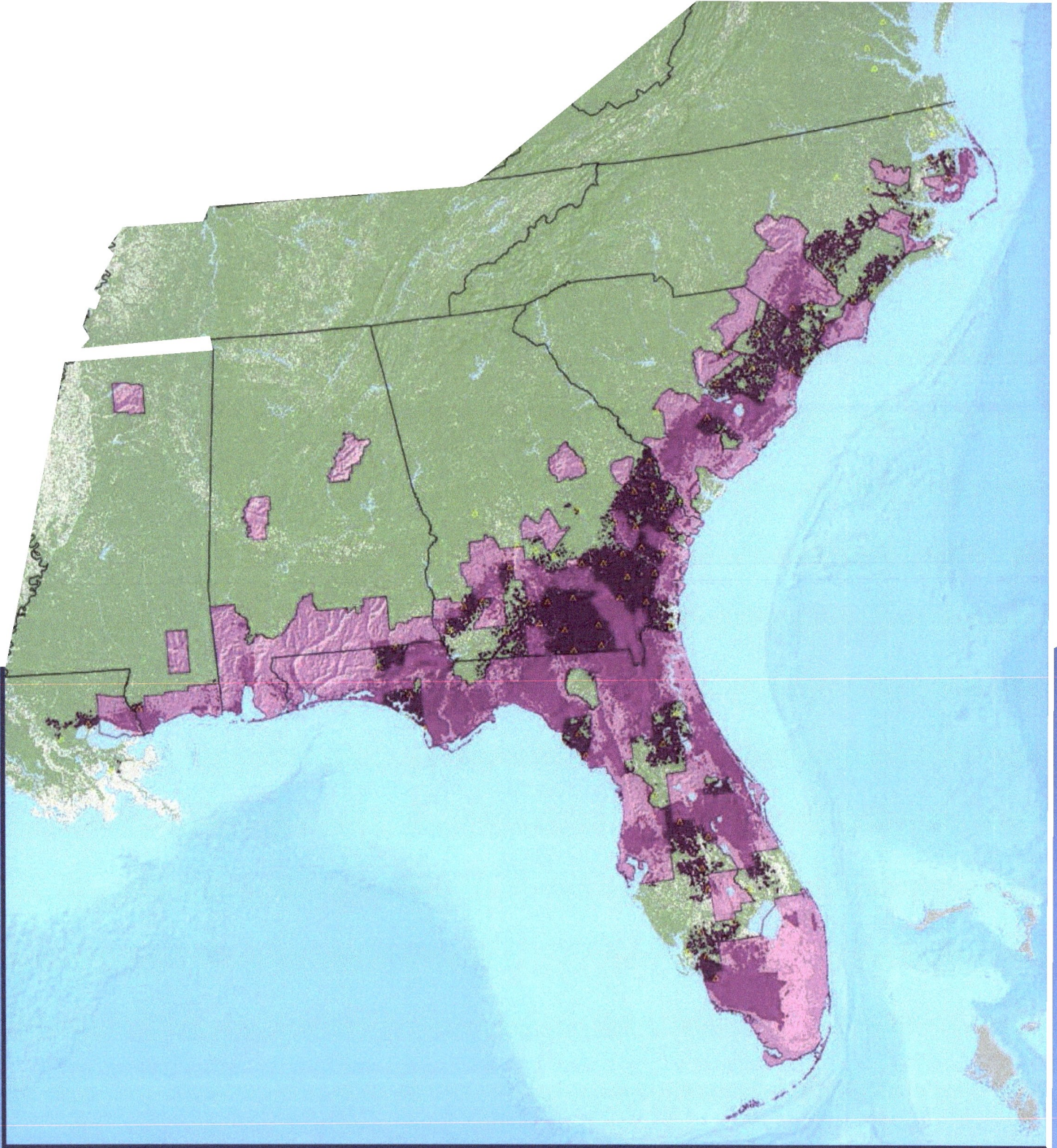

pondcypress

Taxodium ascendens Brongn.

Cupressaceae – Cypress Family

FIA Code **222** – Plants Symbol **TAAS**

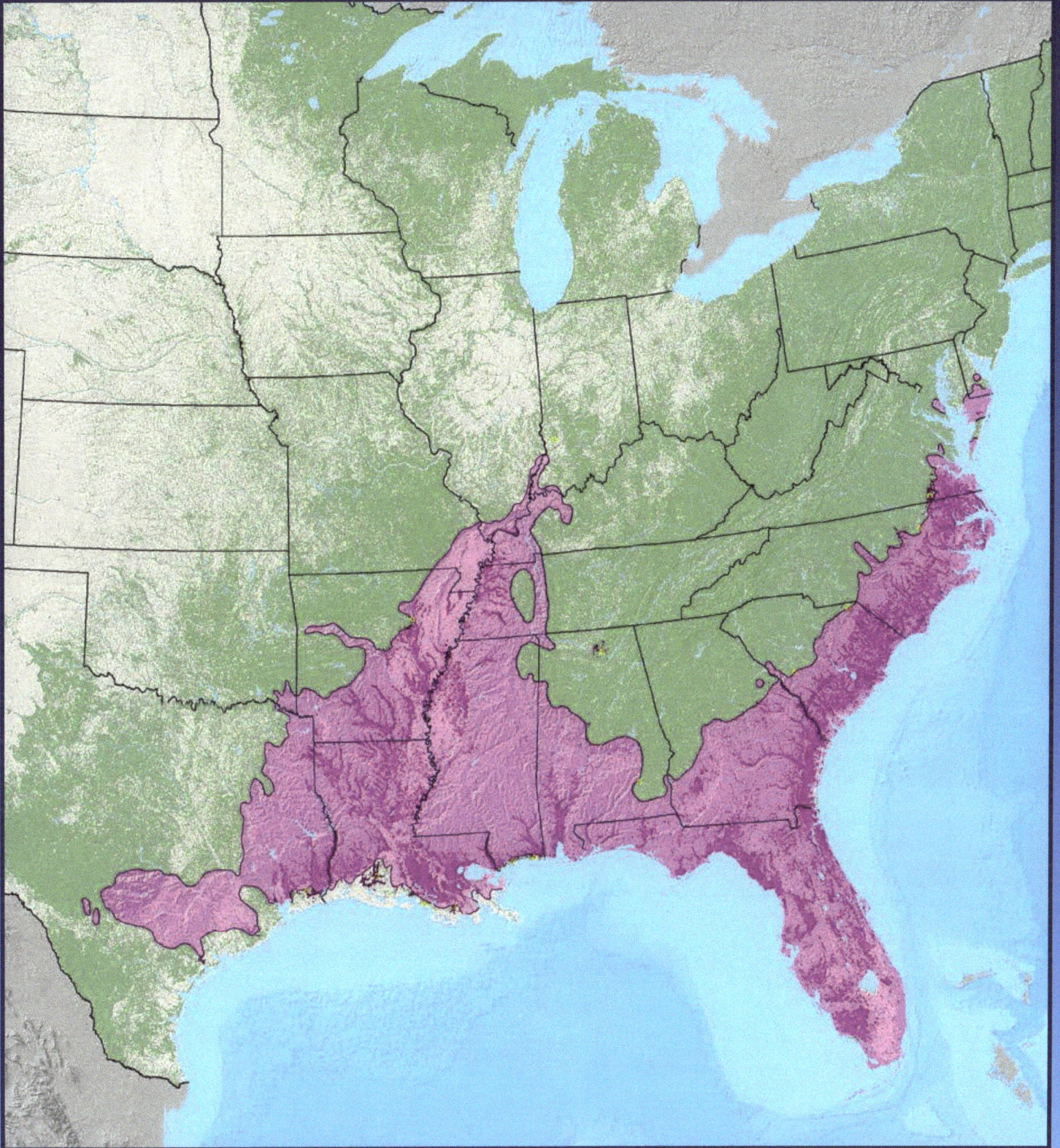

baldcypress

baldcypress
Taxodium distichum (L.) Rich.
Cupressaceae – Cypress Family
FIA Code **221** – Plants Symbol **TADI2**

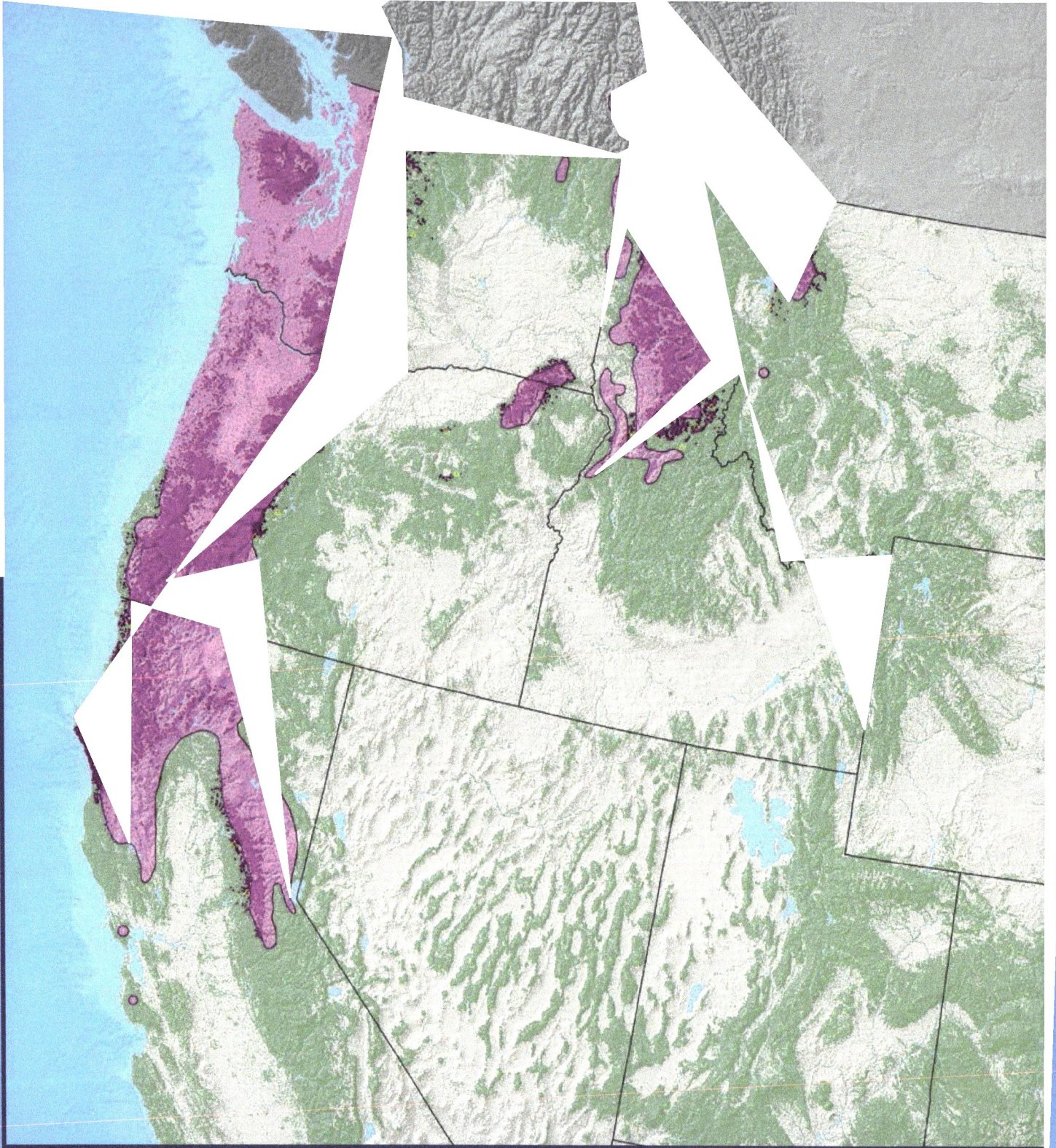

Pacific yew
Taxus brevifolia Nutt.
Taxaceae – Yew Family
FIA Code **231** – Plants Symbol **TABR2**

National Individual Tree Species Atlas

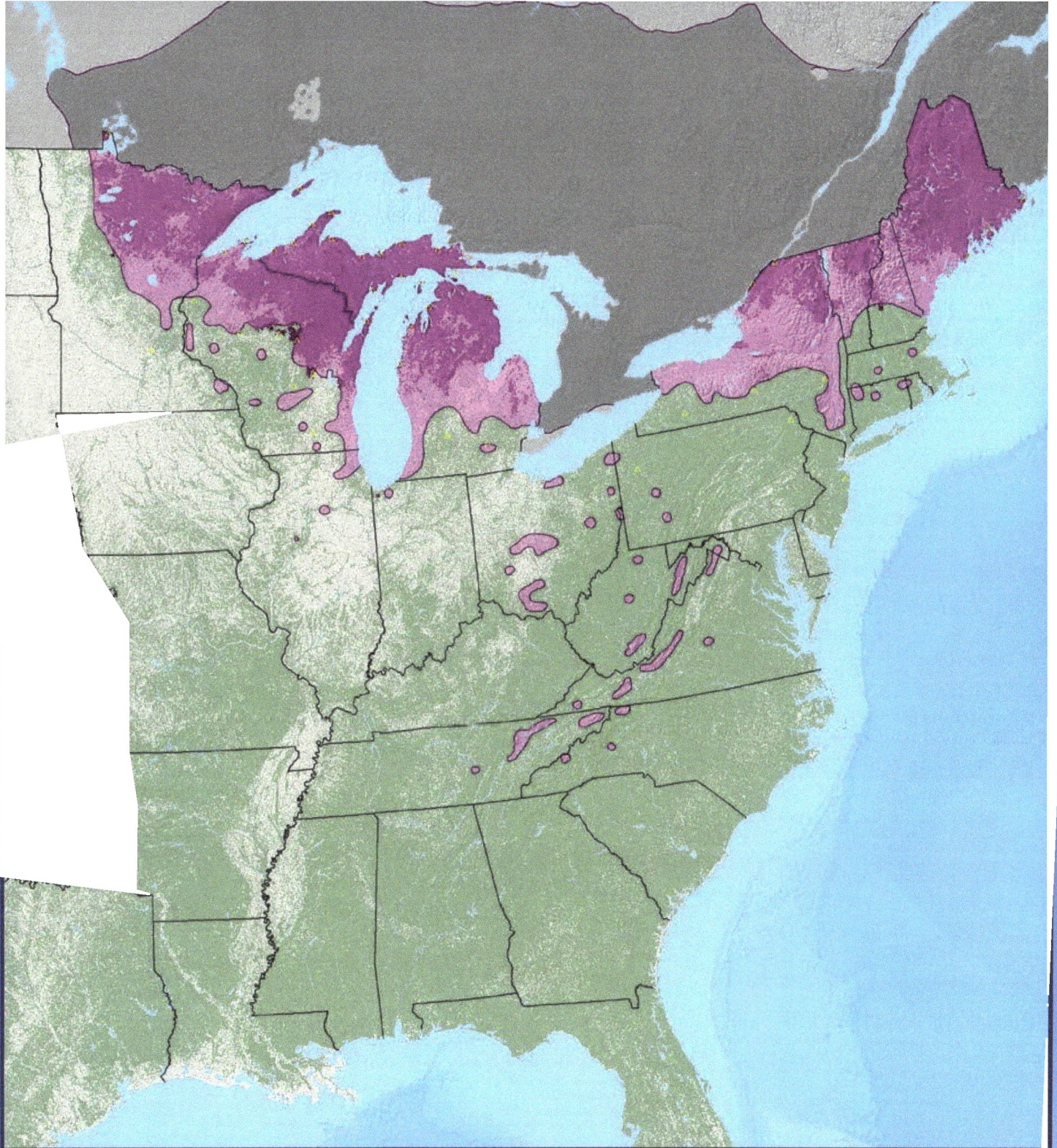

northern white-cedar
Thuja occidentalis L.
Cupressaceae – Cypress Family
FIA Code **241** – Plants Symbol **THOC2**

National Individual Tree Species Atlas

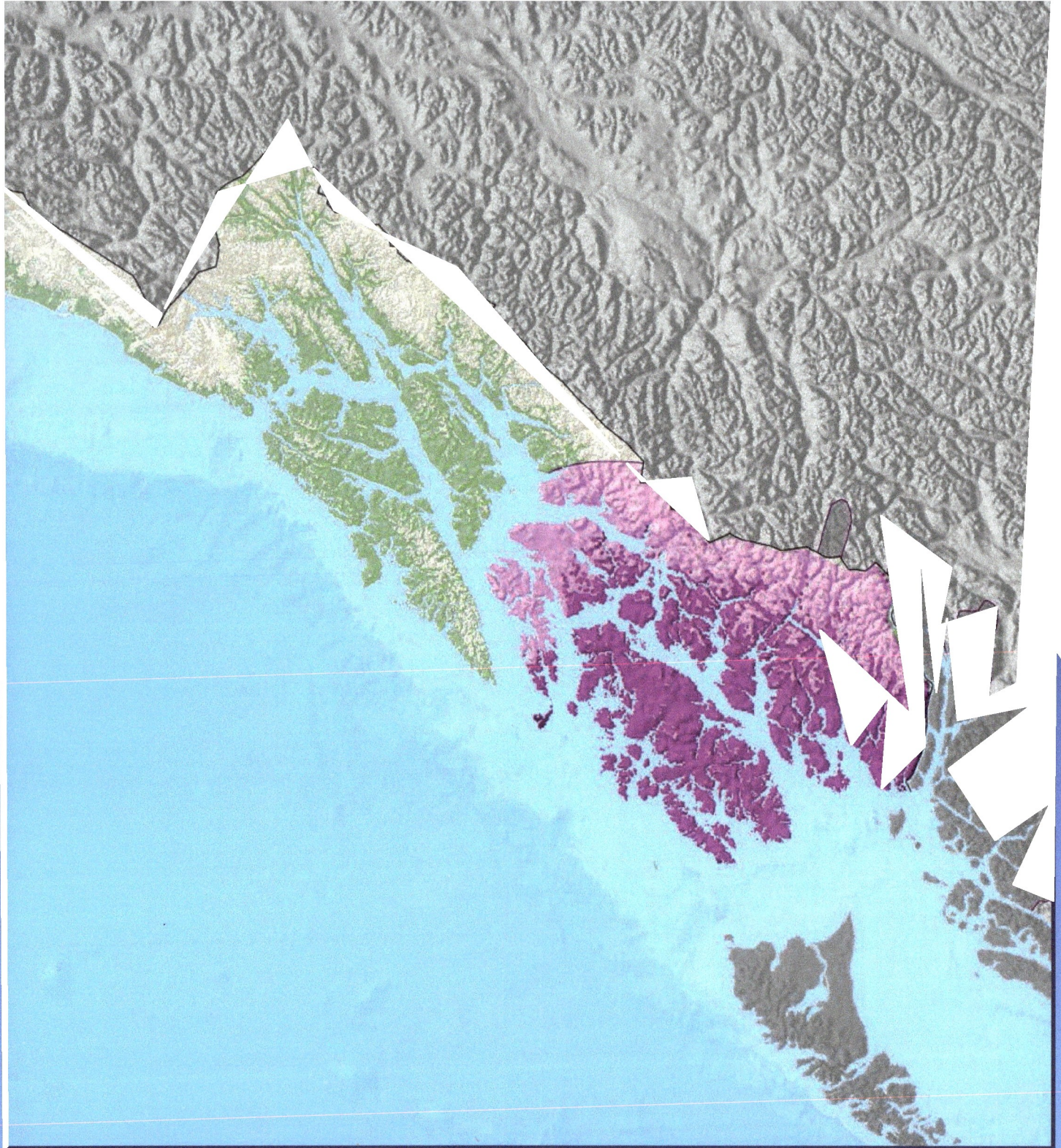

western redcedar

Thuja plicata Donn ex D. Don

Cupressaceae – Cypress Family

FIA Code **242** – Plants Symbol **THPL**

National Individual Tree Species Atlas

western redcedar

Thuja plicata Donn ex D. Don

Cupressaceae – Cypress Family

FIA Code **242** – Plants Symbol **THPL**

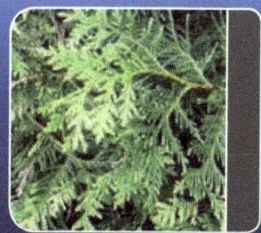

National Individual Tree Species Atlas

California torrey

Torreya californica Torr.

Taxaceae – Yew Family

FIA Code **251** – Plants Symbol **TOCA**

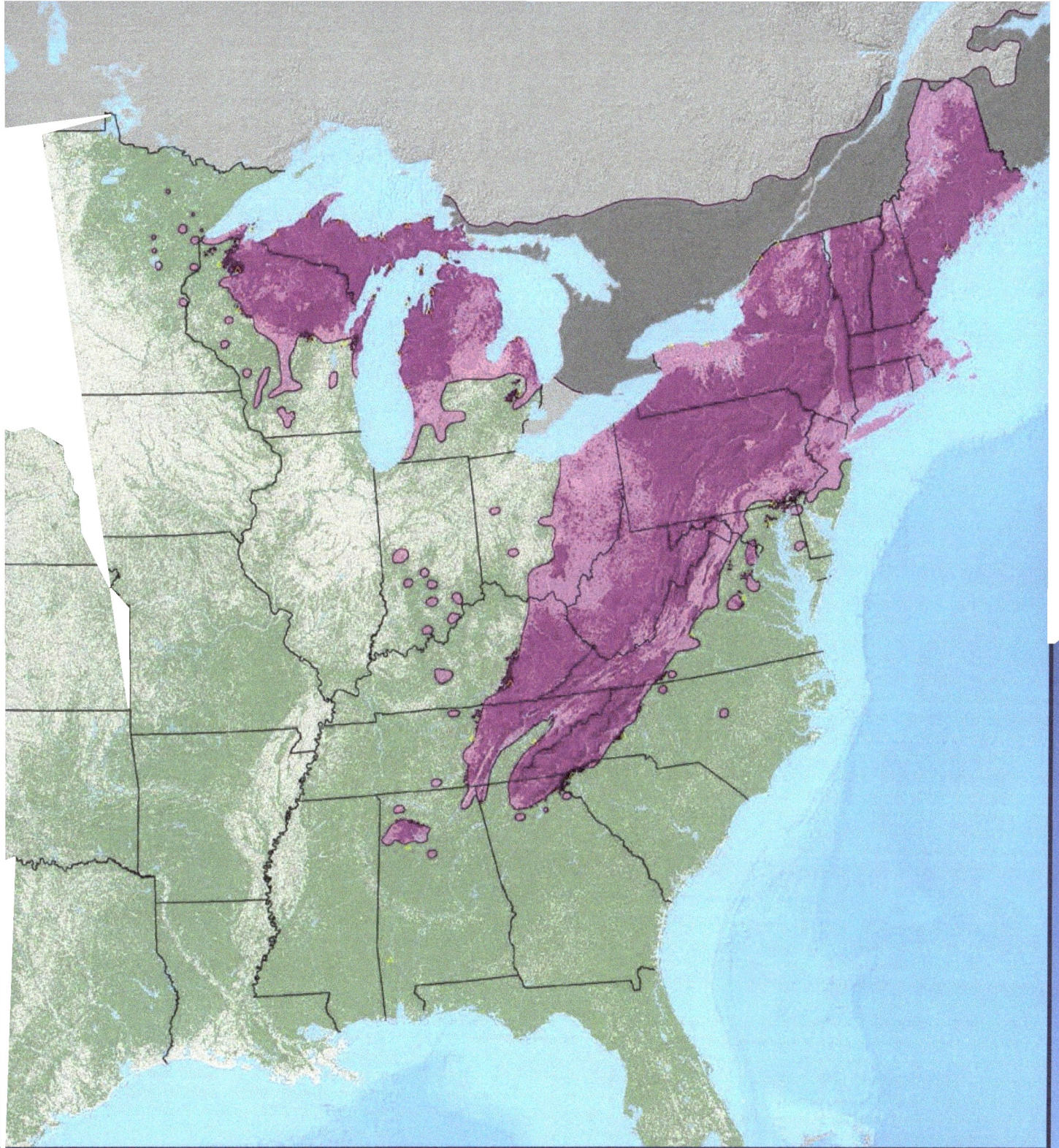

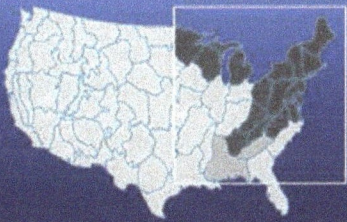

eastern hemlock

Tsuga canadensis (L.) Carrière

Pinaceae – Pine Family

FIA Code **261** – Plants Symbol **TSCA**

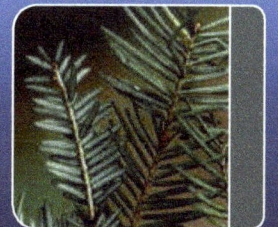

National Individual Tree Species Atlas

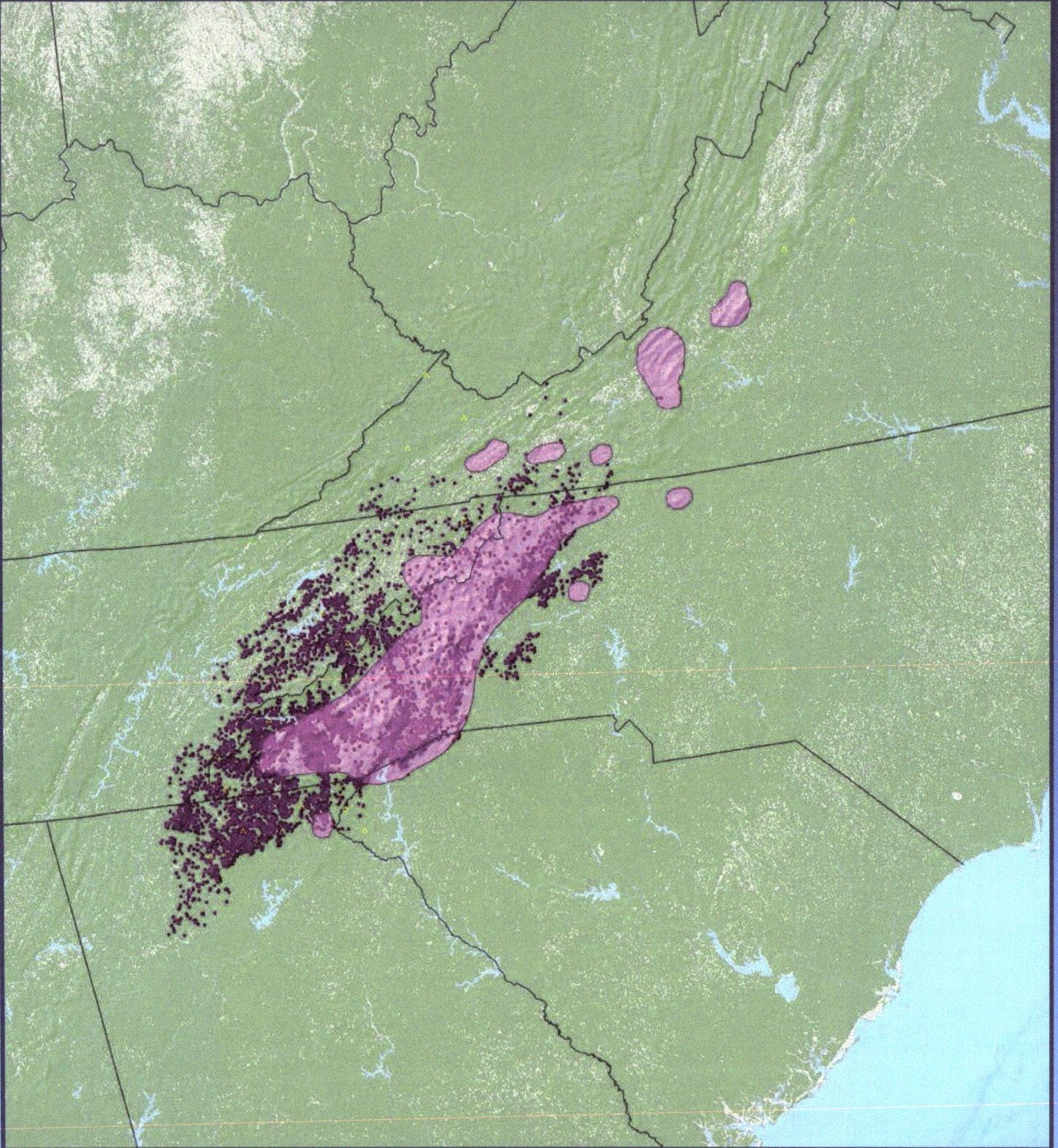

Carolina hemlock

Carolina hemlock

Tsuga caroliniana Engelm.

Pinaceae – Pine Family

FIA Code **262** – Plants Symbol **TSCA2**

National Individual Tree Species Atlas

western hemlock
Tsuga heterophylla (Raf.) Sarg.
Pinaceae – Pine Family
FIA Code **263** – Plants Symbol **TSHE**

western hemlock

Tsuga heterophylla (Raf.) Sarg.

Pinaceae – Pine Family

FIA Code **263** – Plants Symbol **TSHE**

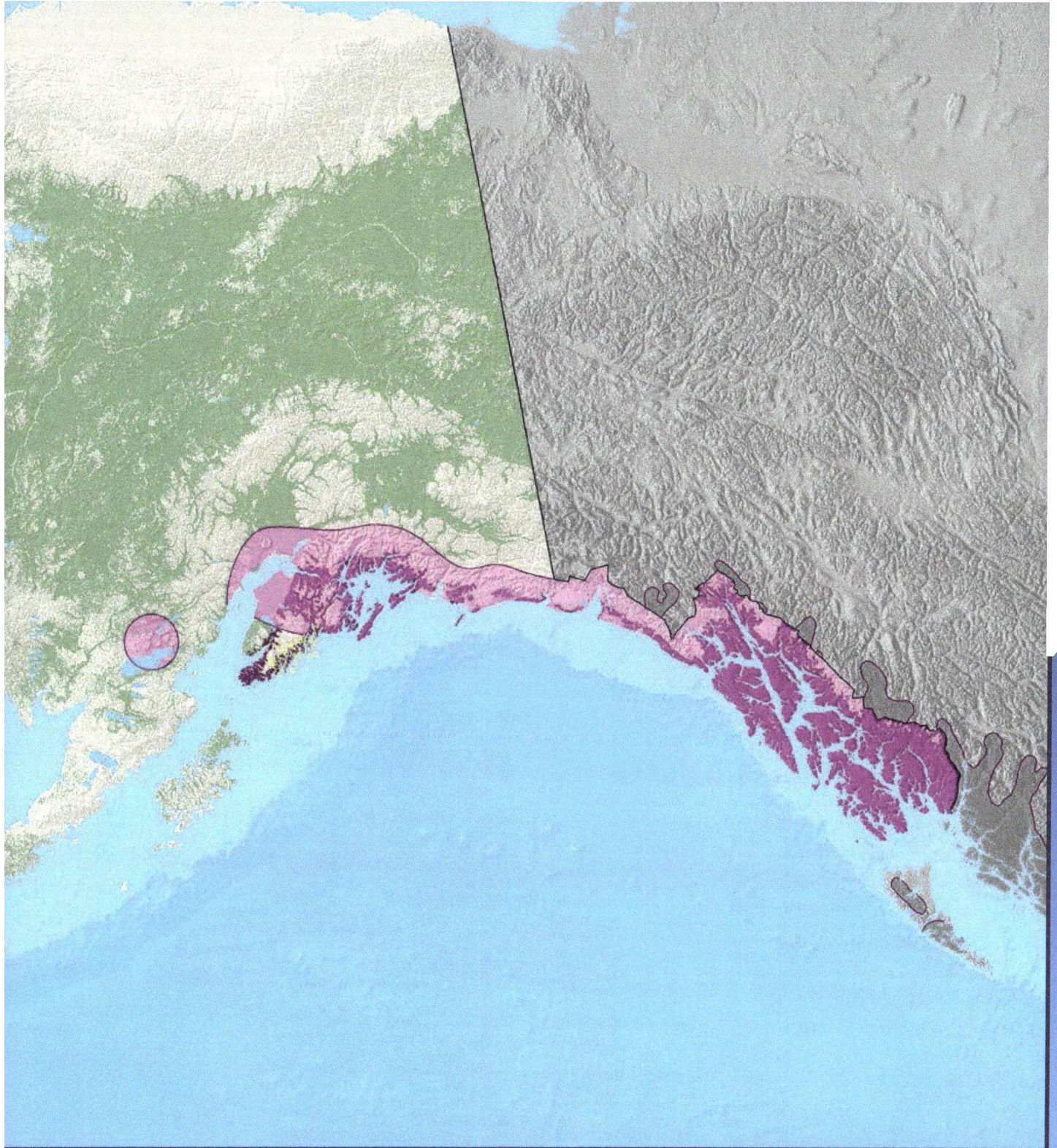

mountain hemlock

Tsuga mertensiana (Bong.) Carrière

Pinaceae – Pine Family

FIA Code **264** – Plants Symbol **TSME**

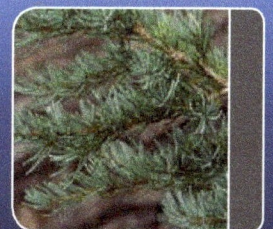

National Individual Tree Species Atlas

mountain hemlock
Tsuga mertensiana (Bong.) Carrière
Pinaceae – Pine Family
FIA Code **264** – Plants Symbol **TSME**

National Individual Tree Species Atlas

Hardwoods

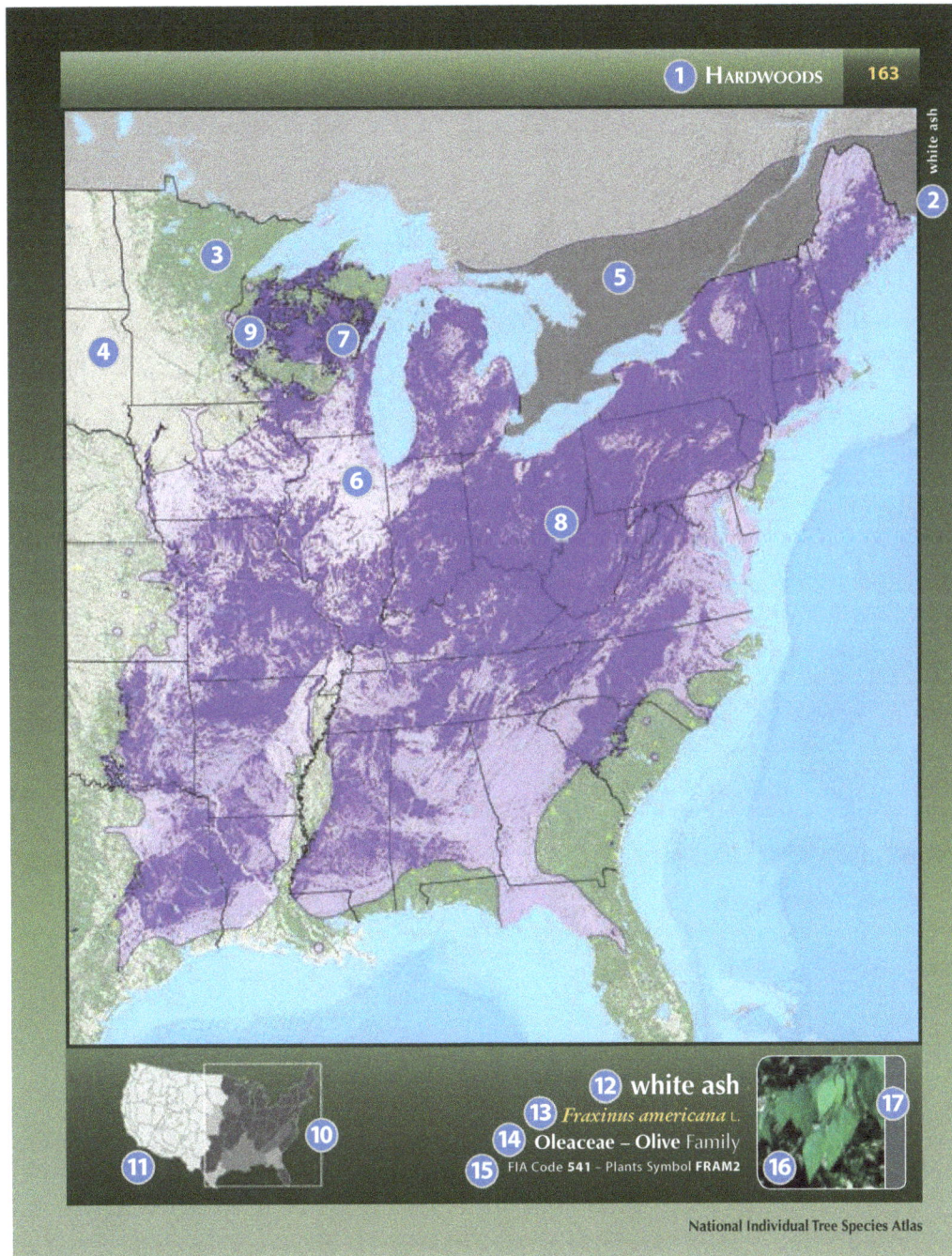

HOW TO READ THE MAPS

1. **Division**
2. **Common name quick reference**
3. **Treed areas**
4. **Non-treed areas**
5. *Mapped non U.S. extent*
6. *Mapped Atlas*
7. *Modeled Atlas*
8. **Coincidence**
9. ⚠ **Perturbed FIA plots**
10. **USGS map zones**
11. **Map extent with model correlations** (x100)
    - 81 -100 Strongest fit
    - 61 - 80
    - 41 - 60
    - 21 - 40
    - 0 - 20
    - Not modeled
    - No data available
12. **Common name**
13. **Scientific name and authority**
14. **Family name**
15. **FIA code and Plants symbol**
16. **Leaf**
17. **Color chip**

1 HARDWOODS 163

2 white ash

12 white ash
13 *Fraxinus americana* L.
14 Oleaceae – Olive Family
15 FIA Code **541** – Plants Symbol **FRAM2**

National Individual Tree Species Atlas

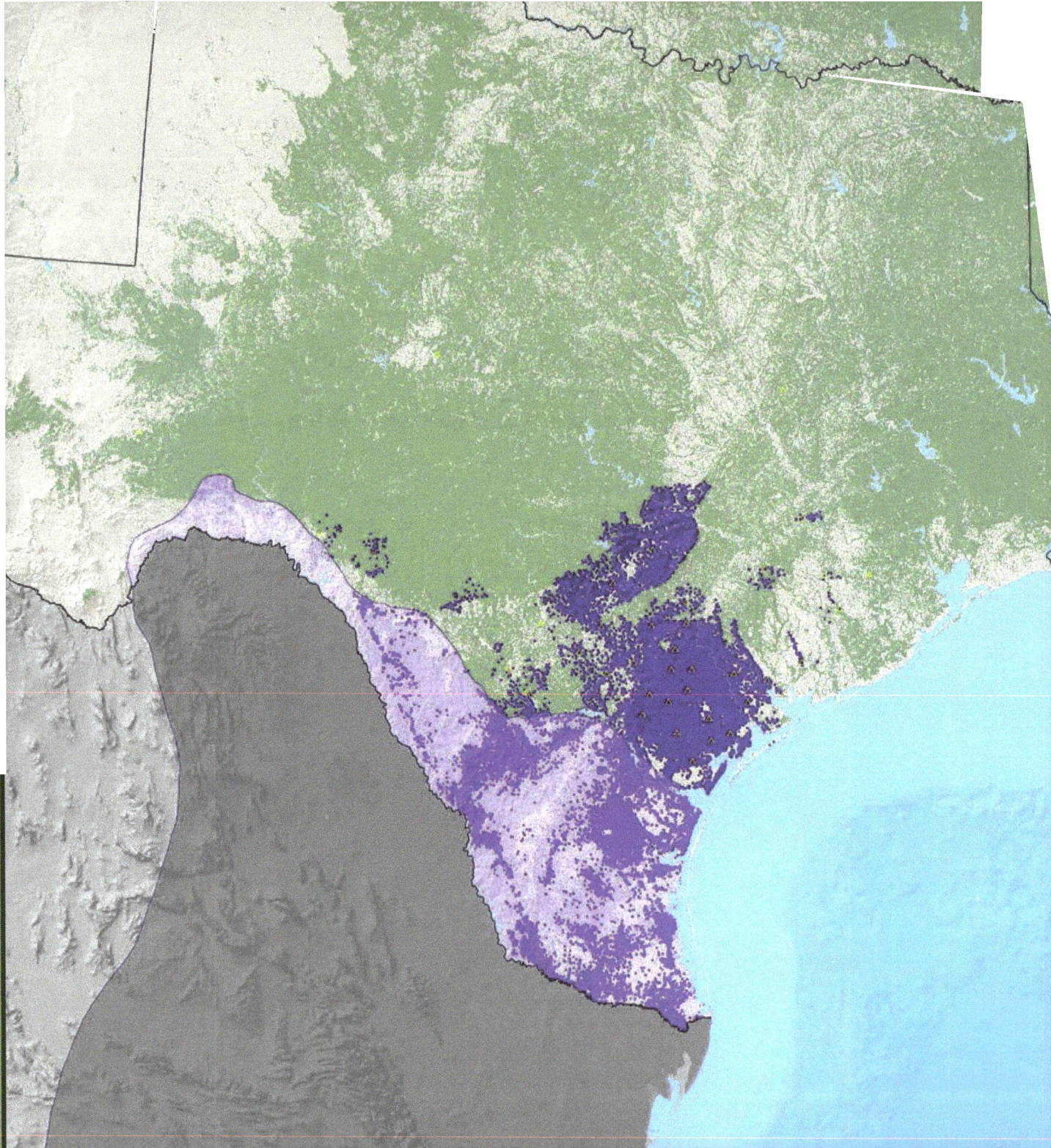

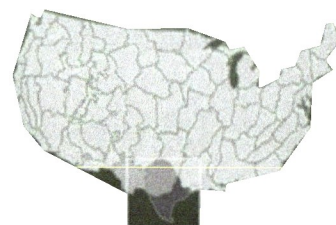

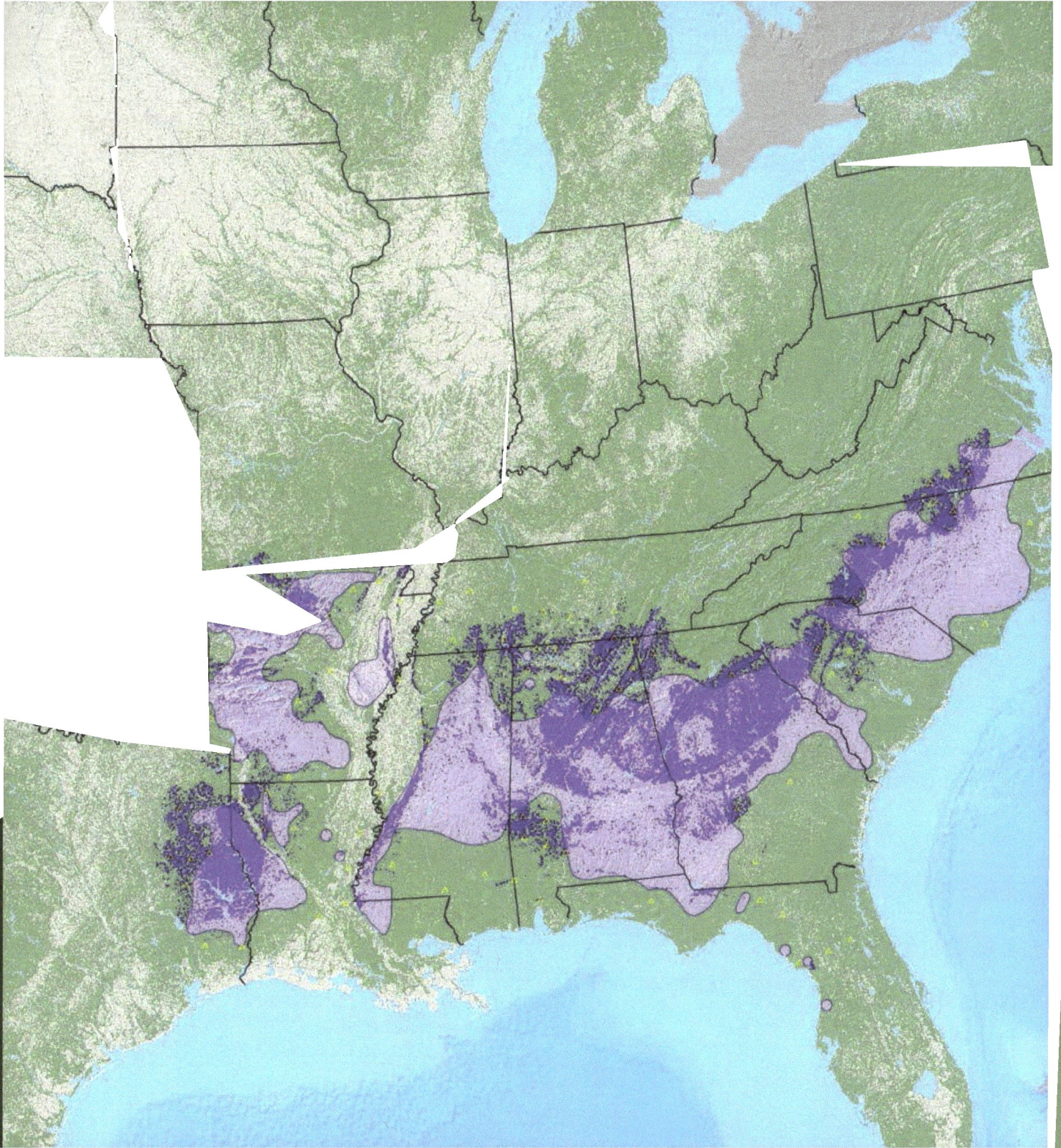
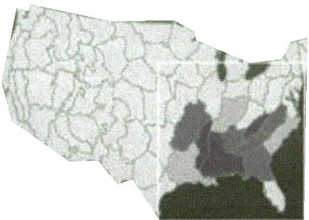

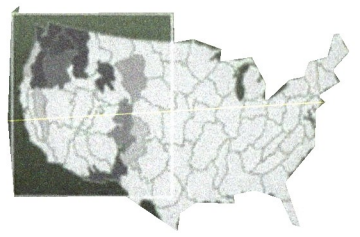

National Individual Tree Species Atlas

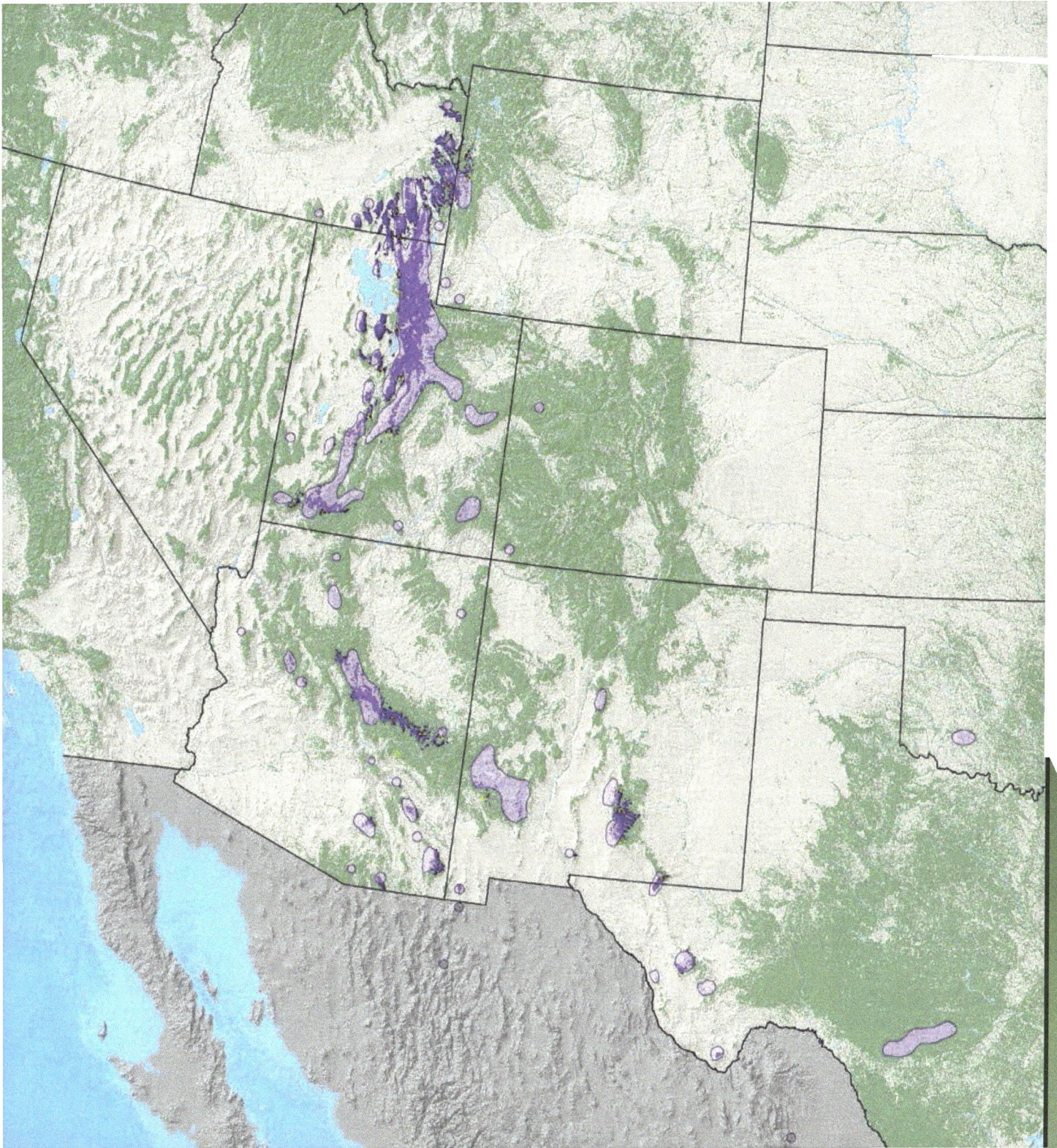

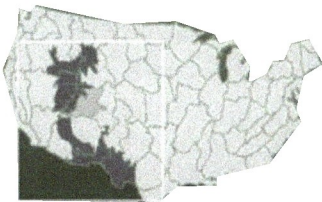

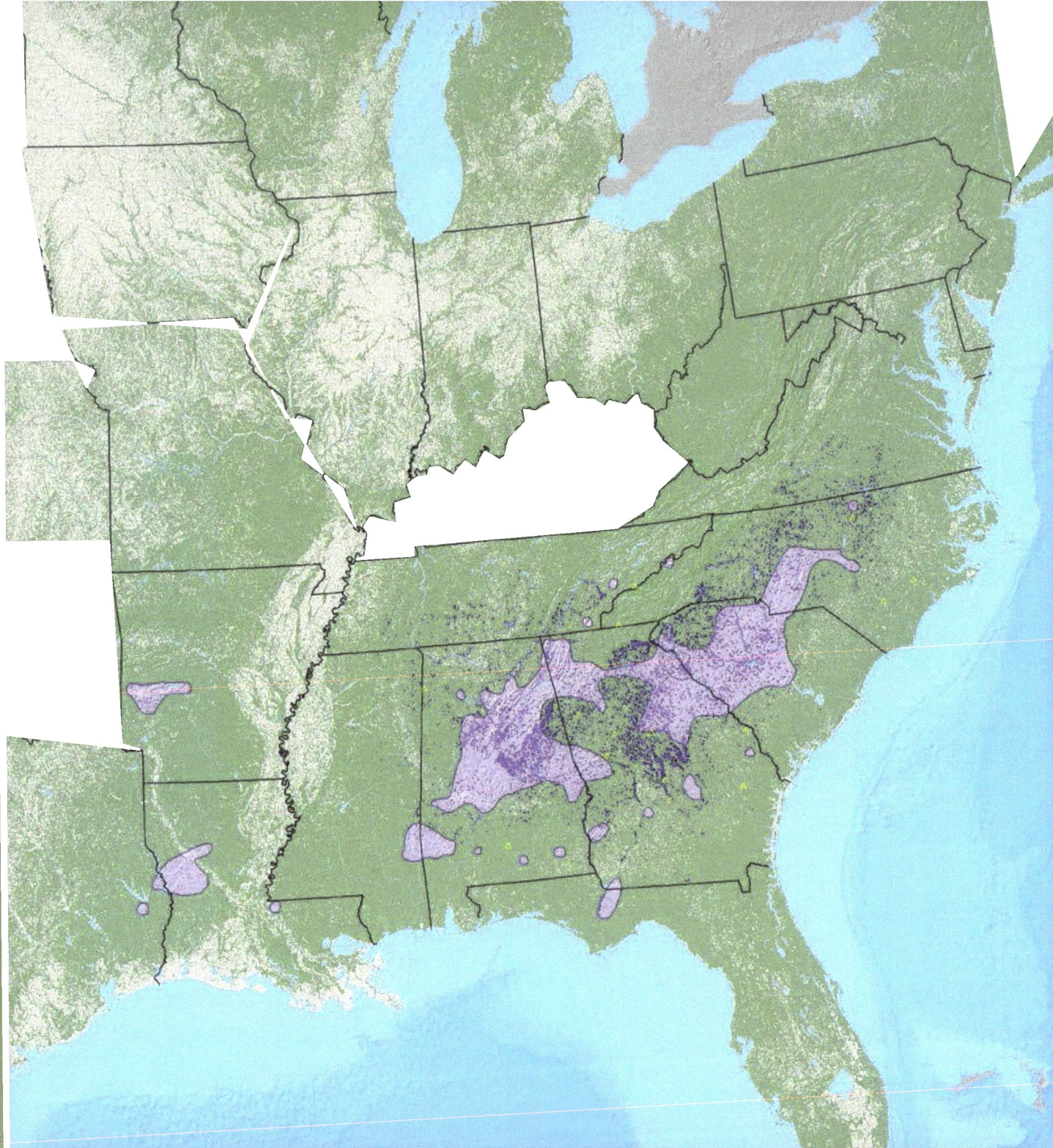

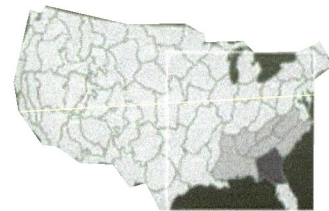

National Individual Tree Species Atlas

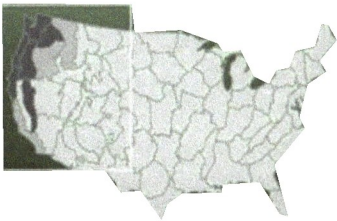

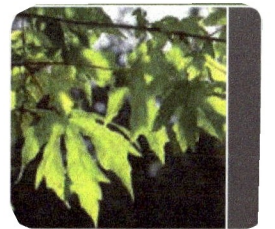

National Individual Tree Species Atlas

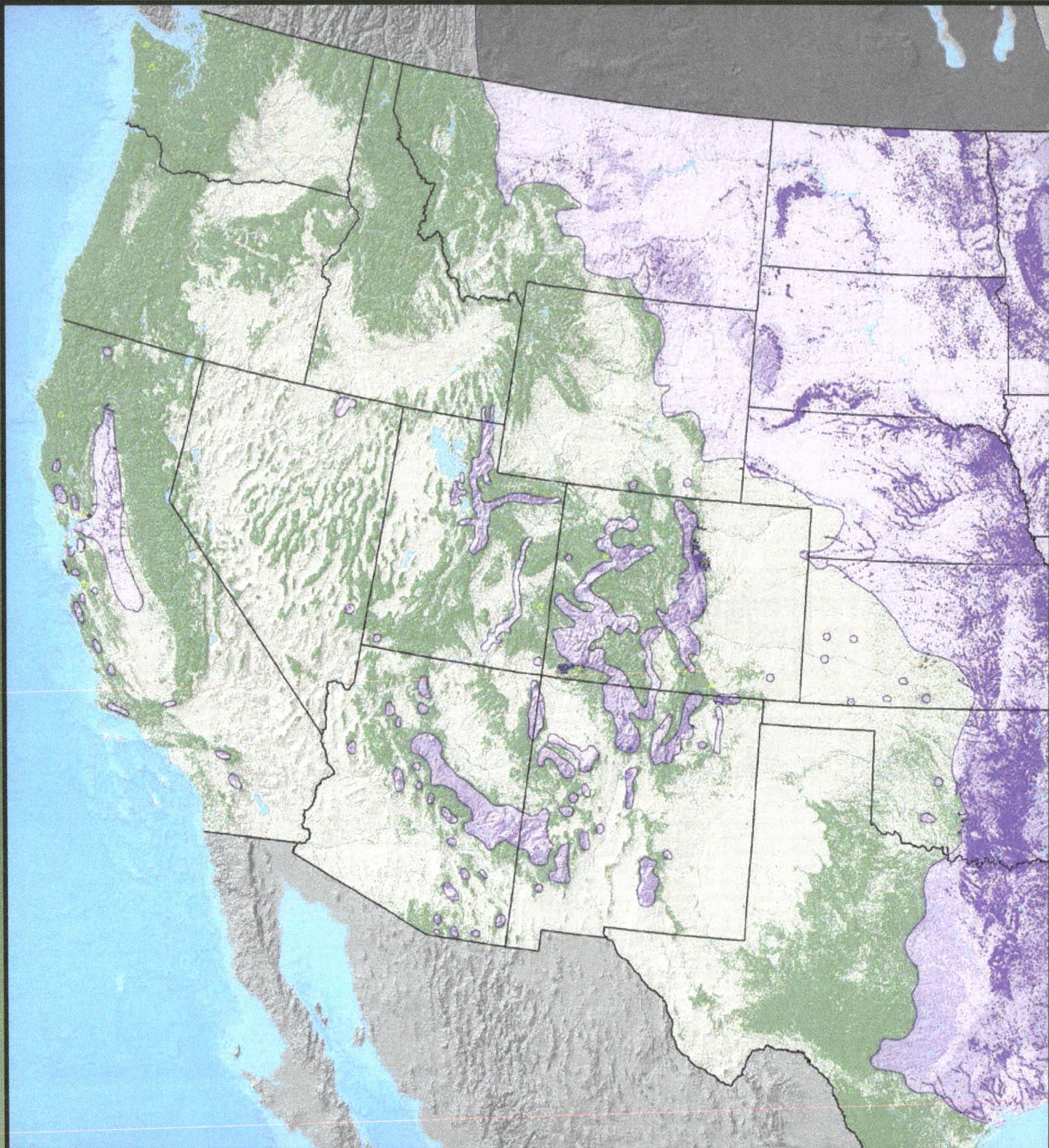

boxelder
Acer negundo L.
Aceraceae – Maple Family
FIA Code **313** – Plants Symbol **ACNE2**

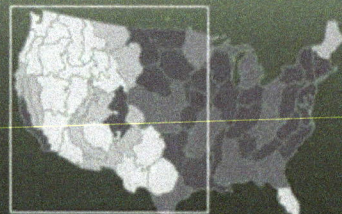

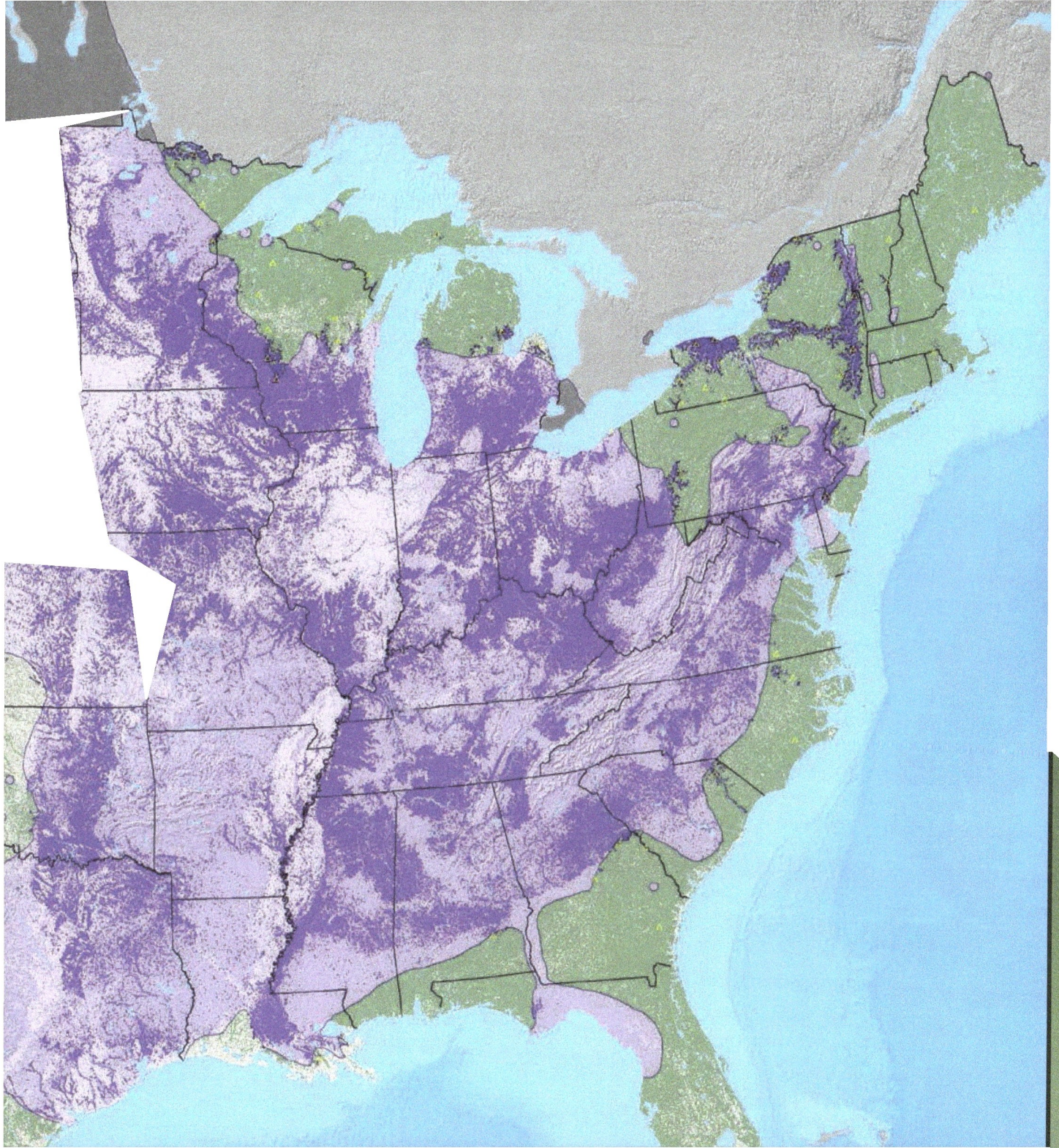
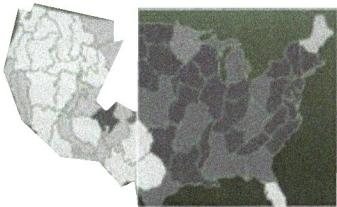
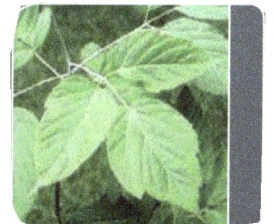

National Individual Tree Species Atlas

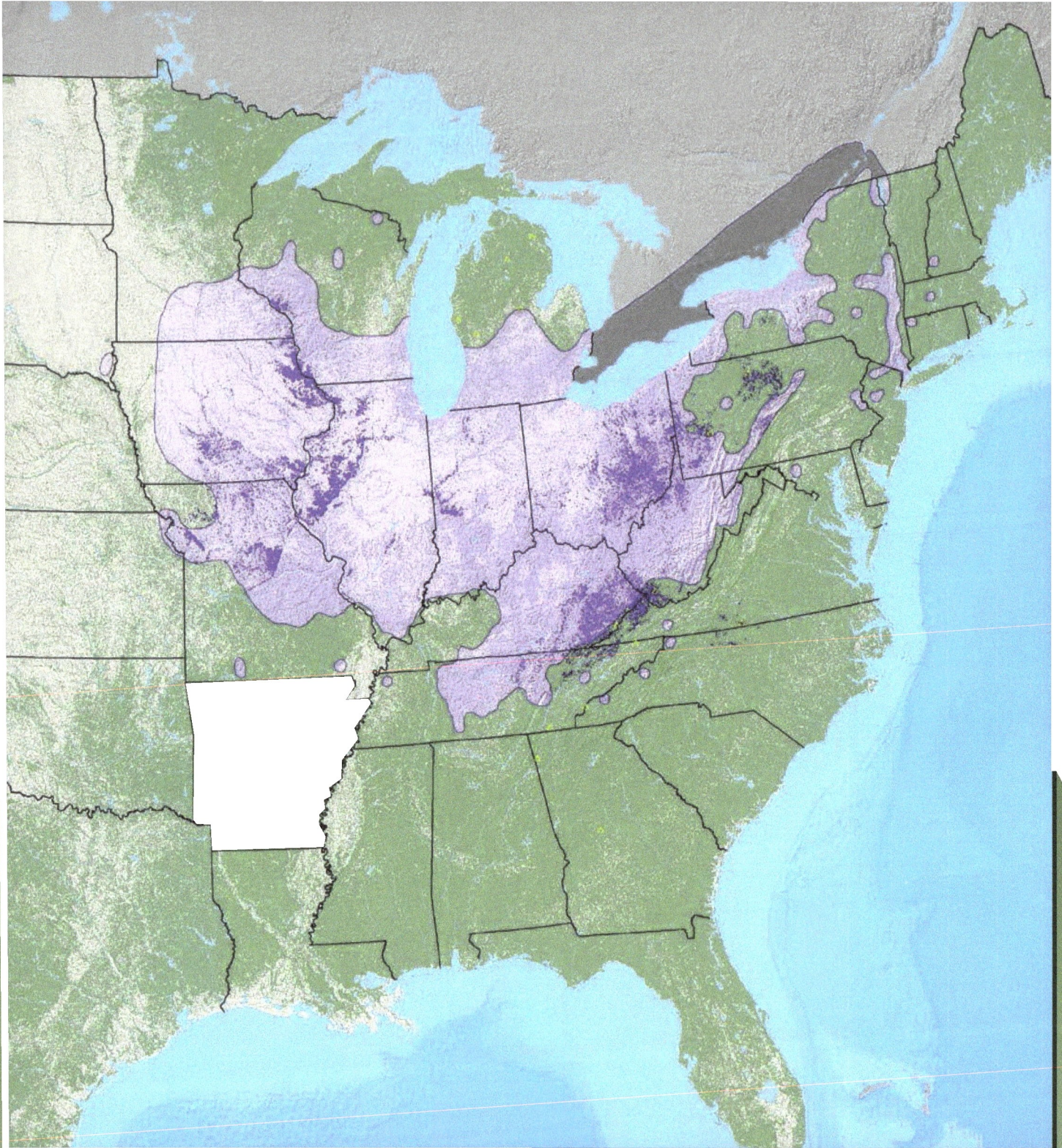

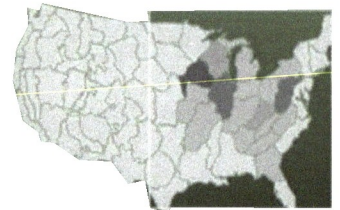

National Individual Tree Species Atlas

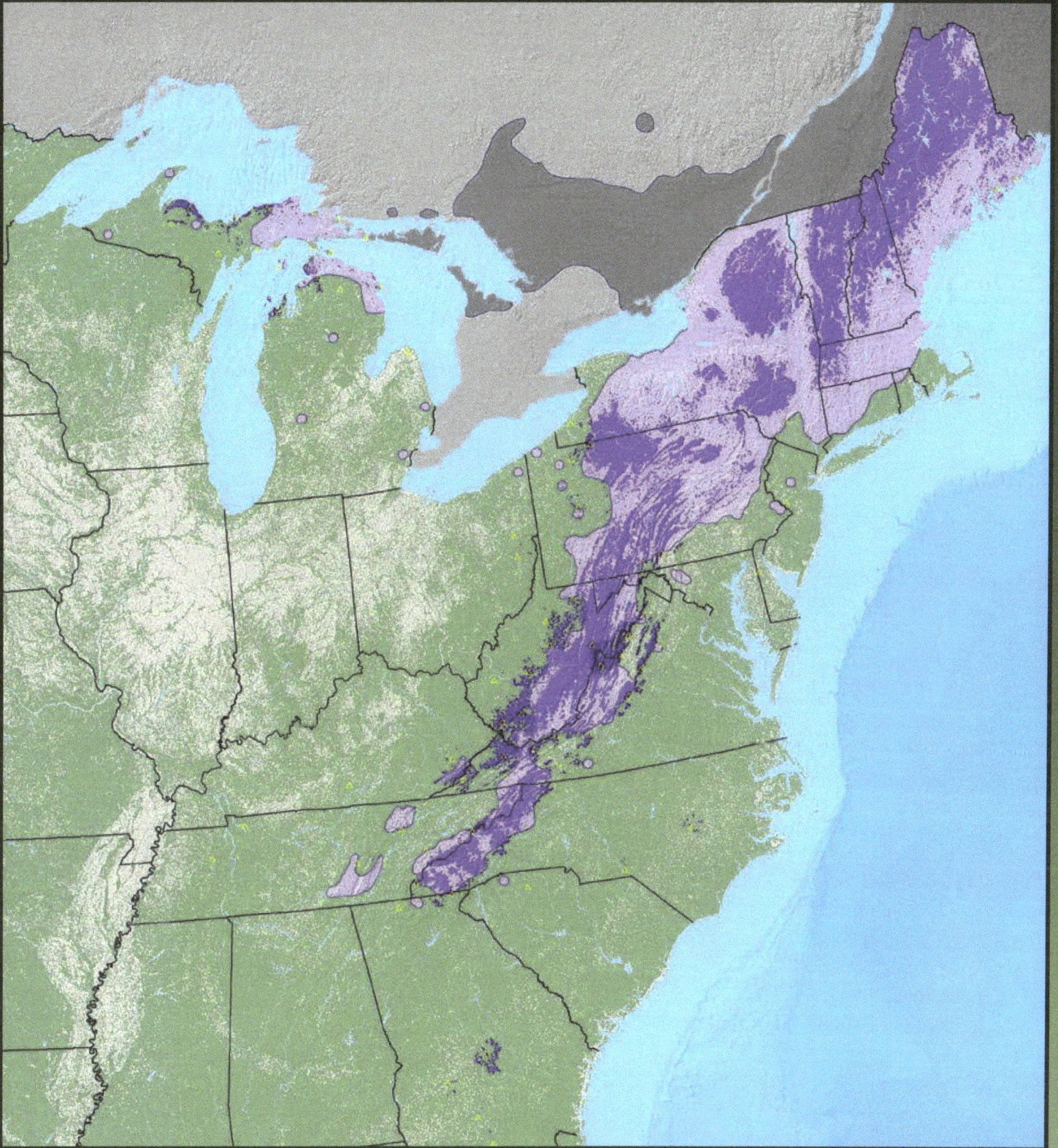

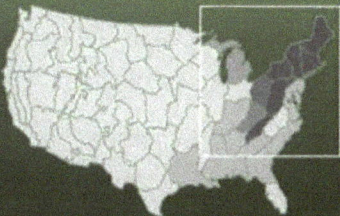

striped maple

striped maple
Acer pensylvanicum L.
Aceraceae – Maple Family
FIA Code **315** – Plants Symbol **ACPE**

National Individual Tree Species Atlas

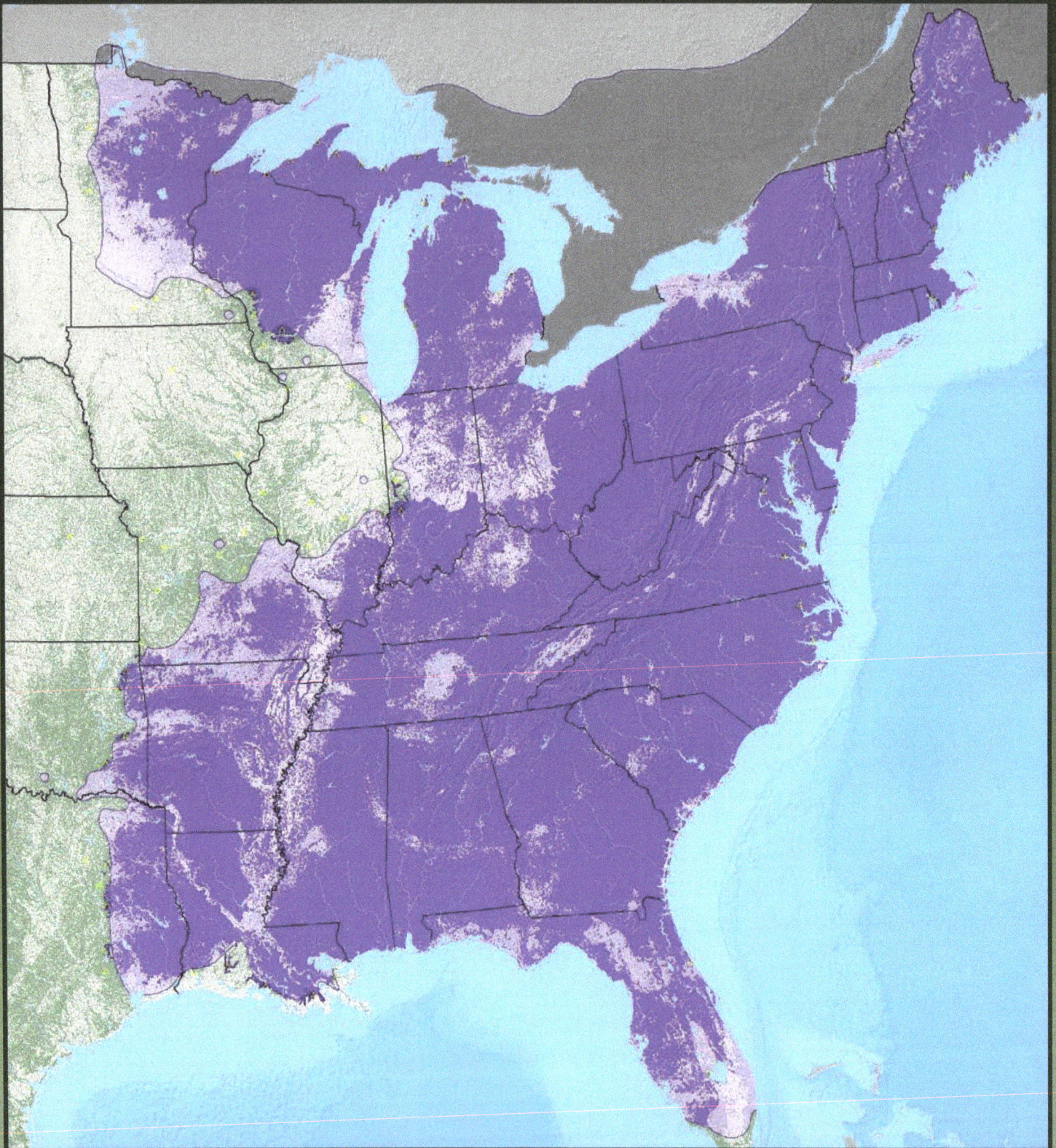

red maple

red maple
Acer rubrum L.
Aceraceae – Maple Family
FIA Code **316** – Plants Symbol **ACRU**

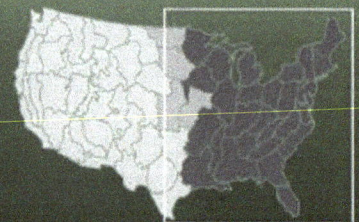

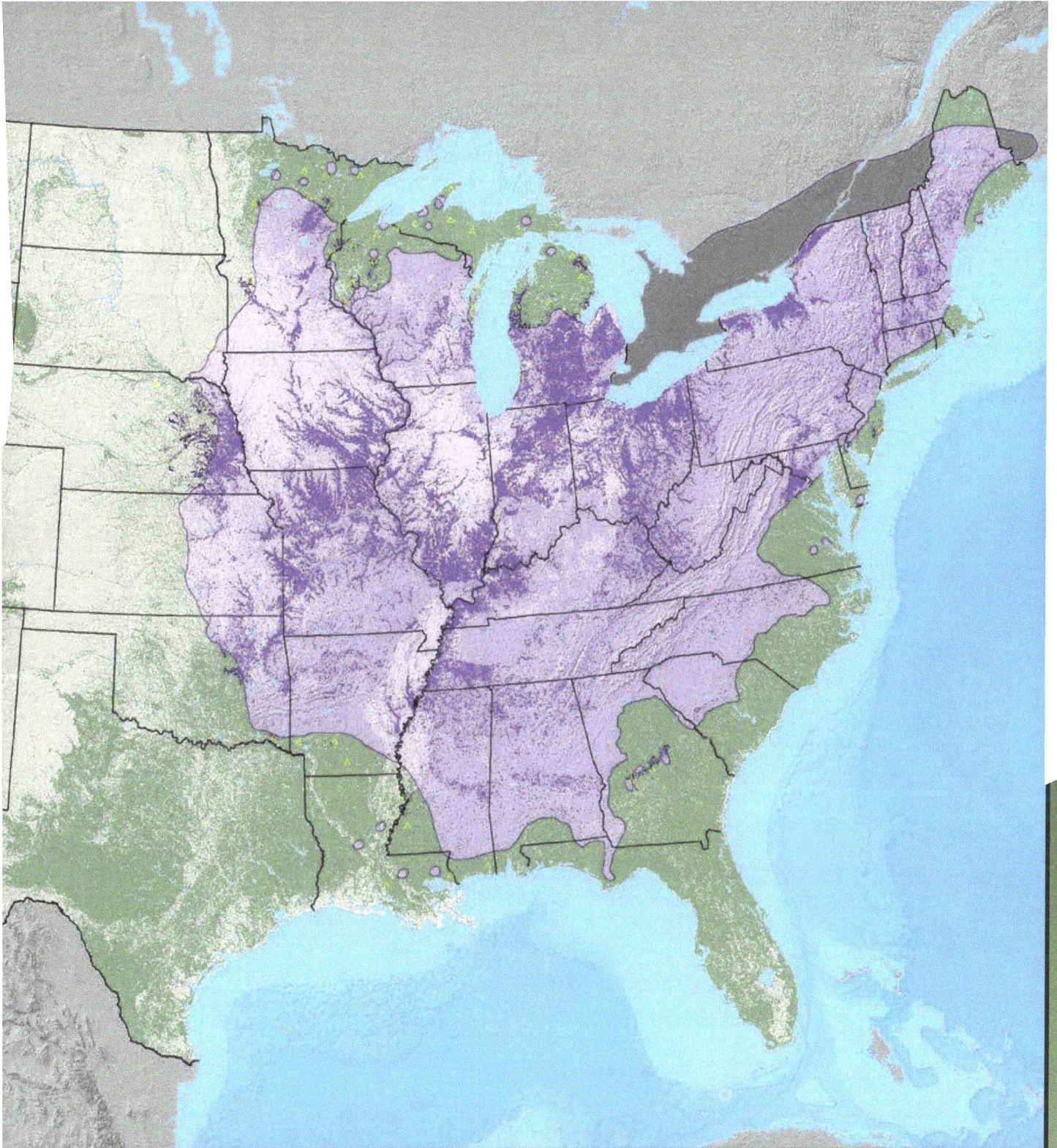

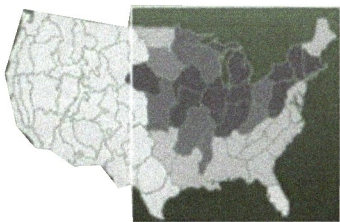

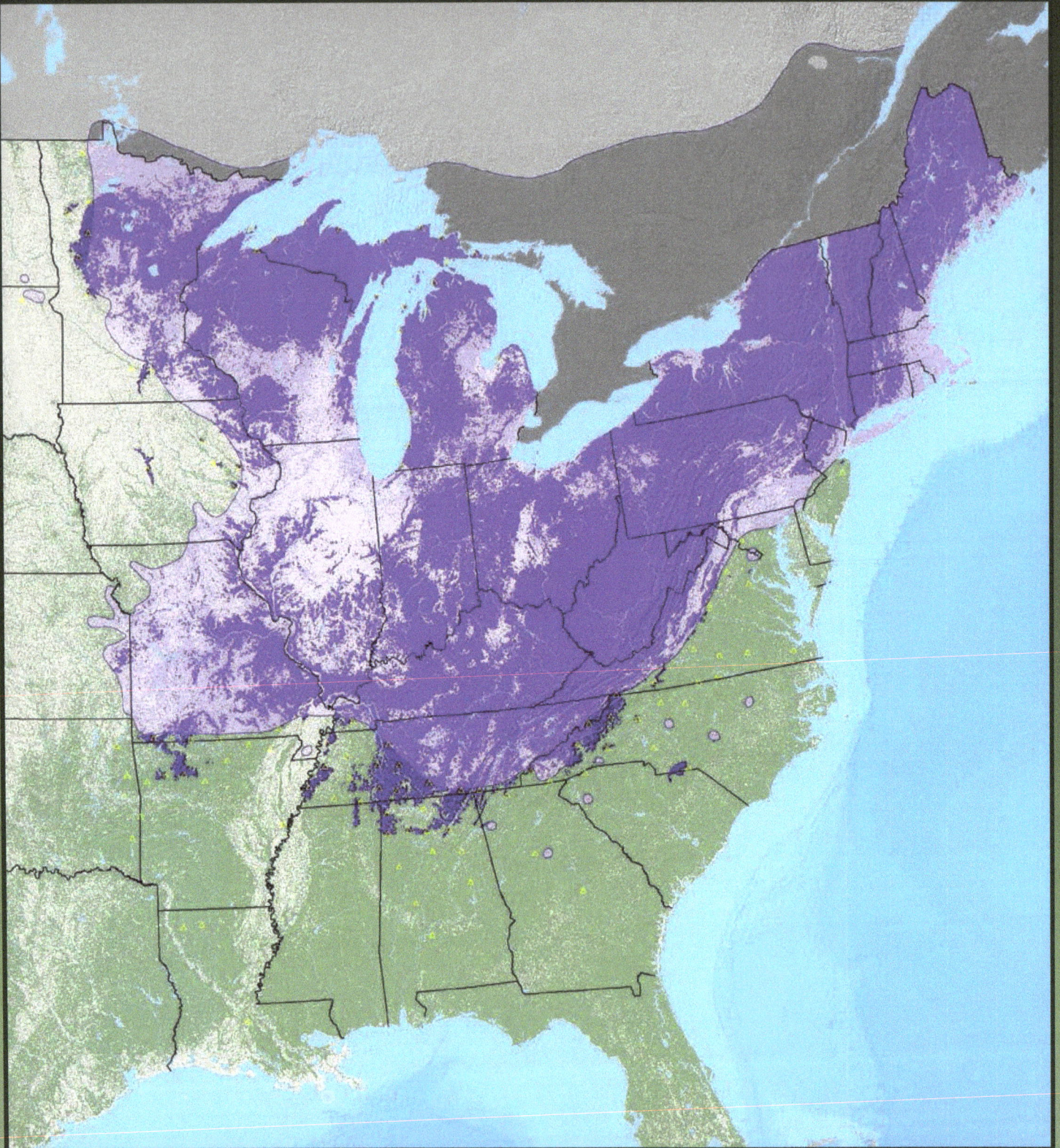

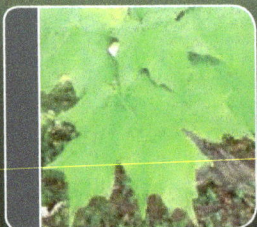

sugar maple

sugar maple
Acer saccharum Marsh.

Aceraceae – Maple Family

FIA Code **318** – Plants Symbol **ACSA3**

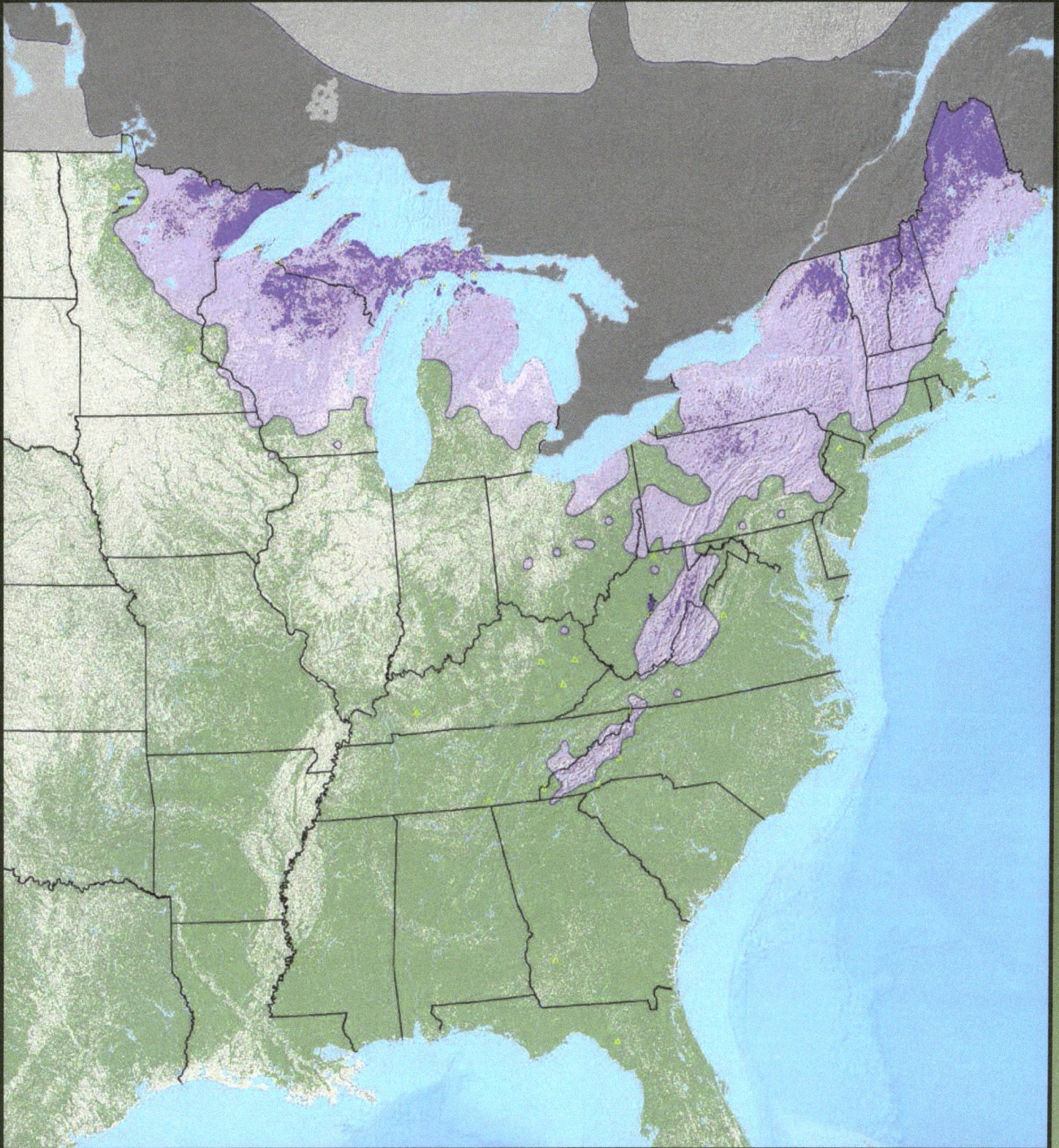

mountain maple

mountain maple
Acer spicatum Lam.
Aceraceae – Maple Family
FIA Code **319** – Plants Symbol **ACSP2**

National Individual Tree Species Atlas

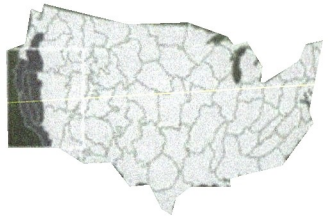

National Individual Tree Species Atlas

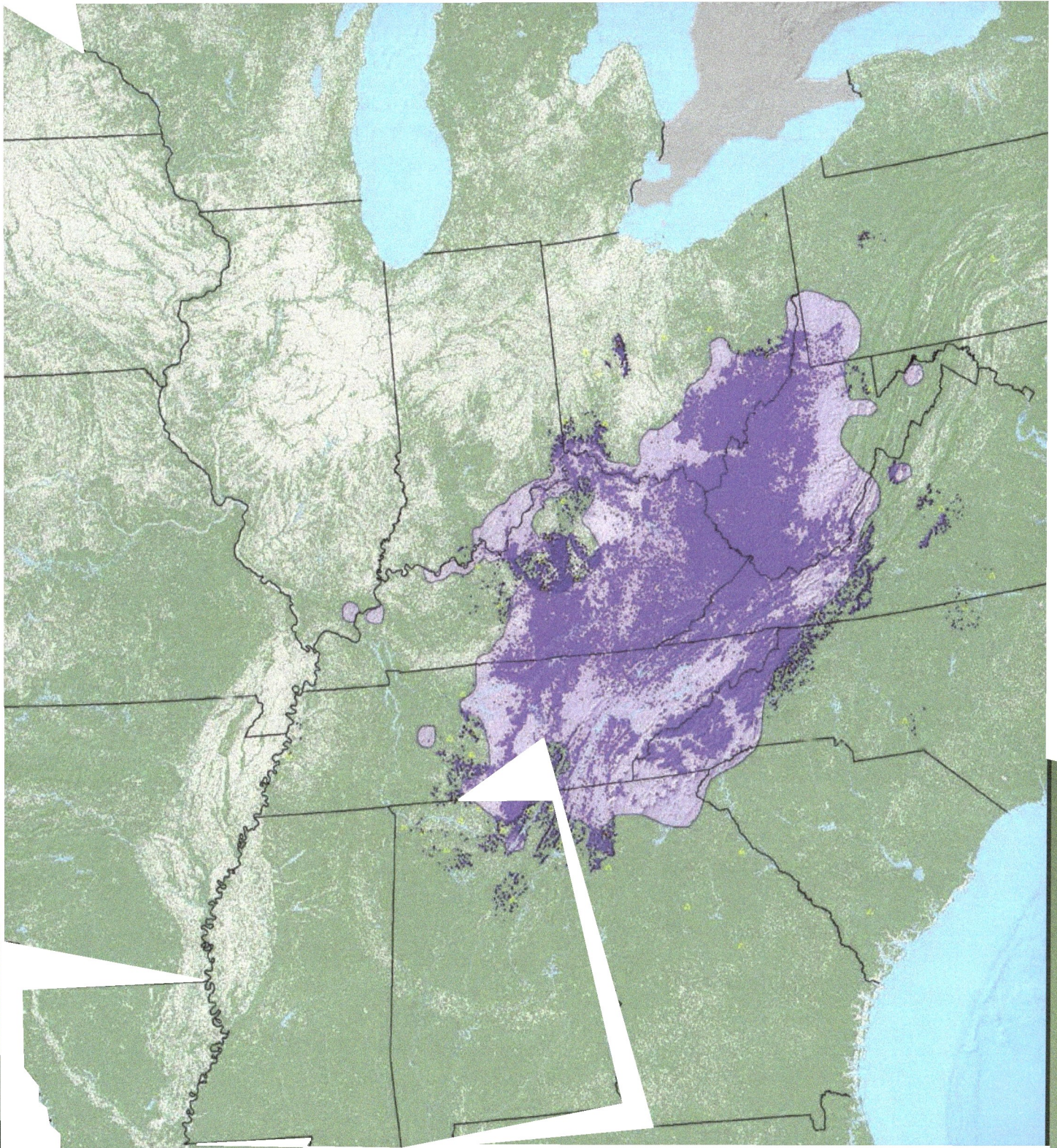

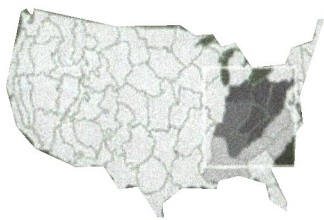

National Individual Tree Species Atlas

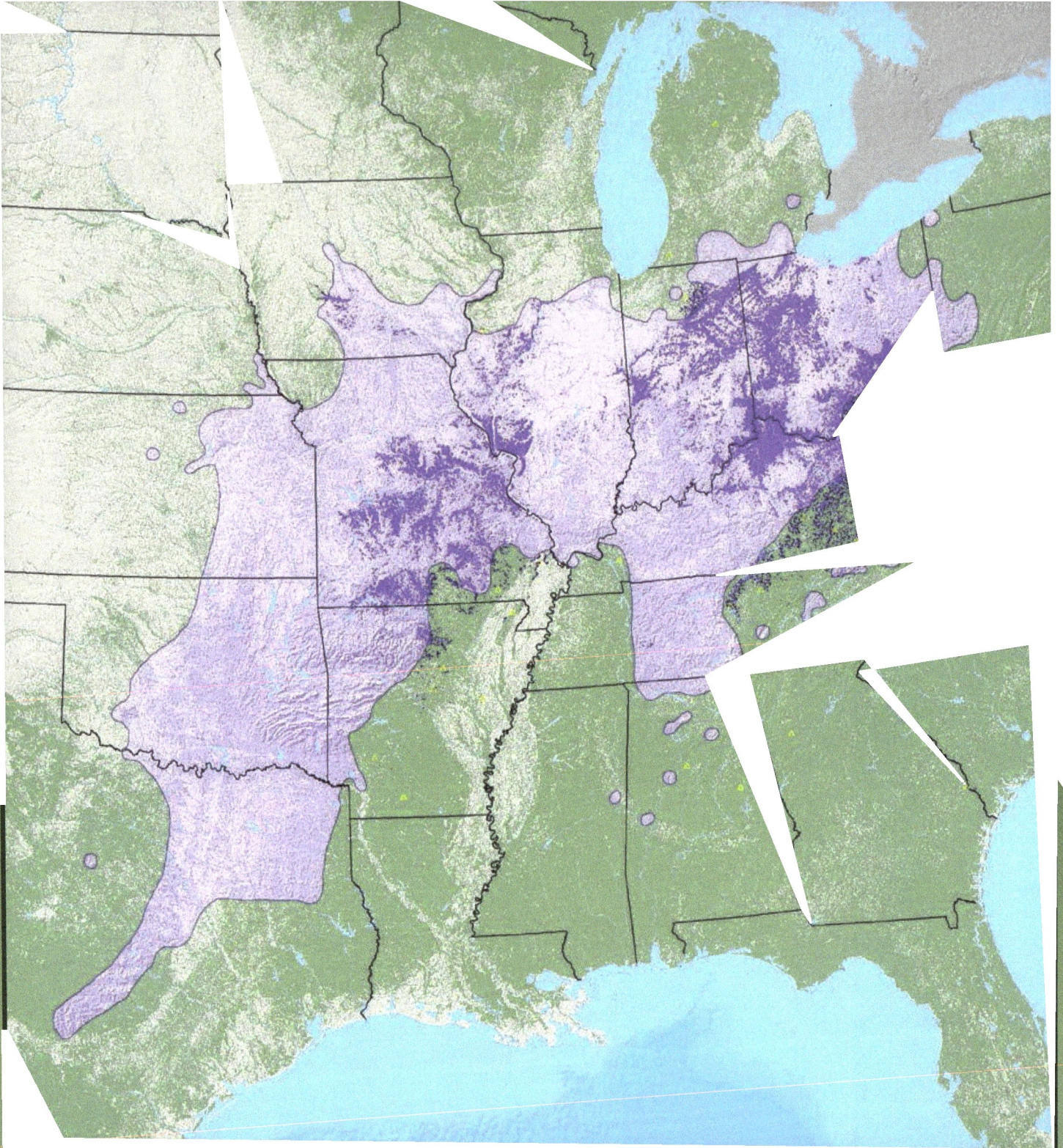

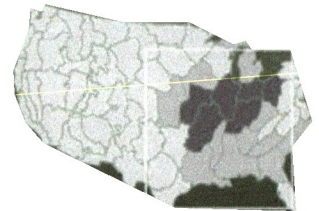

National Individual Tree Species Atlas

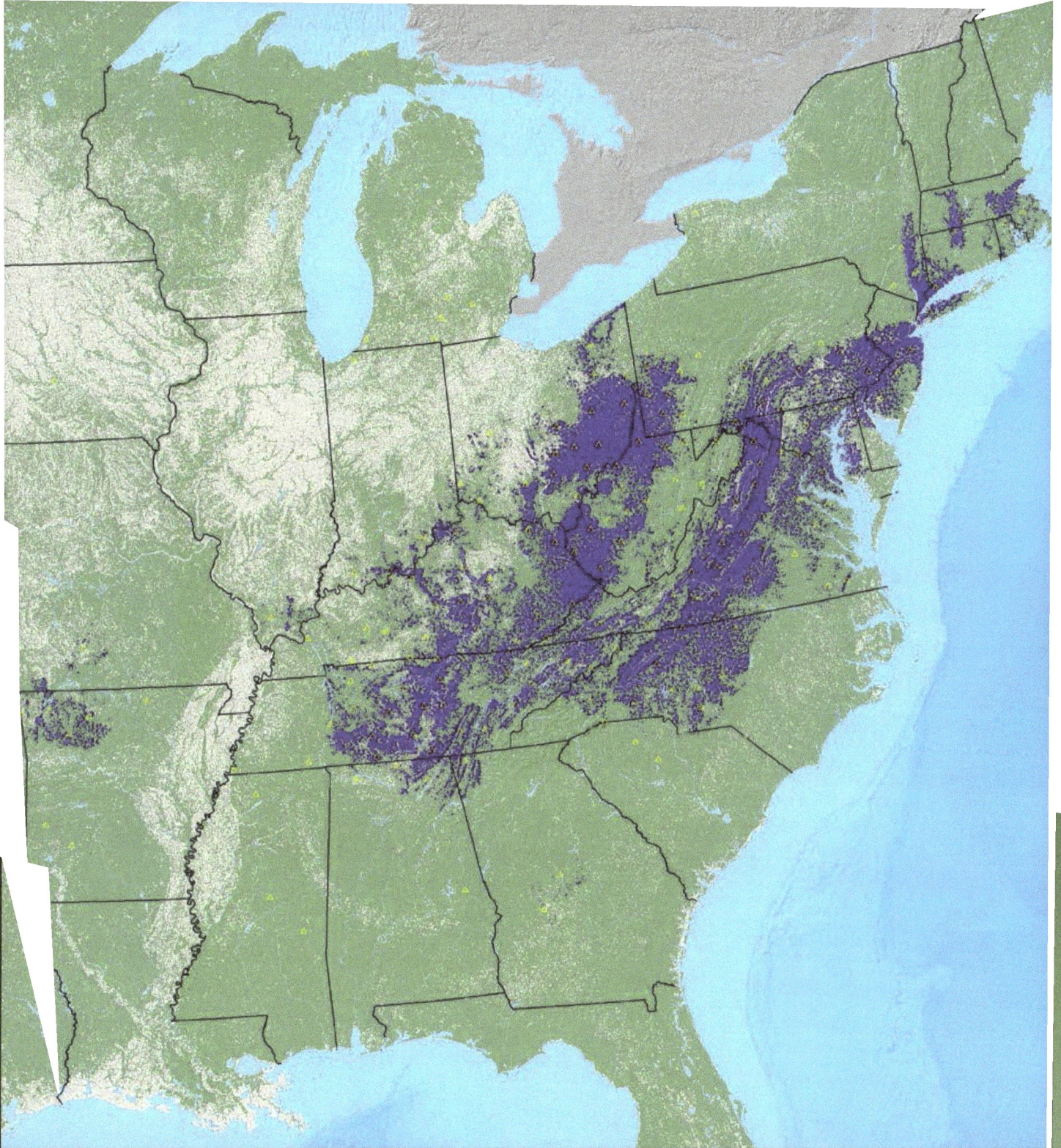

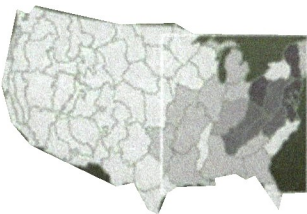

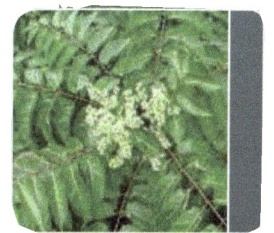

National Individual Tree Species Atlas

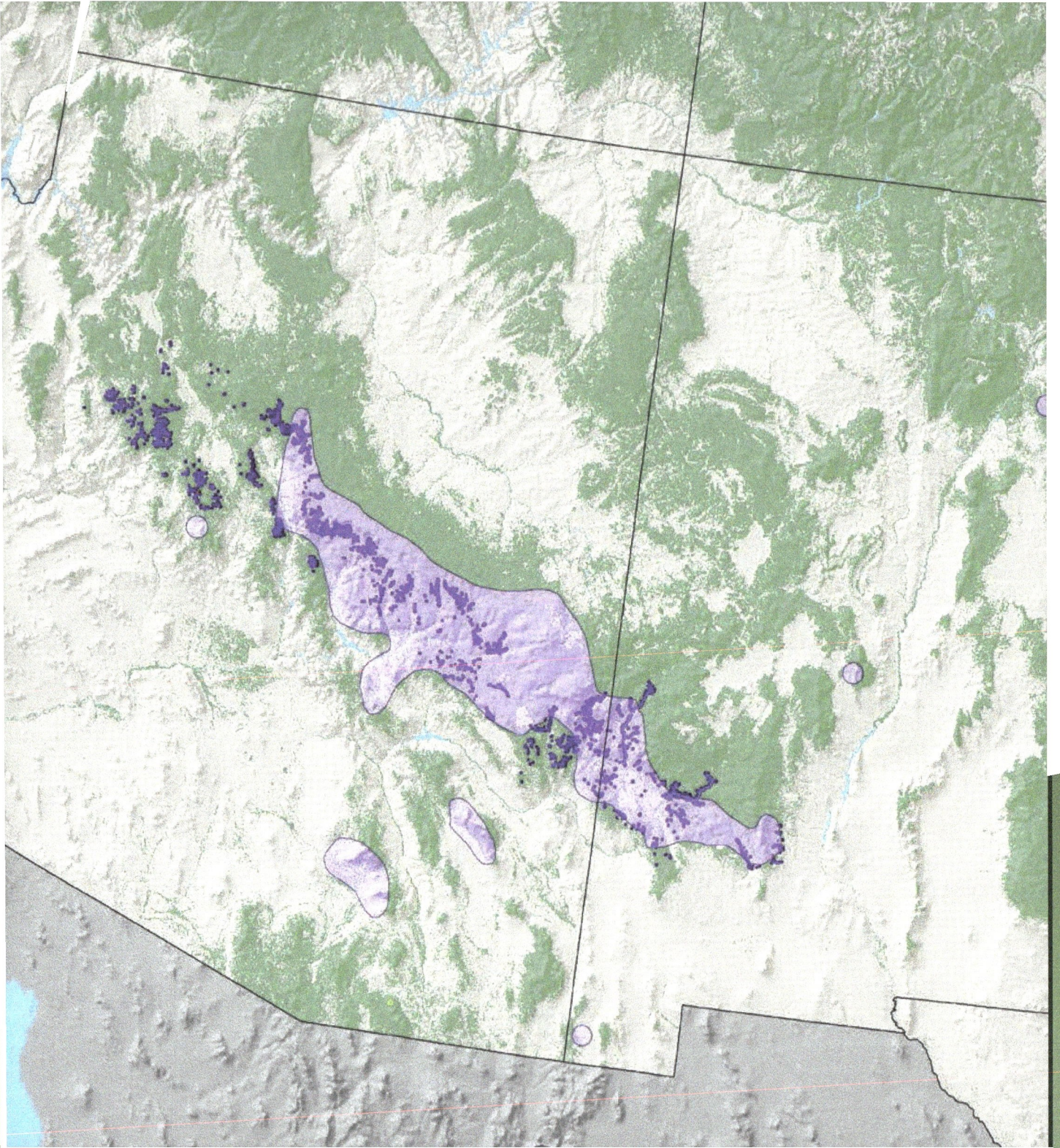

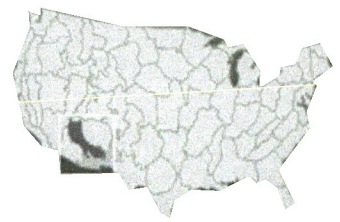

National Individual Tree Species Atlas

white alder

white alder
Alnus rhombifolia Nutt.
Betulaceae – Birch Family
FIA Code **352** – Plants Symbol **ALRH2**

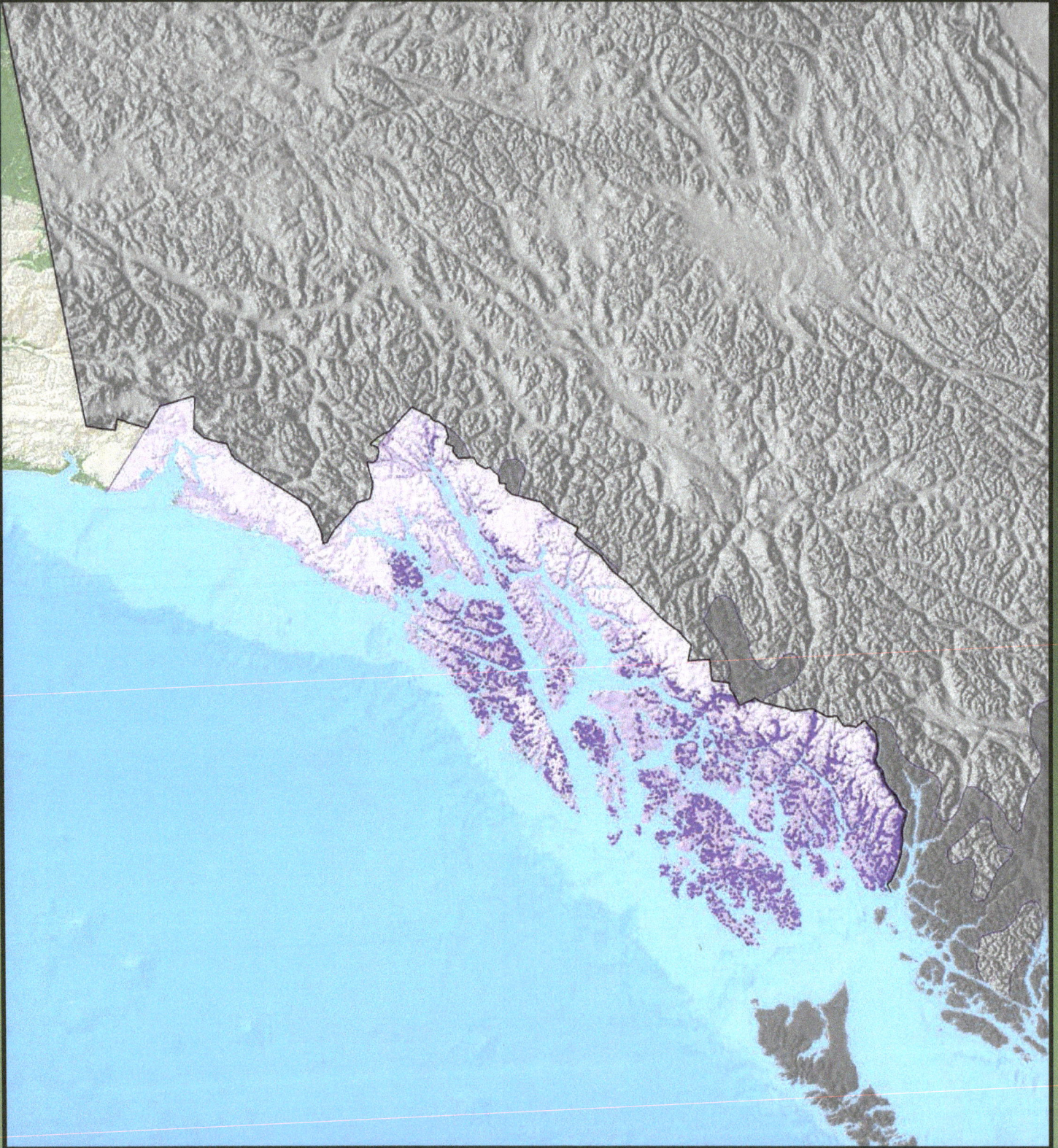

red alder
Alnus rubra Bong.
Betulaceae – Birch Family
FIA Code **351** – Plants Symbol **ALRU2**

National Individual Tree Species Atlas

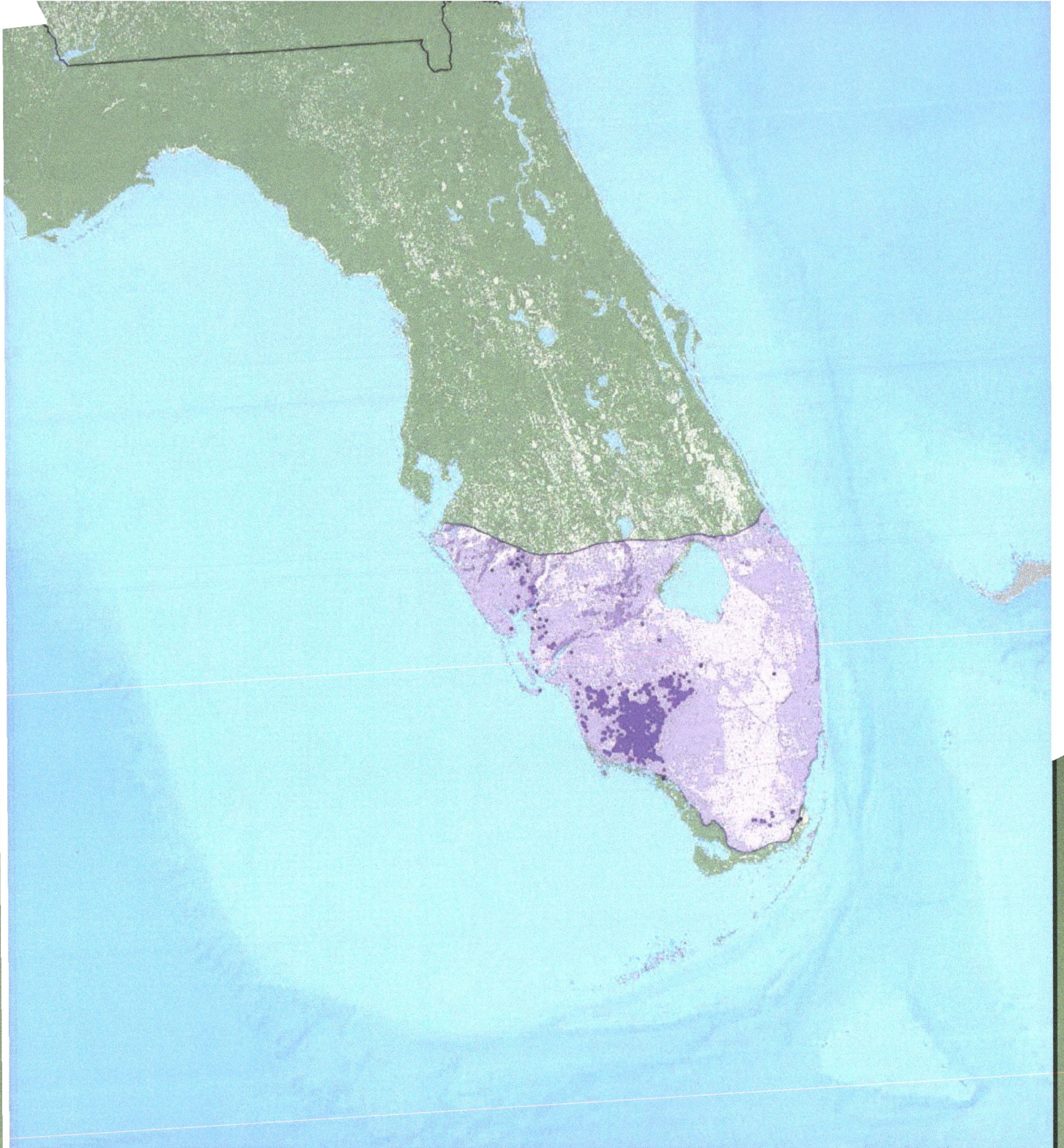

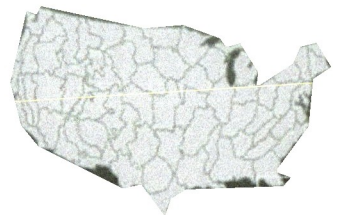

National Individual Tree Species Atlas

National Individual Tree Species Atlas

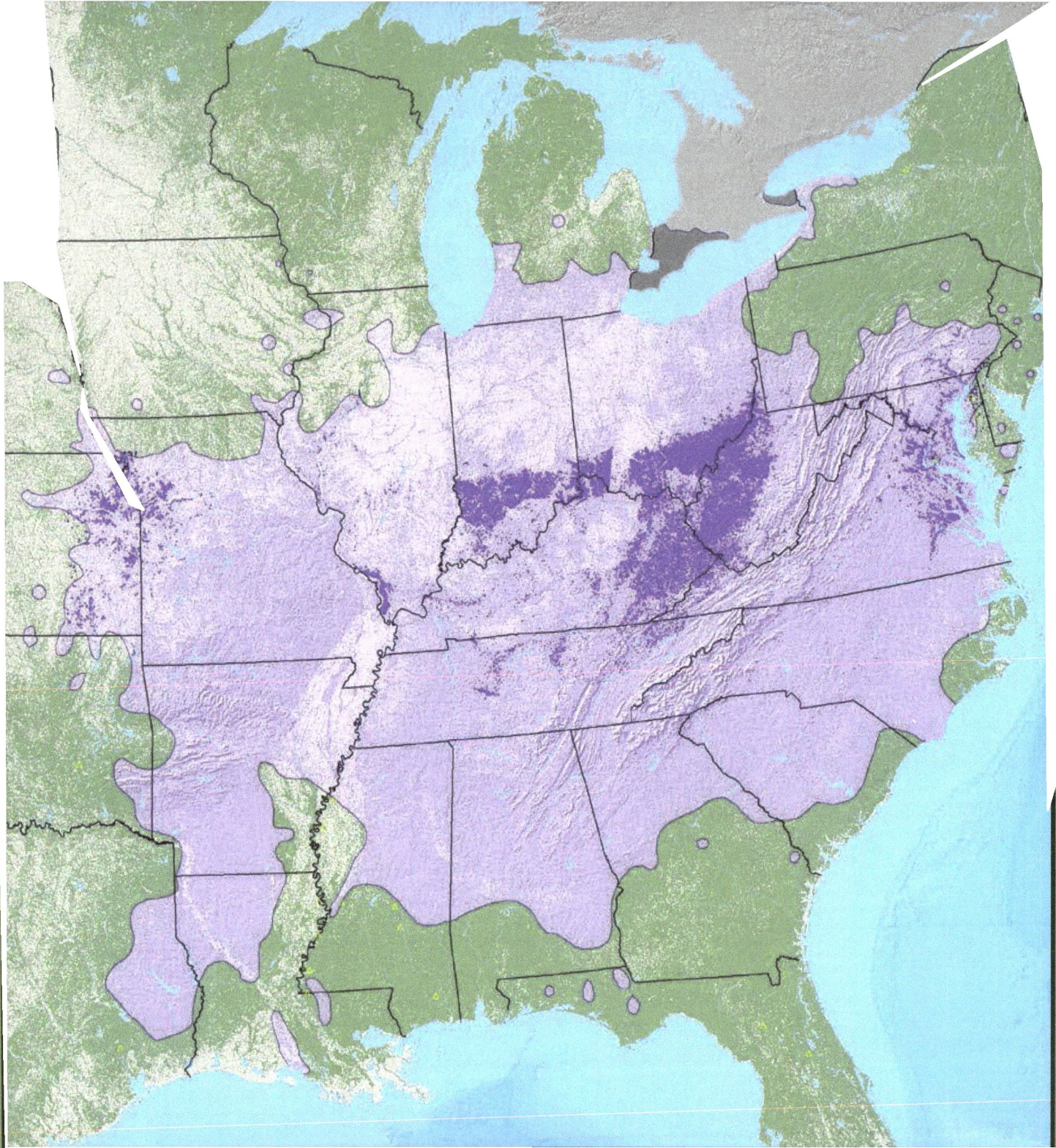

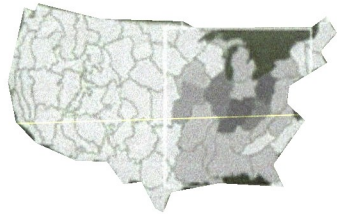

National Individual Tree Species Atlas

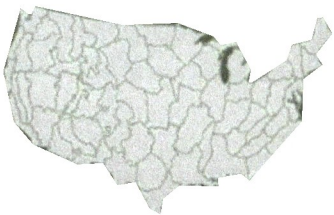

National Individual Tree Species Atlas

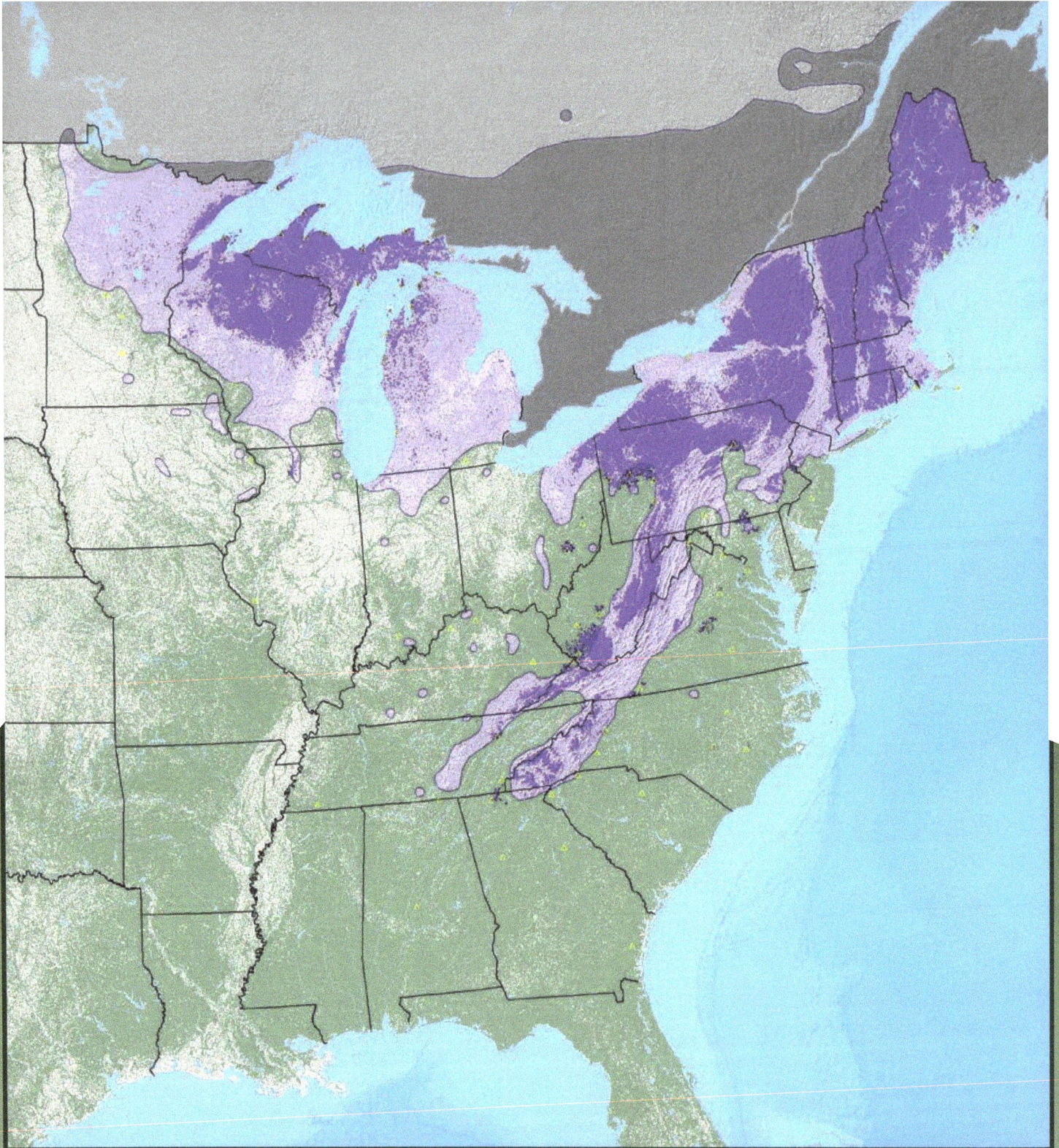

yellow birch
Betula alleghaniensis Britton
Betulaceae – Birch Family
FIA Code **371** – Plants Symbol **BEAL2**

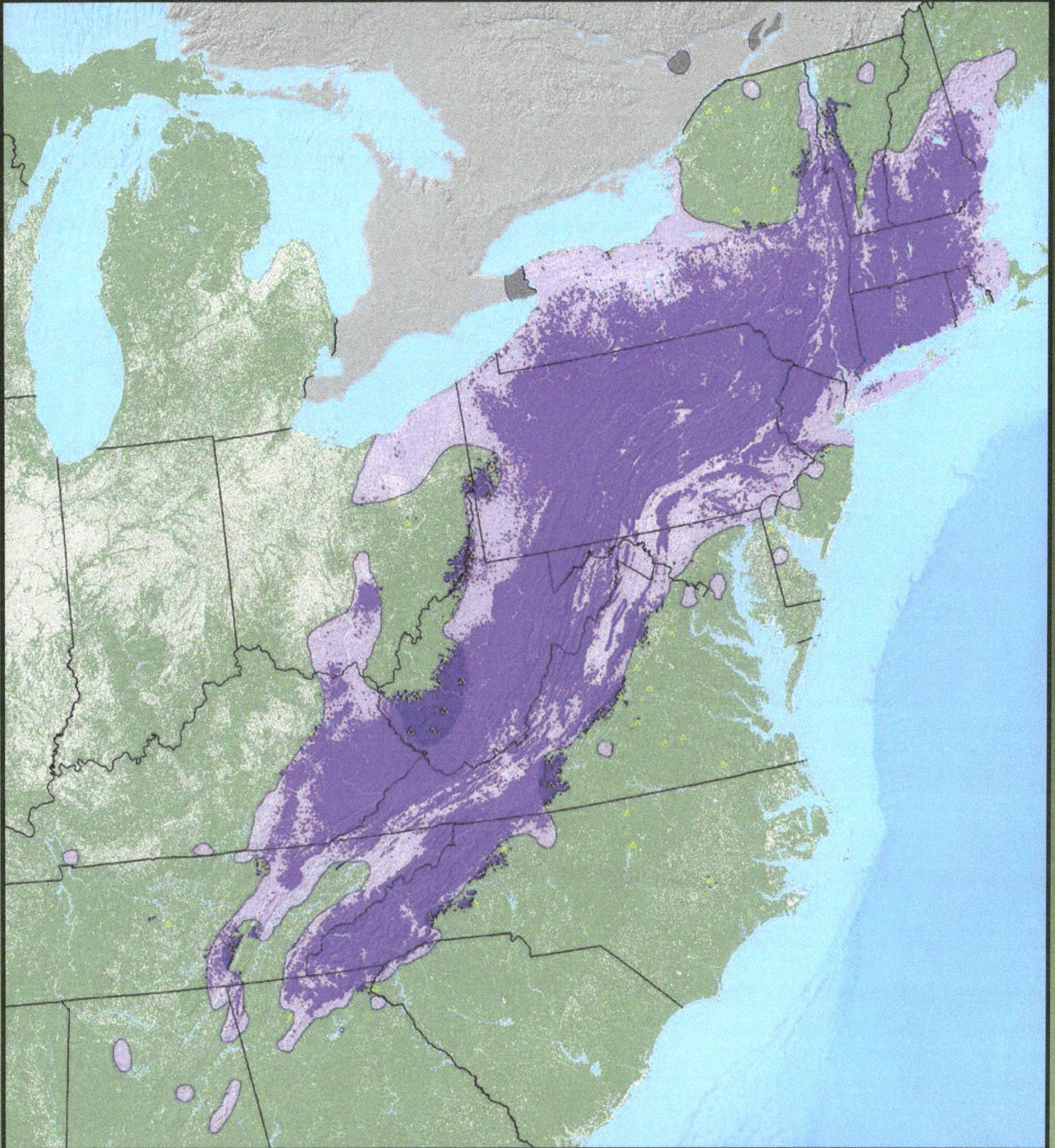

sweet birch

sweet birch
Betula lenta L.
Betulaceae – Birch Family
FIA Code **372** – Plants Symbol **BELE**

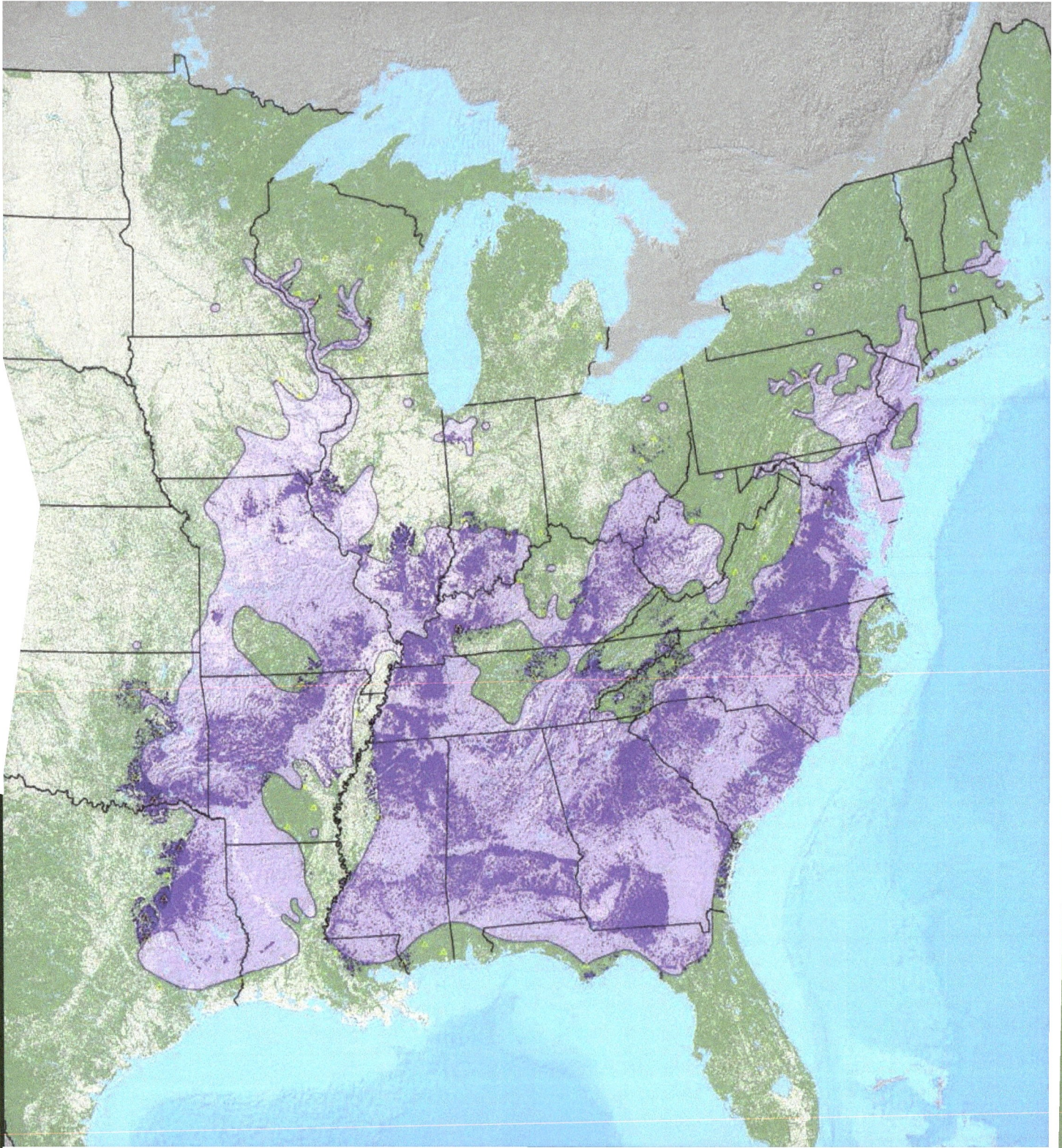

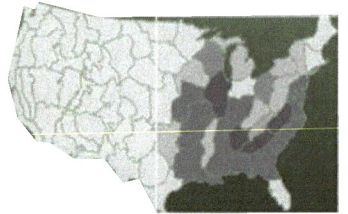

National Individual Tree Species Atlas

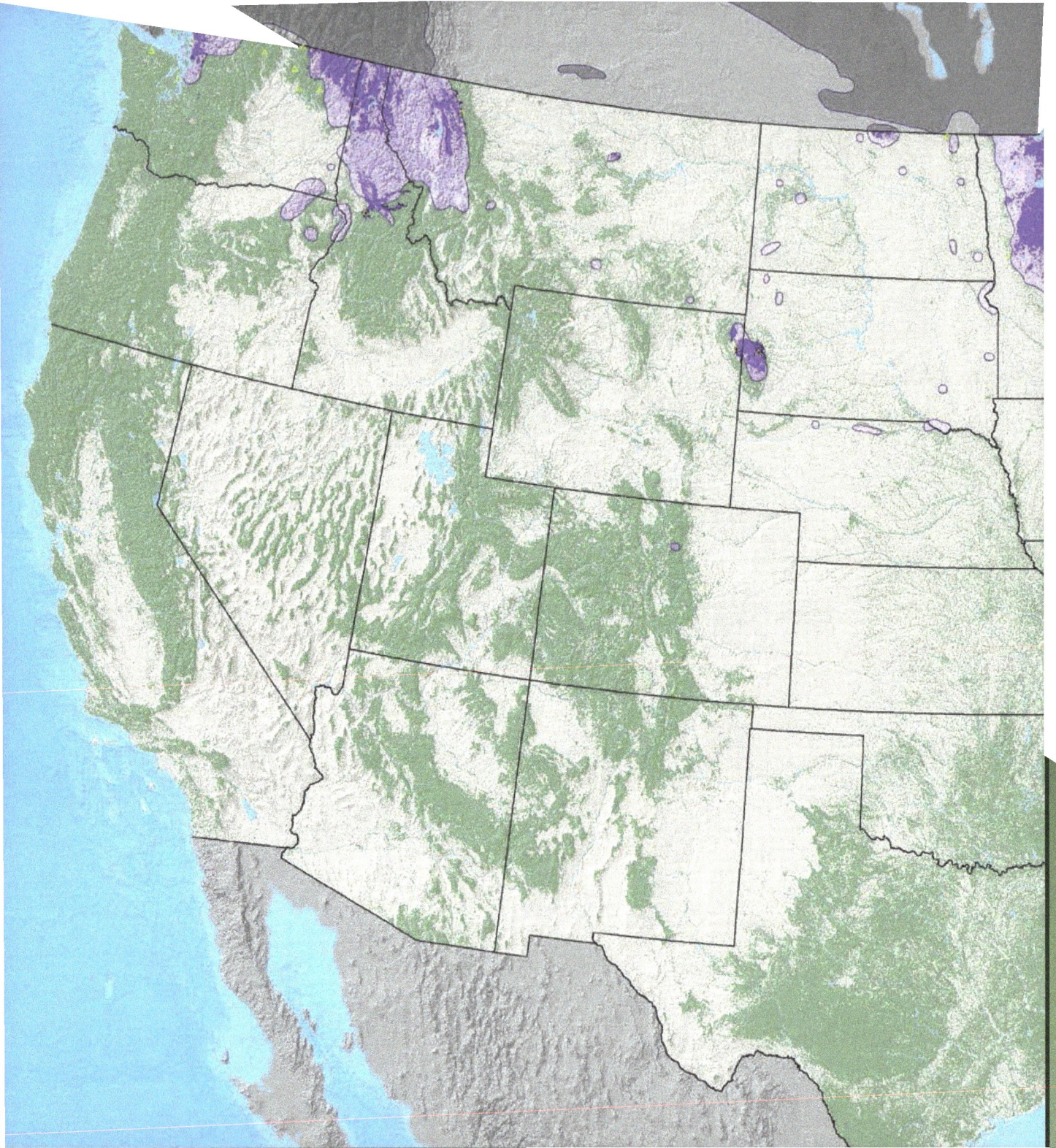

National Individual Tree Species Atlas

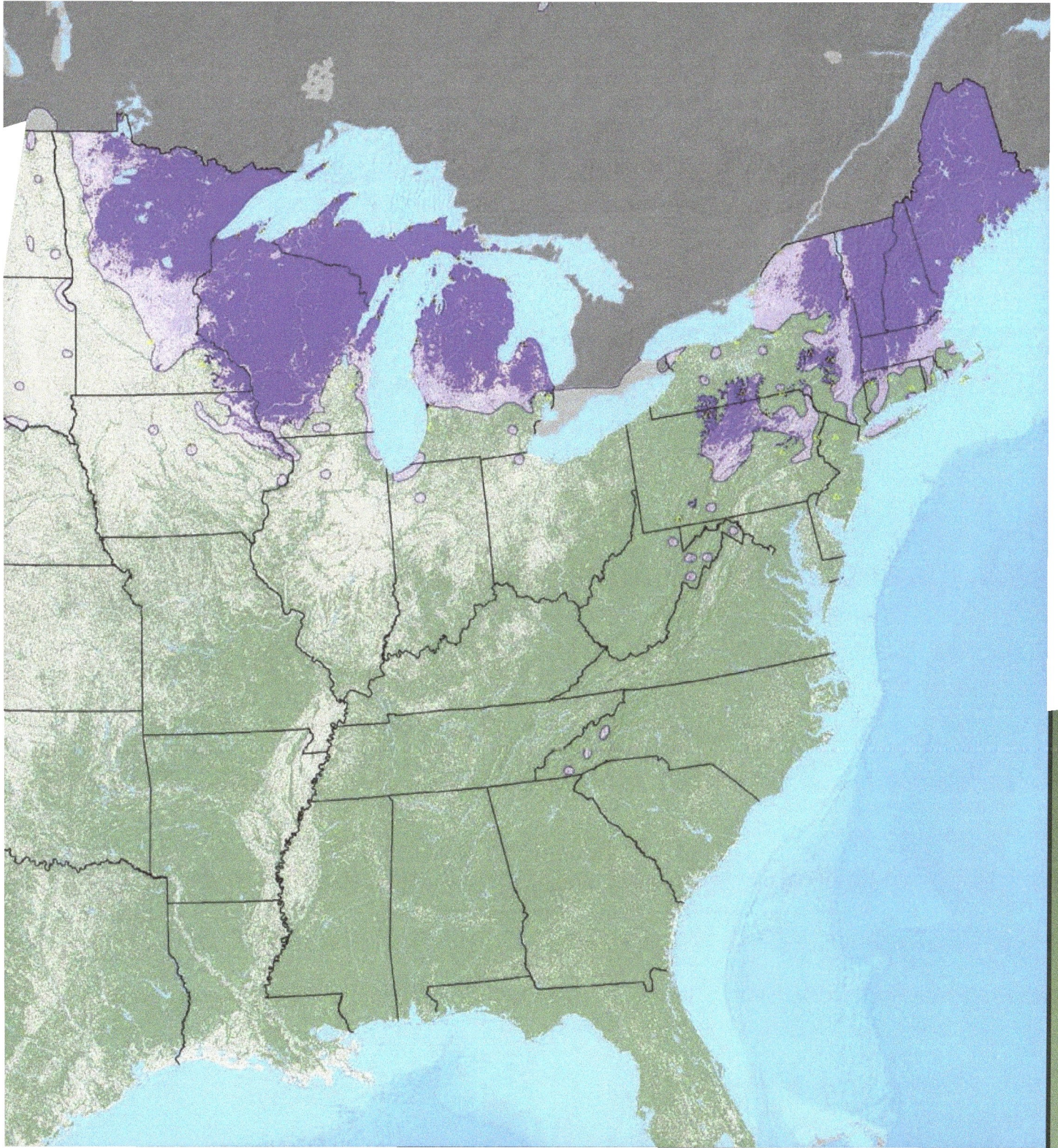

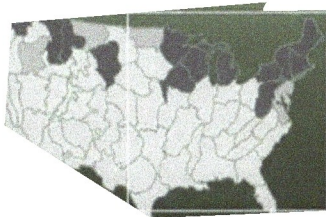

National Individual Tree Species Atlas

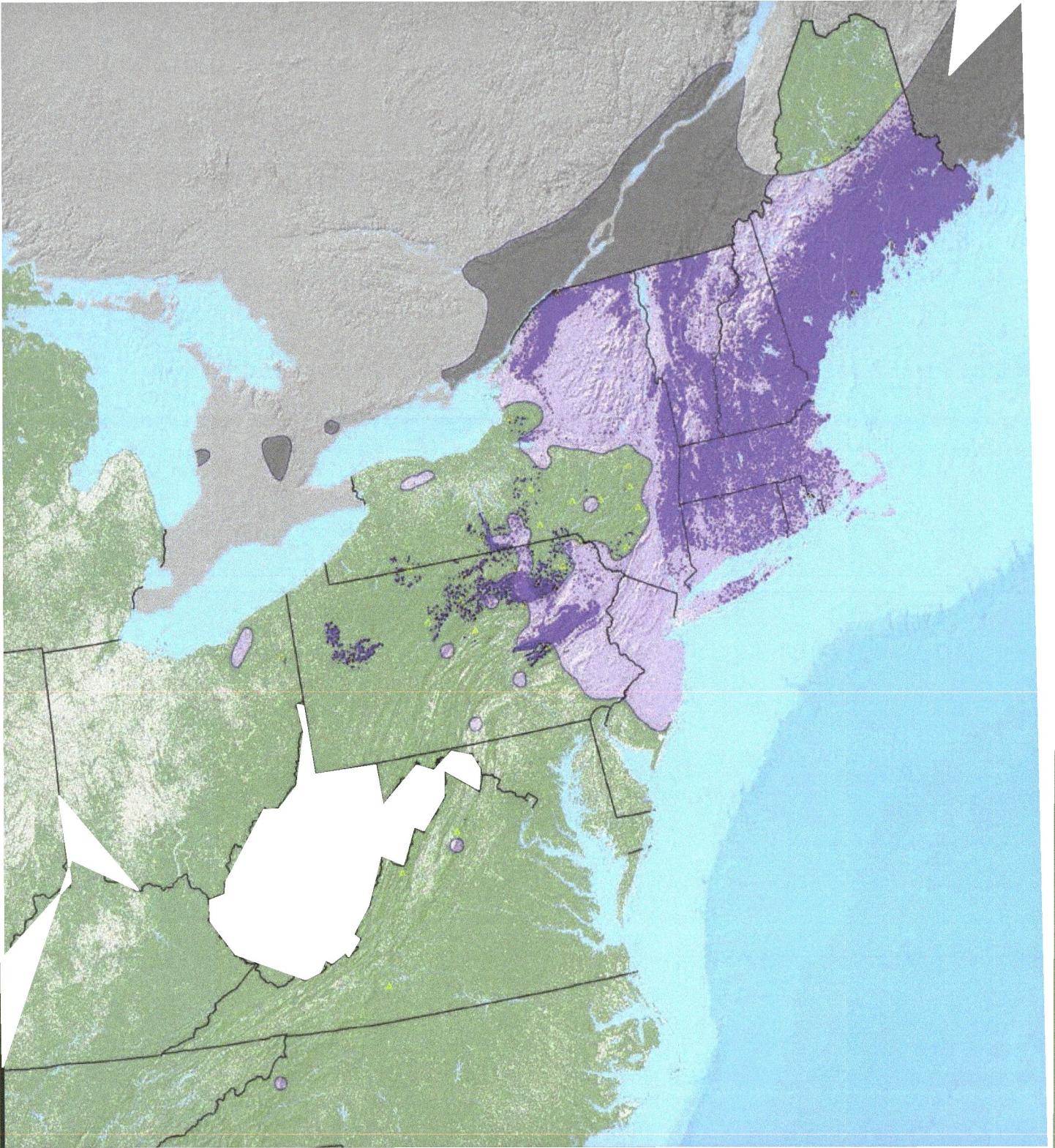

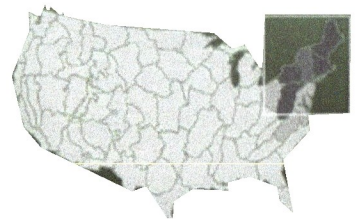

National Individual Tree Species Atlas

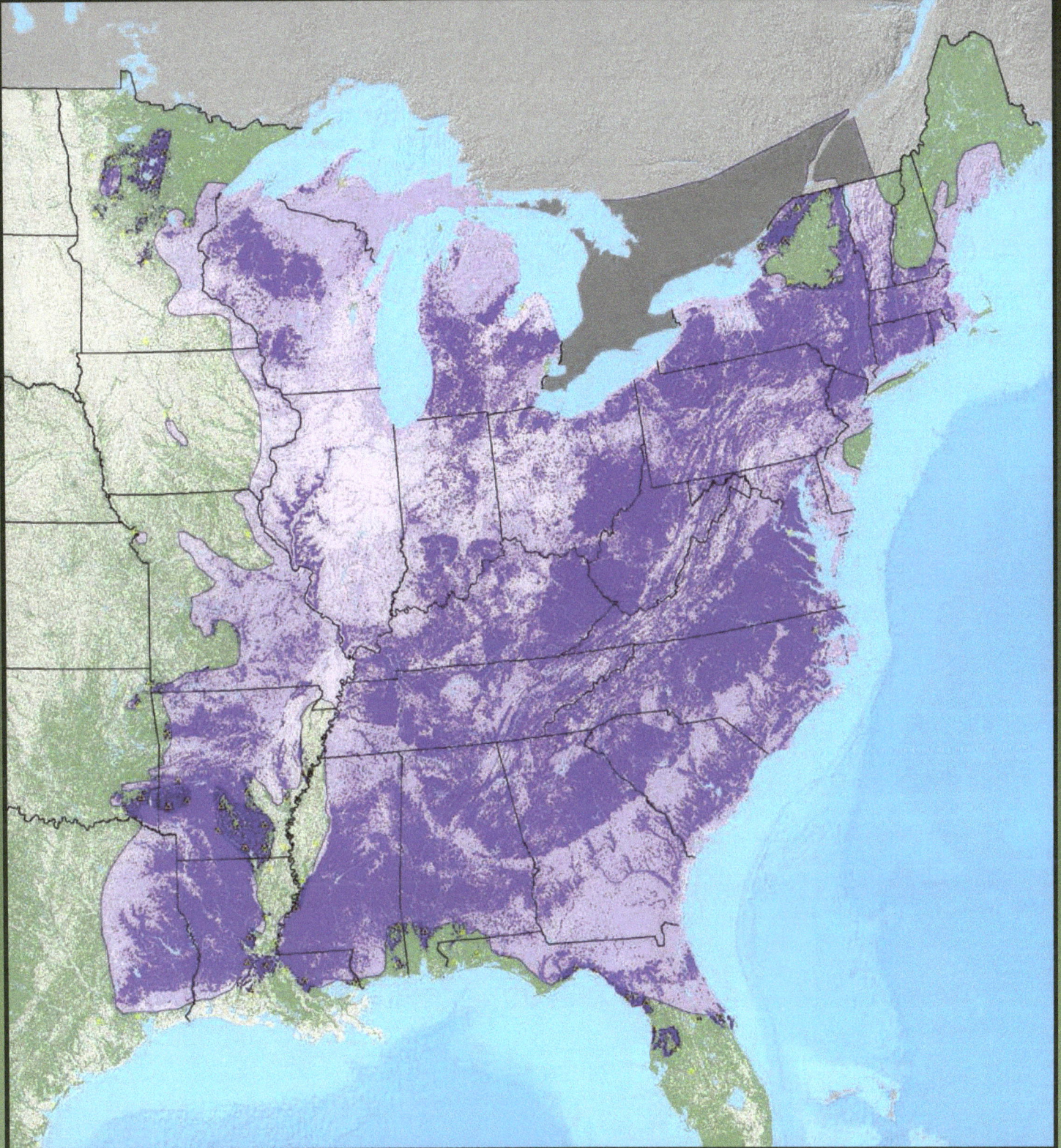

American hornbeam

American hornbeam
Carpinus caroliniana Walter
Betulaceae – Birch Family
FIA Code **391** – Plants Symbol **CACA18**

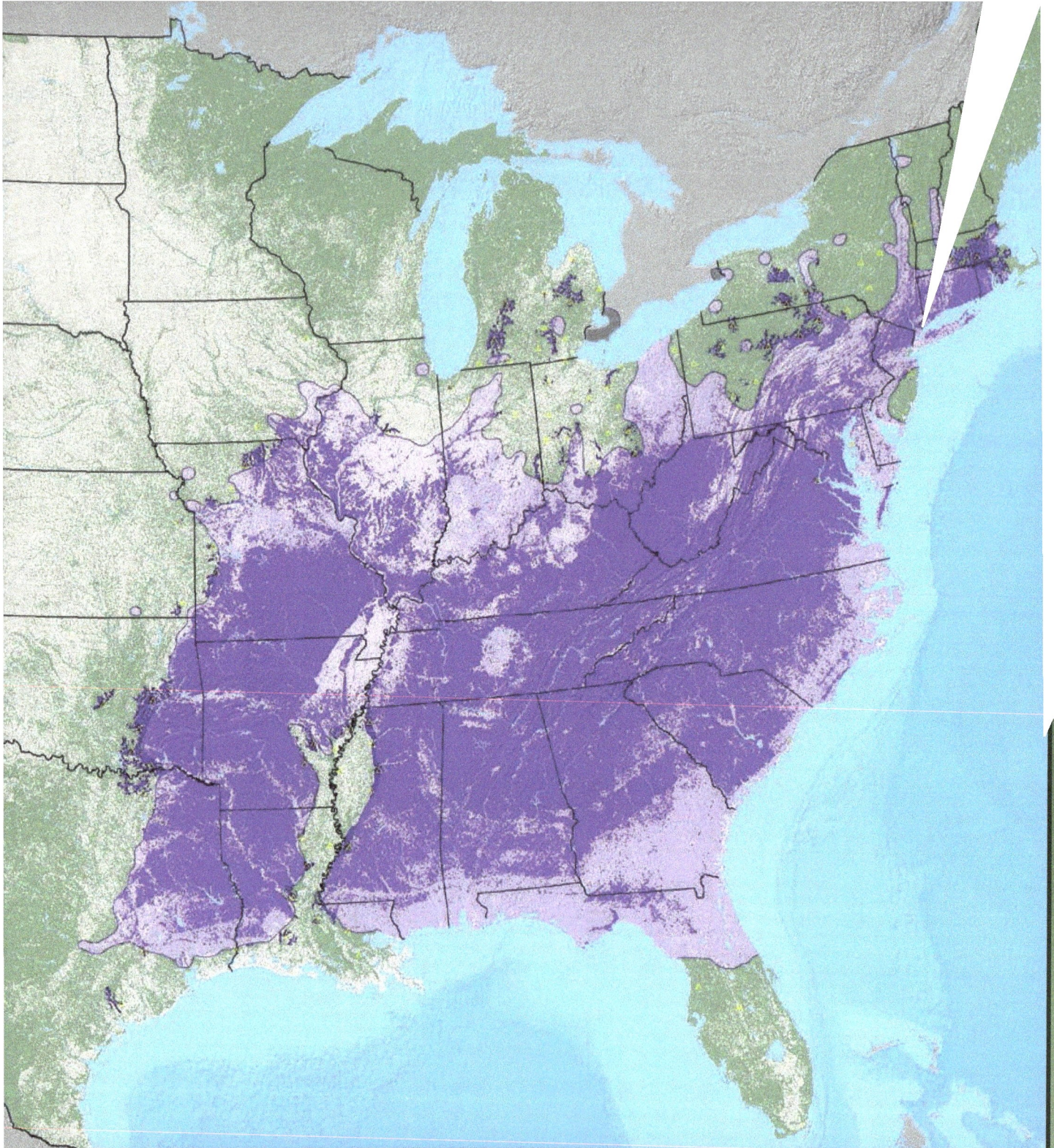

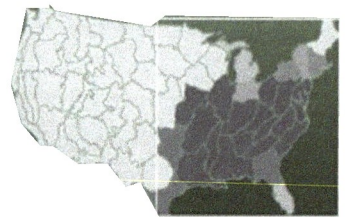

National Individual Tree Species Atlas

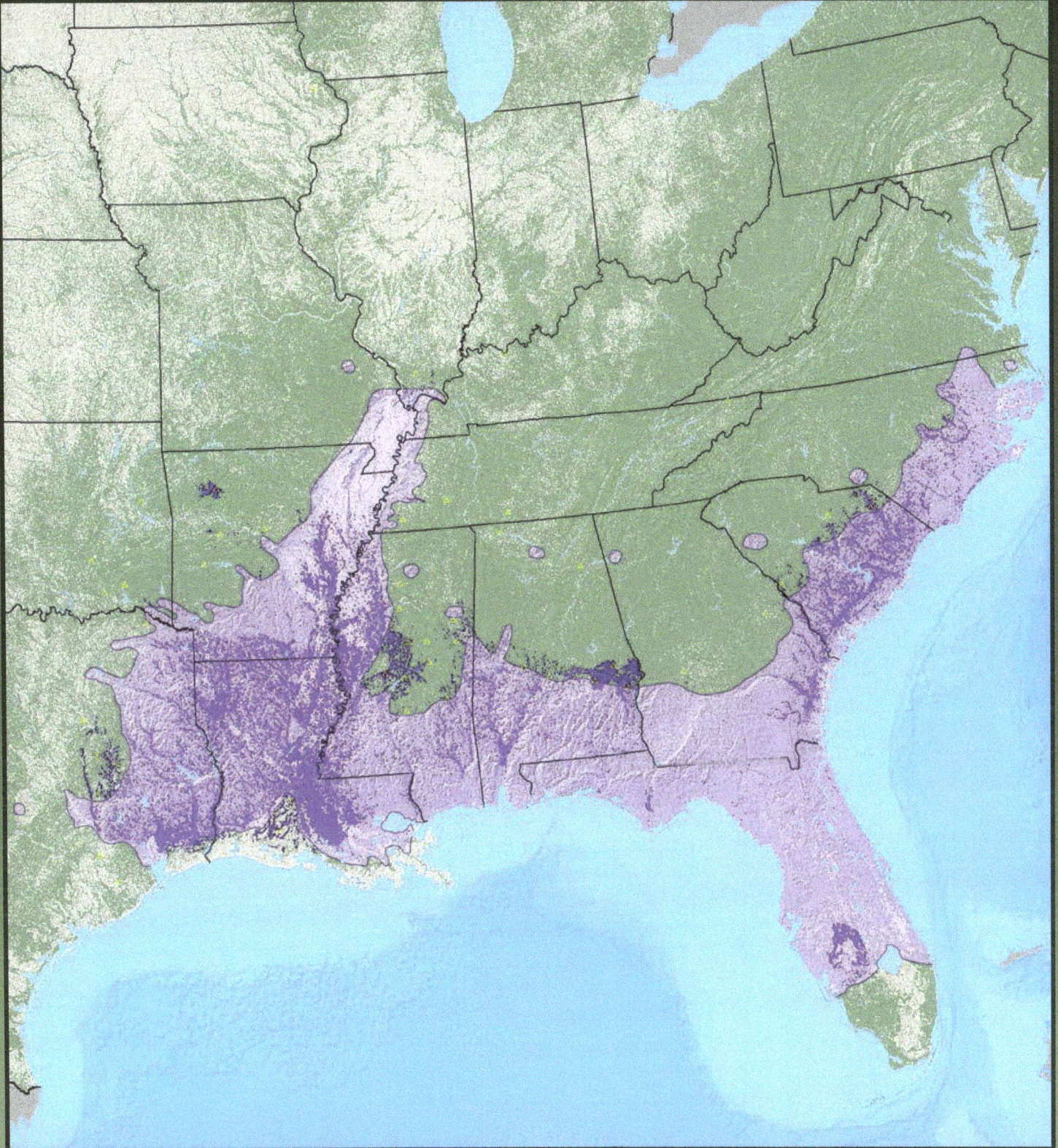

water hickory

water hickory

Carya aquatica (Michx. f.) Nutt.

Juglandaceae – Walnut Family

FIA Code **401** – Plants Symbol **CAAQ2**

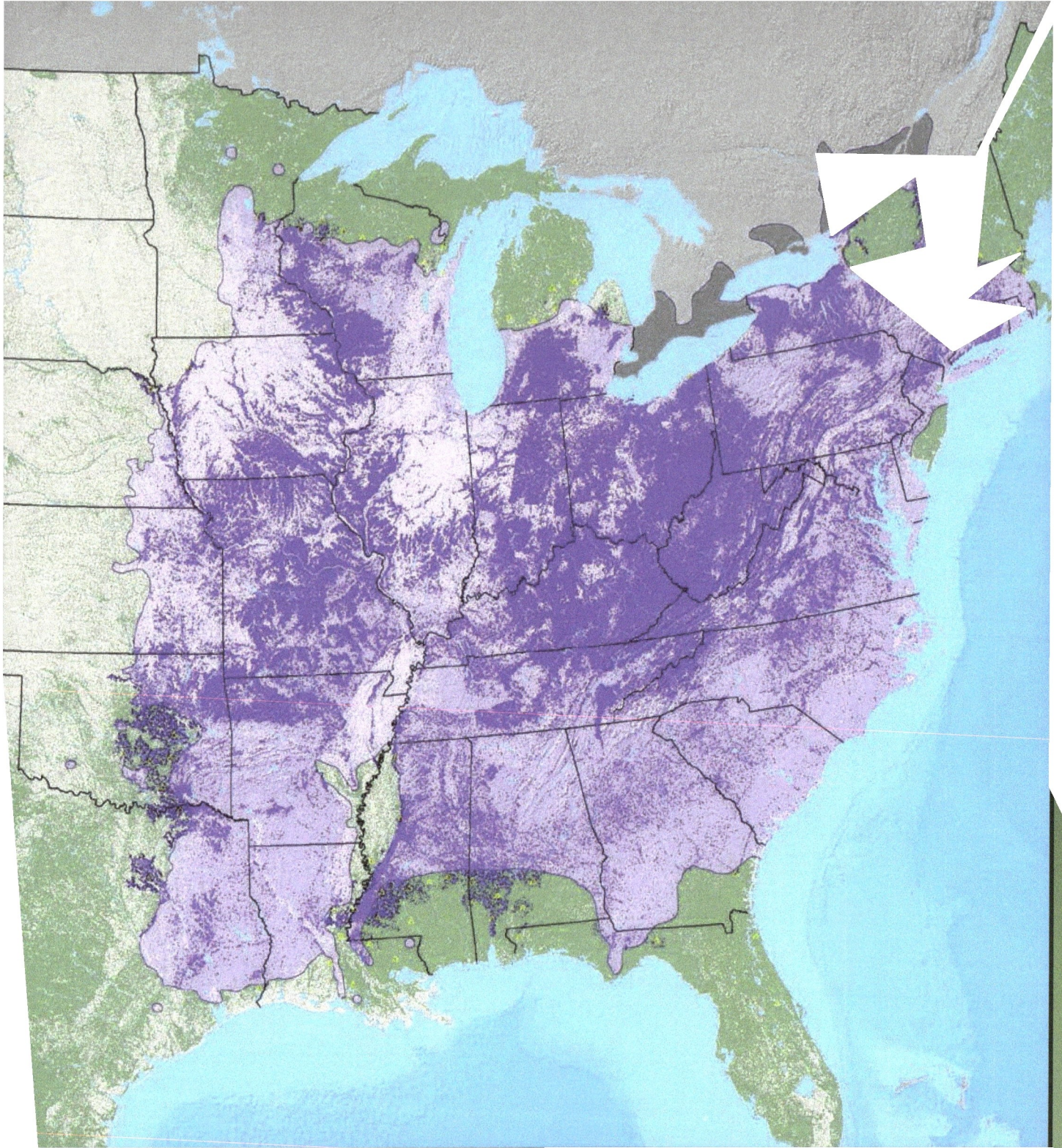

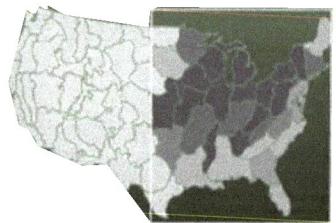

National Individual Tree Species Atlas

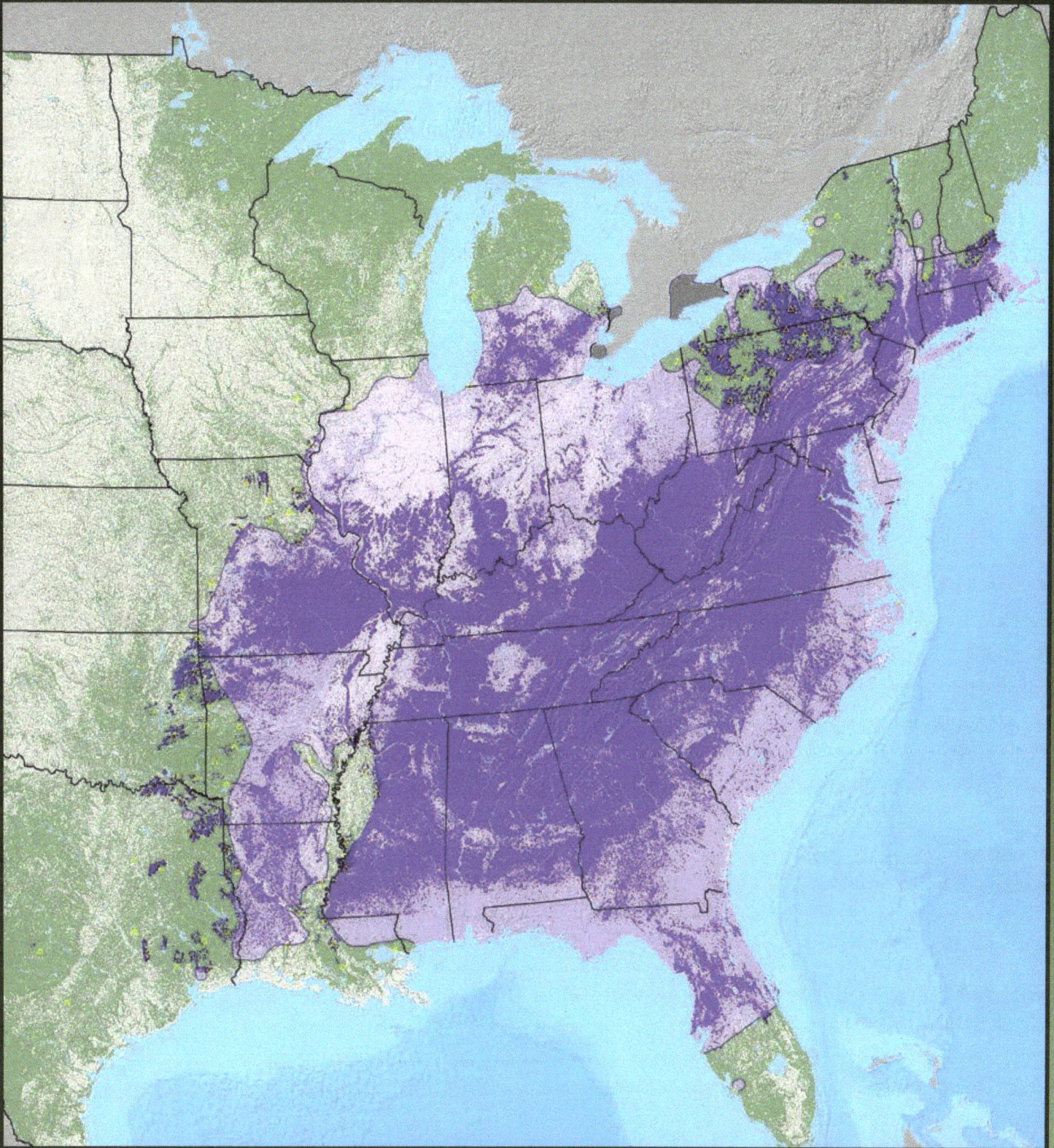

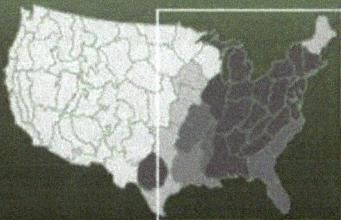

pignut hickory

pignut hickory
Carya glabra (Mill.) Sweet
Juglandaceae – Walnut Family
FIA Code **403** – Plants Symbol **CAGL8**

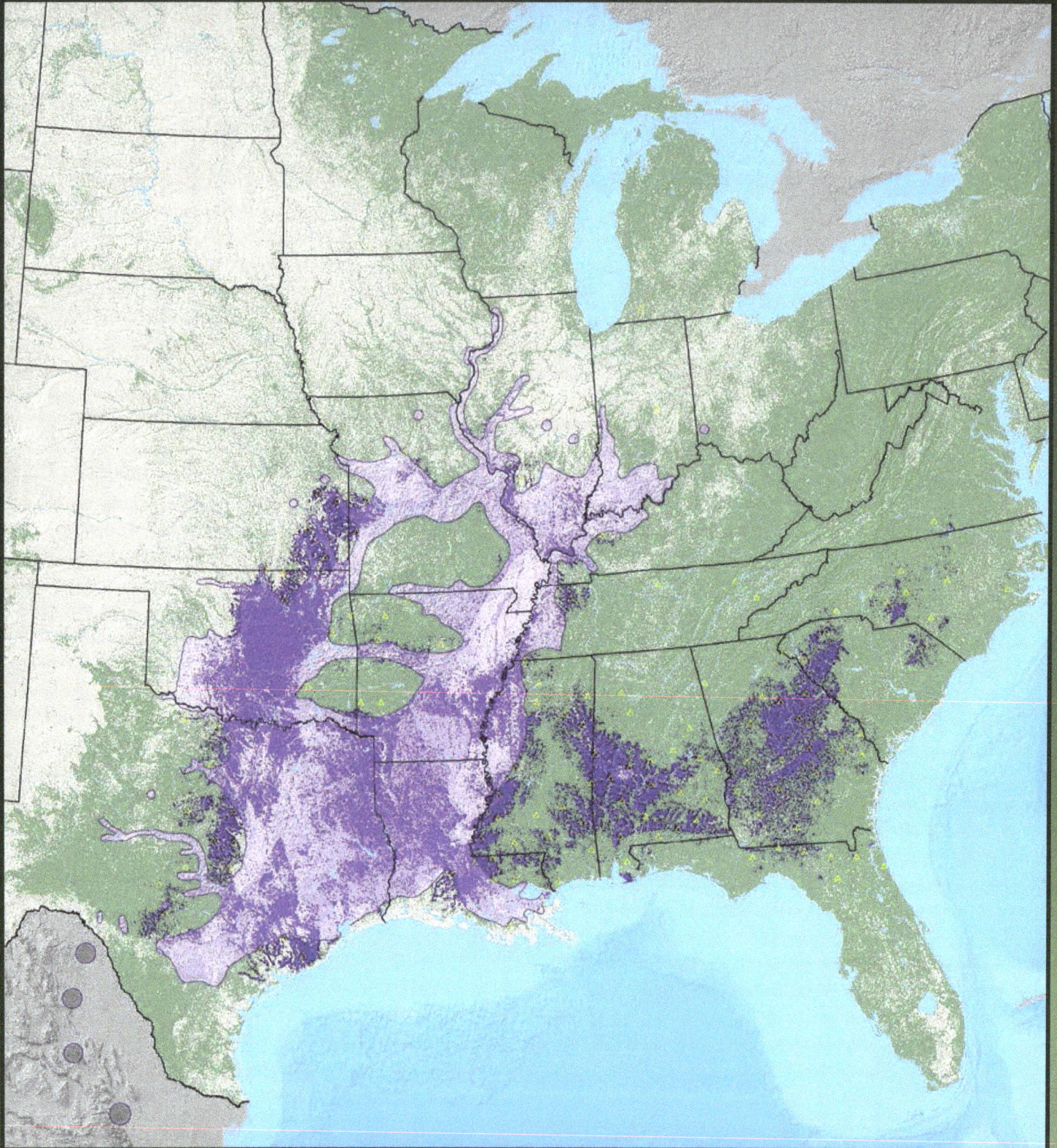

pecan

pecan
Carya illinoinensis (Wangenh.) K. Koch

Juglandaceae – Walnut Family

FIA Code **404** – Plants Symbol **CAIL2**

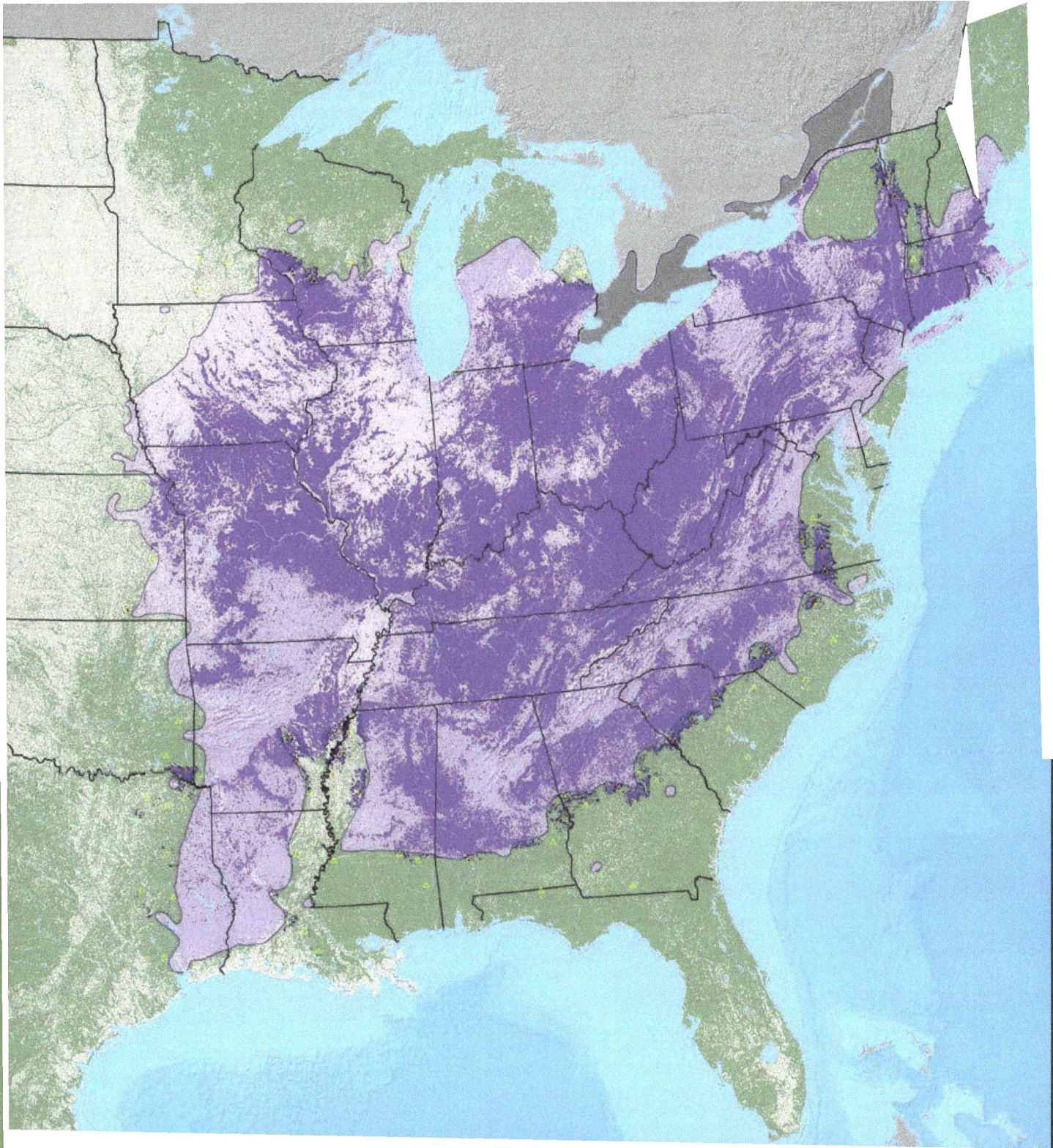

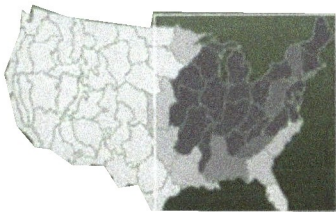

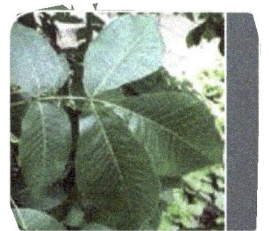

National Individual Tree Species Atlas

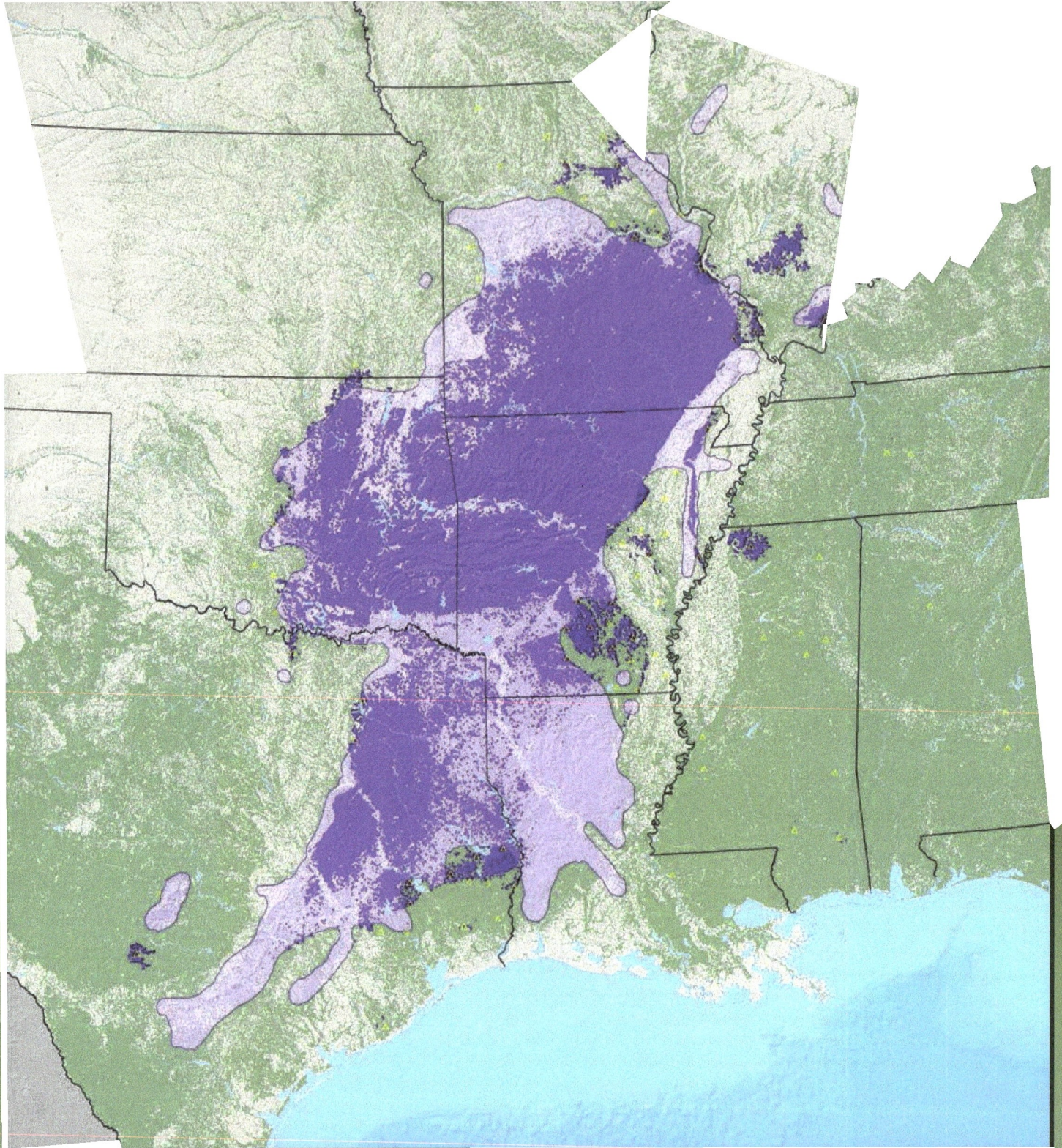

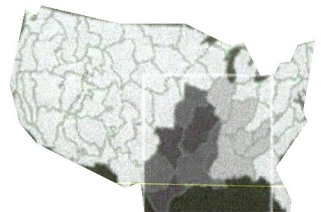

National Individual Tree Species Atlas

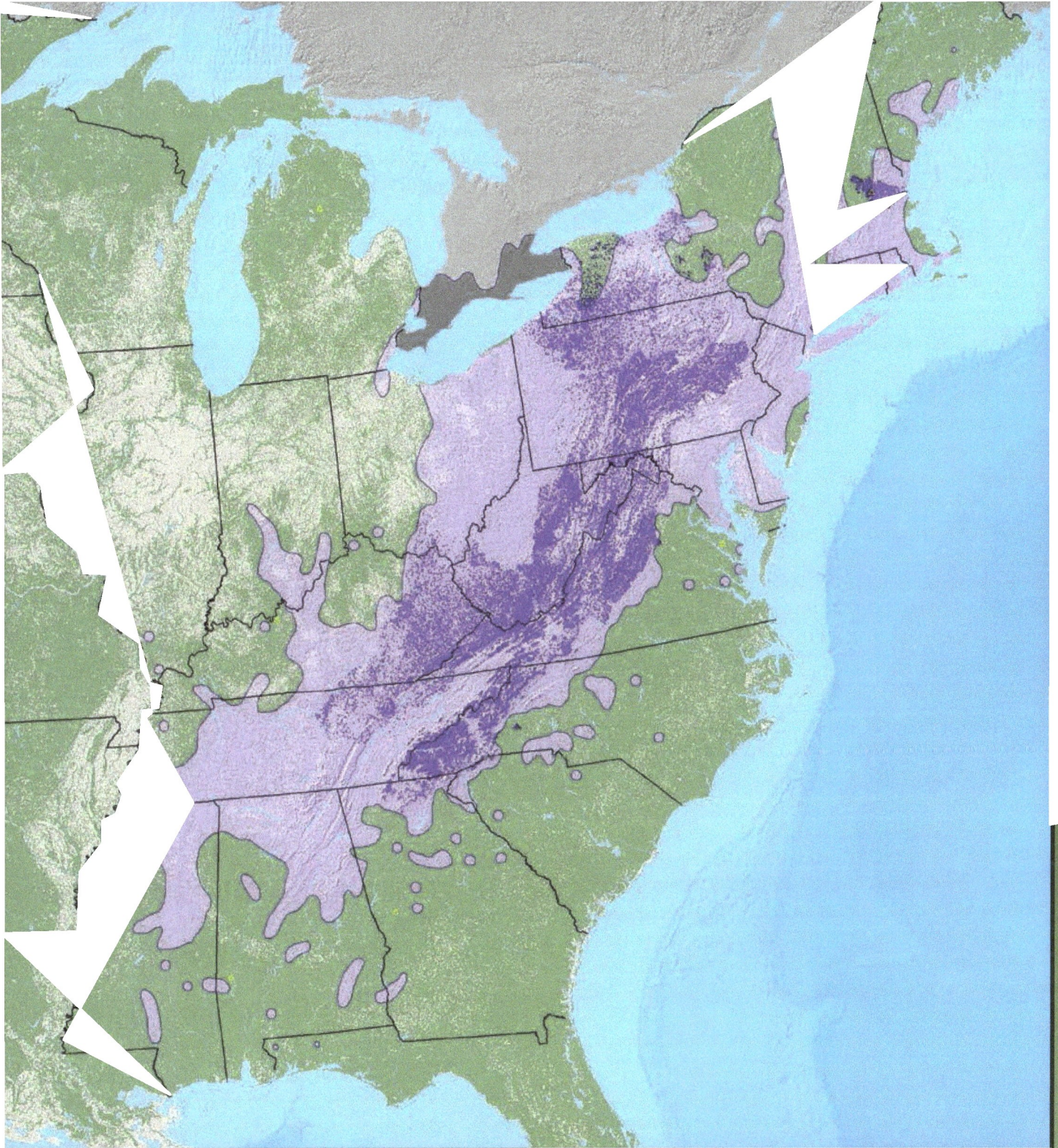

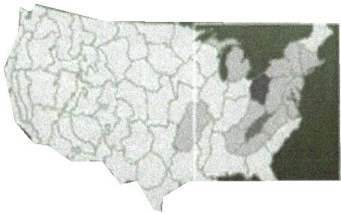

National Individual Tree Species Atlas

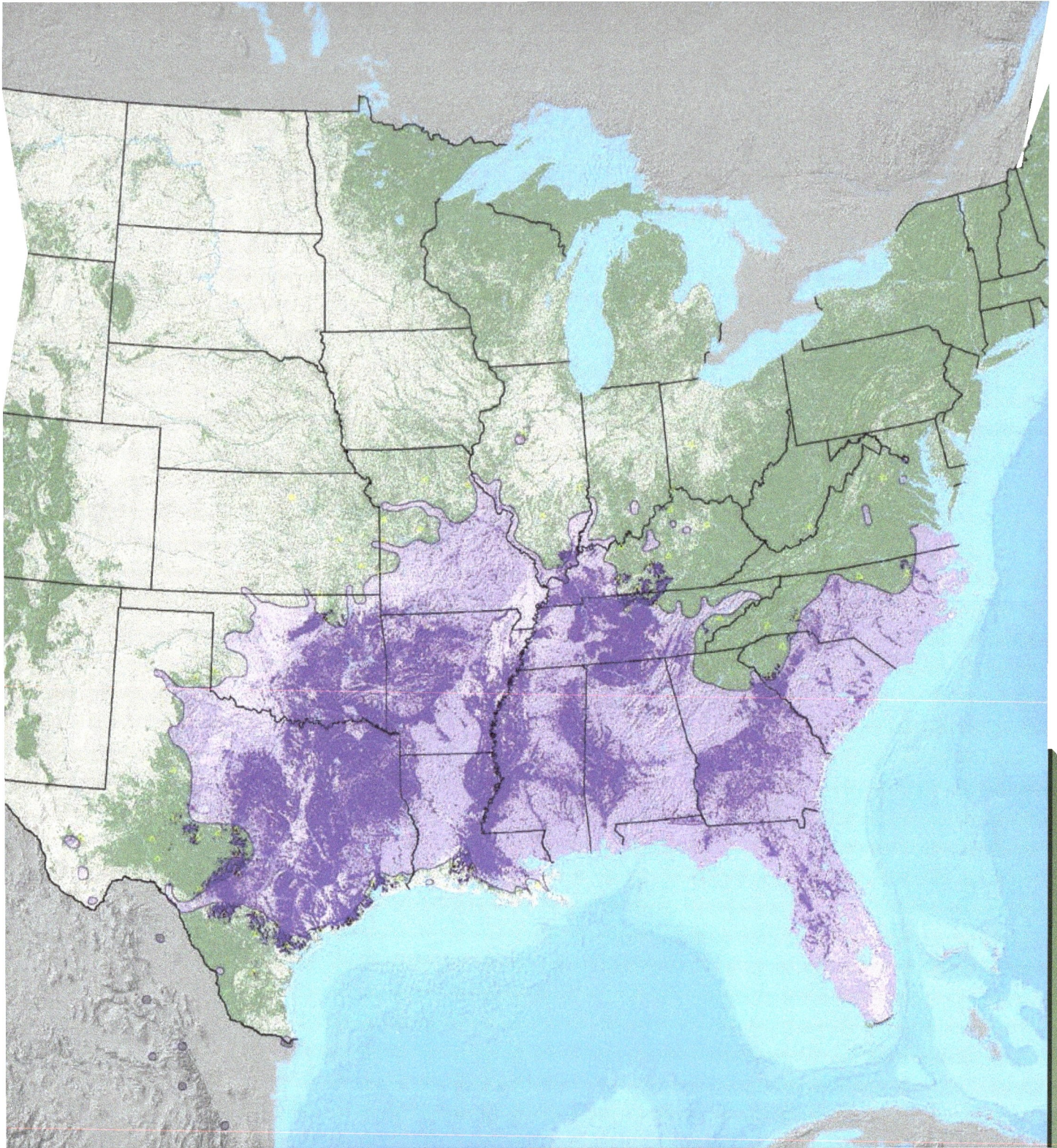

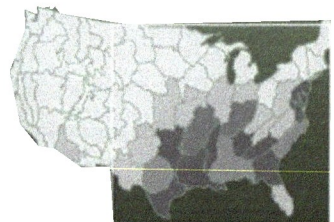

National Individual Tree Species Atlas

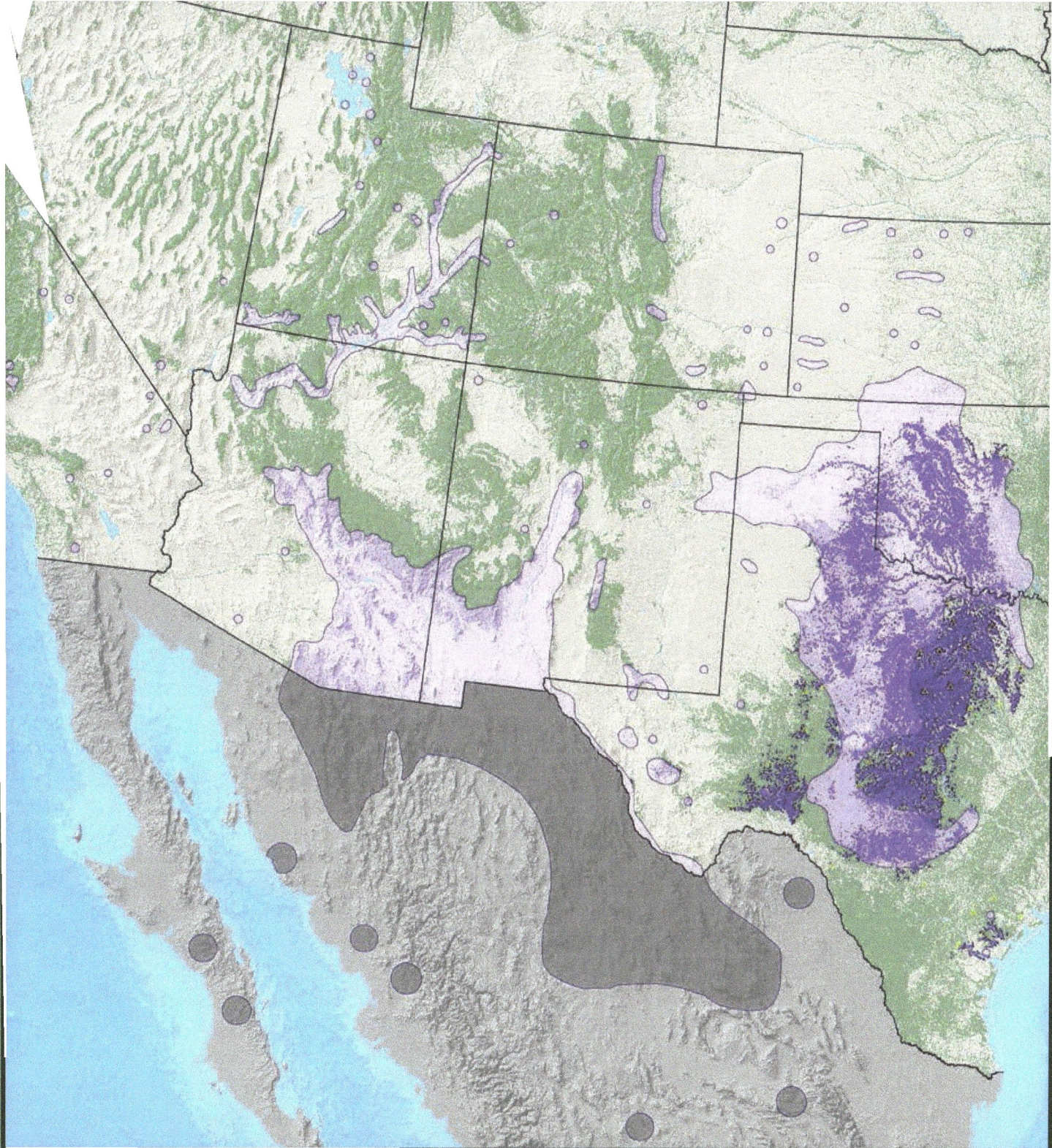

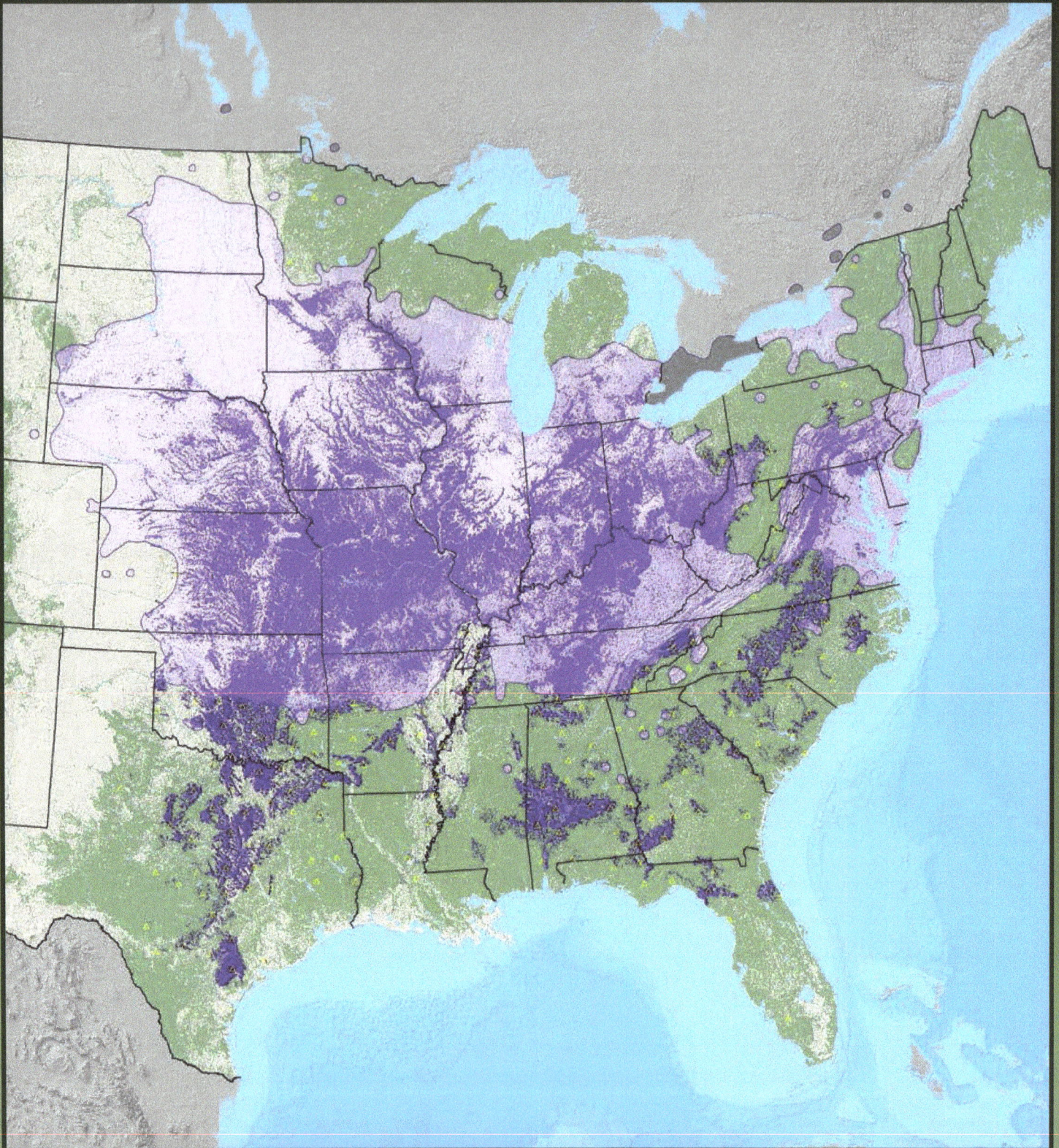

hackberry
Celtis occidentalis L.
Ulmaceae – Elm Family
FIA Code **462** – Plants Symbol **CEOC**

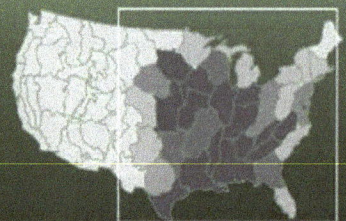

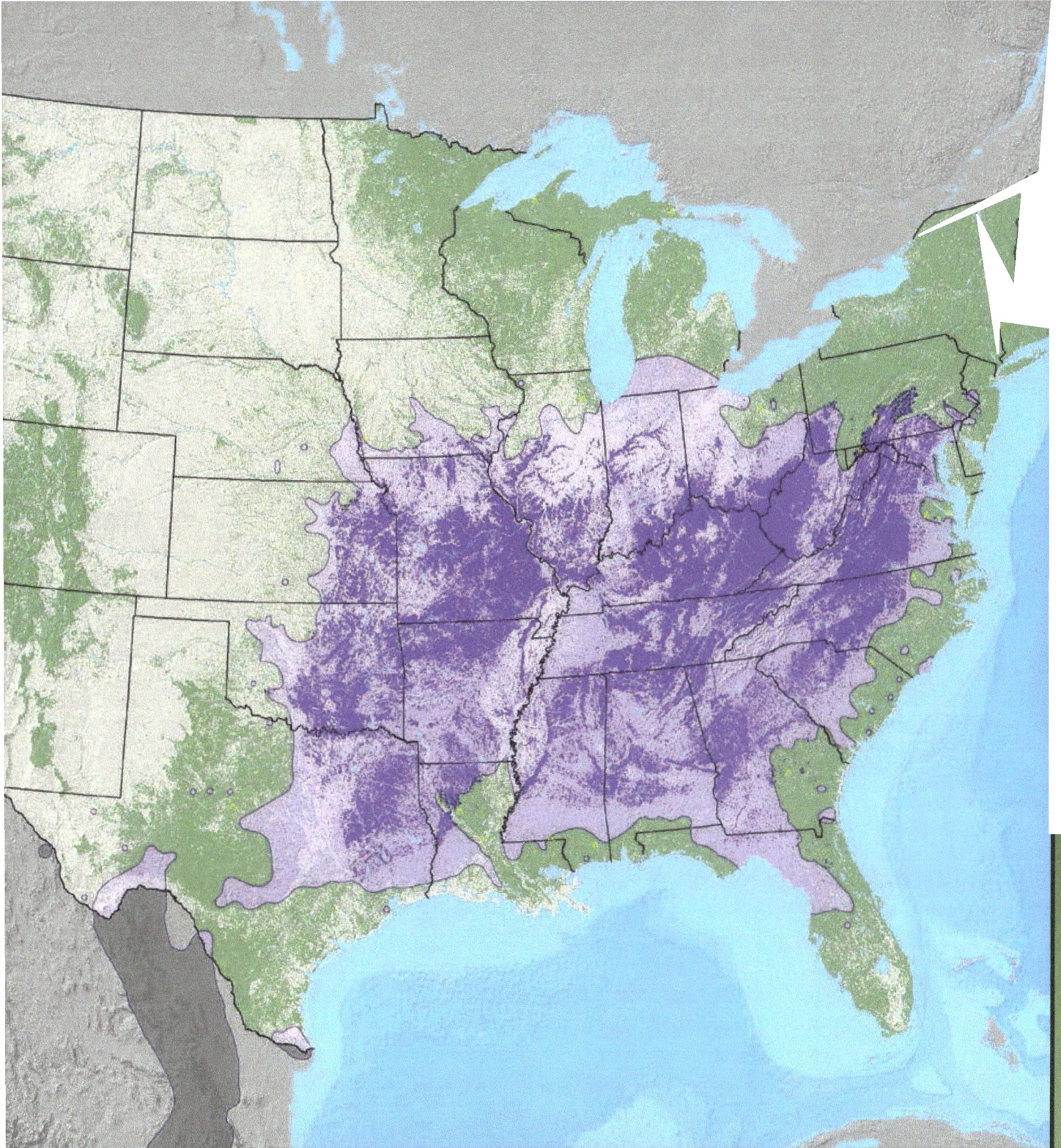

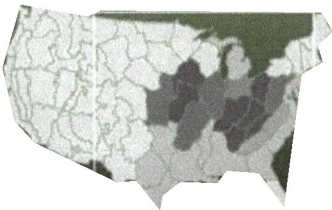

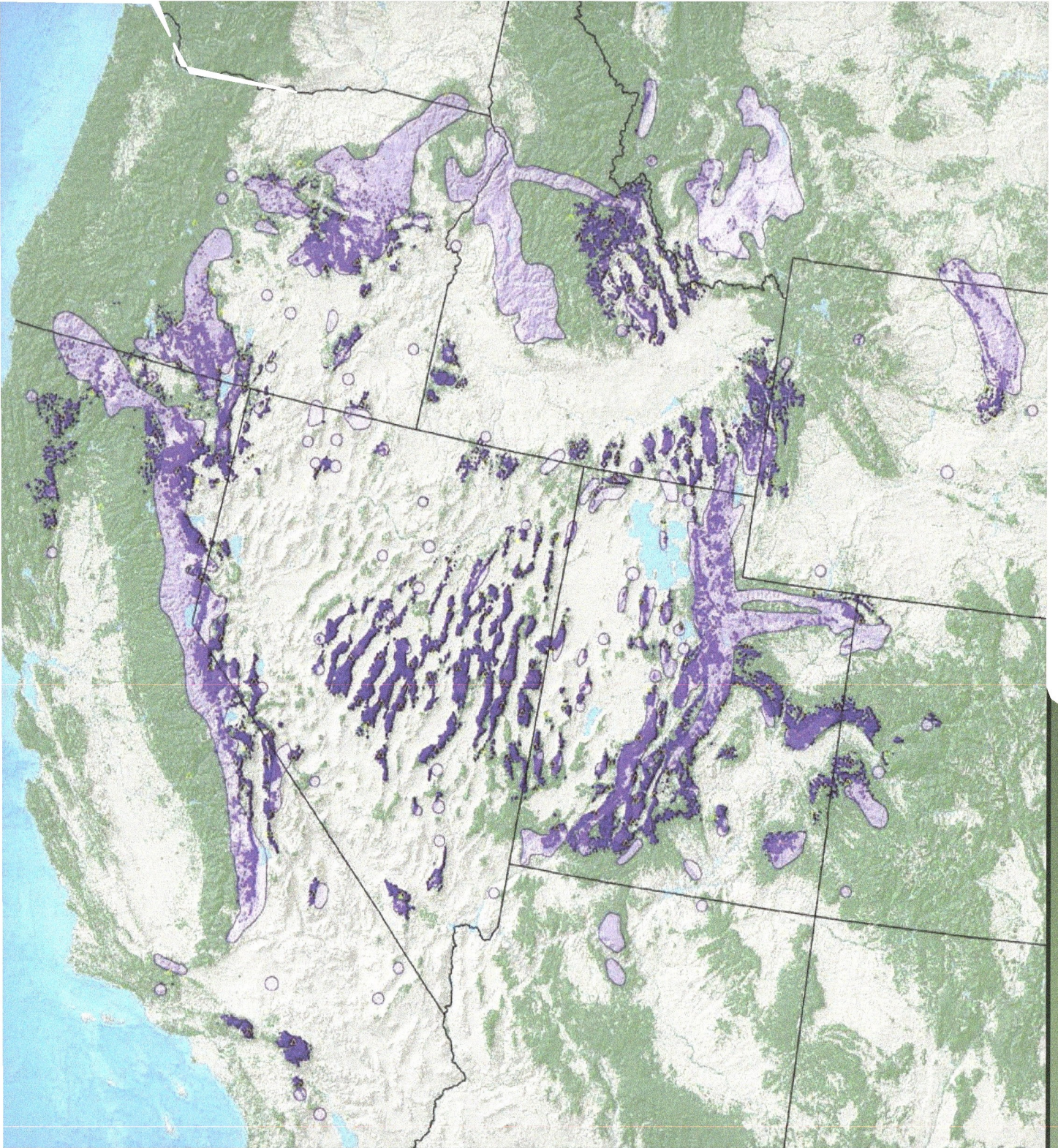

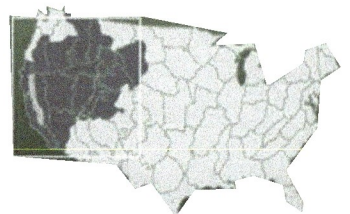

National Individual Tree Species Atlas

giant chinkapin

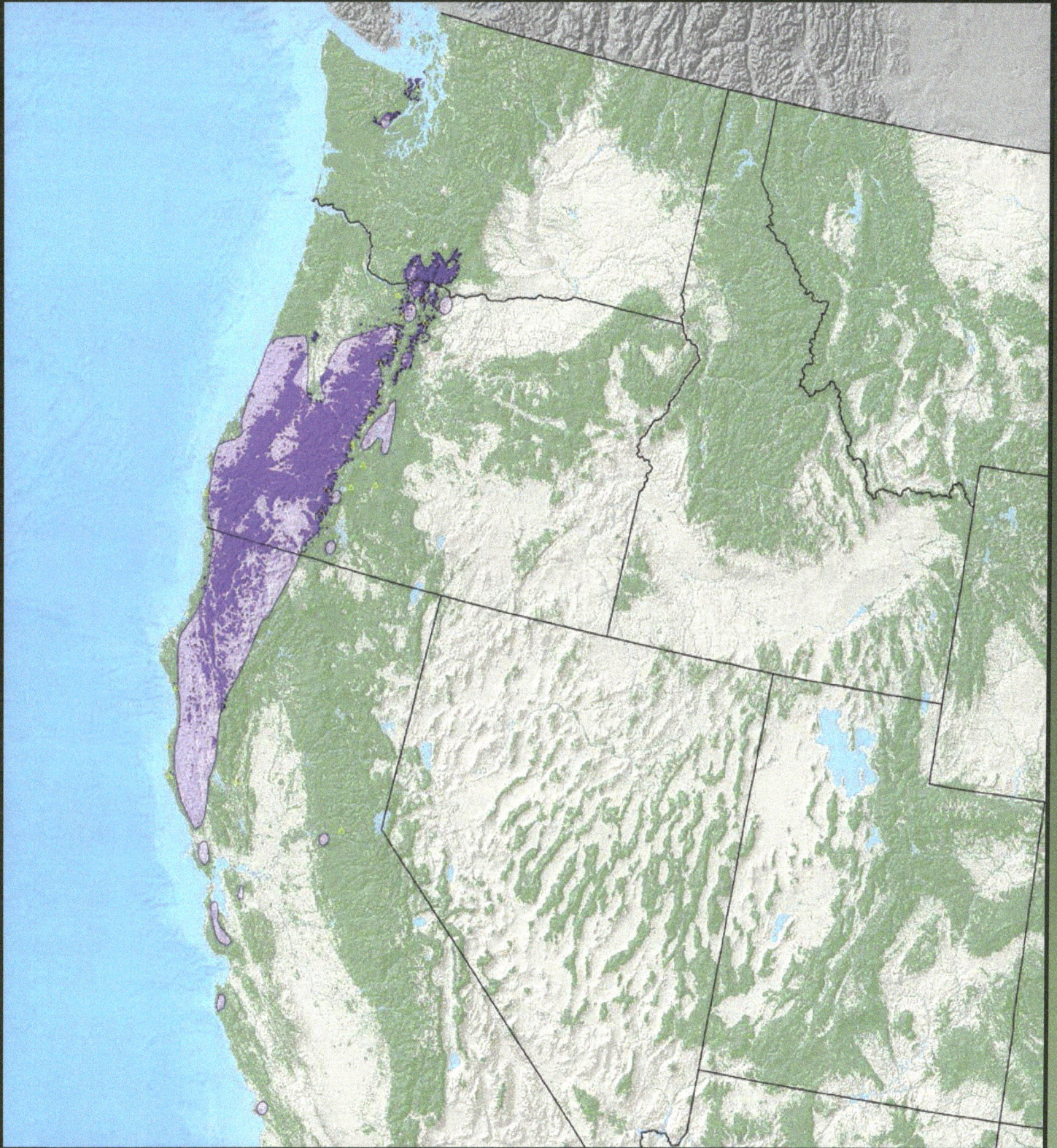

giant chinkapin
Chrysolepis chrysophylla (Douglas ex Hook.) Hjelmqvist
var. *chrysophylla*
Fagaceae – Beech Family
FIA Code **431** – Plants Symbol **CHCHC4**

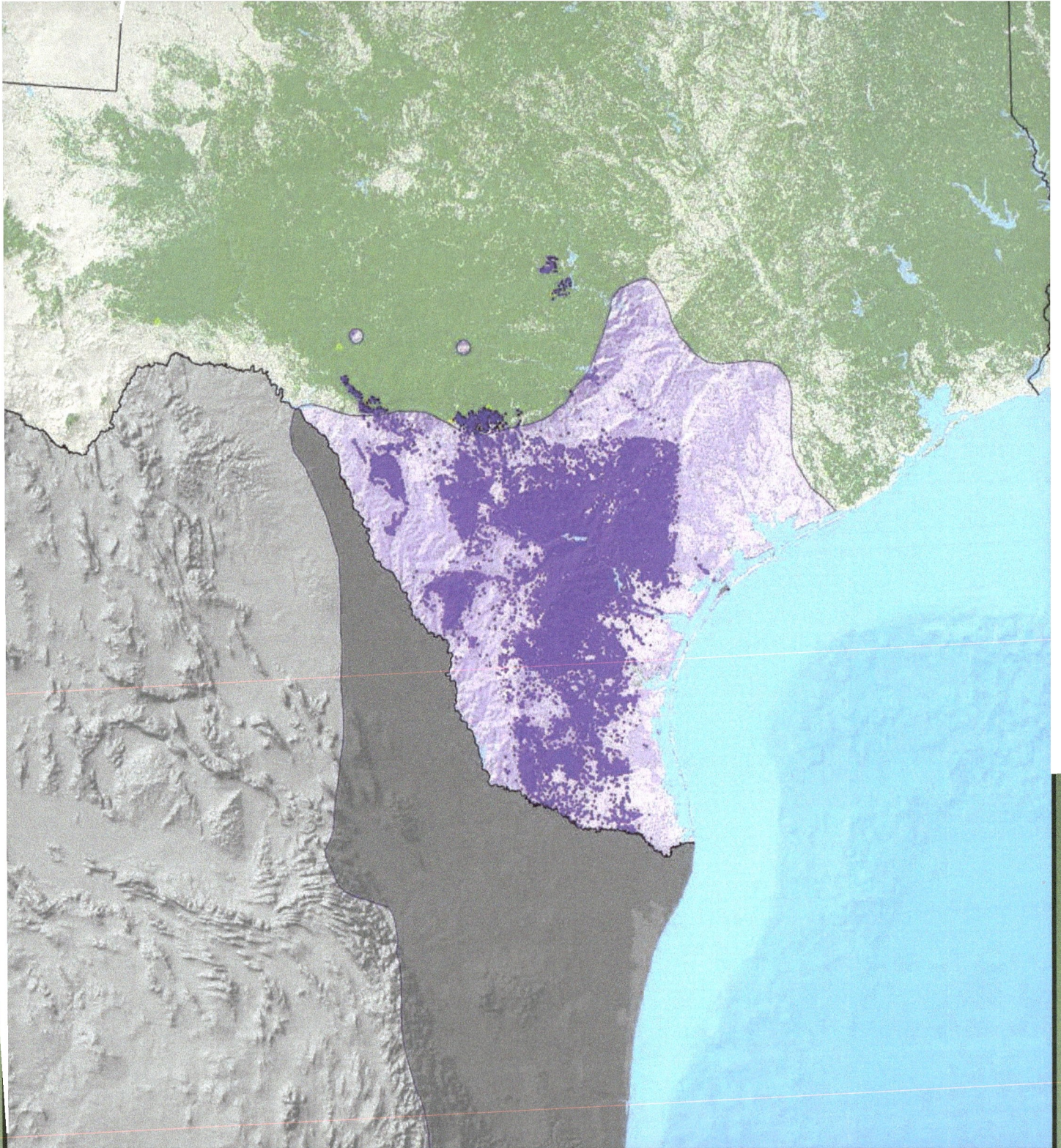

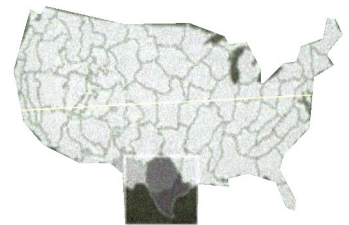

National Individual Tree Species Atlas

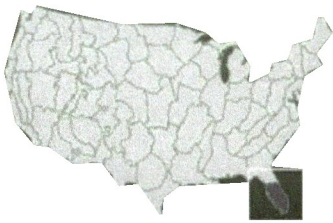

National Individual Tree Species Atlas

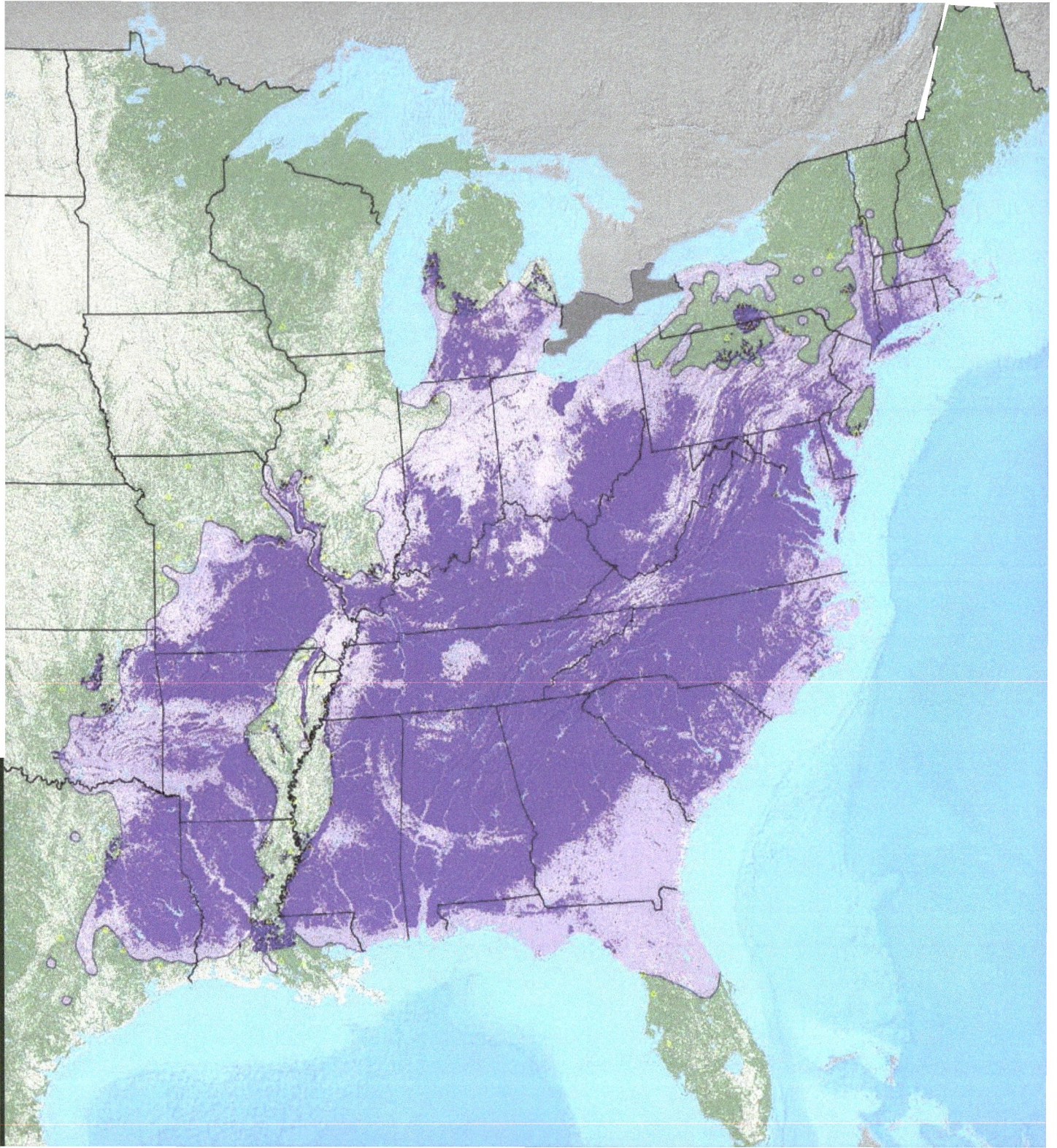

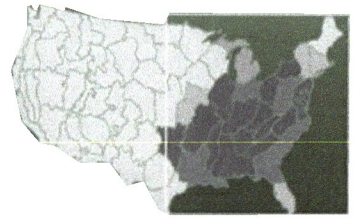

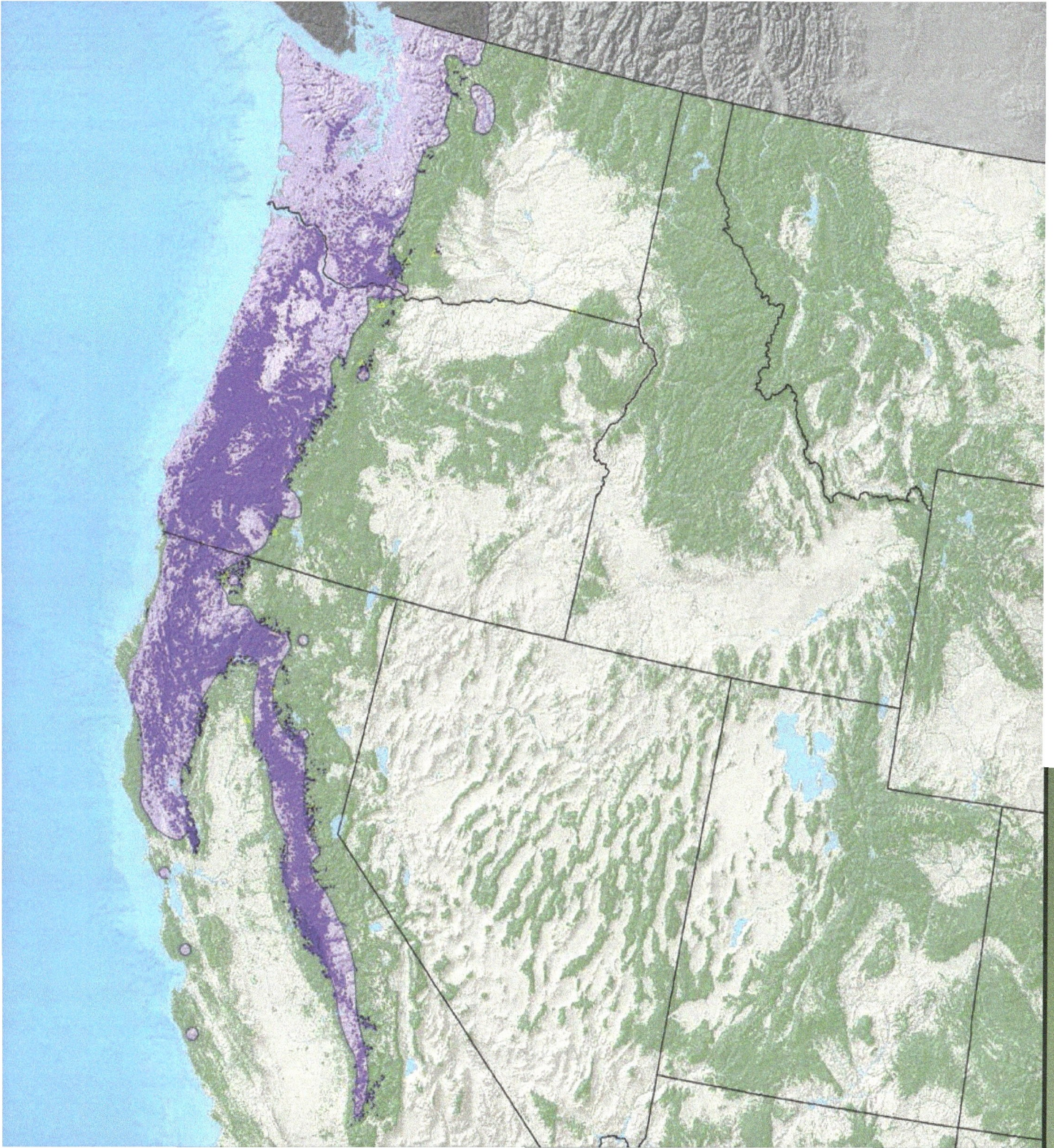
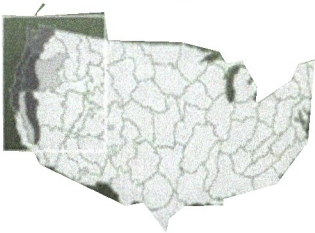

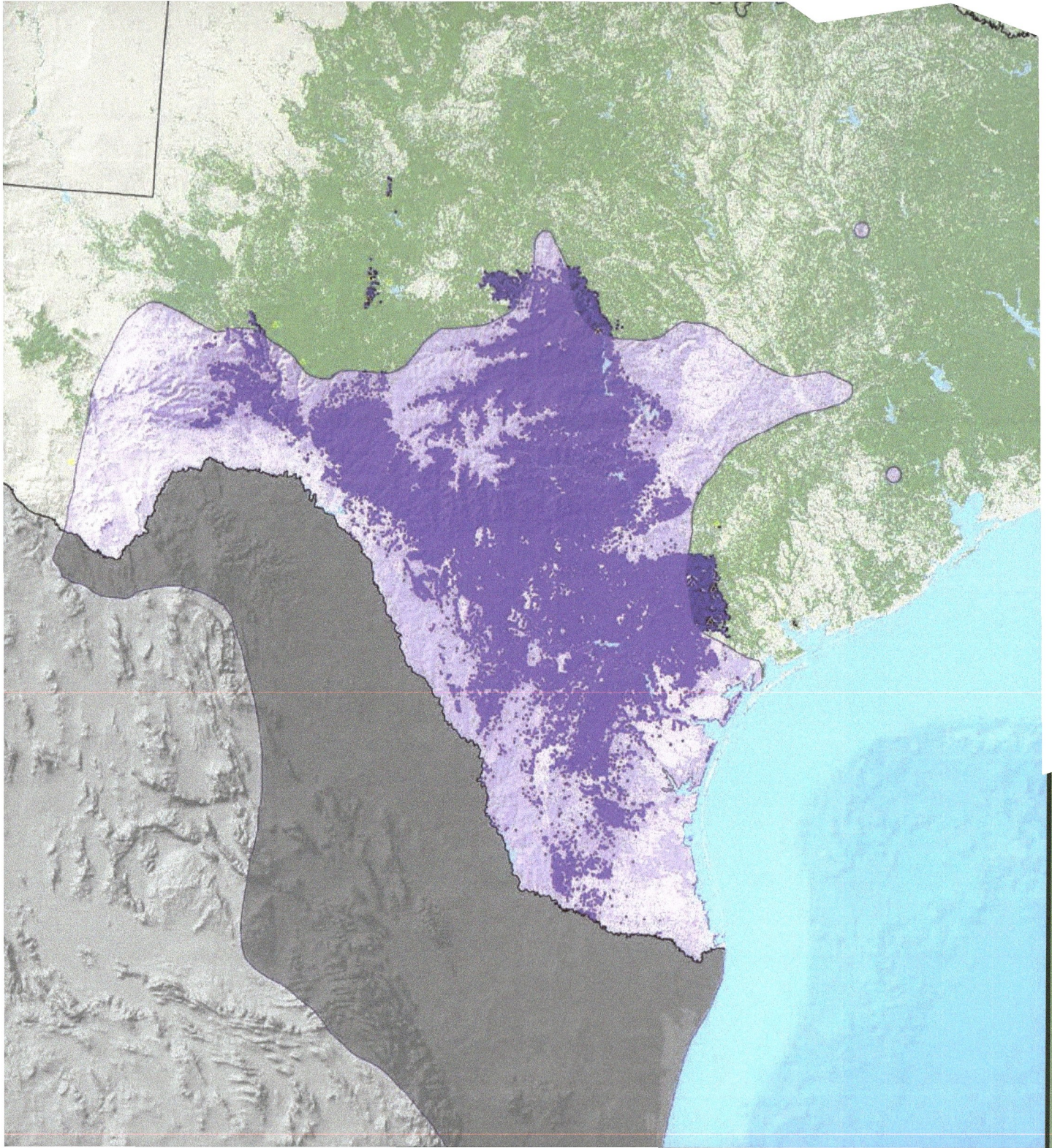

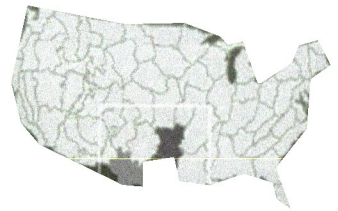

National Individual Tree Species Atlas

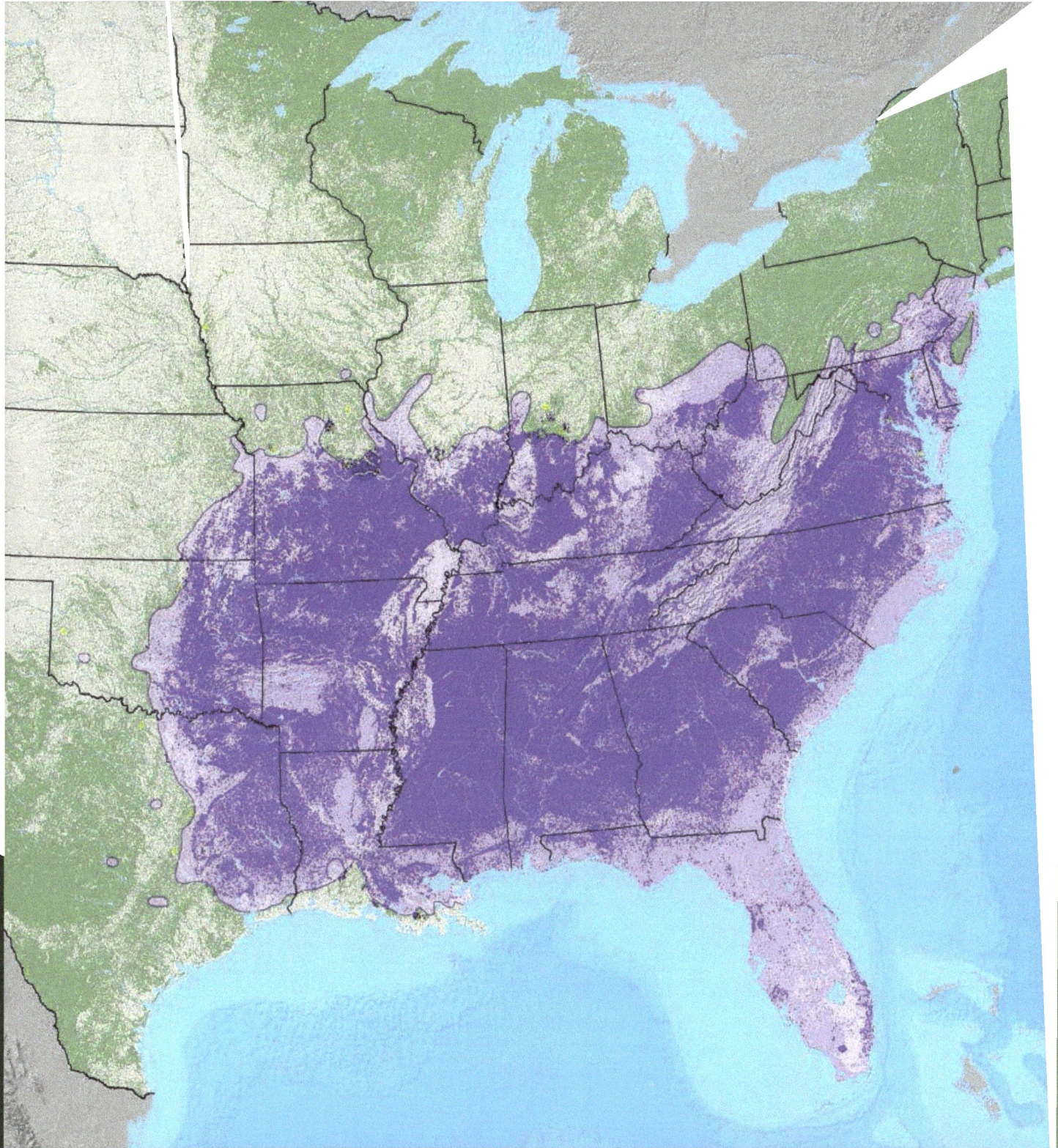

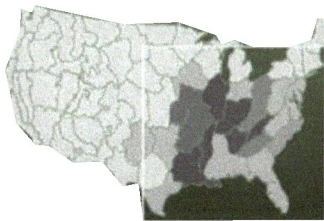

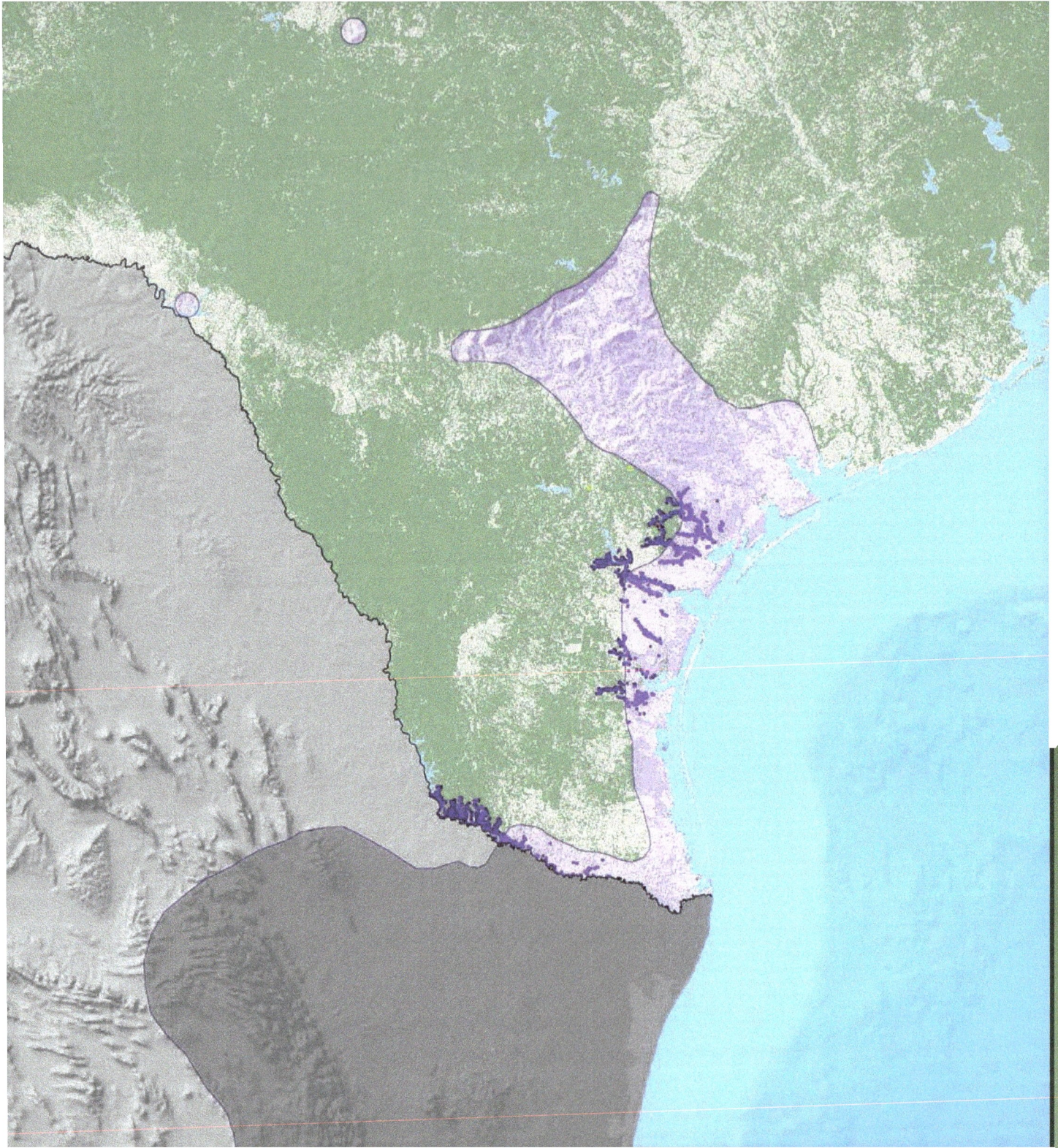

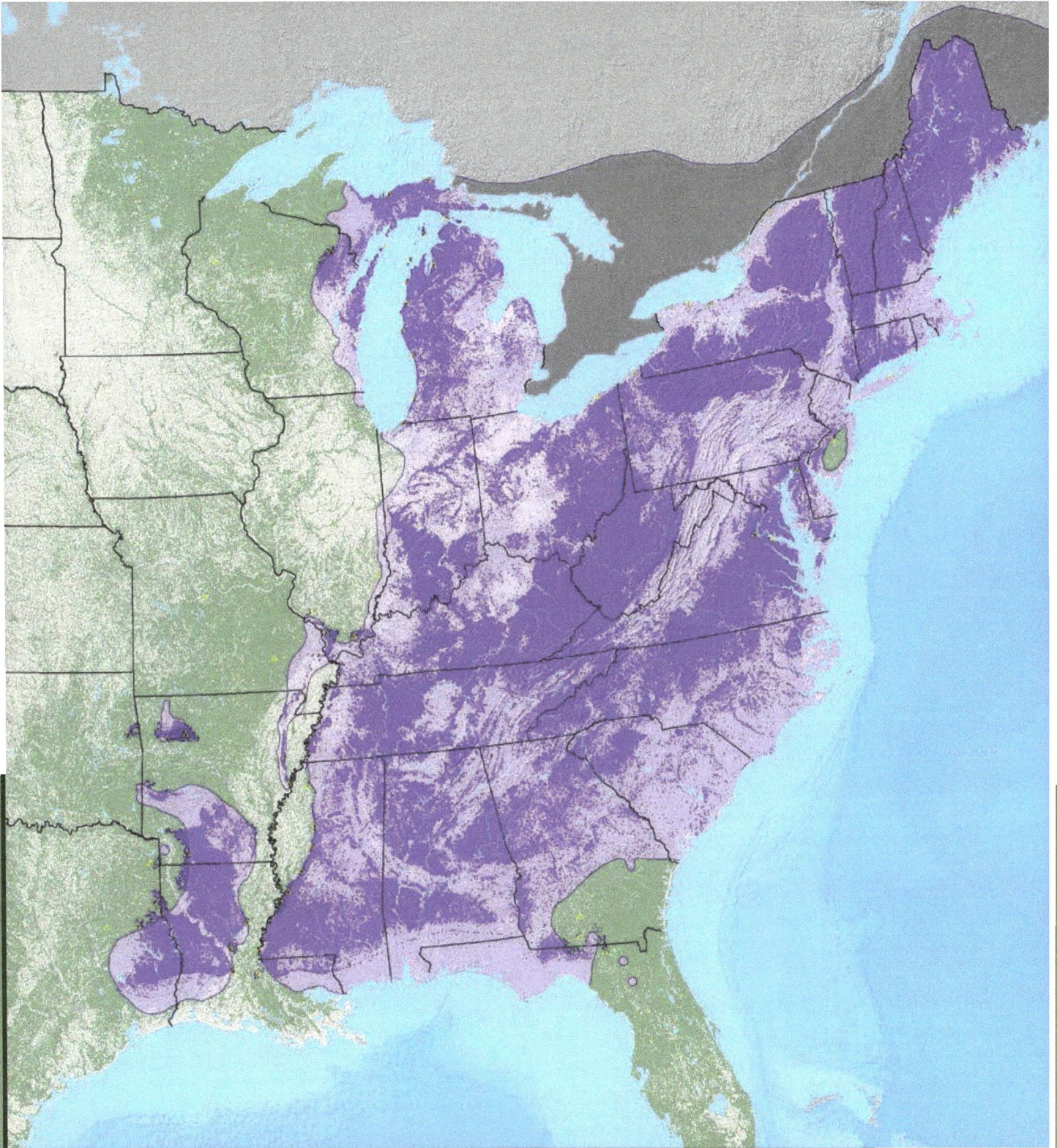

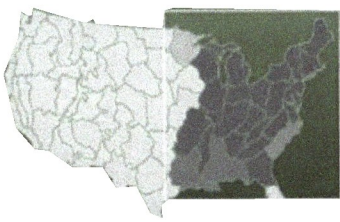

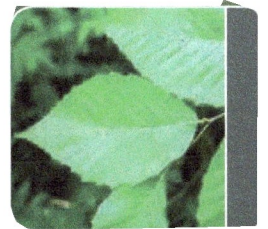

National Individual Tree Species Atlas

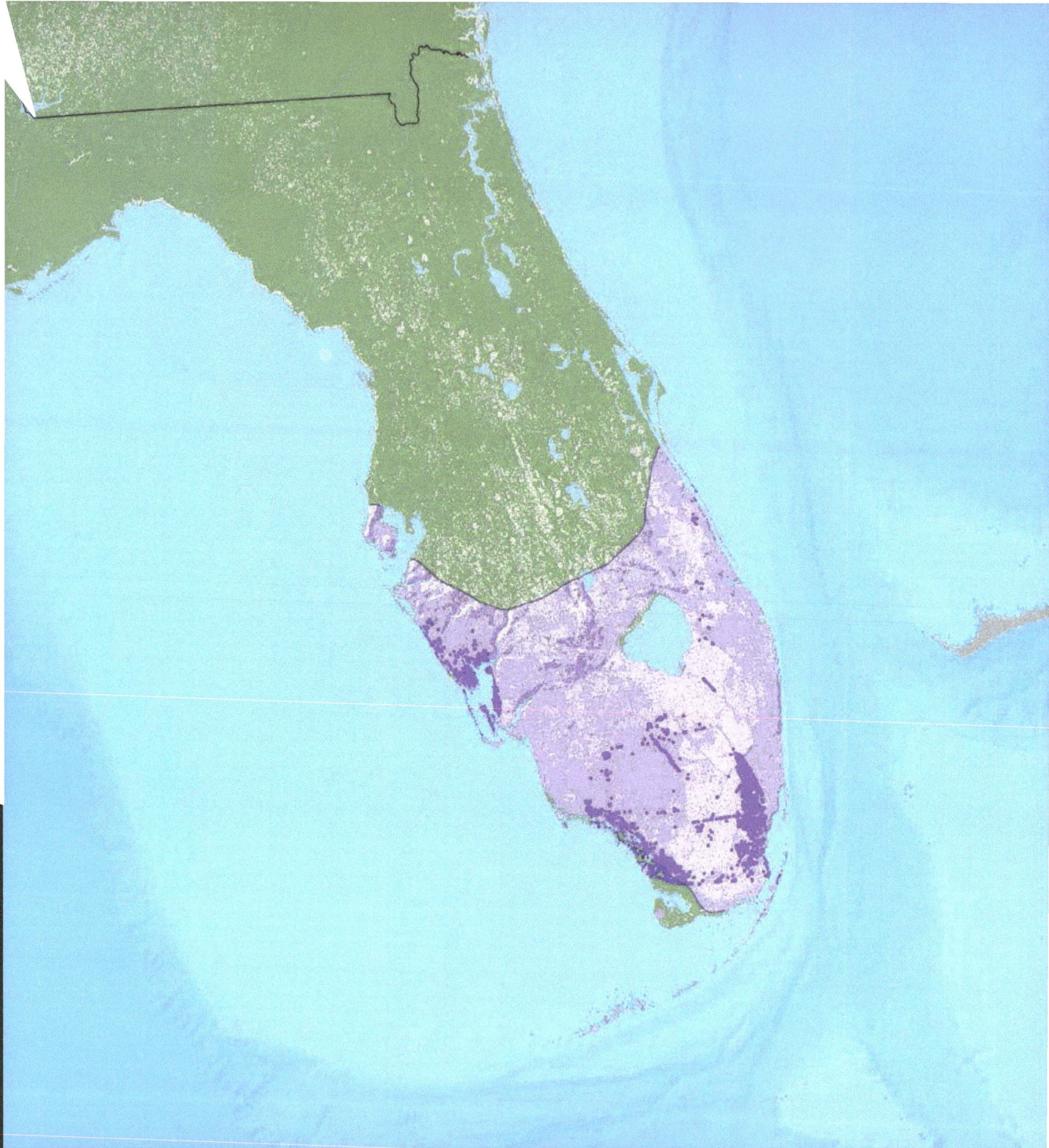

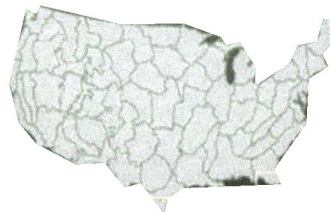

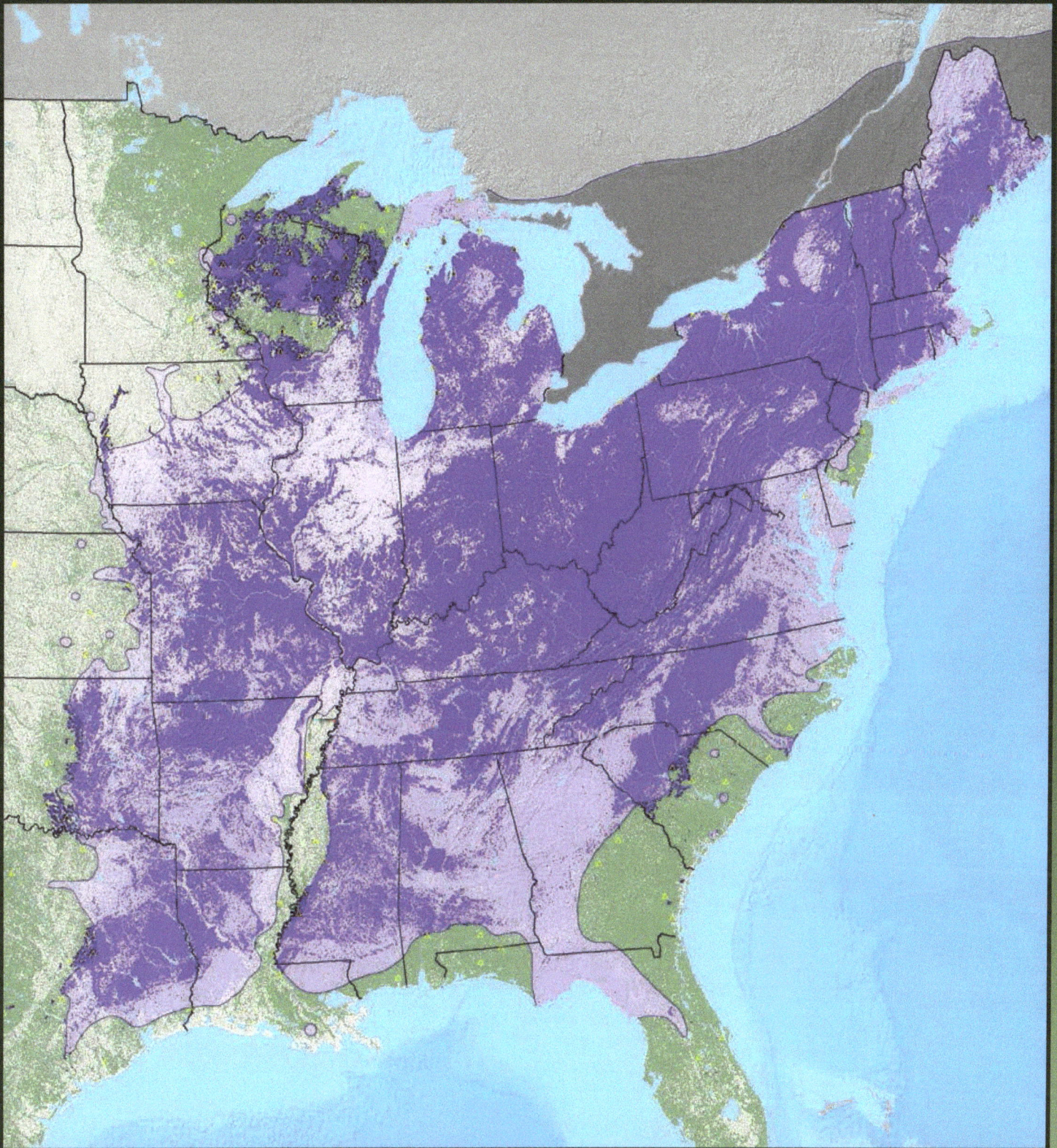

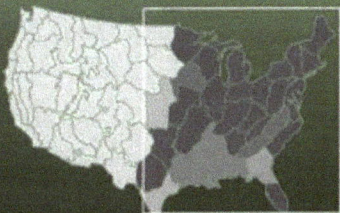

white ash

white ash
Fraxinus americana L.
Oleaceae – Olive Family
FIA Code **541** – Plants Symbol **FRAM2**

National Individual Tree Species Atlas

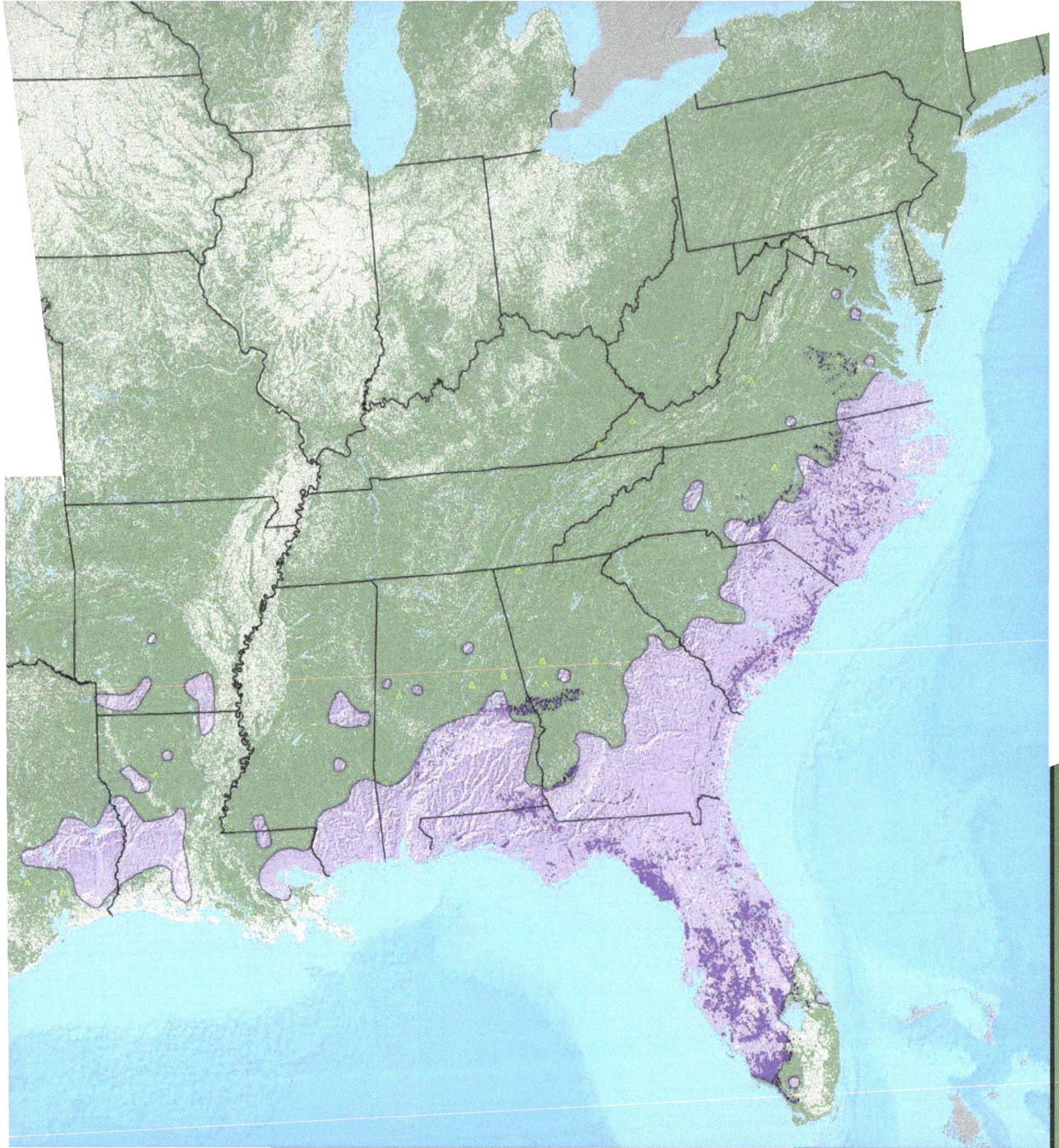

National Individual Tree Species Atlas

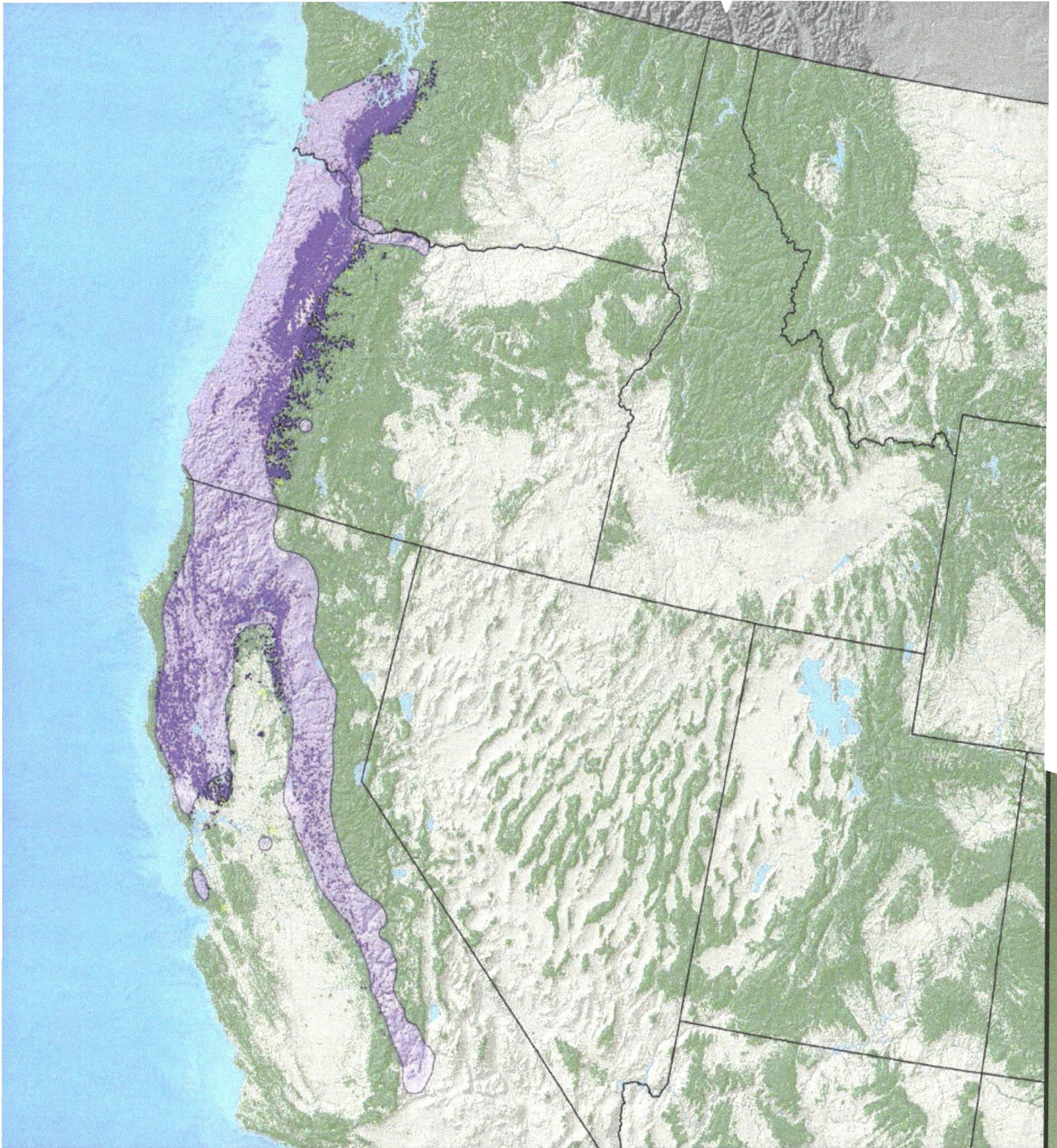

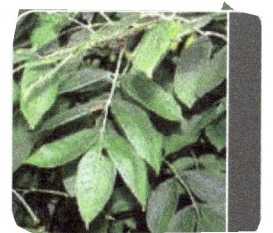

National Individual Tree Species Atlas

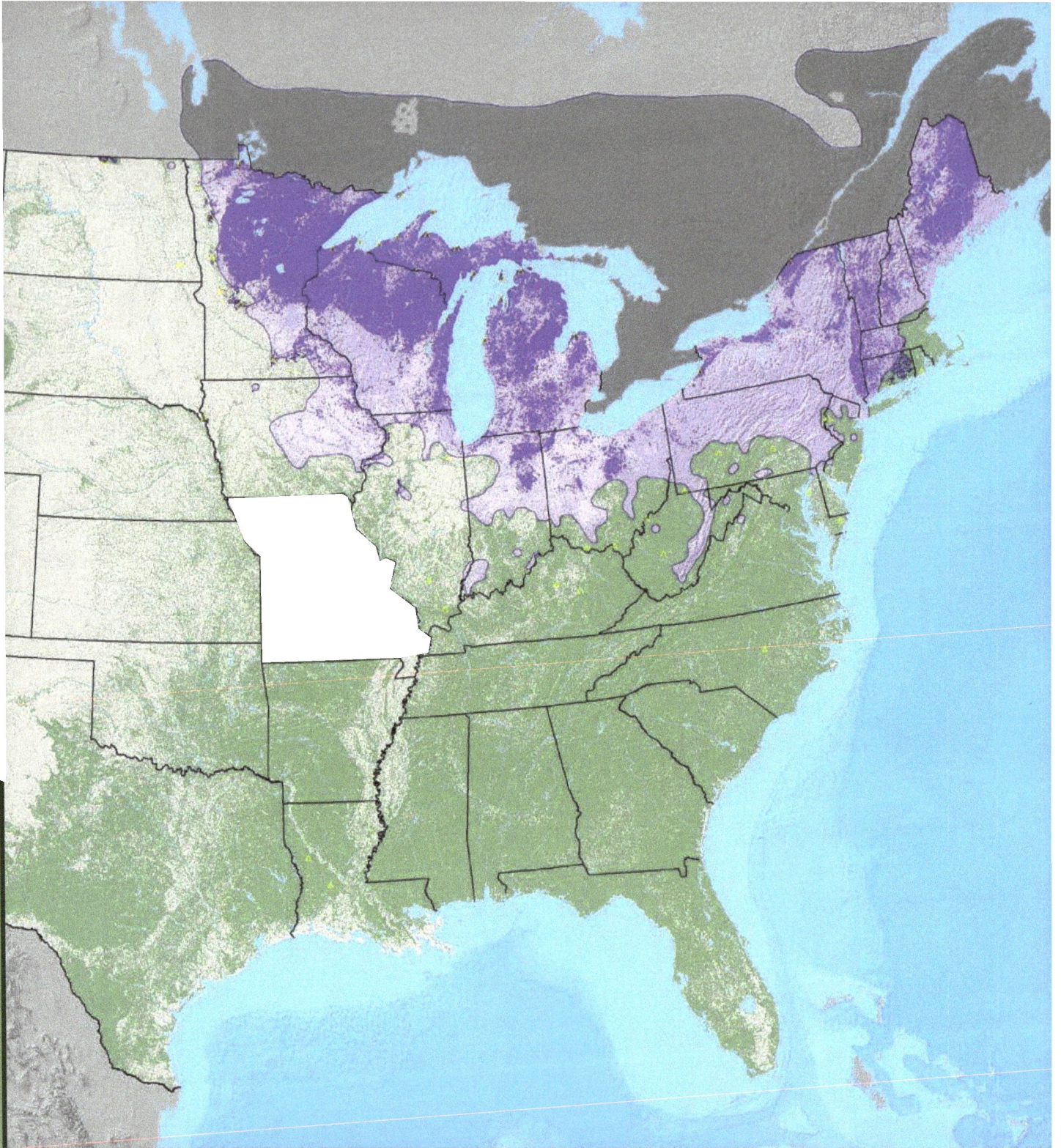

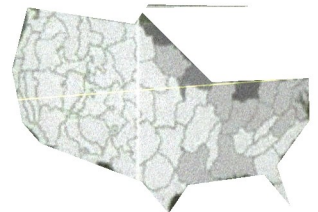

National Individual Tree Species Atlas

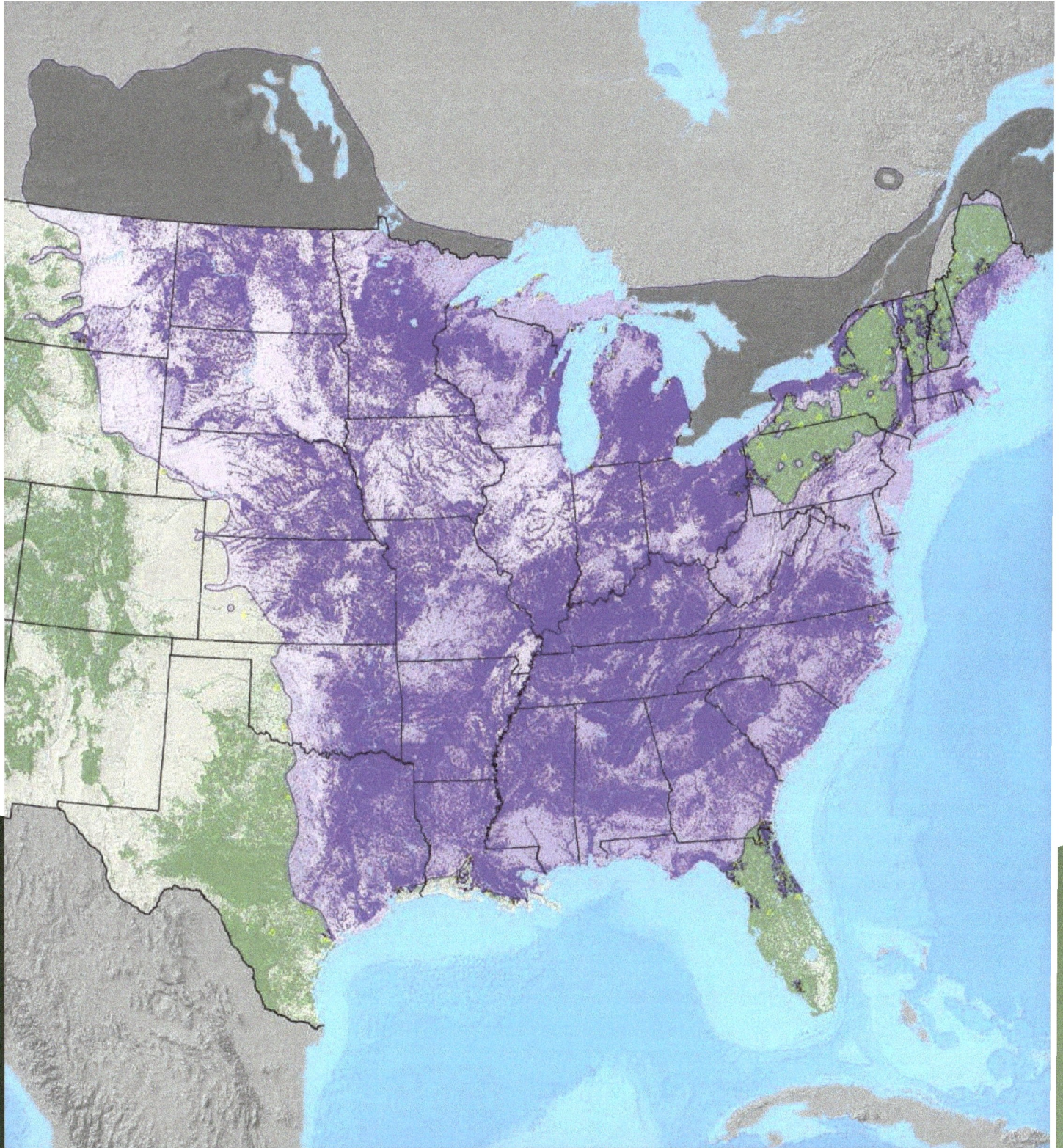

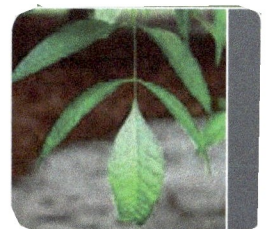

National Individual Tree Species Atlas

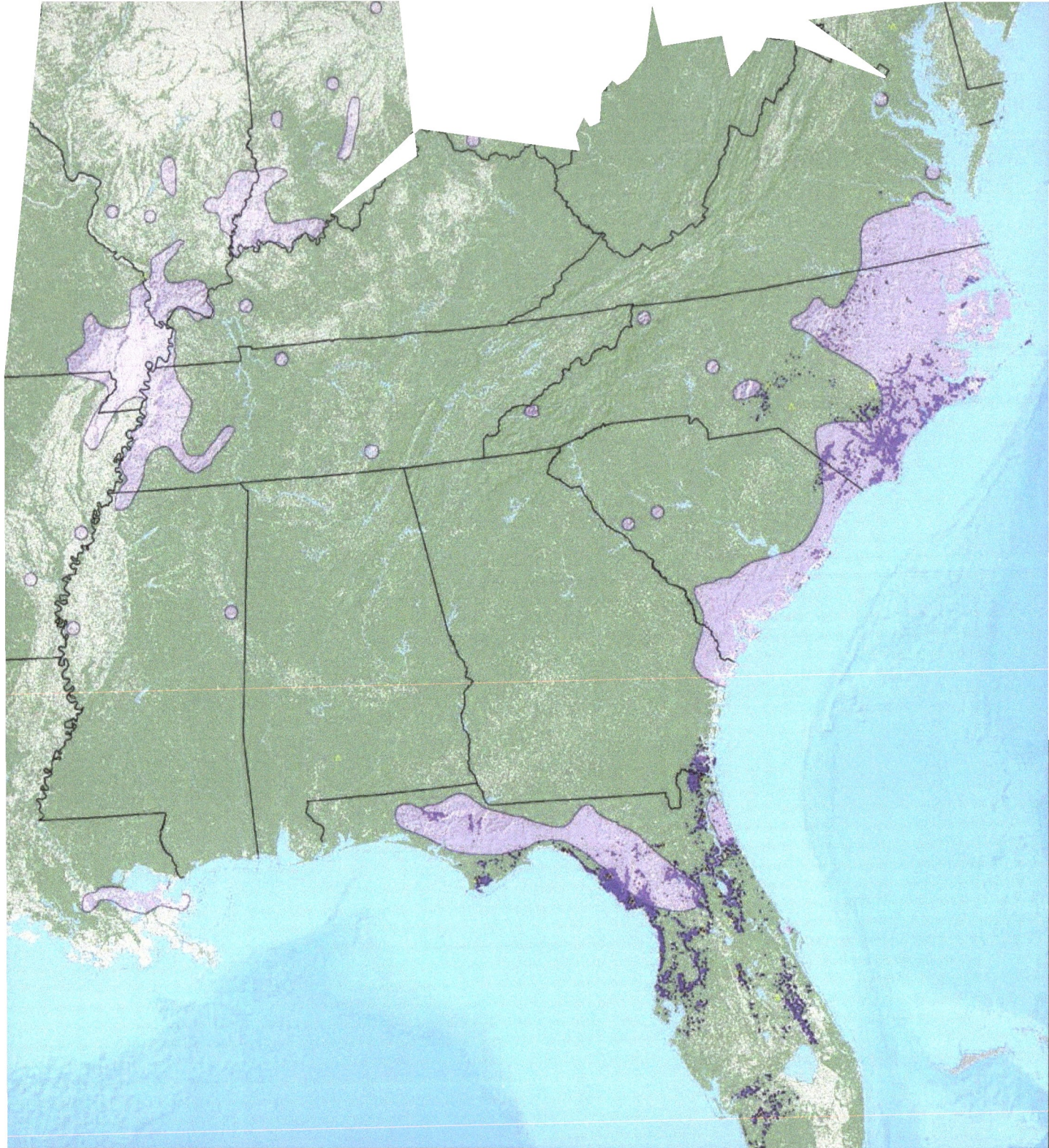

National Individual Tree Species Atlas

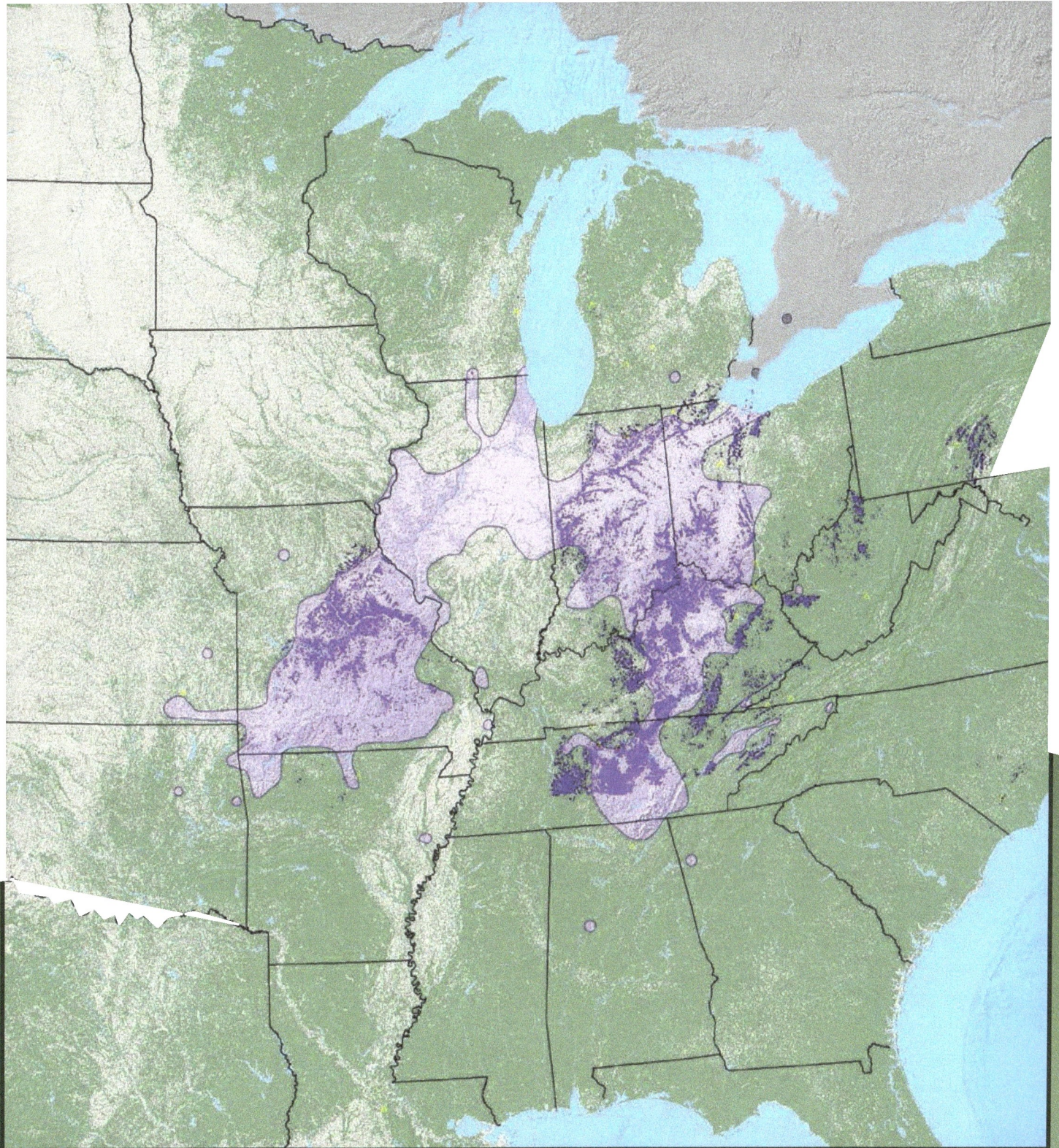

blue ash

Fraxinus quadrangulata Michx.

Oleaceae – Olive Family

FIA Code **546** – Plants Symbol **FRQU**

National Individual Tree Species Atlas

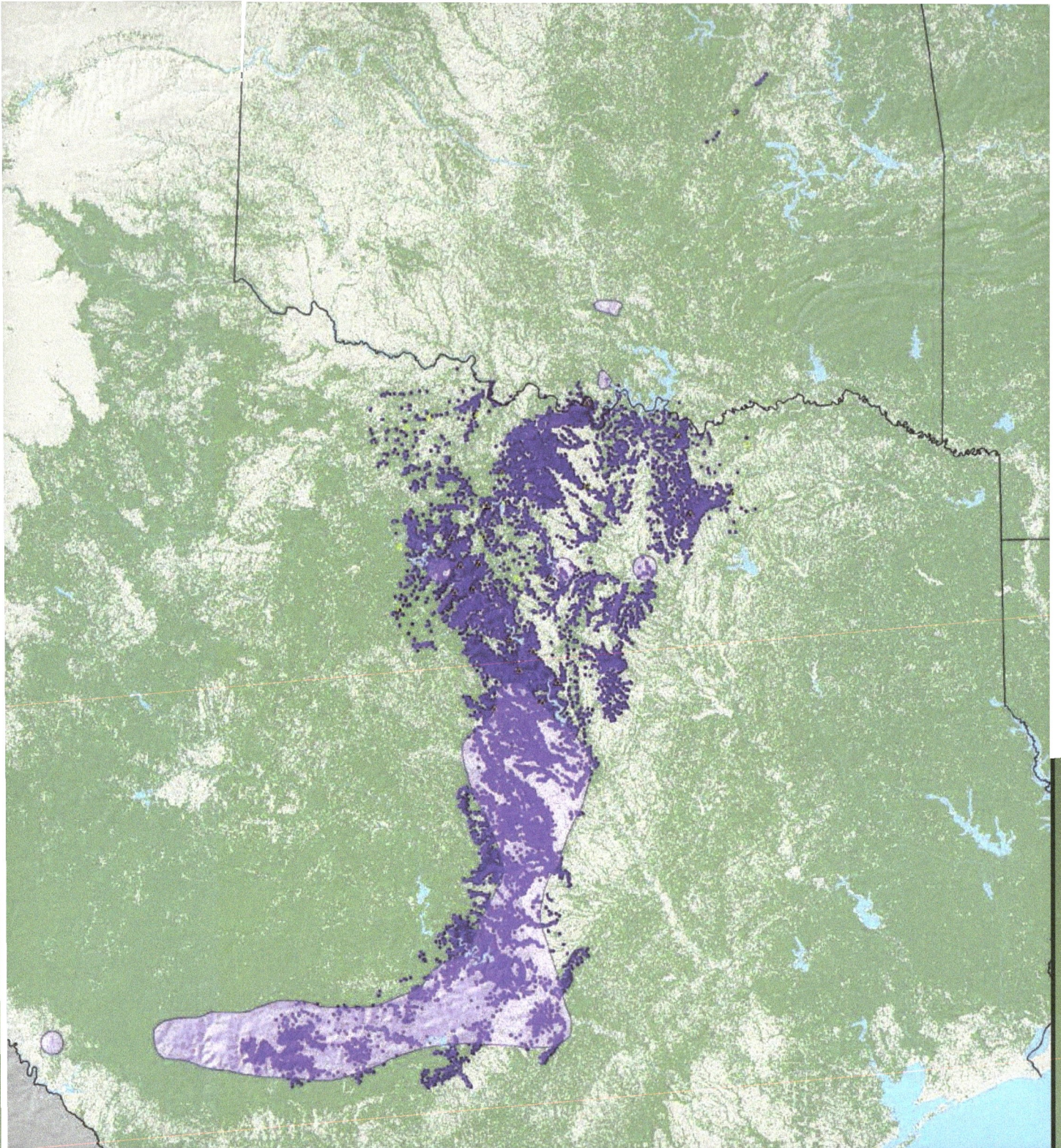

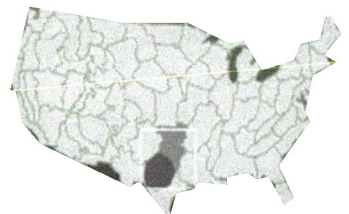

National Individual Tree Species Atlas

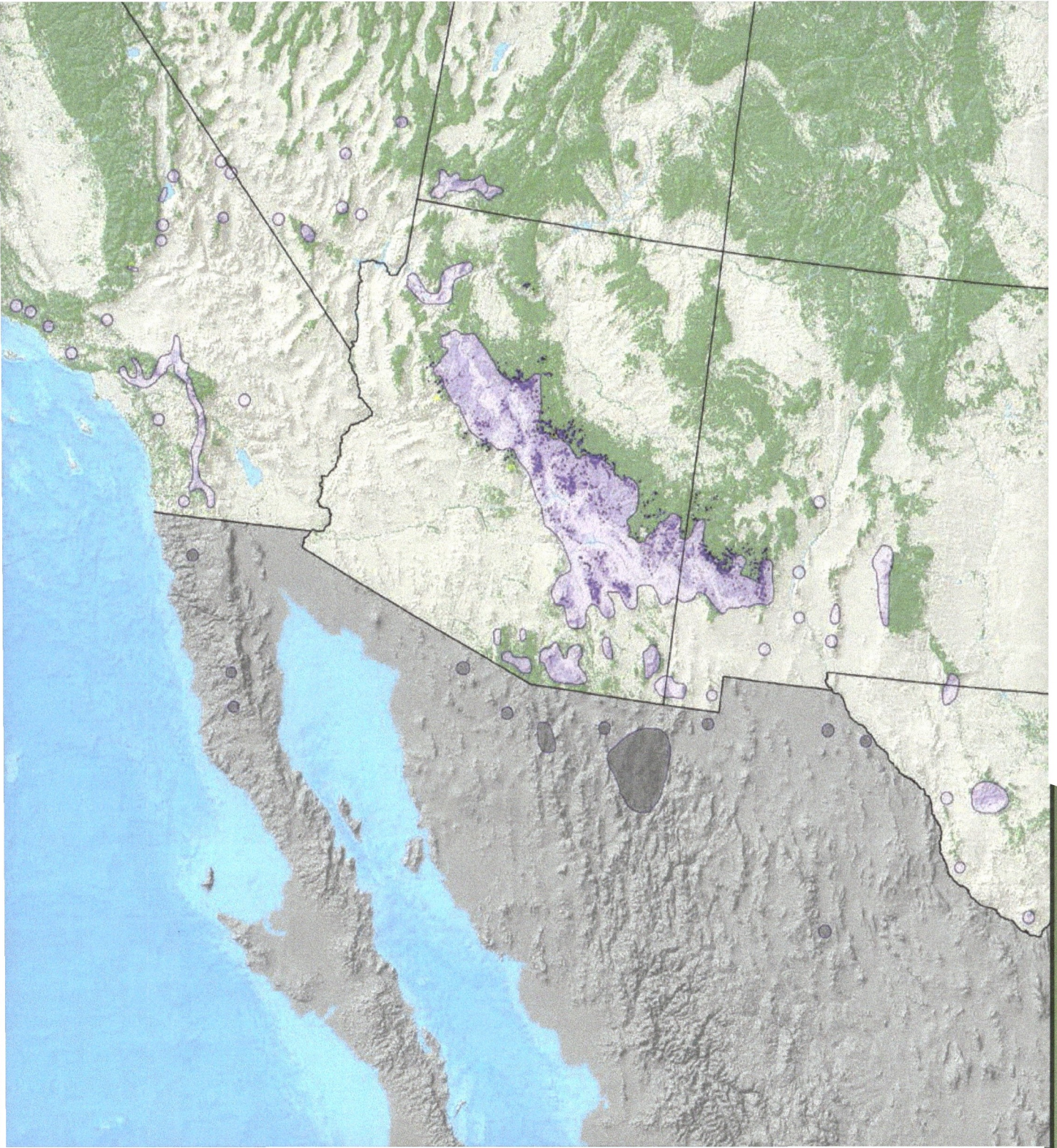

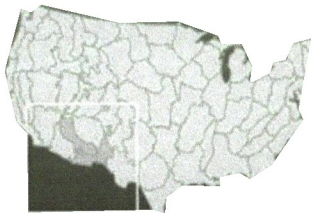

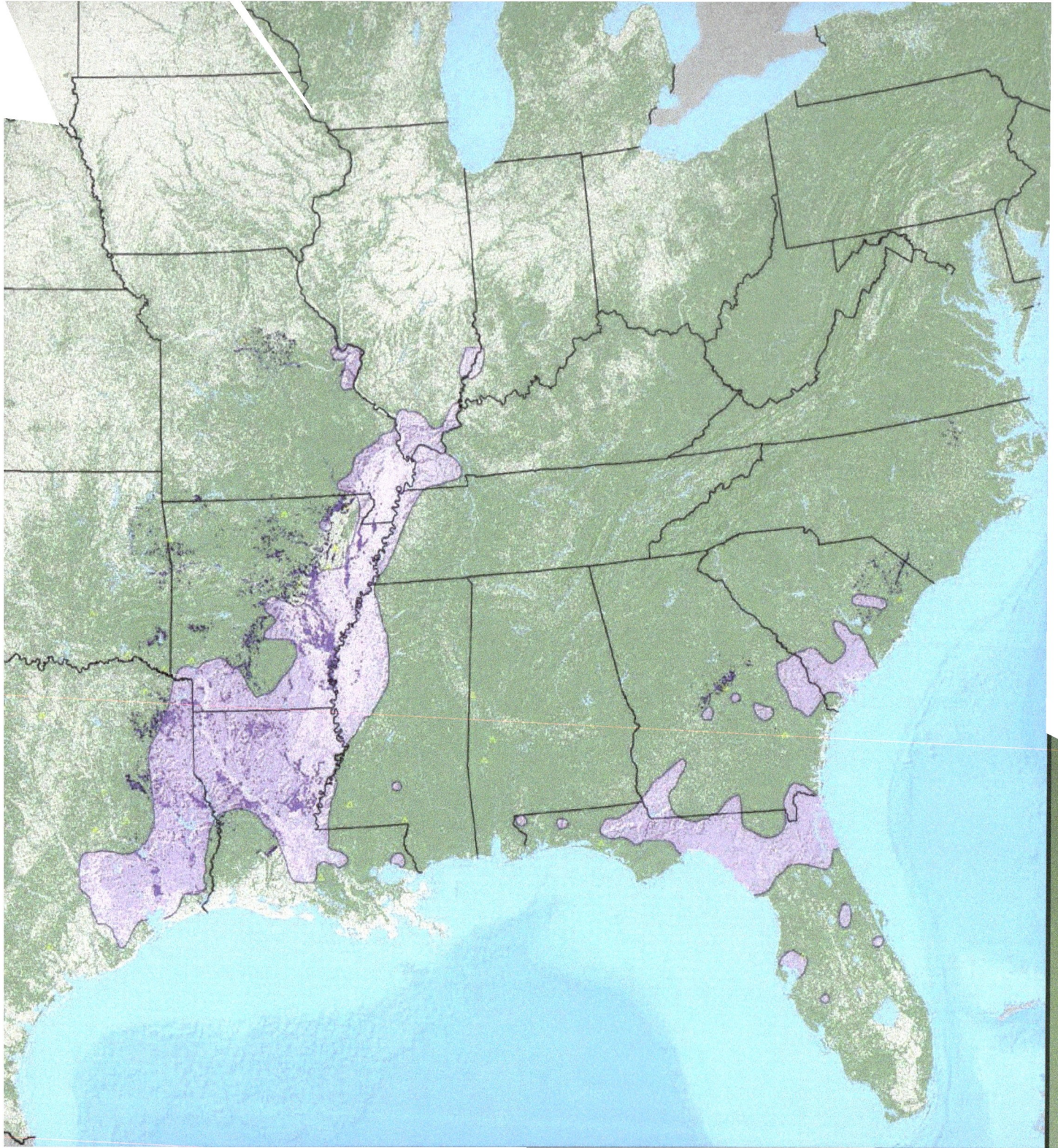

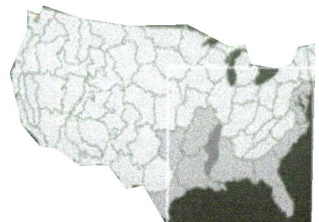

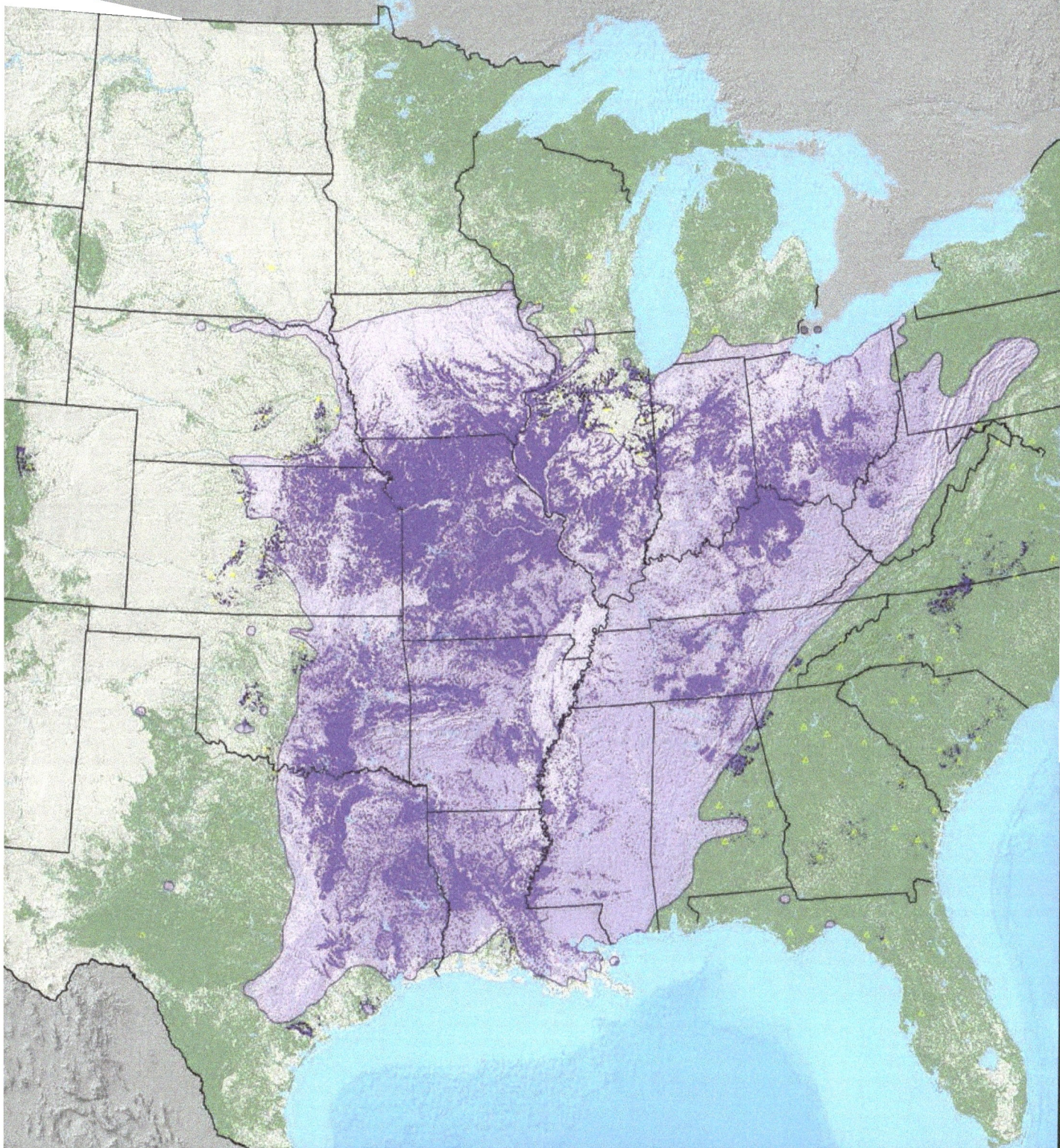

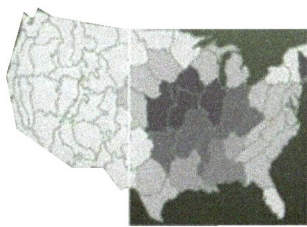

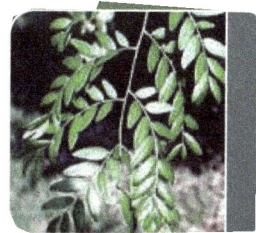

National Individual Tree Species Atlas

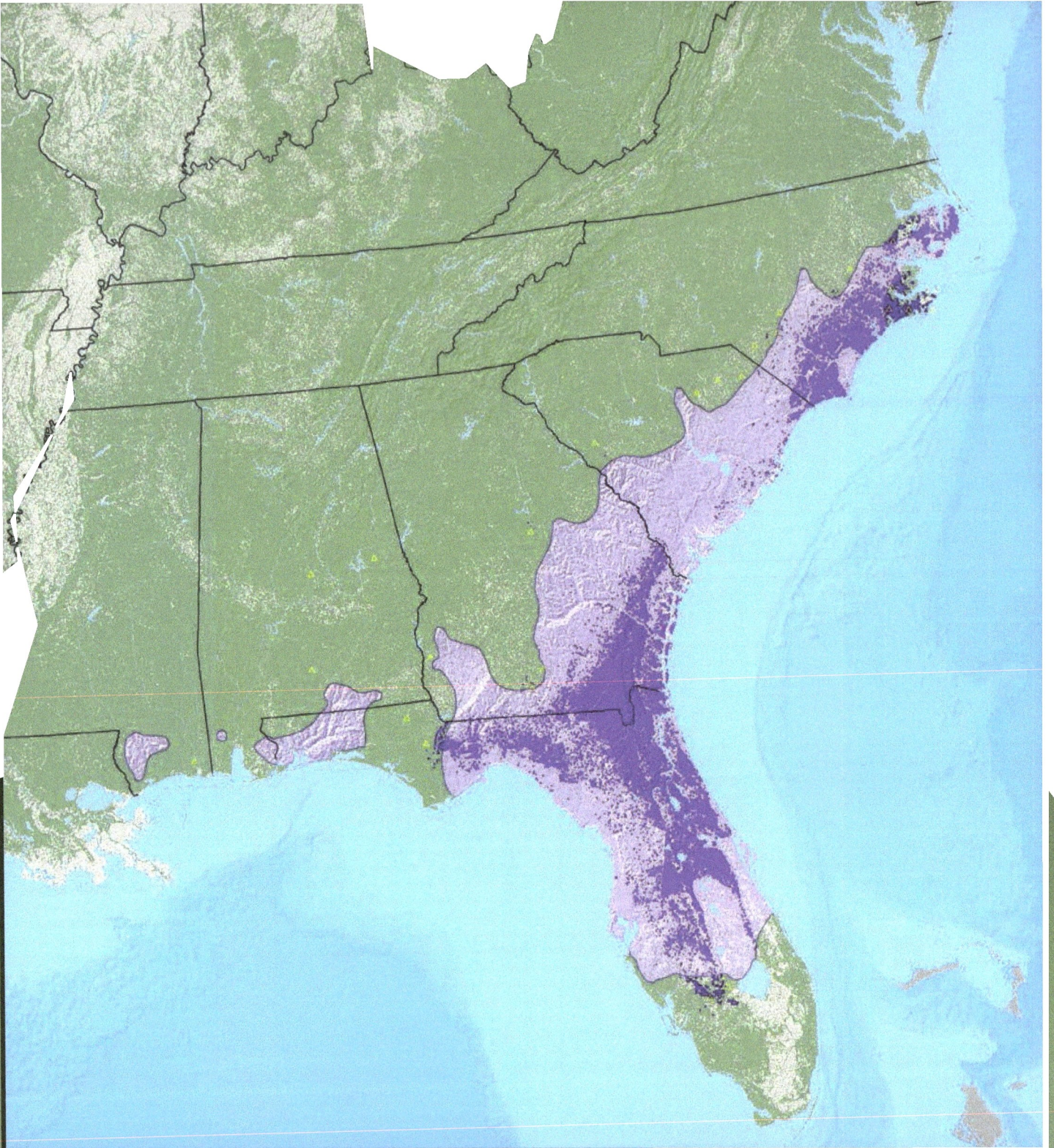

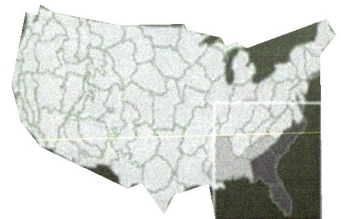

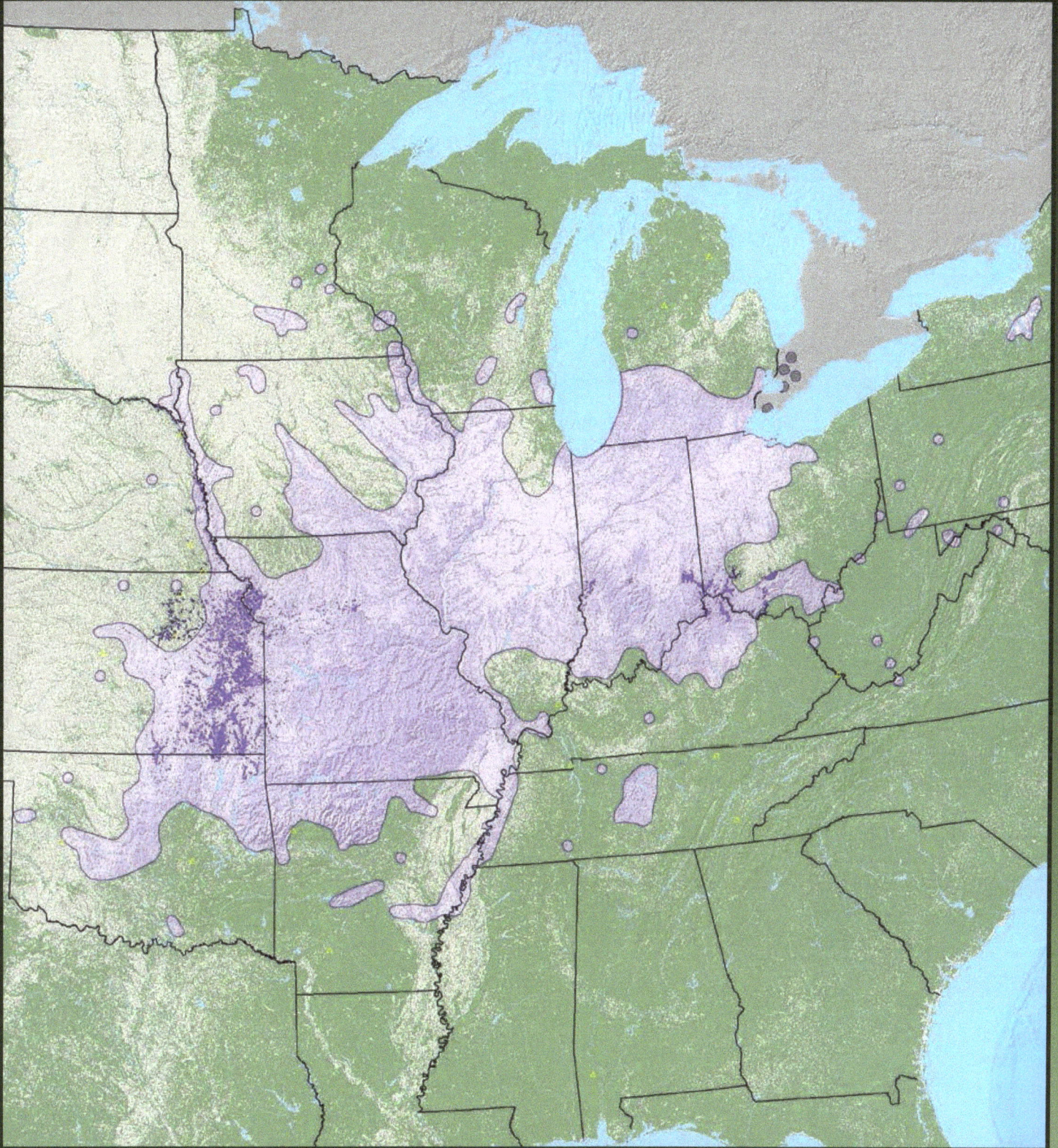

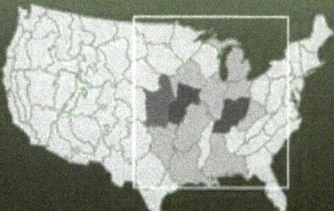

Kentucky coffeetree

Gymnocladus dioicus (L.) K. Koch

Fabaceae – Pea Family

FIA Code **571** – Plants Symbol **GYDI**

National Individual Tree Species Atlas

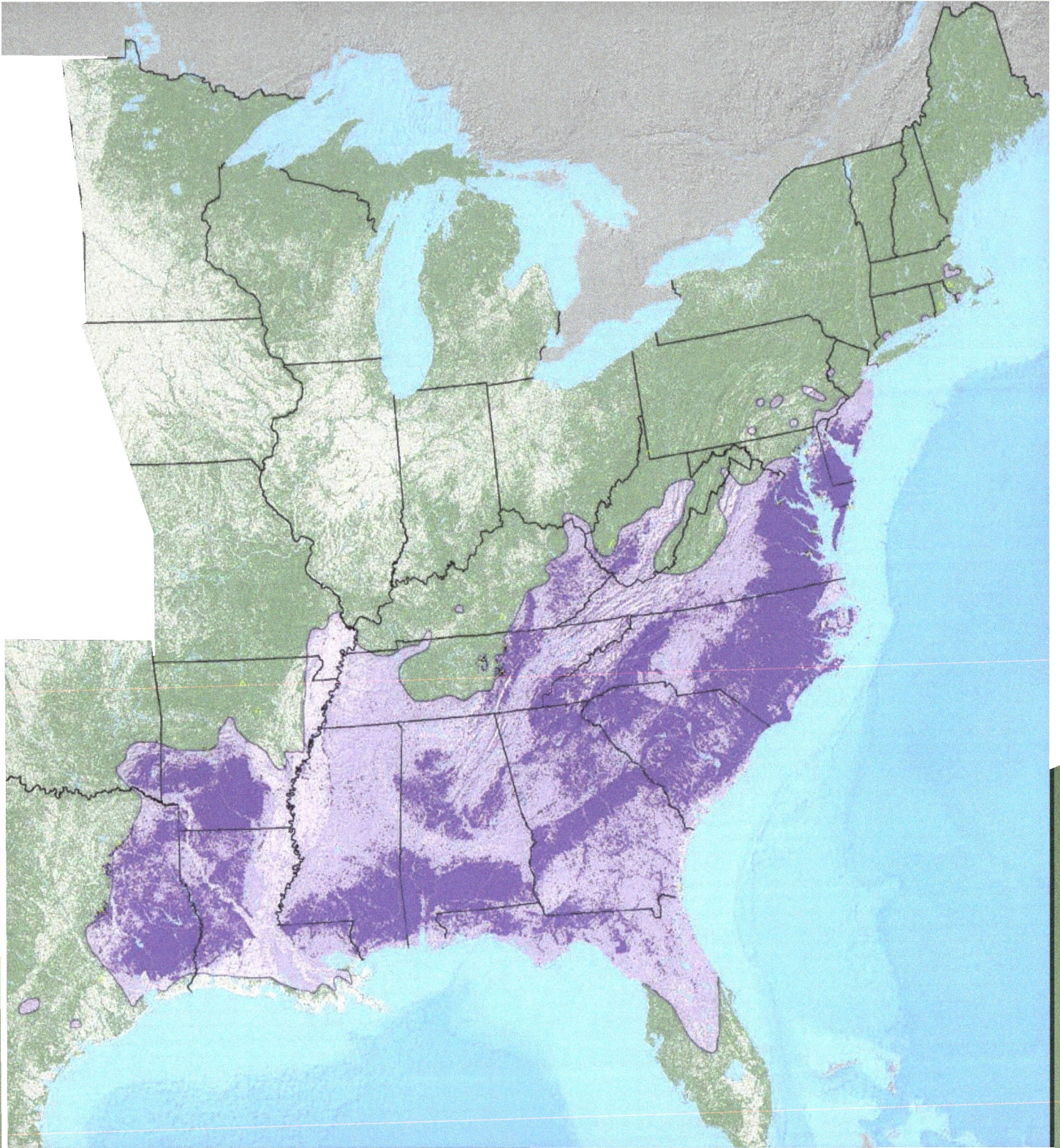

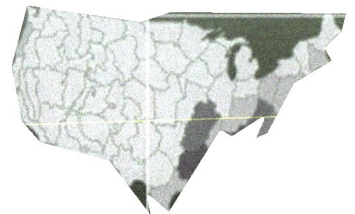

National Individual Tree Species Atlas

National Individual Tree Species Atlas

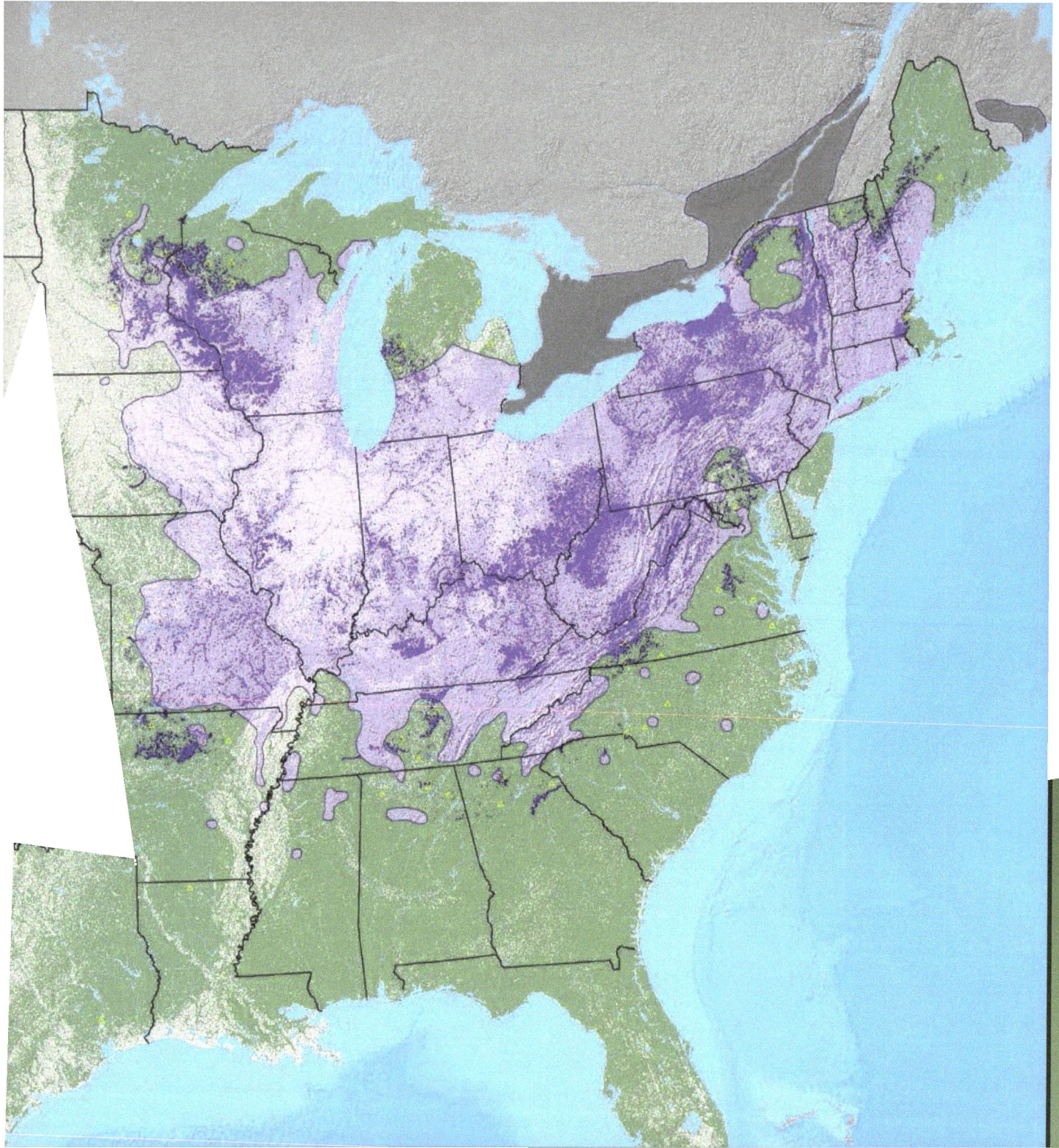

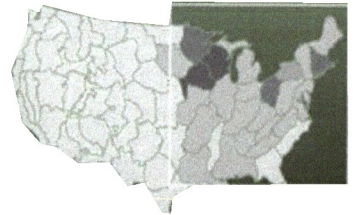

National Individual Tree Species Atlas

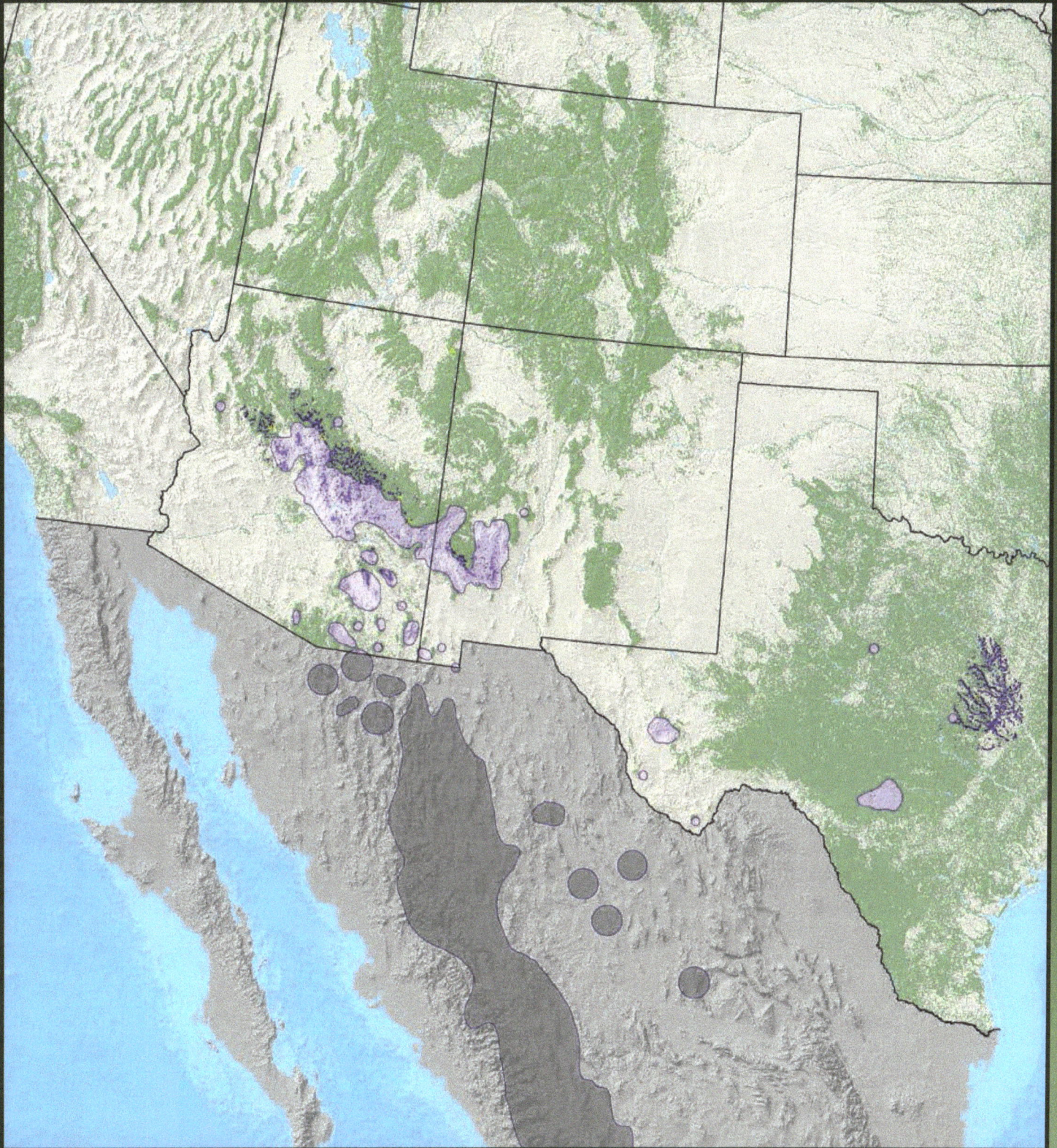

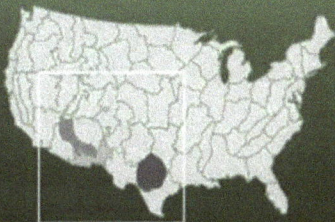

Arizona walnut

Arizona walnut
Juglans major (Torr.) Heller
Juglandaceae – Walnut Family
FIA Code **606** – Plants Symbol **JUMA**

National Individual Tree Species Atlas

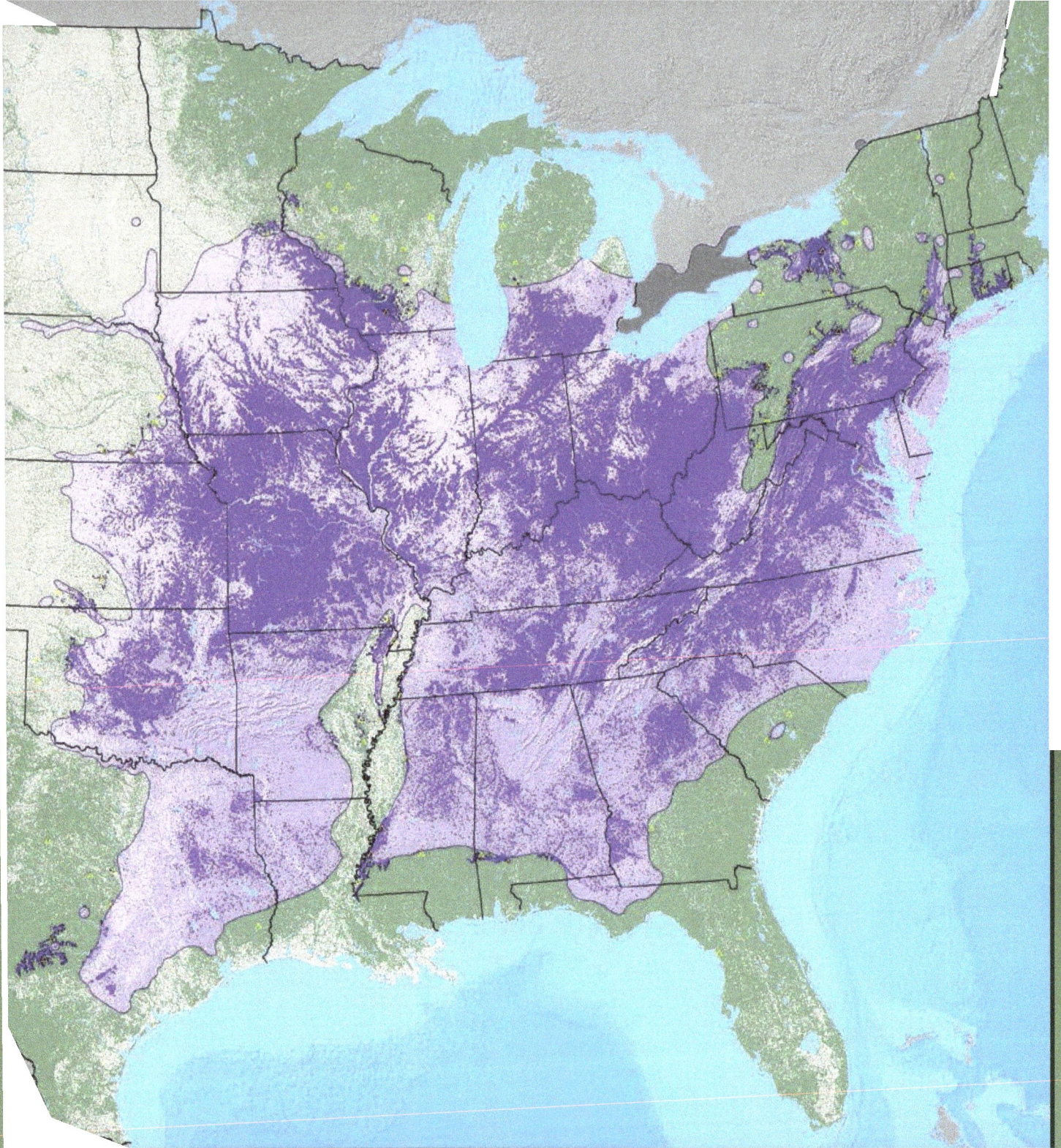

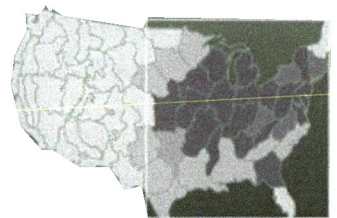

National Individual Tree Species Atlas

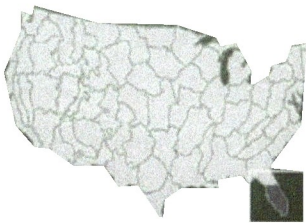
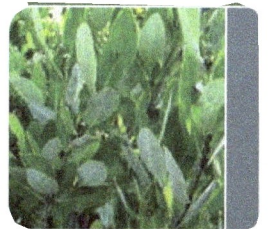

National Individual Tree Species Atlas

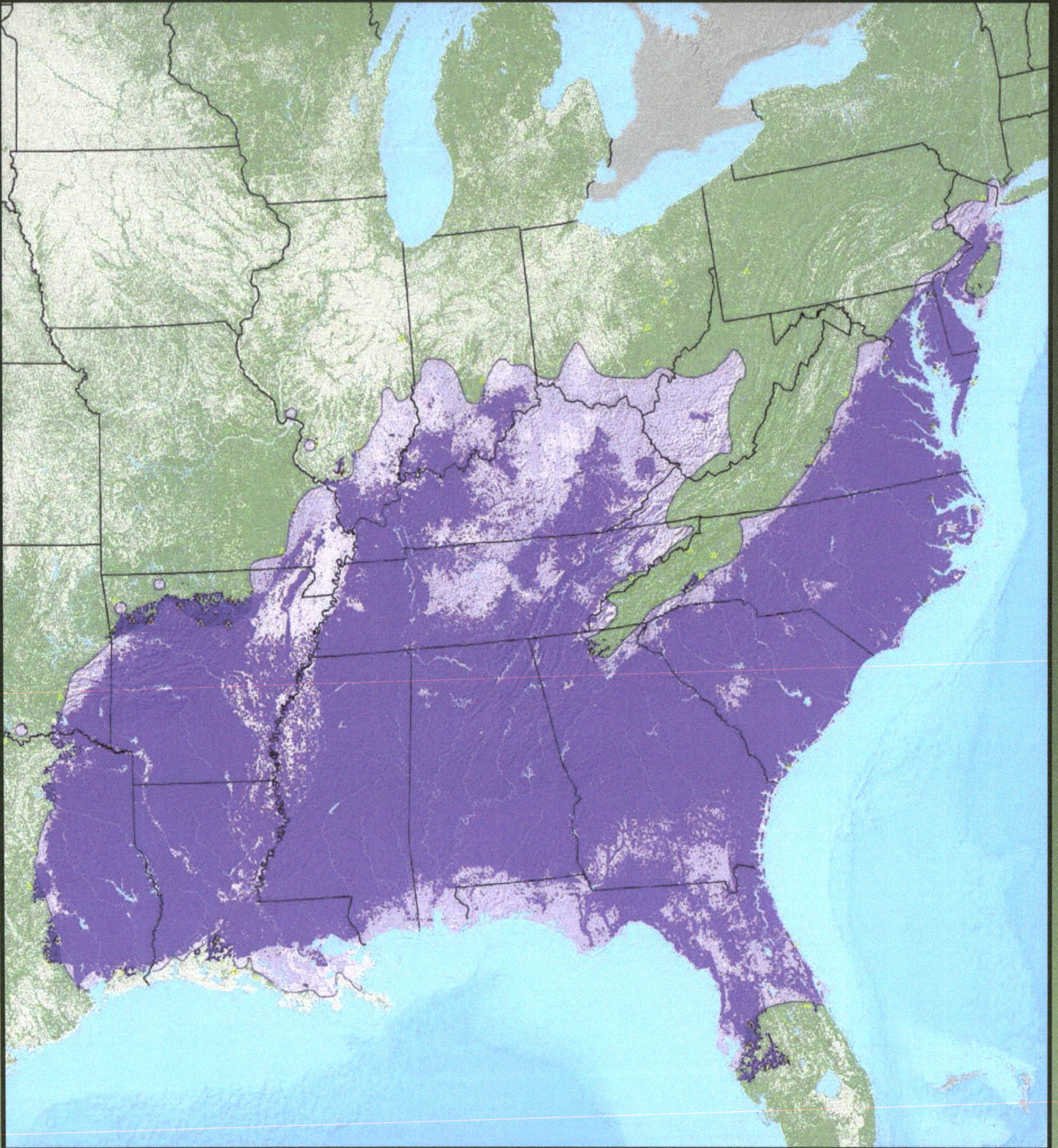

sweetgum

sweetgum
Liquidambar styraciflua L.
Hamamelidaceae – Witch-hazel Family
FIA Code **611** – Plants Symbol **LIST2**

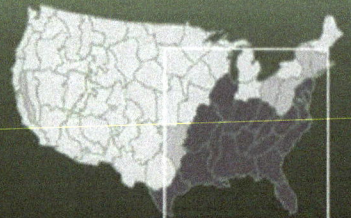

National Individual Tree Species Atlas

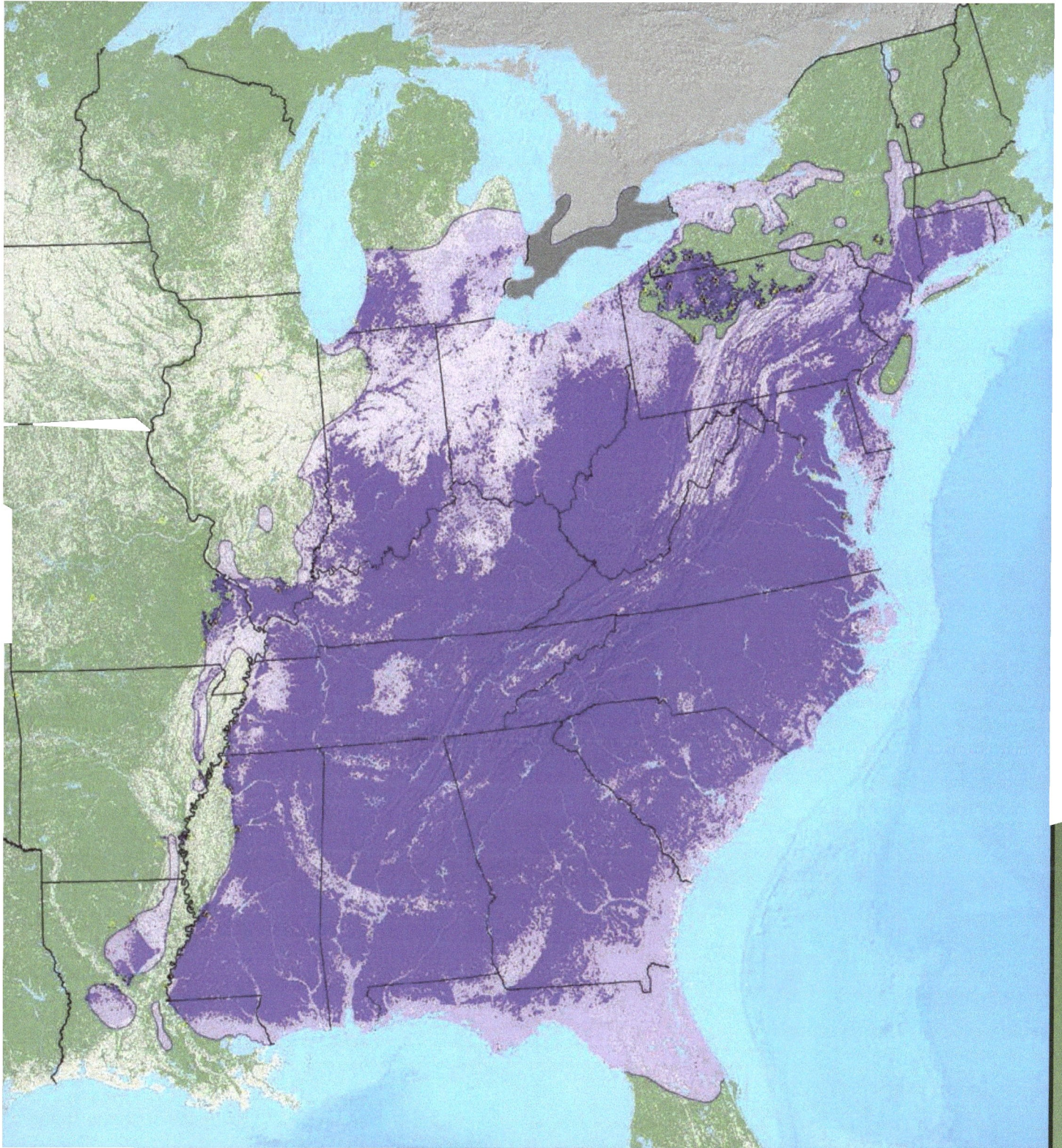

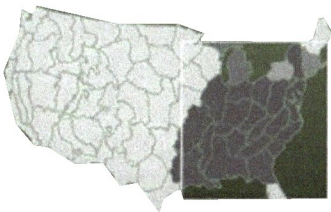

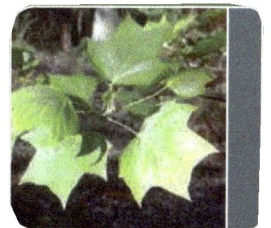

National Individual Tree Species Atlas

National Individual Tree Species Atlas

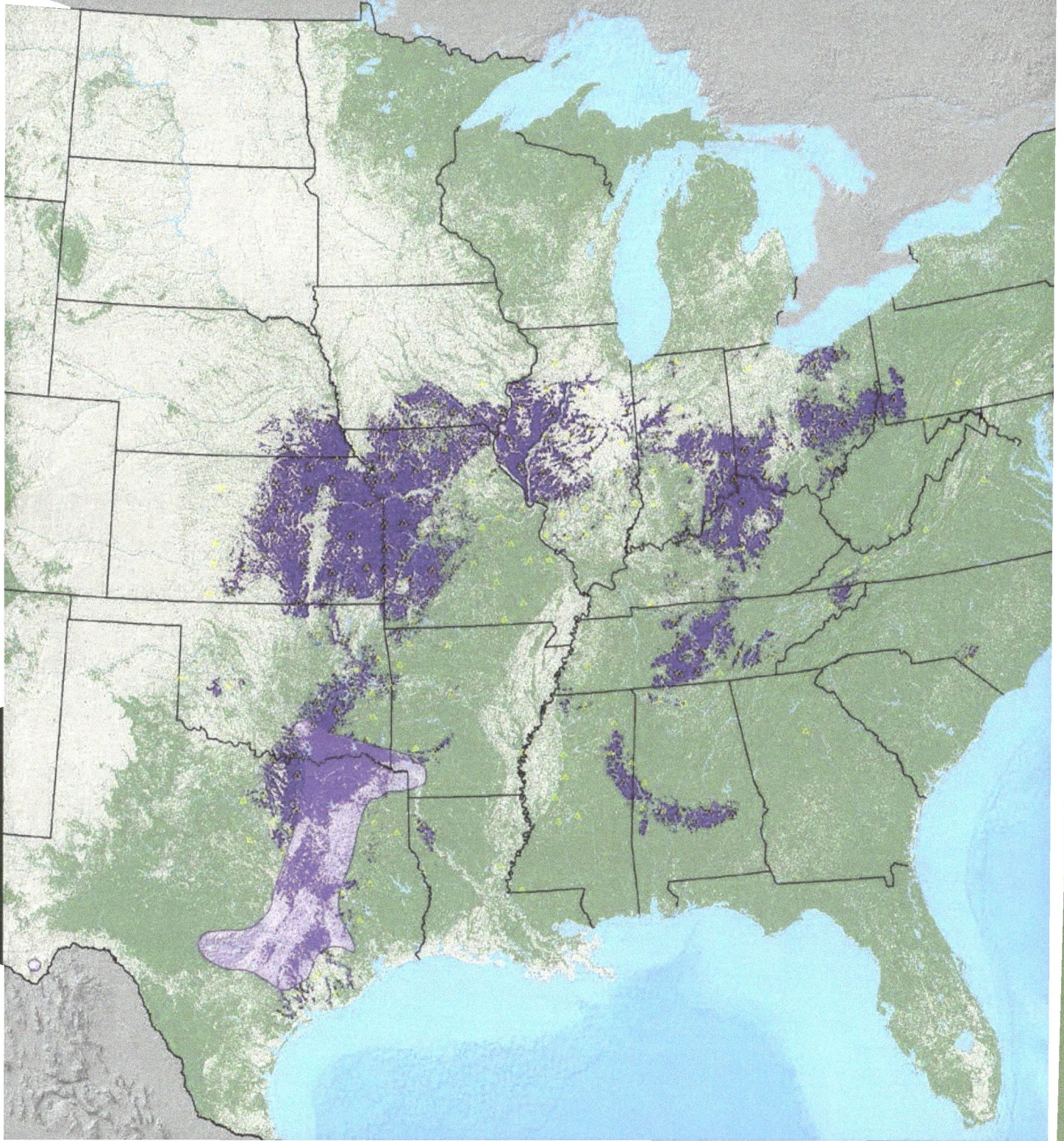

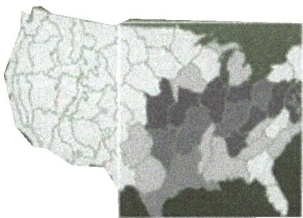

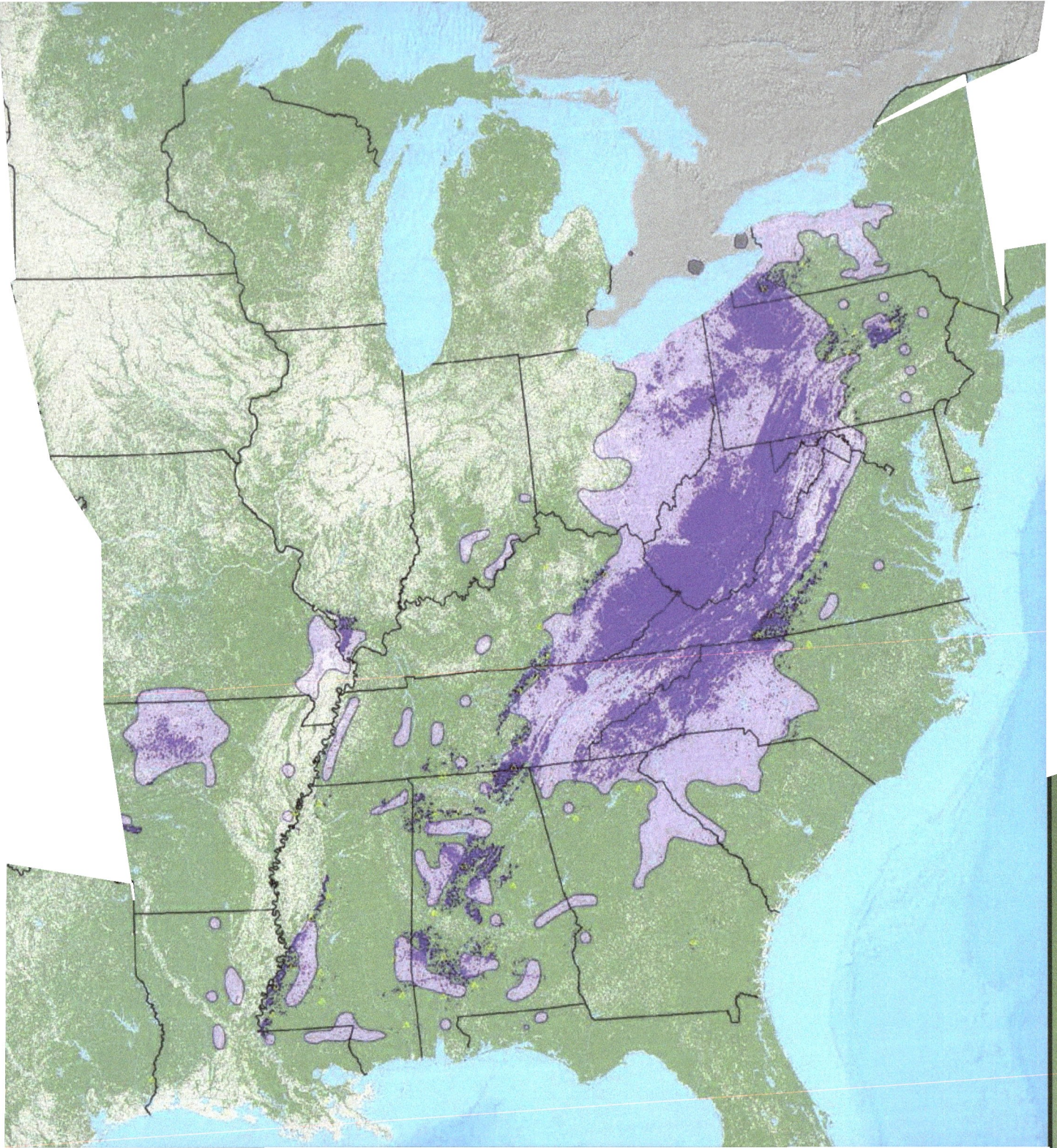

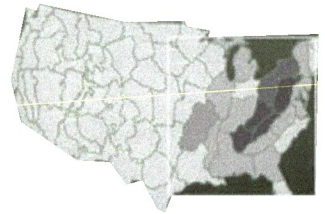

National Individual Tree Species Atlas

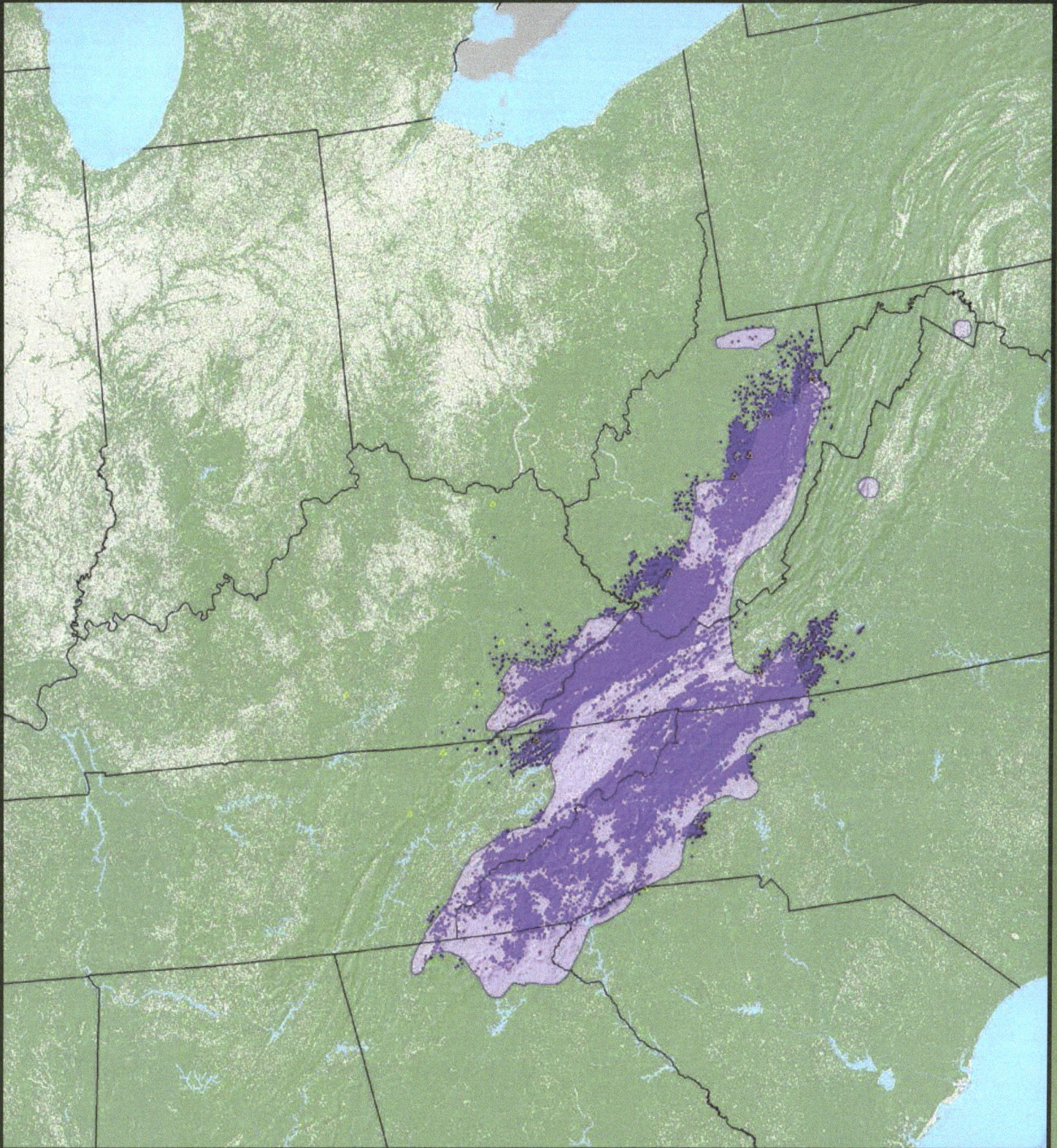

mountain magnolia

mountain magnolia
Magnolia fraseri Walter
Magnoliaceae – Magnolia Family
FIA Code **655** – Plants Symbol **MAFR**

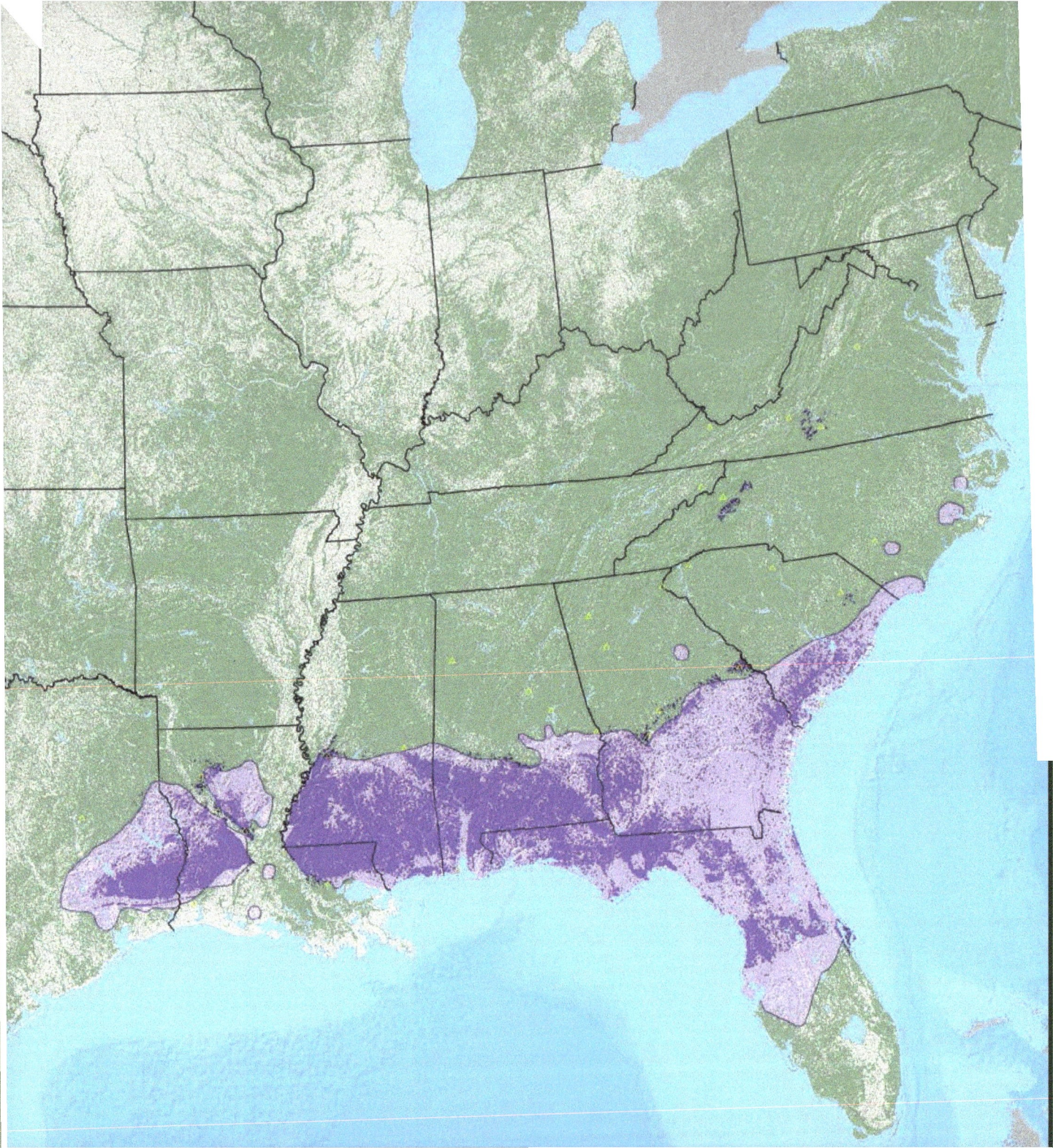

National Individual Tree Species Atlas

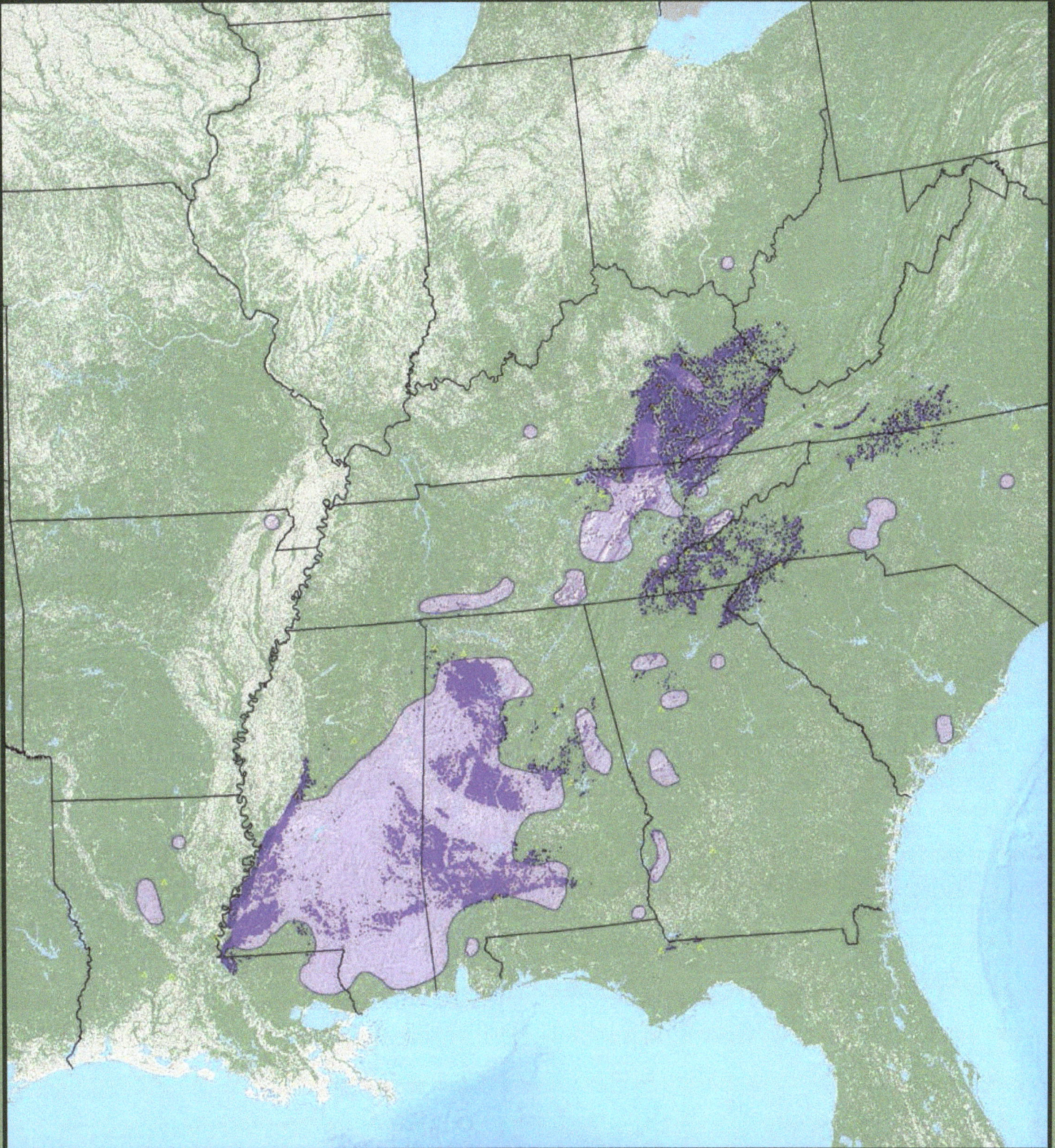

bigleaf magnolia
Magnolia macrophylla Michx.
Magnoliaceae – Magnolia Family
FIA Code **654** – Plants Symbol **MAMA2**

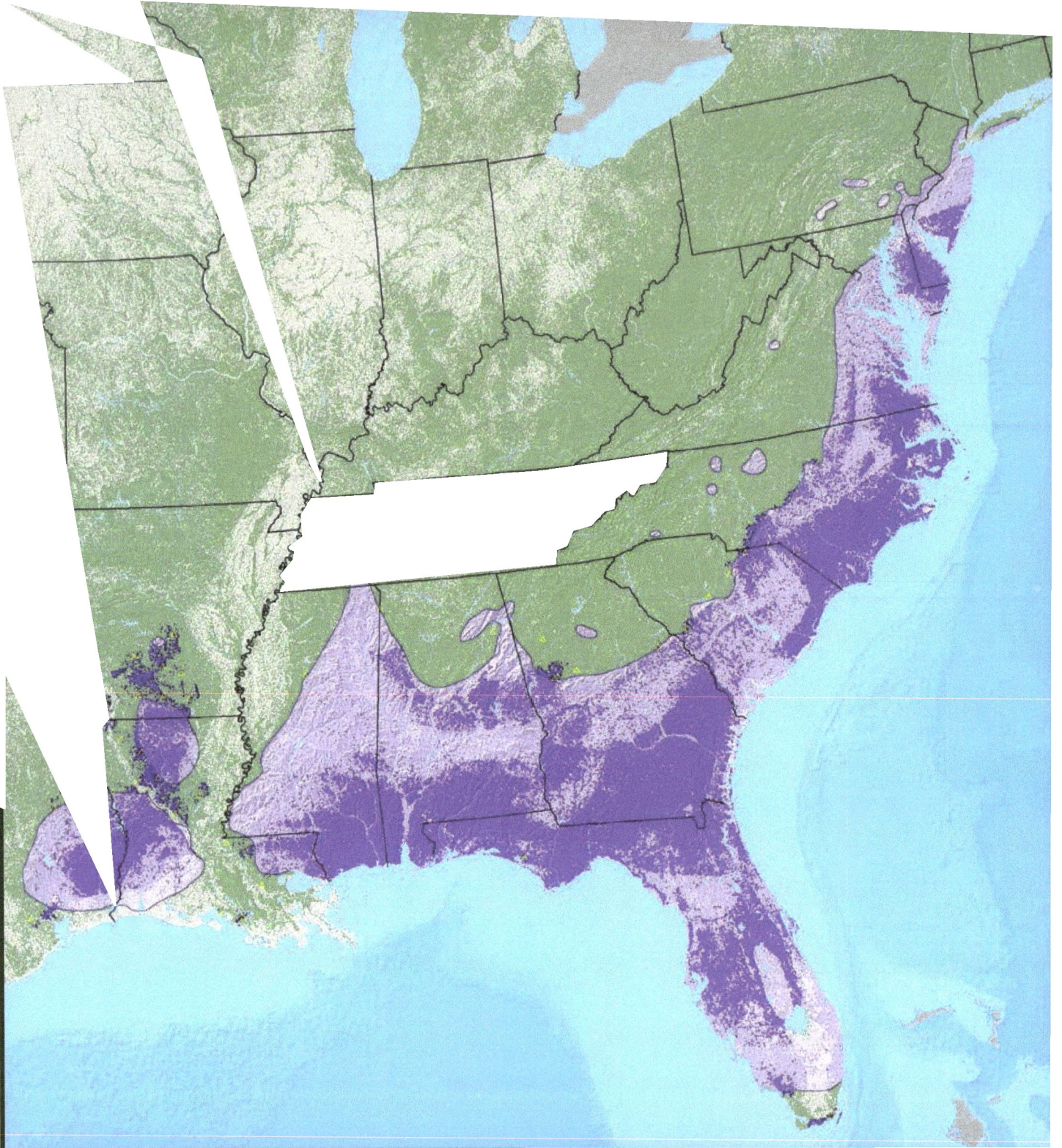

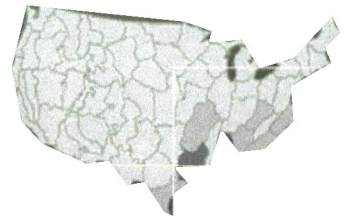

National Individual Tree Species Atlas

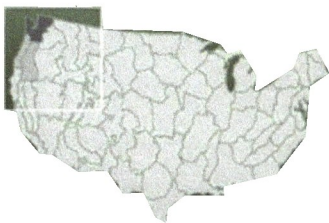

National Individual Tree Species Atlas

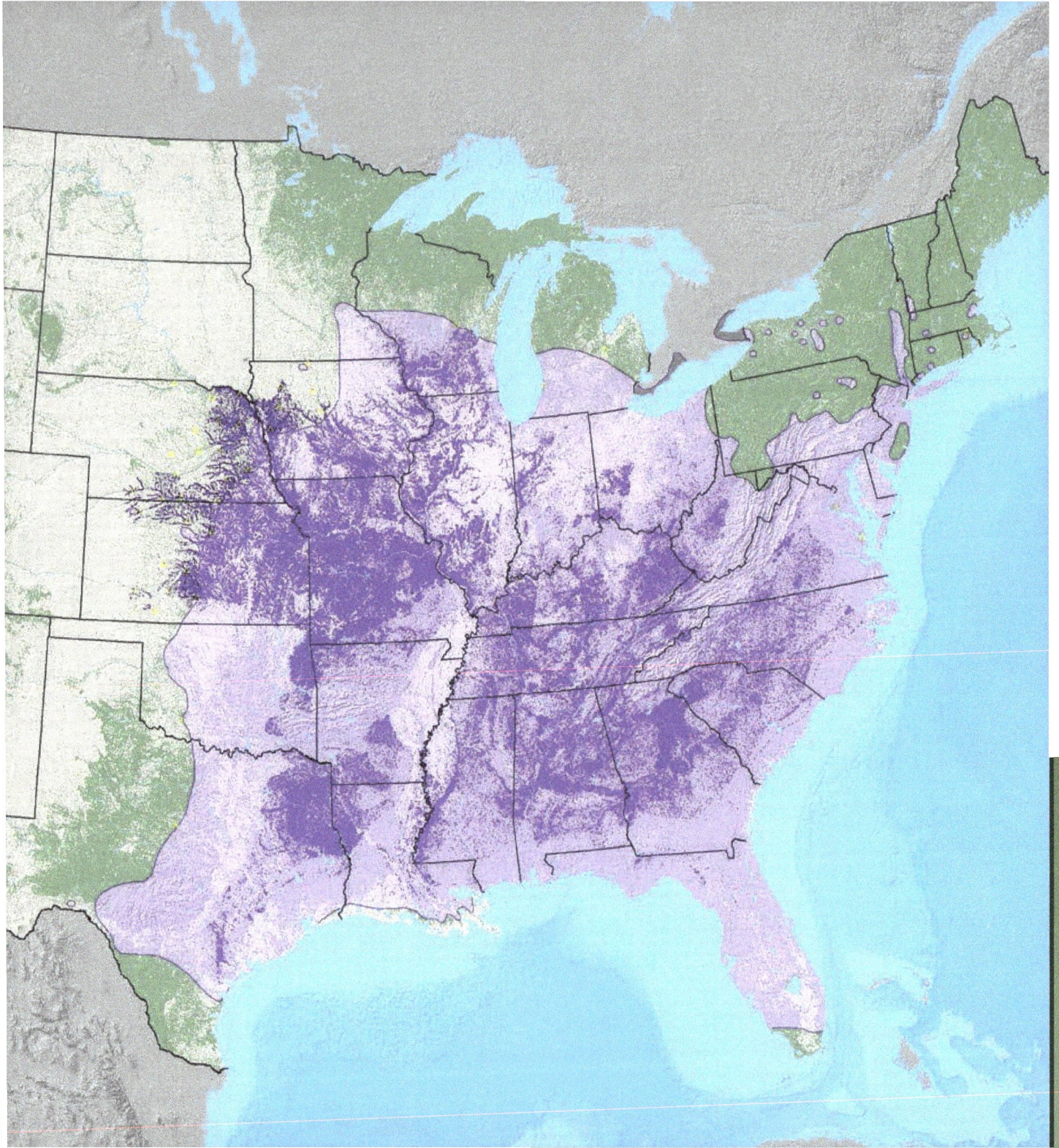

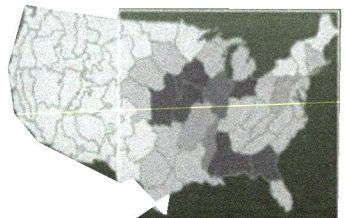

National Individual Tree Species Atlas

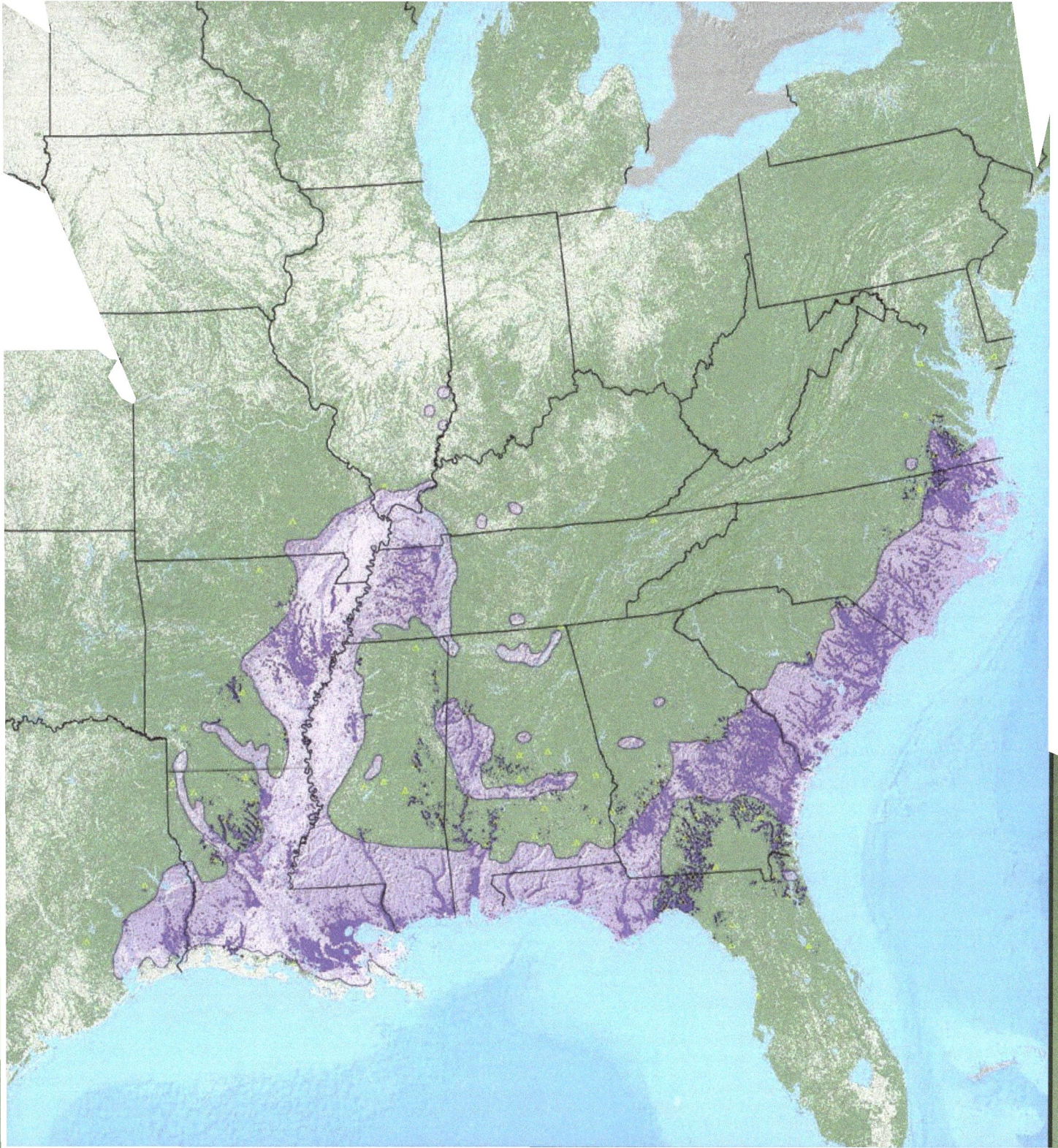

swamp tupelo

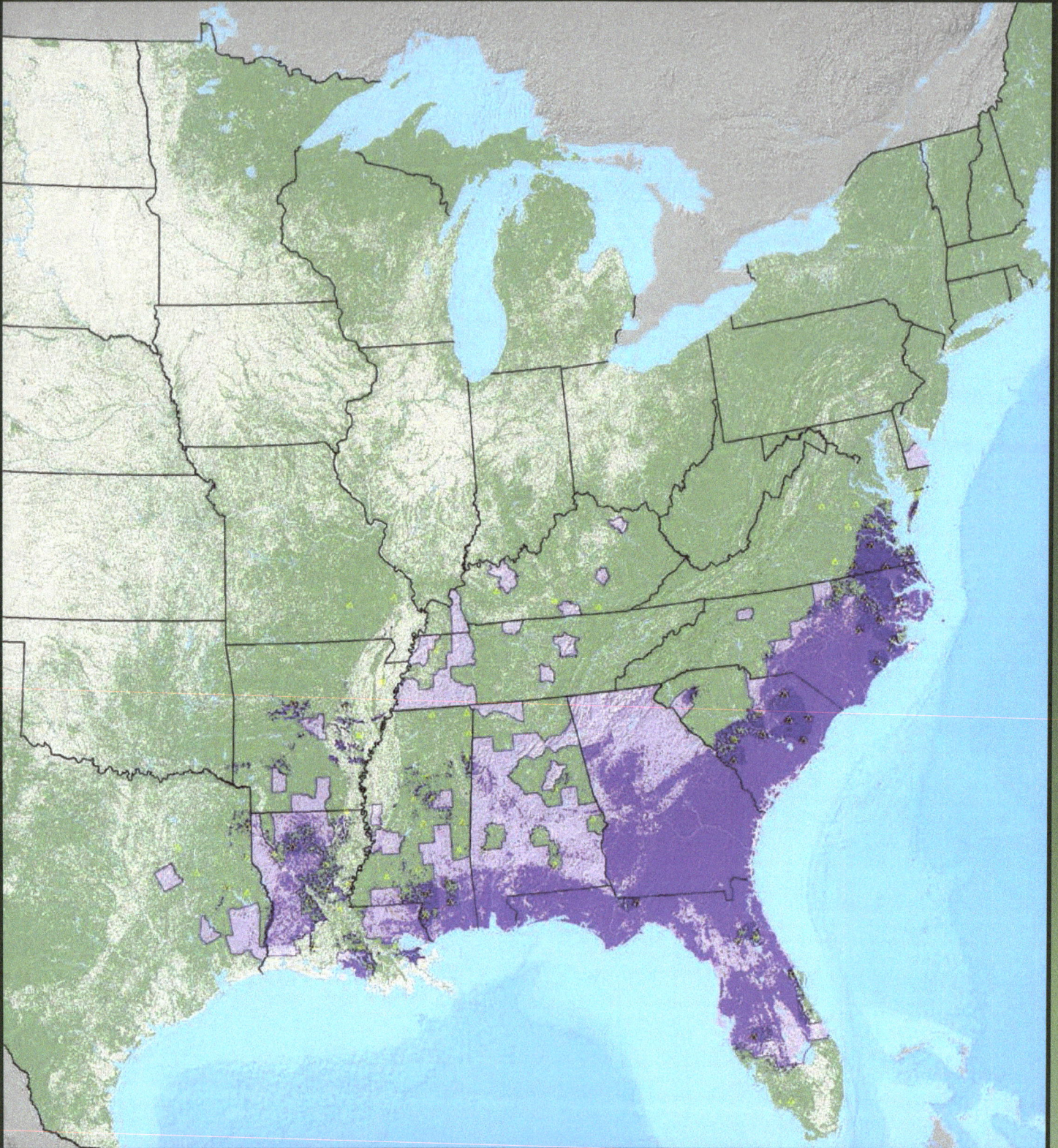

swamp tupelo

Nyssa biflora Walter

Cornaceae – Dogwood Family

FIA Code **694** – Plants Symbol **NYBI**

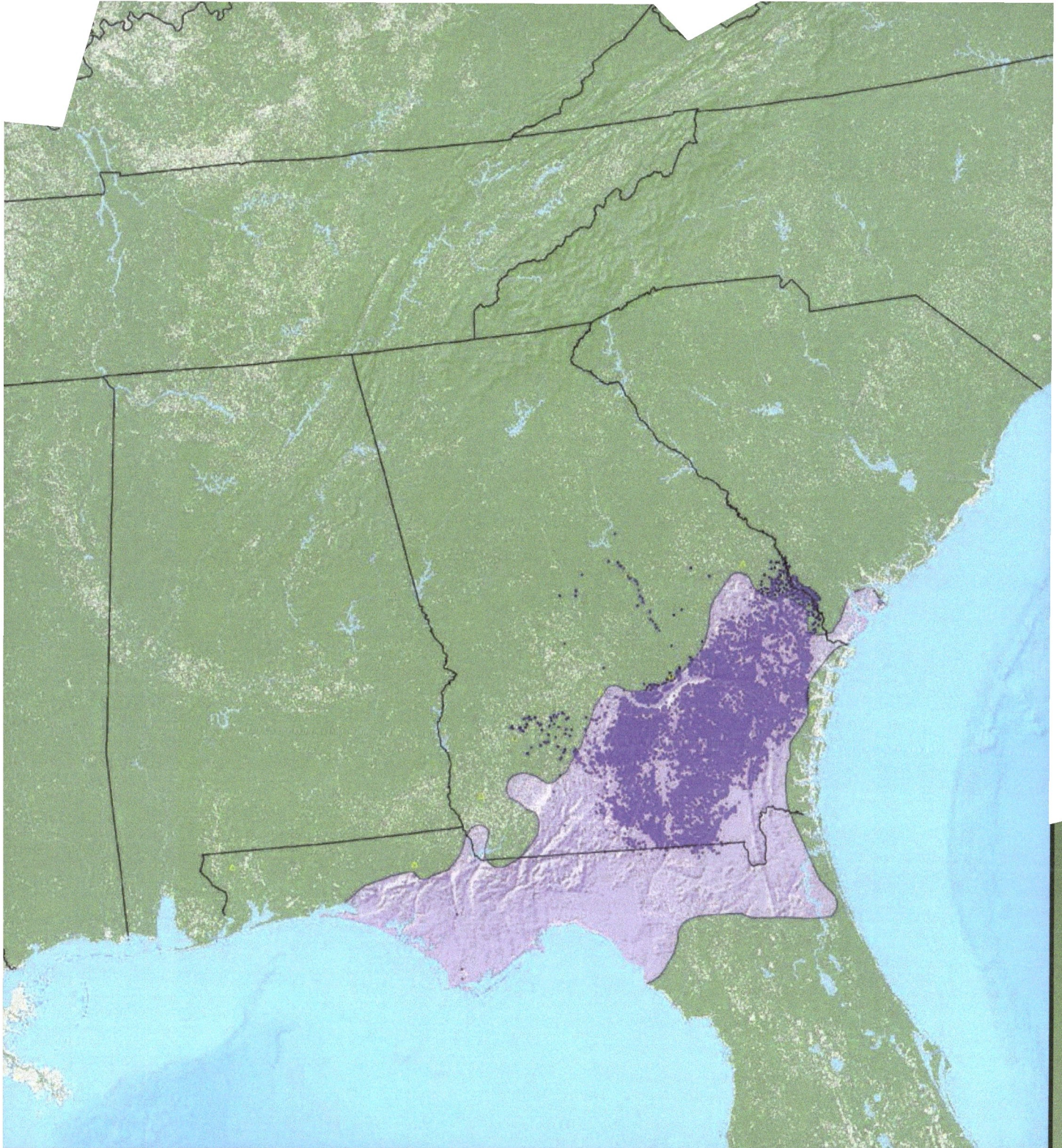
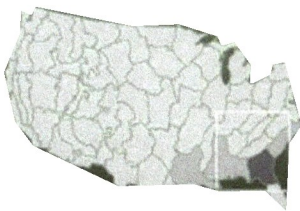

National Individual Tree Species Atlas

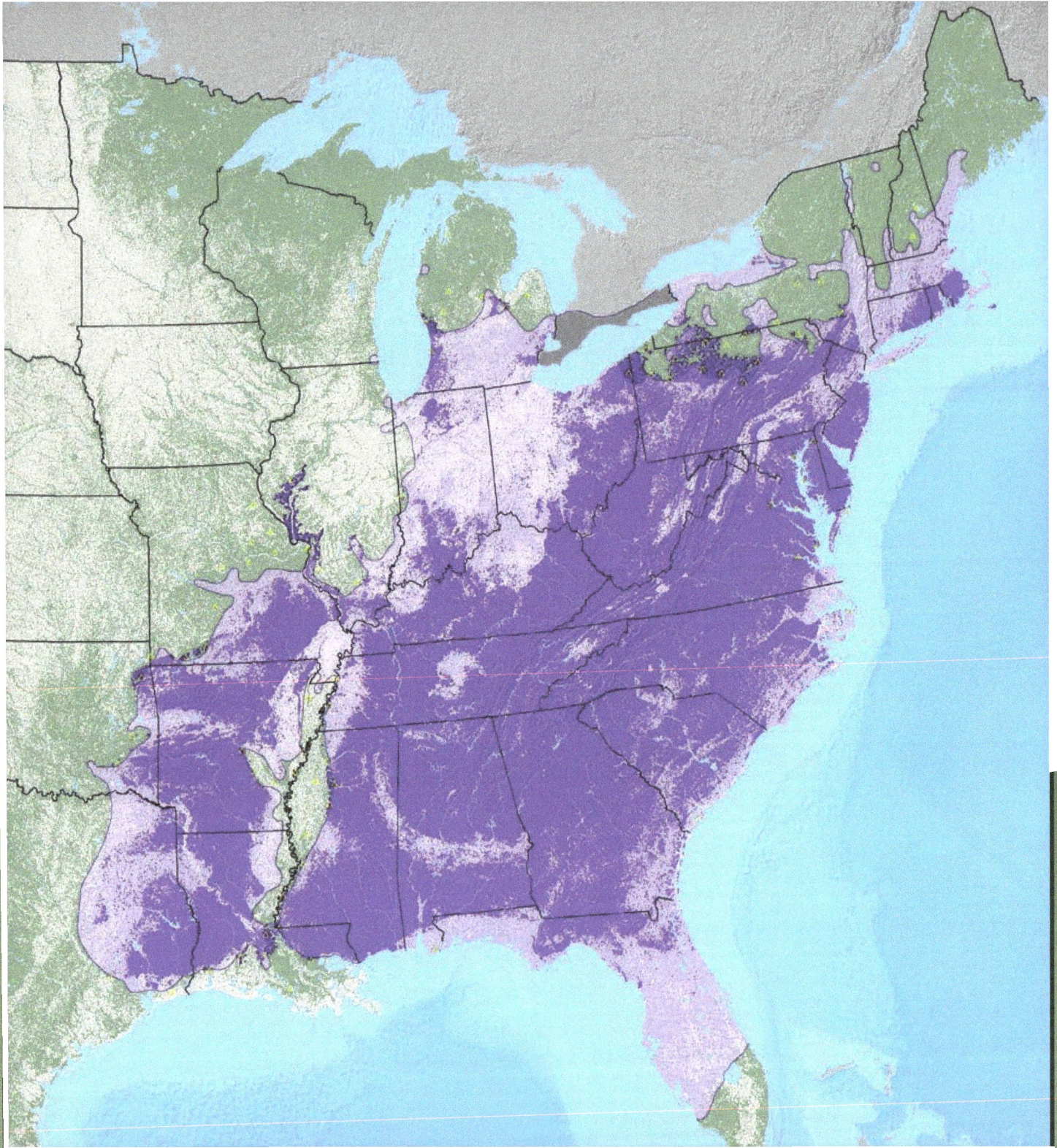

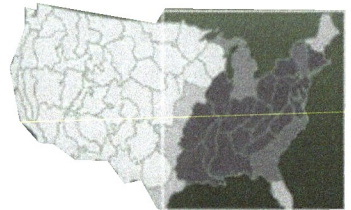

National Individual Tree Species Atlas

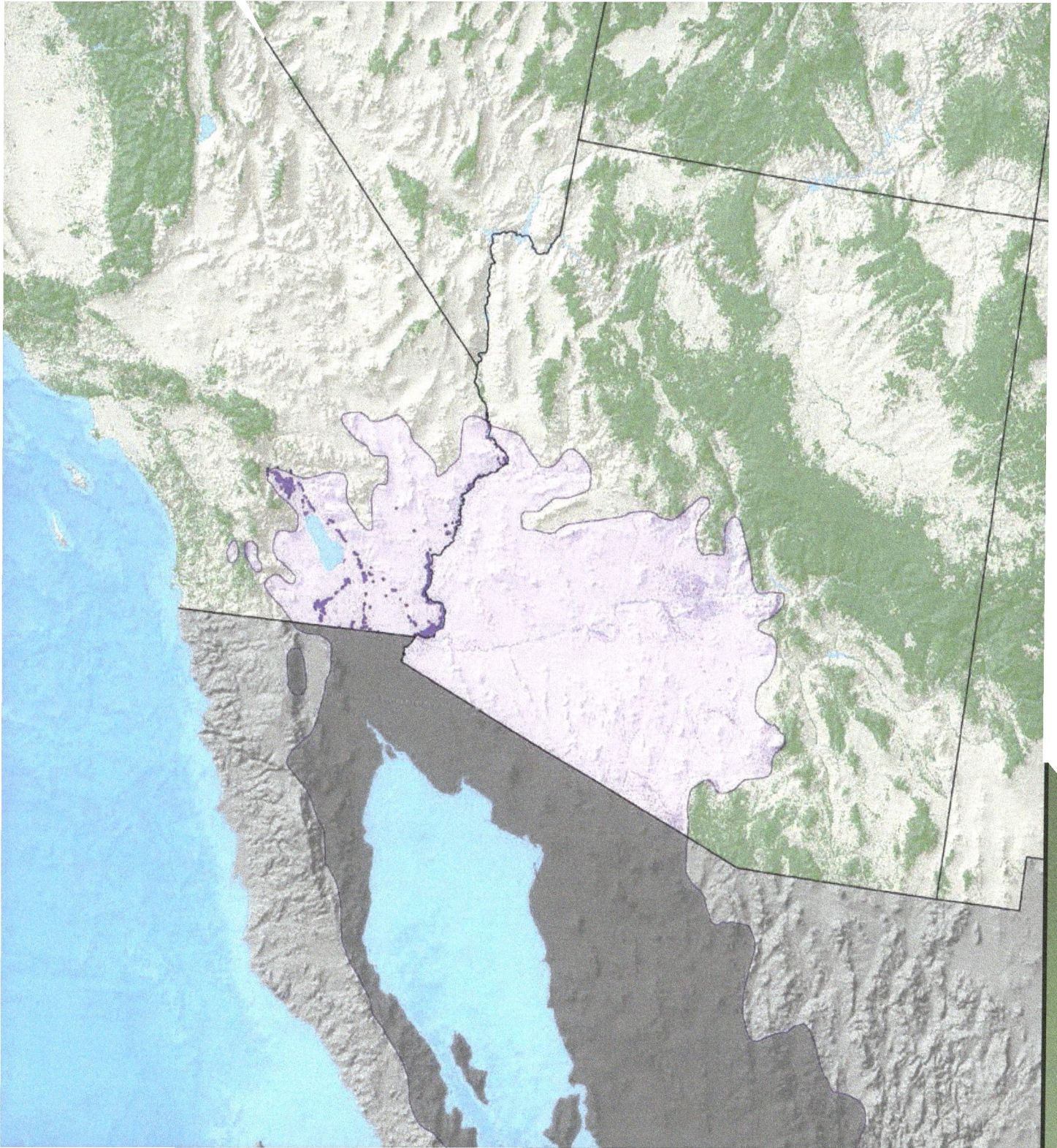
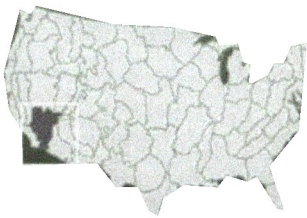

National Individual Tree Species Atlas

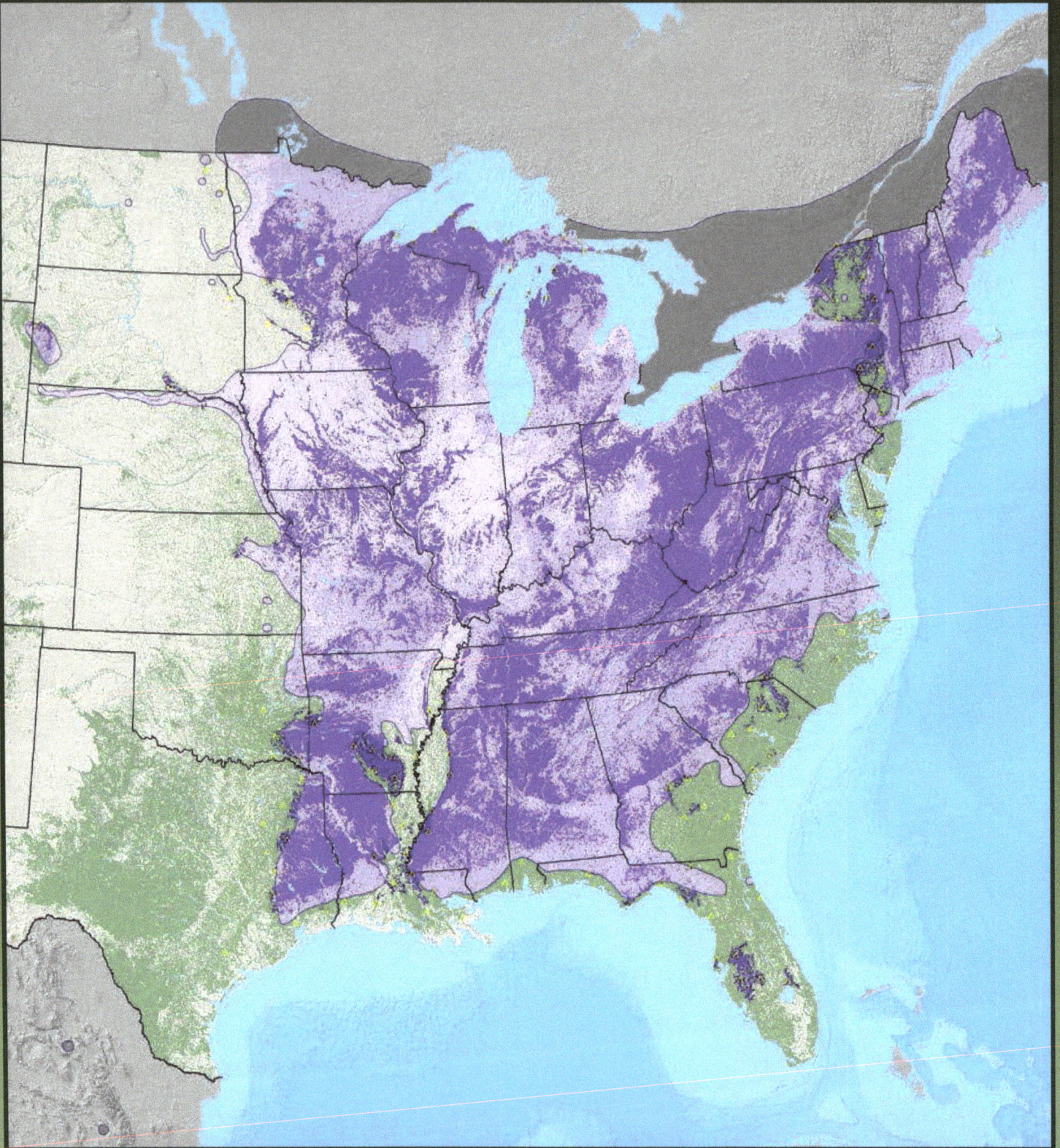

eastern hophornbeam

Ostrya virginiana (Mill.) K. Koch

Betulaceae – Birch Family

FIA Code **701** – Plants Symbol **OSVI**

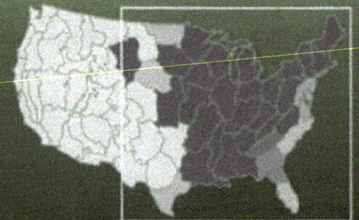

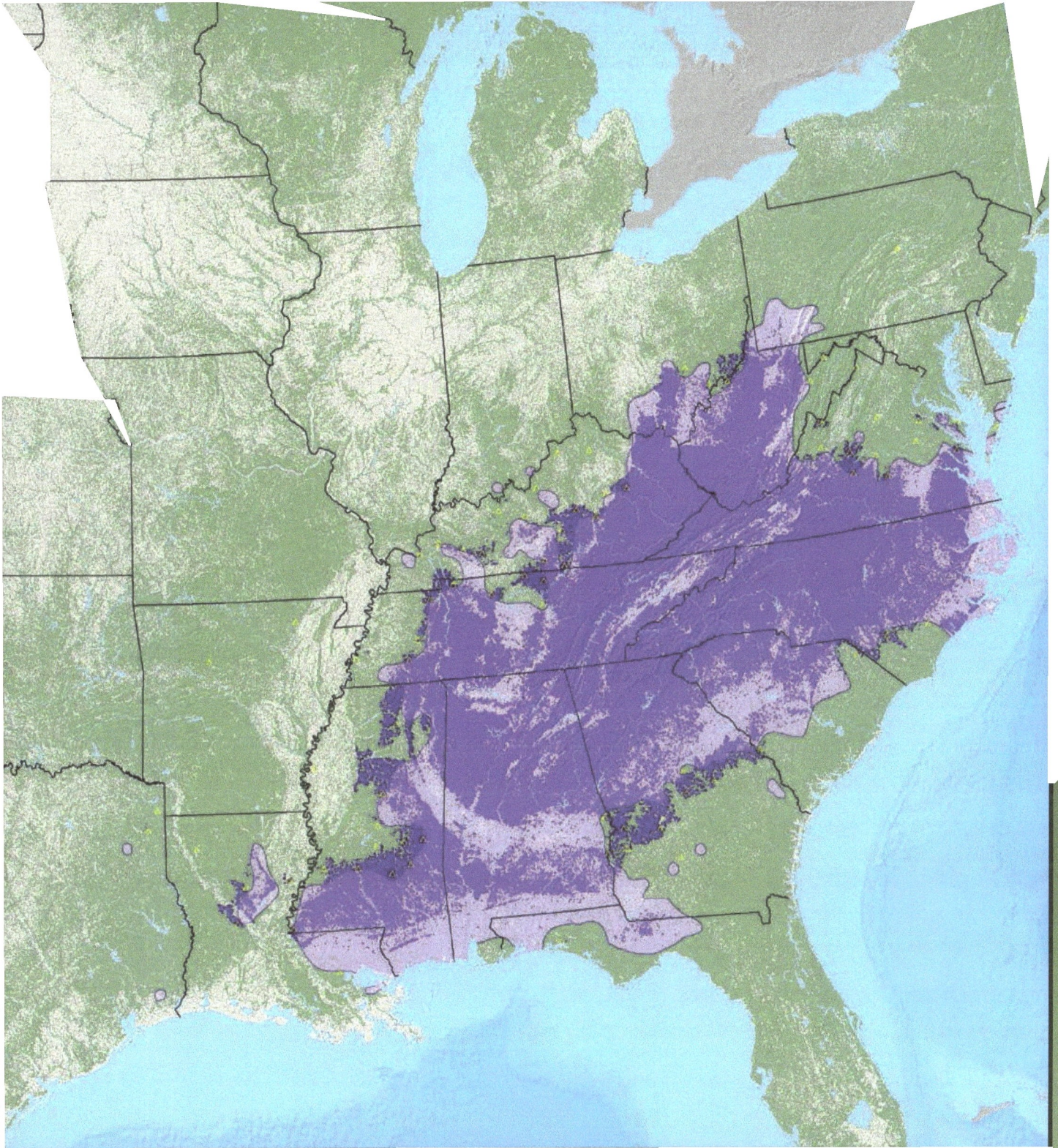

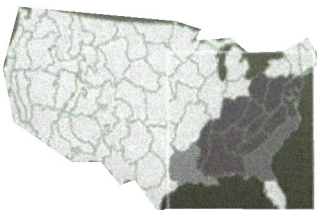

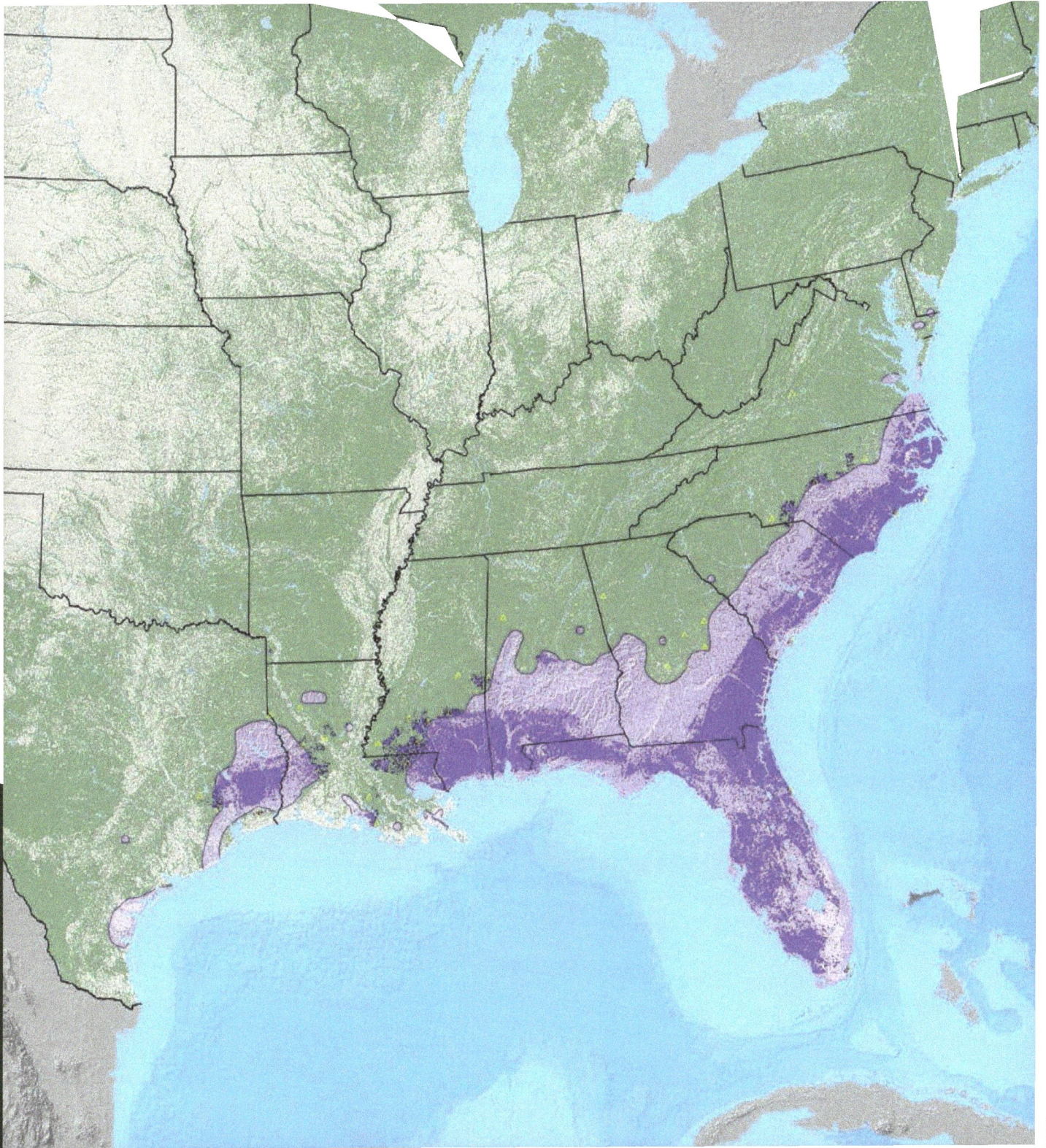

National Individual Tree Species Atlas

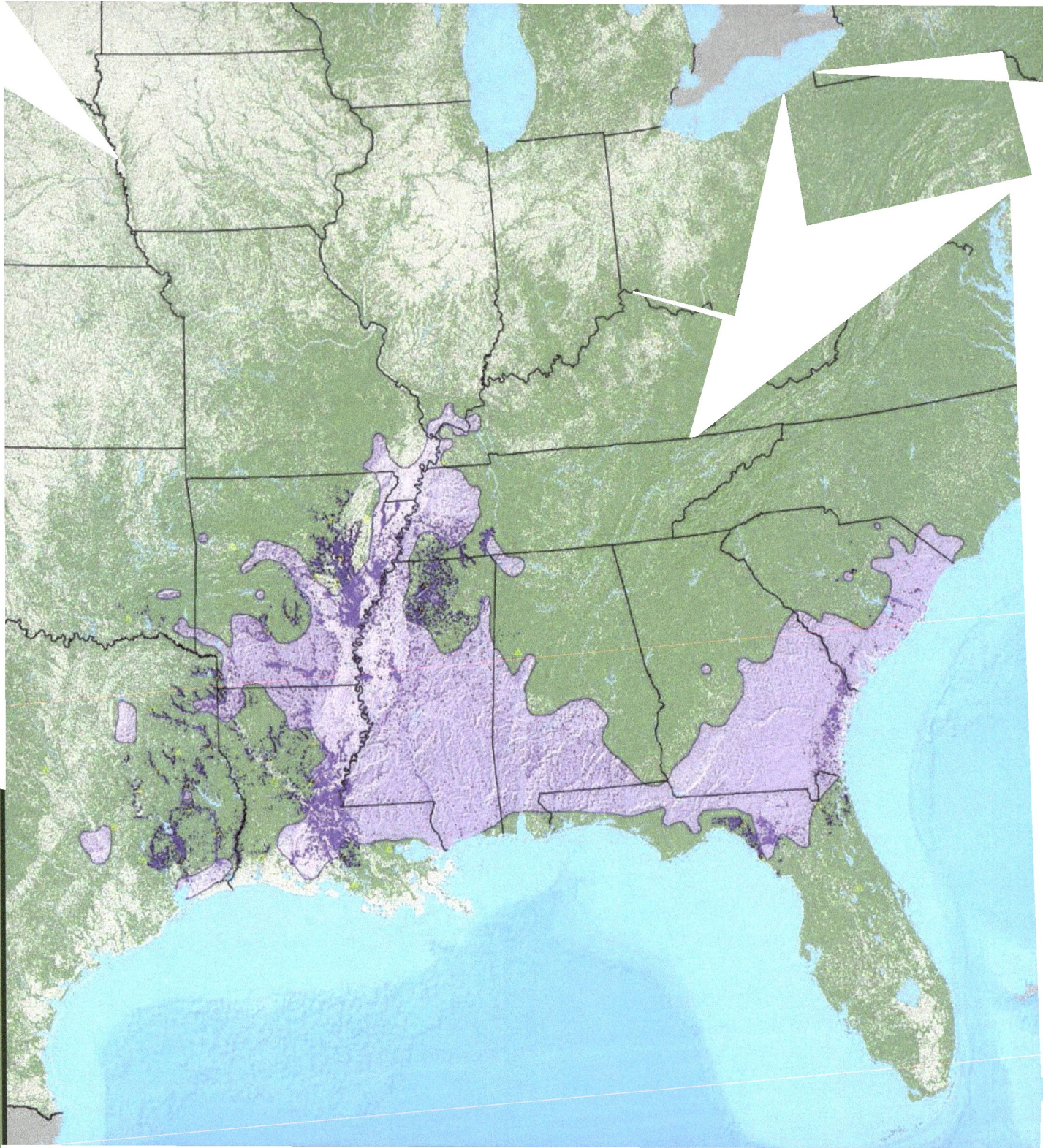

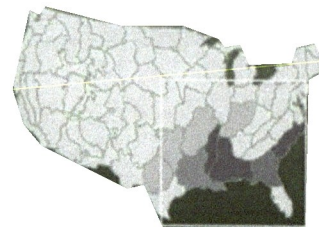

National Individual Tree Species Atlas

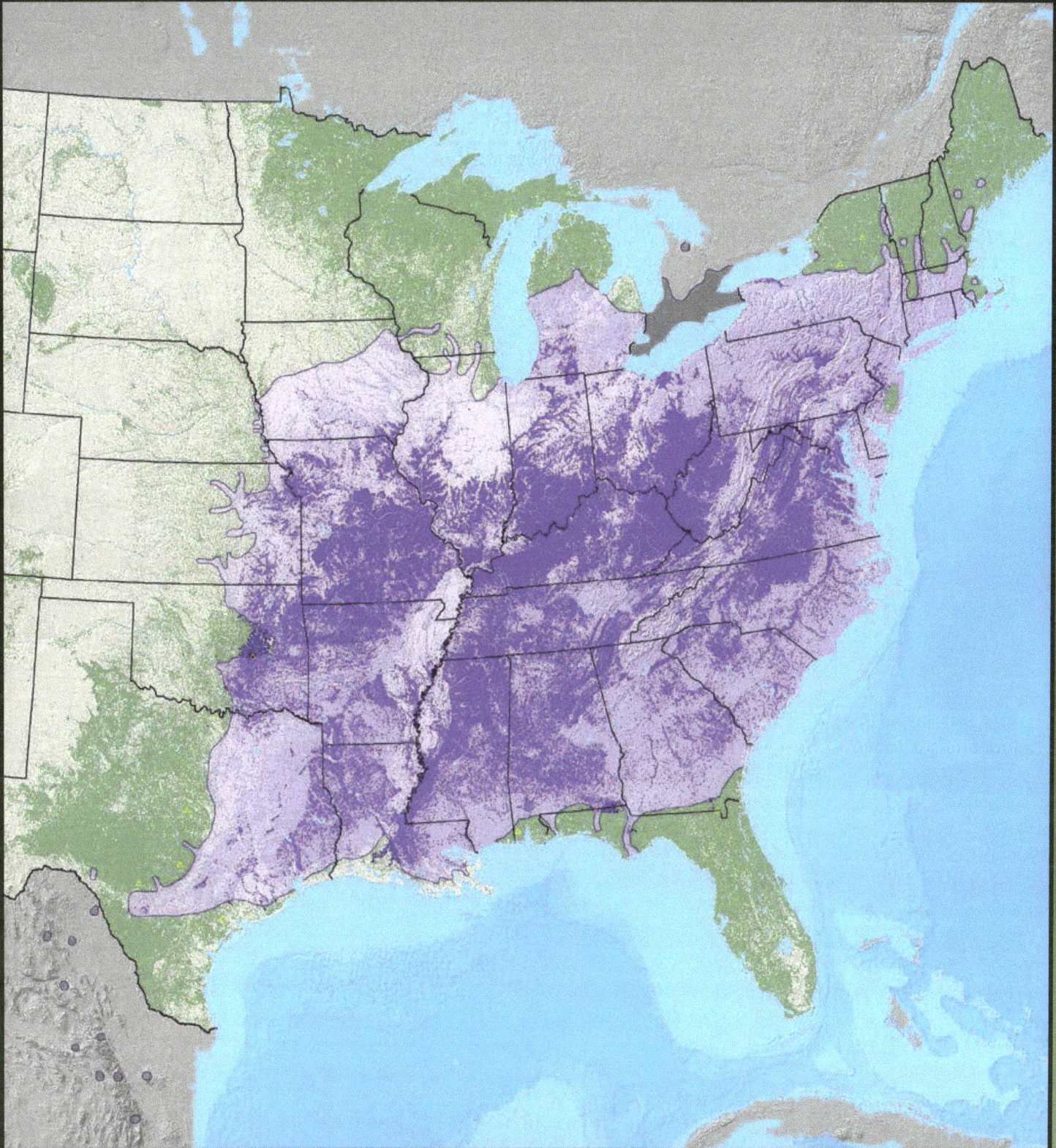

American sycamore

American sycamore
Platanus occidentalis L.
Platanaceae – Plane-tree Family
FIA Code **731** – Plants Symbol **PLOC**

National Individual Tree Species Atlas

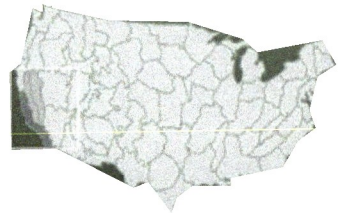

National Individual Tree Species Atlas

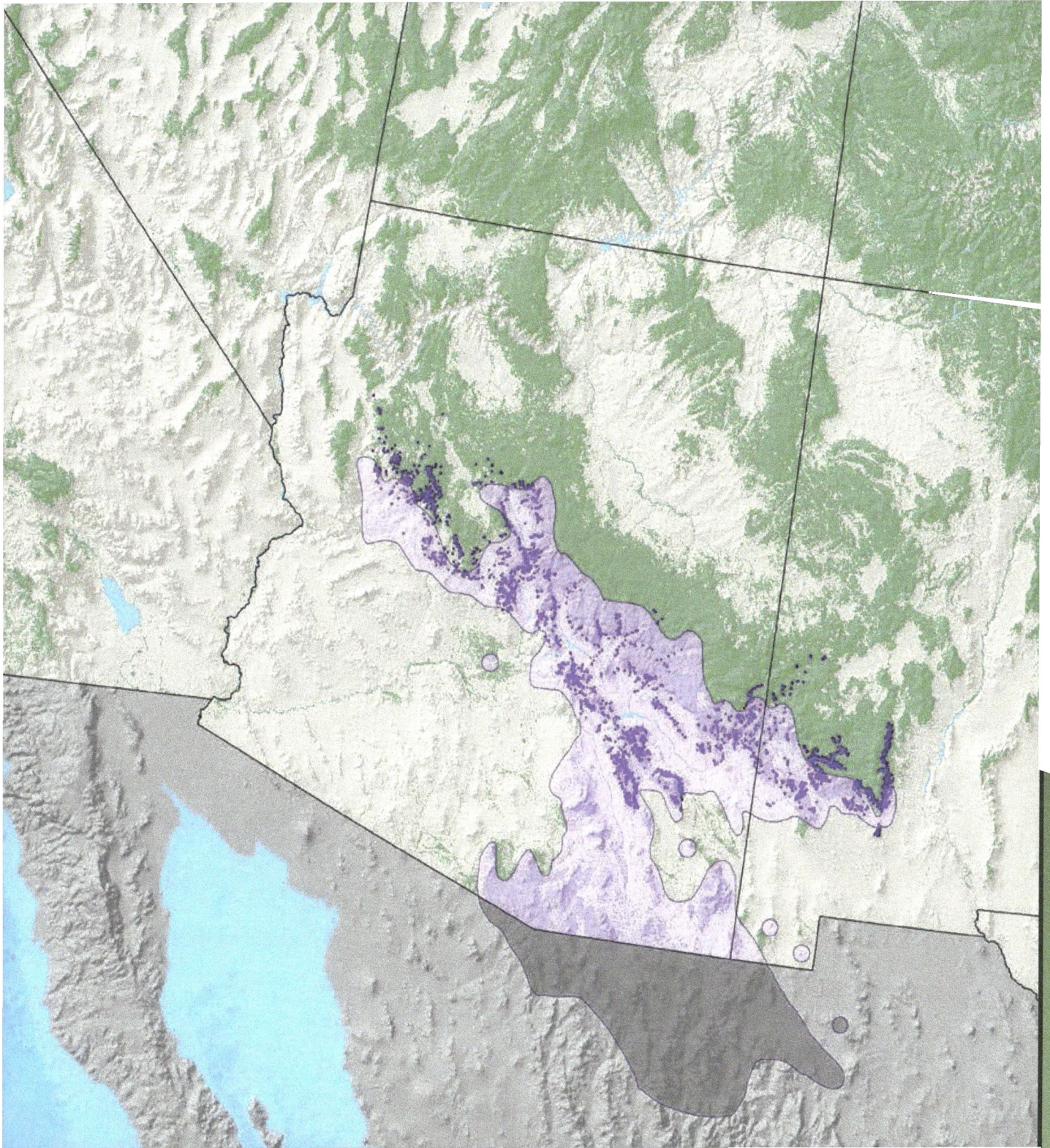
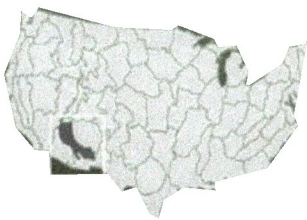
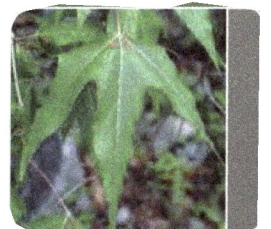

National Individual Tree Species Atlas

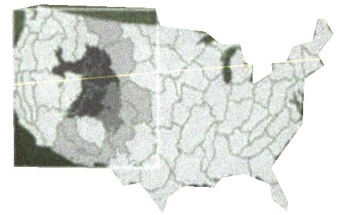

National Individual Tree Species Atlas

National Individual Tree Species Atlas

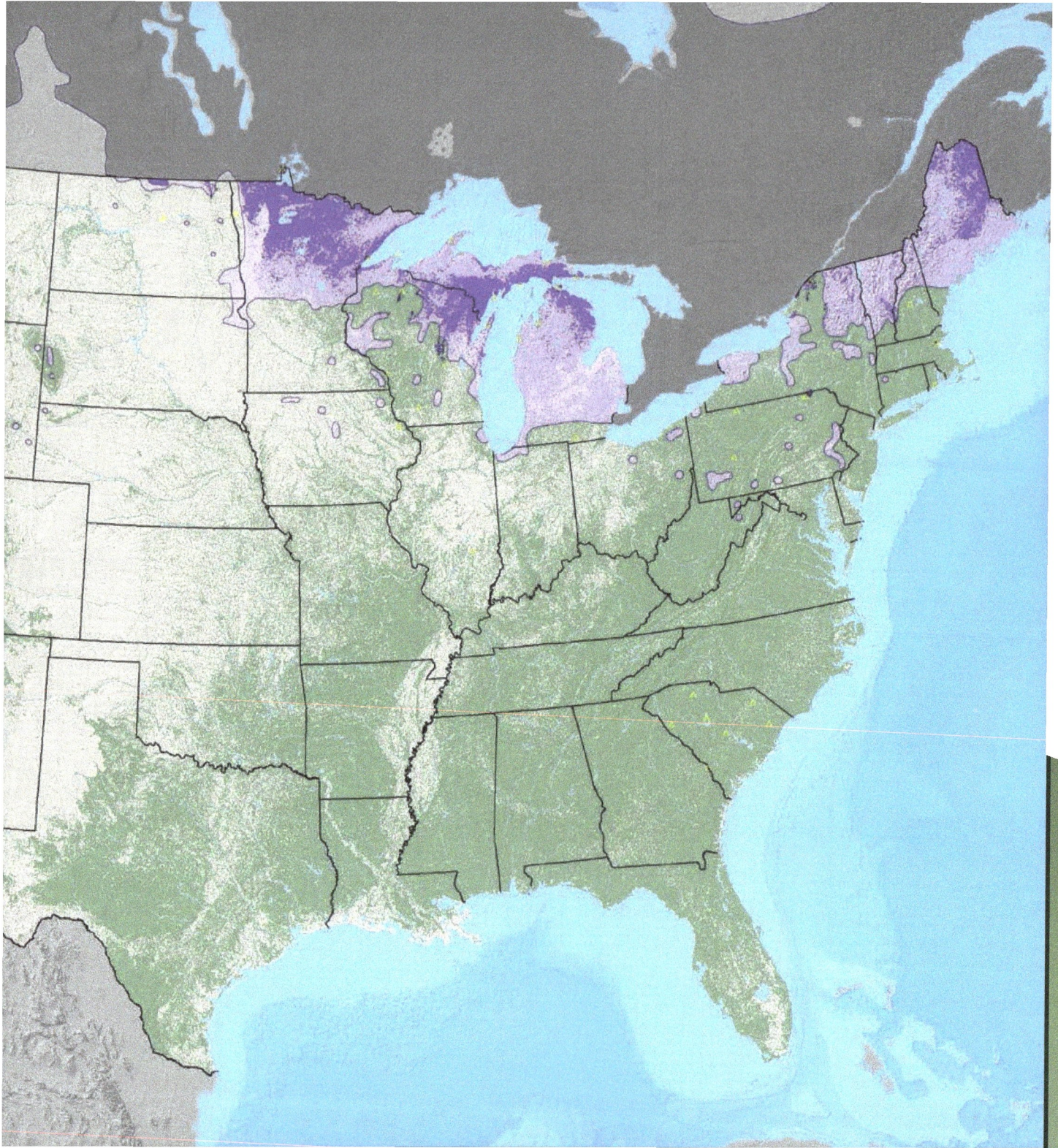

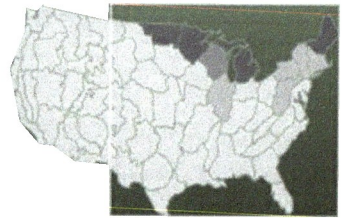

National Individual Tree Species Atlas

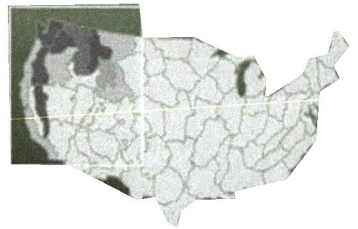

National Individual Tree Species Atlas

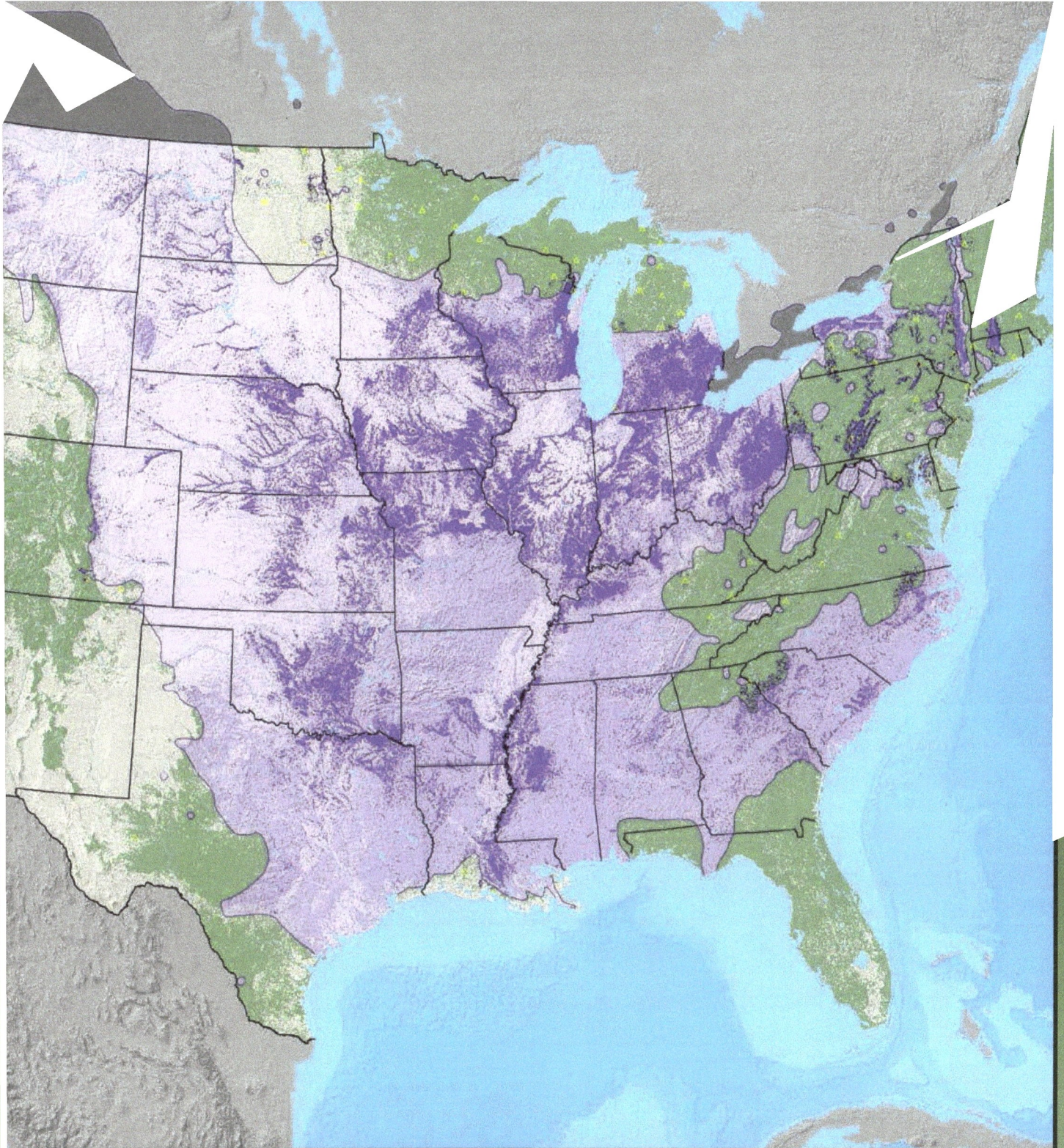

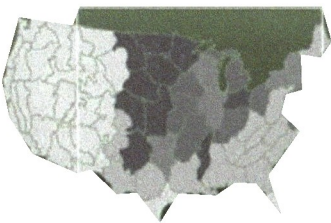

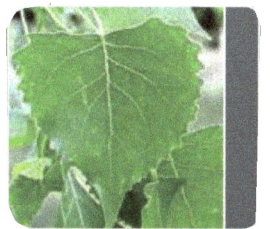

National Individual Tree Species Atlas

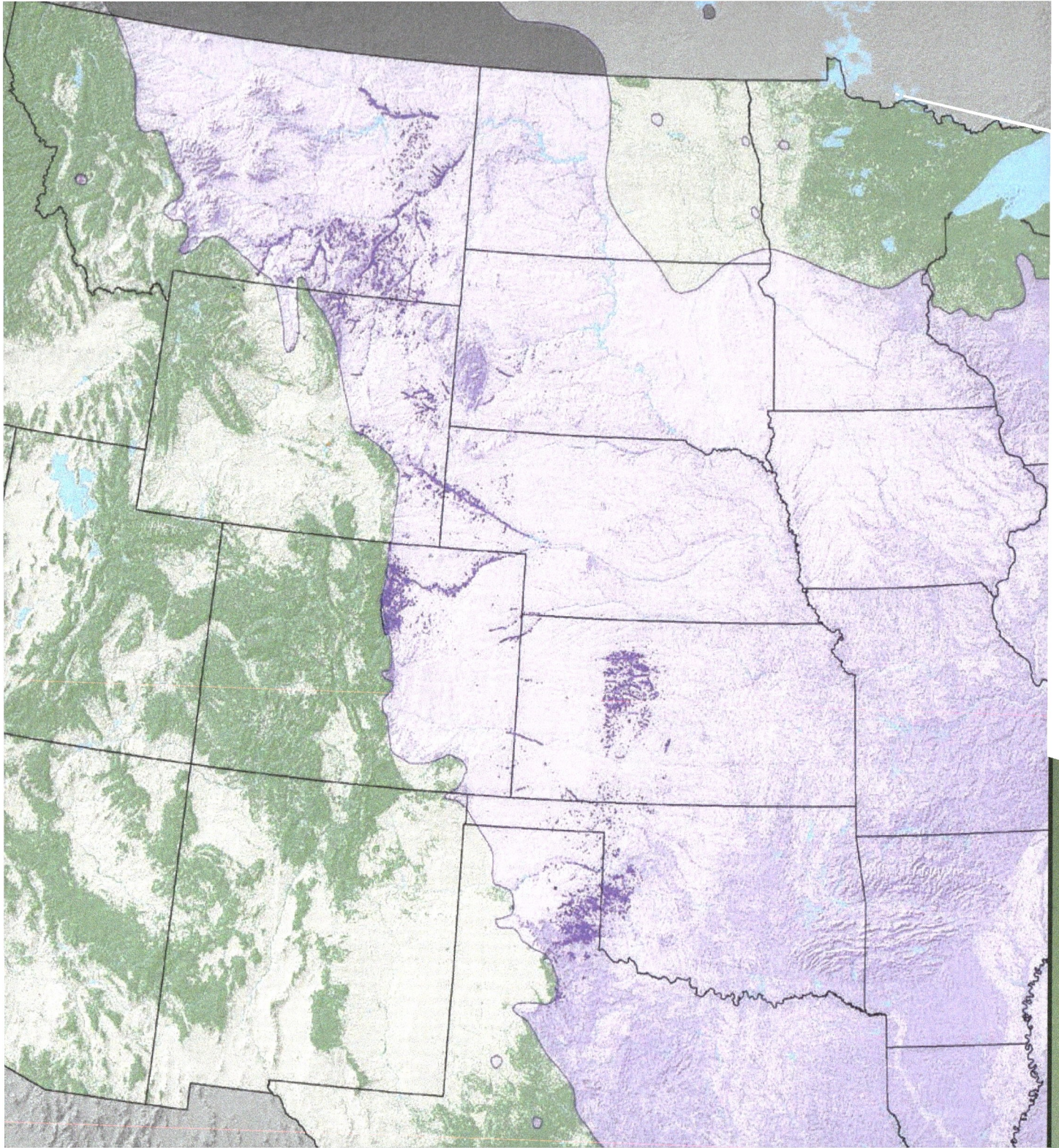

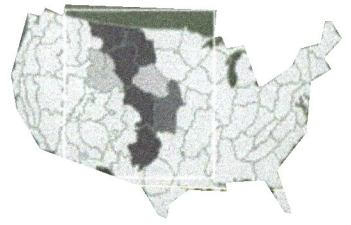

National Individual Tree Species Atlas

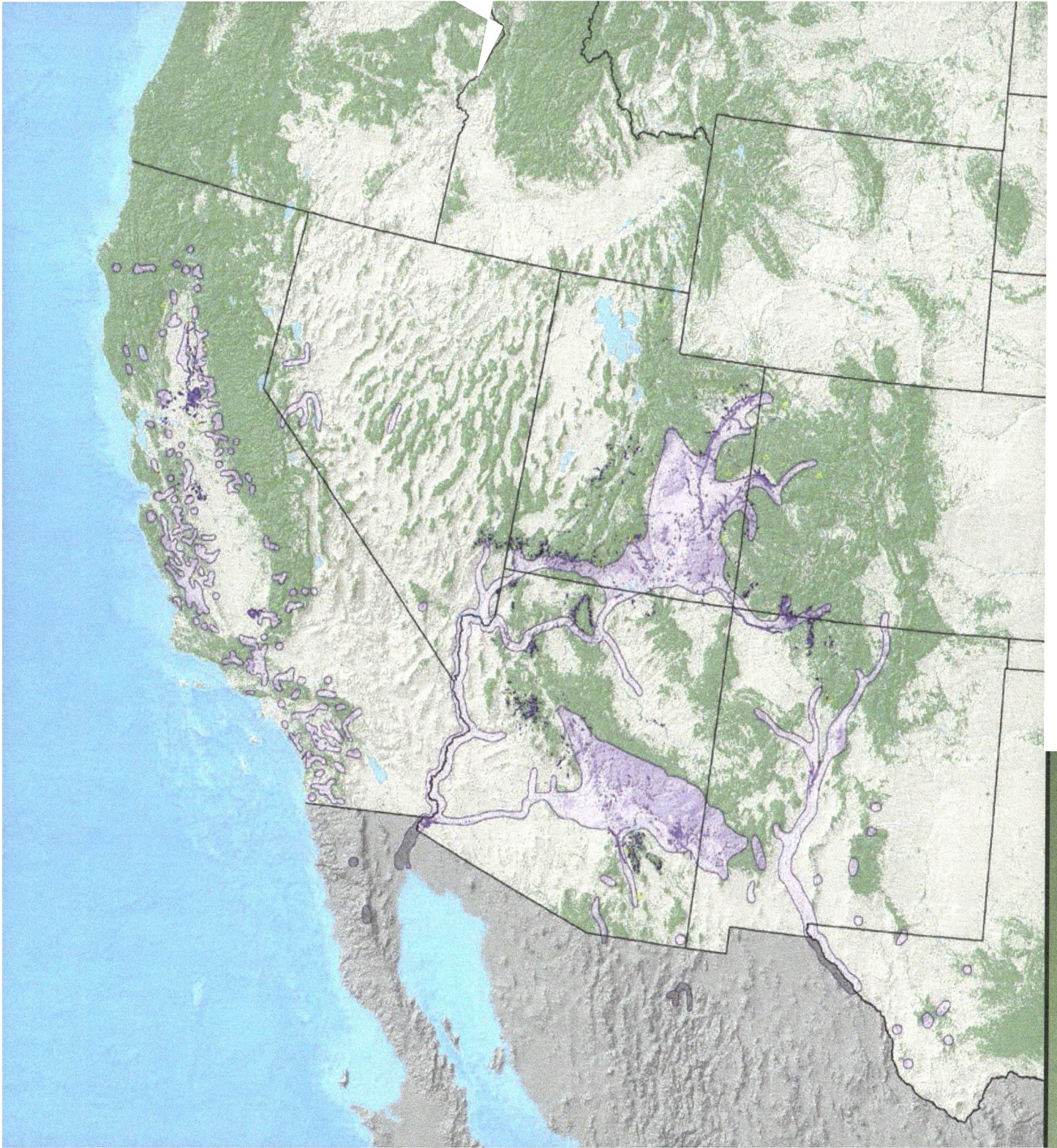
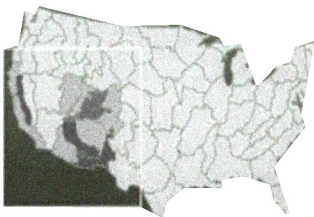
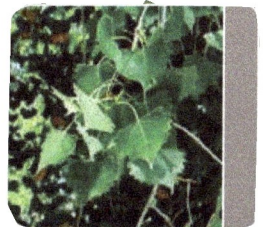

National Individual Tree Species Atlas

bigtooth aspen

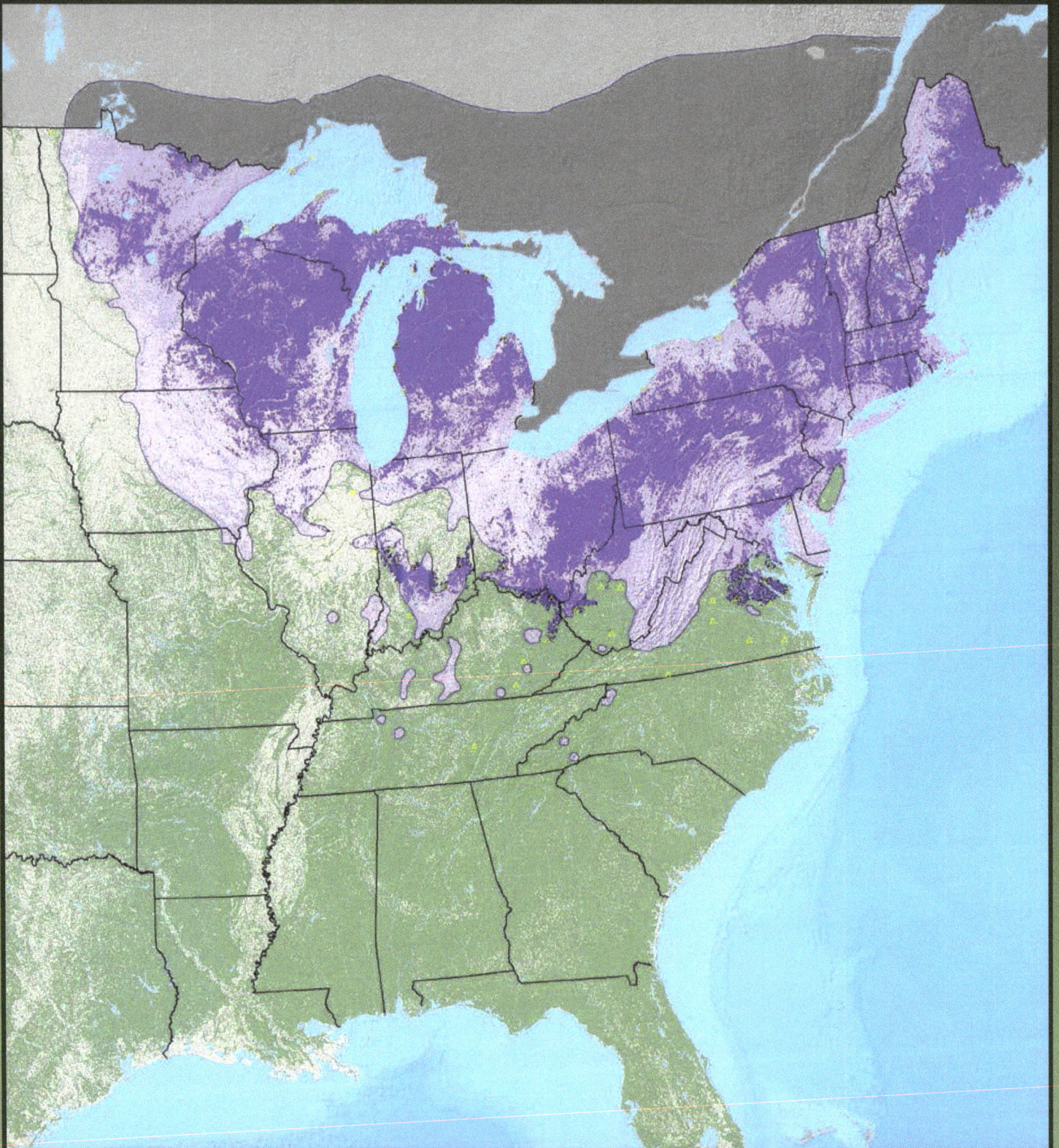

Populus grandidentata Michx.

Salicaceae – Willow Family

FIA Code **743** – Plants Symbol **POGR4**

National Individual Tree Species Atlas

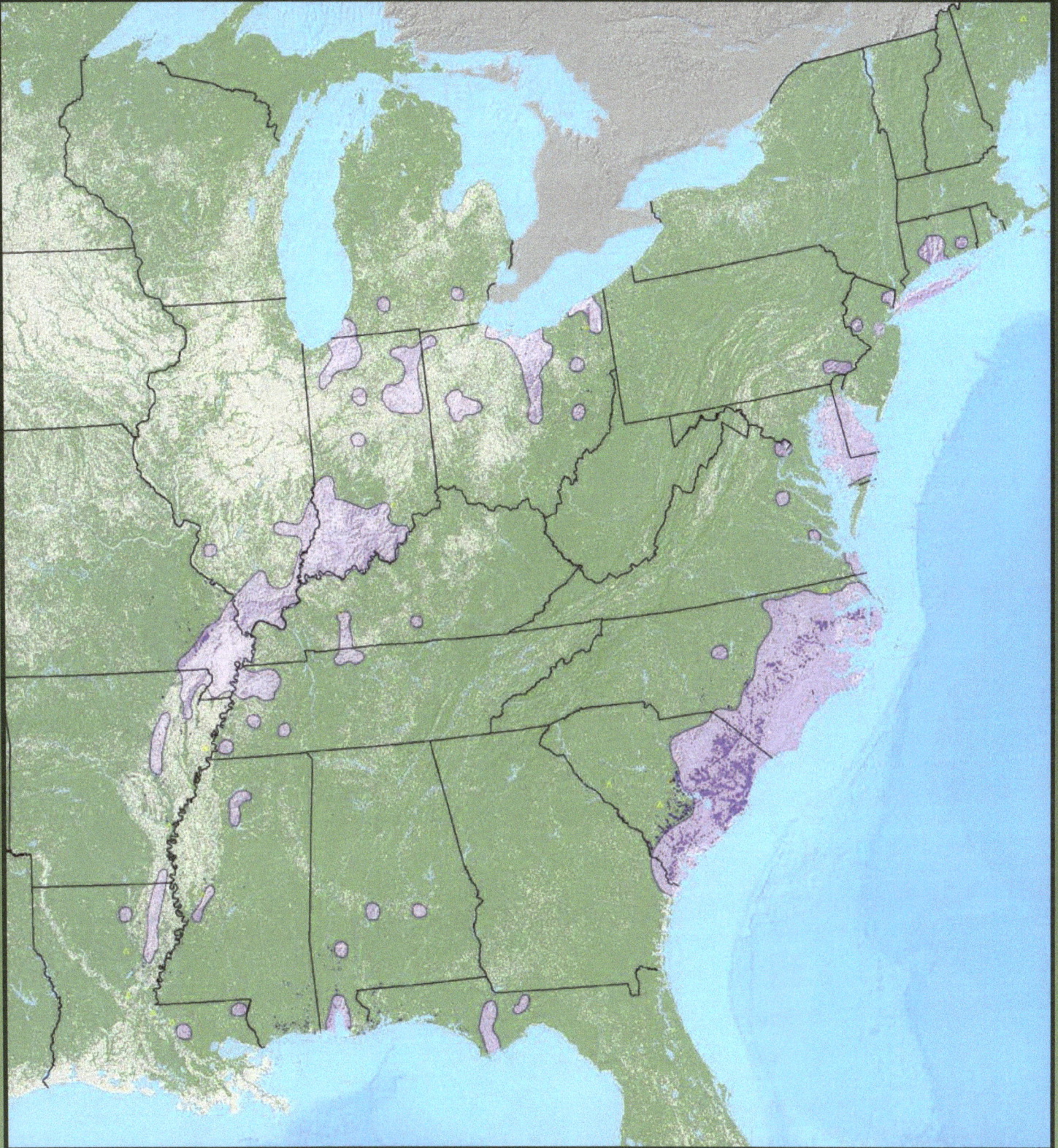

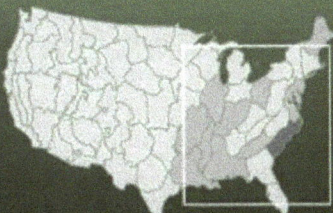

swamp cottonwood

swamp cottonwood
Populus heterophylla L.
Salicaceae – Willow Family
FIA Code **744** – Plants Symbol **POHE4**

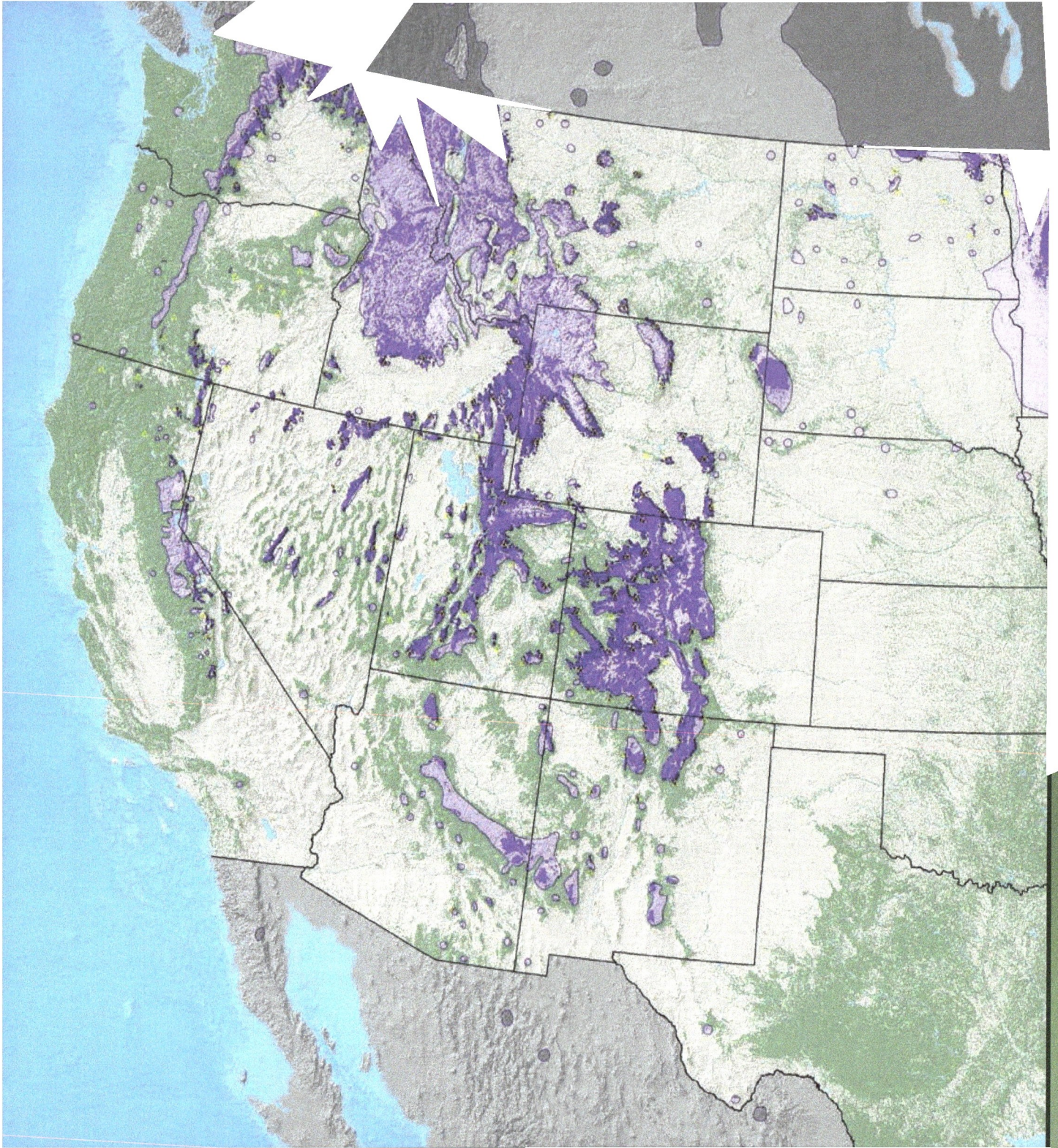

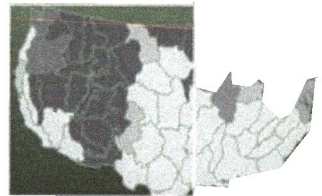

National Individual Tree Species Atlas

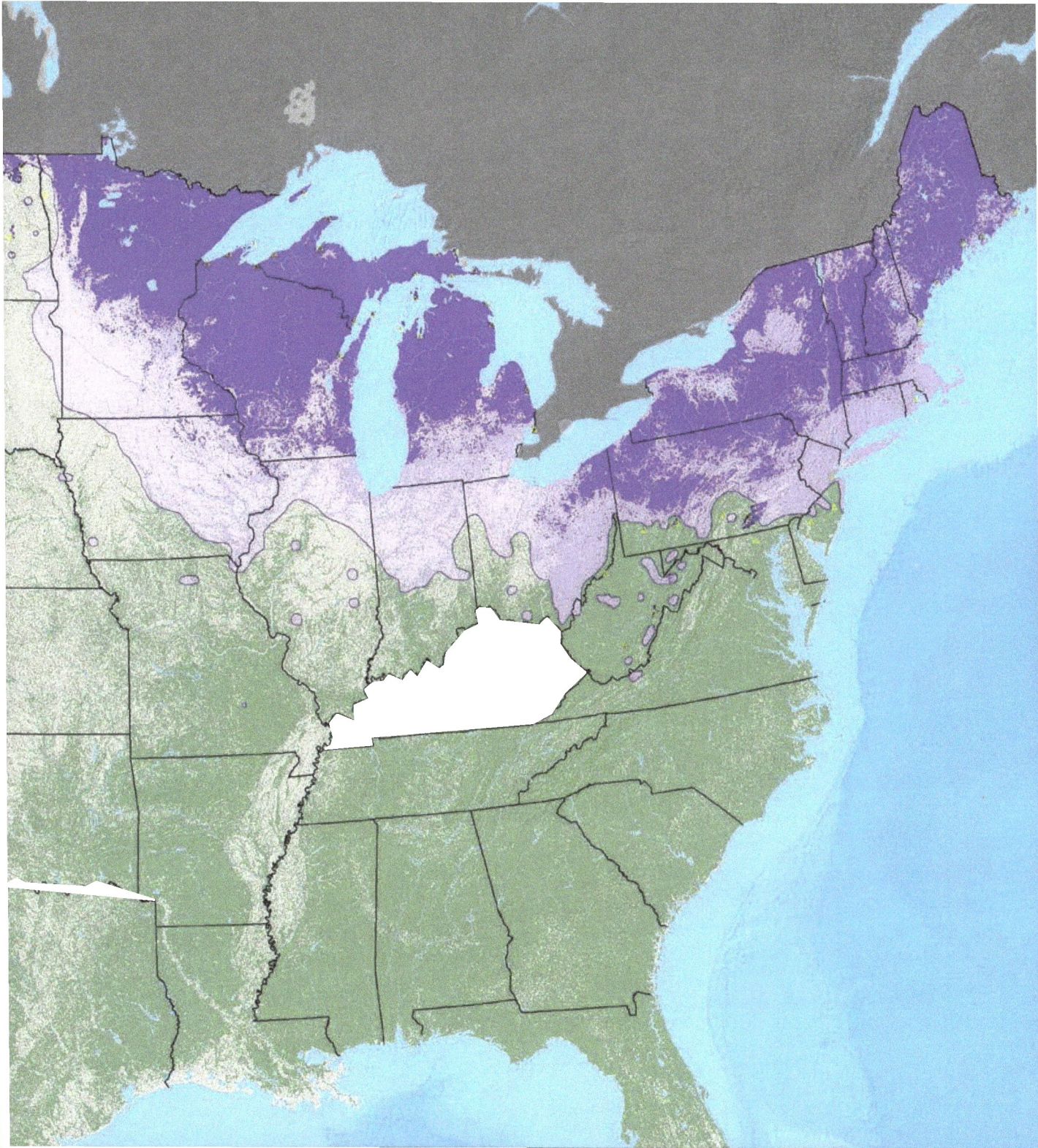

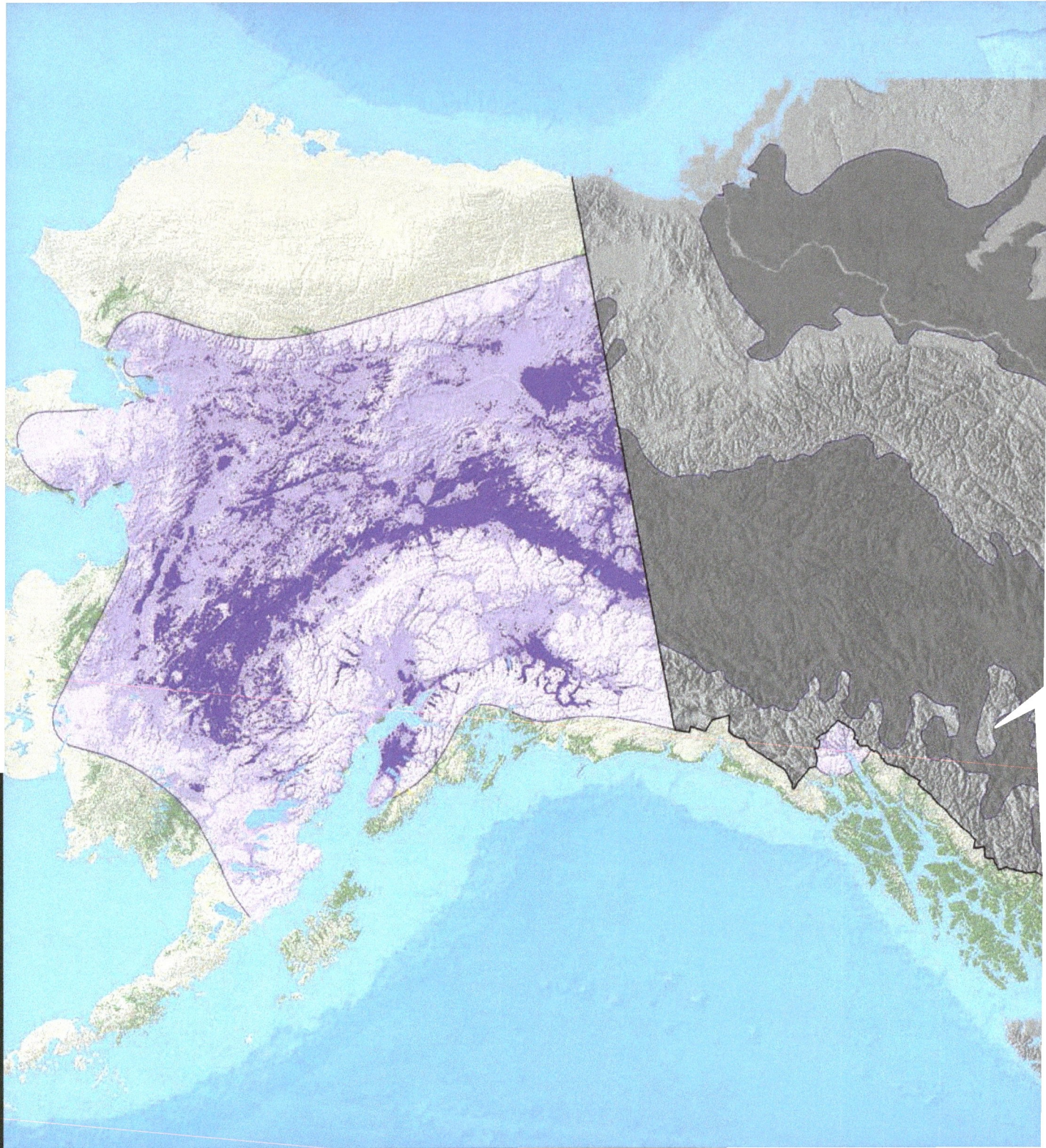

National Individual Tree Species Atlas

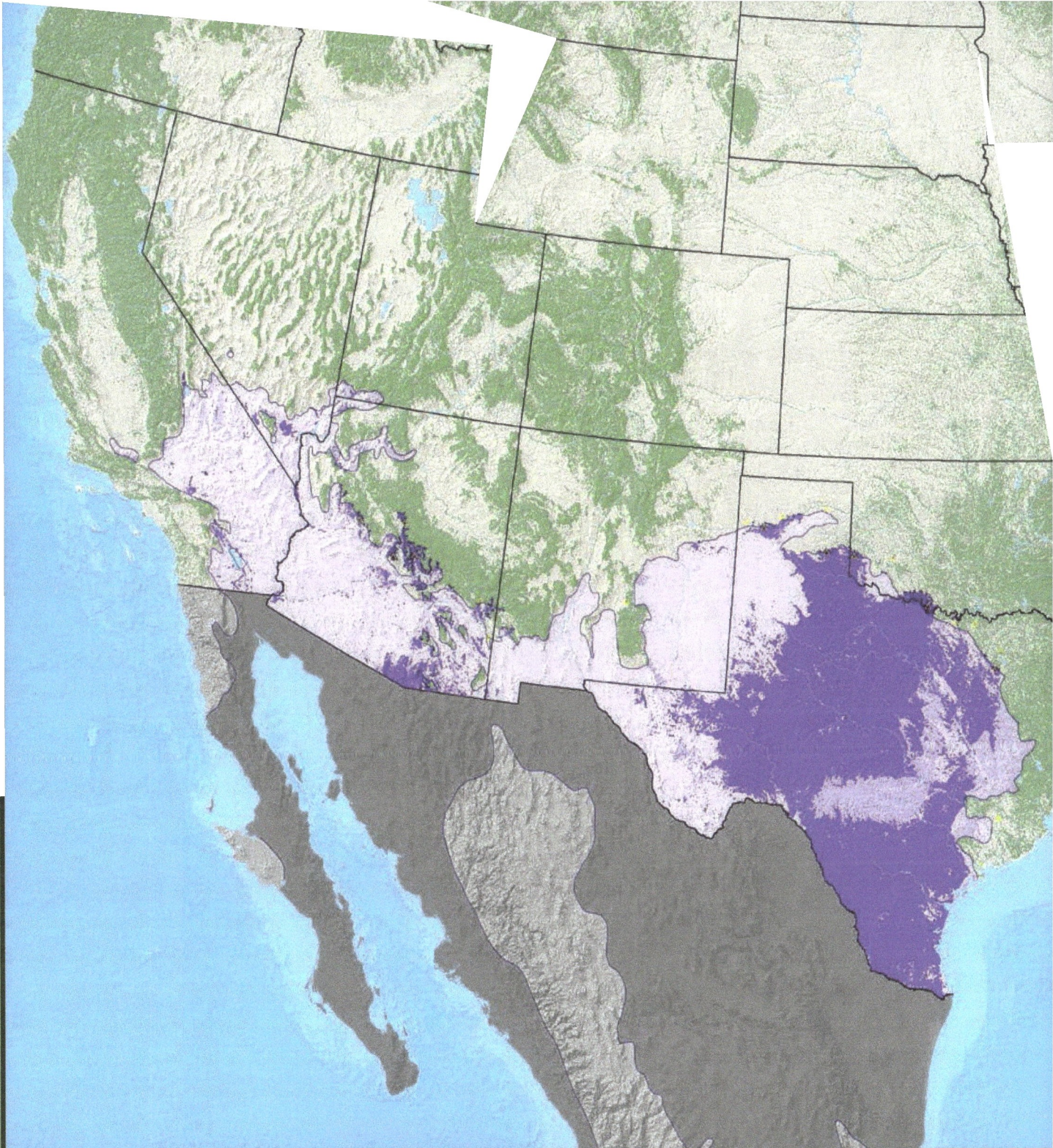

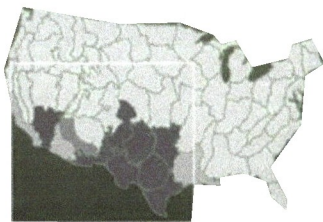

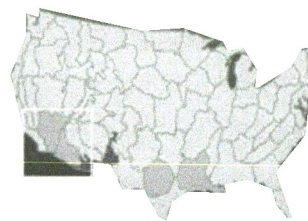

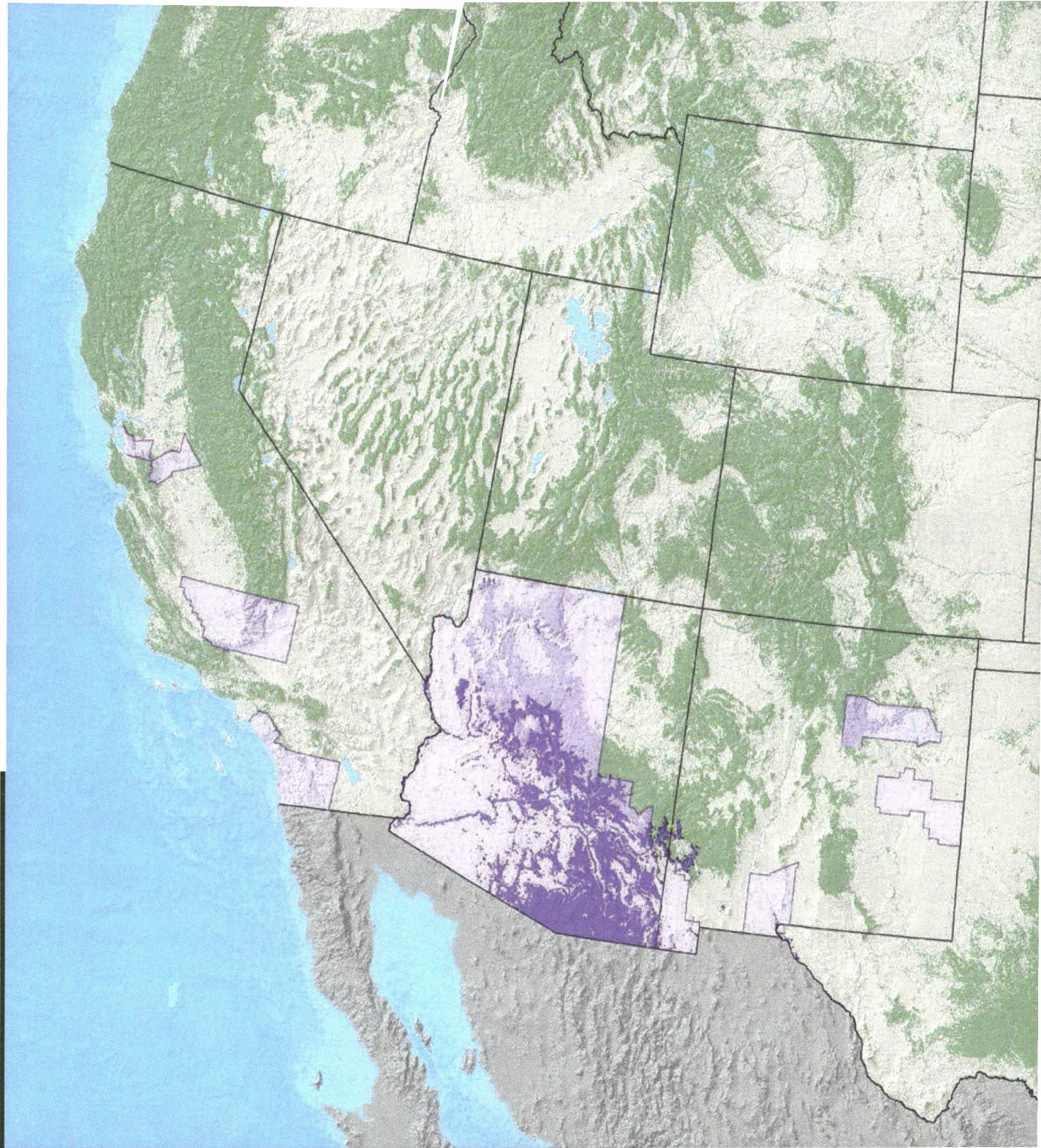

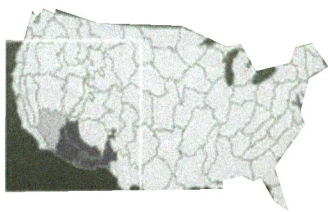

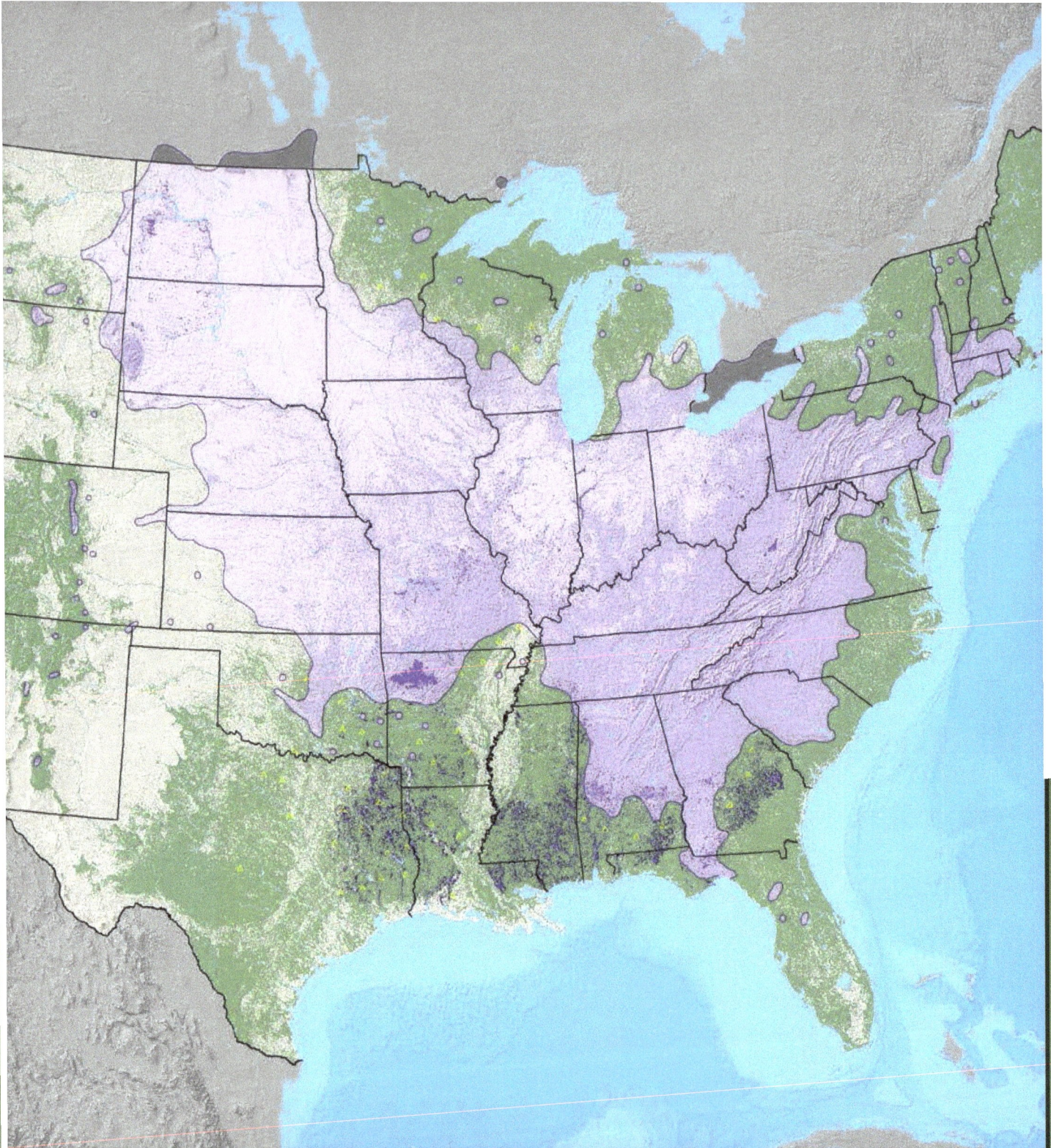

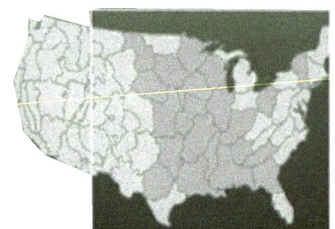

National Individual Tree Species Atlas

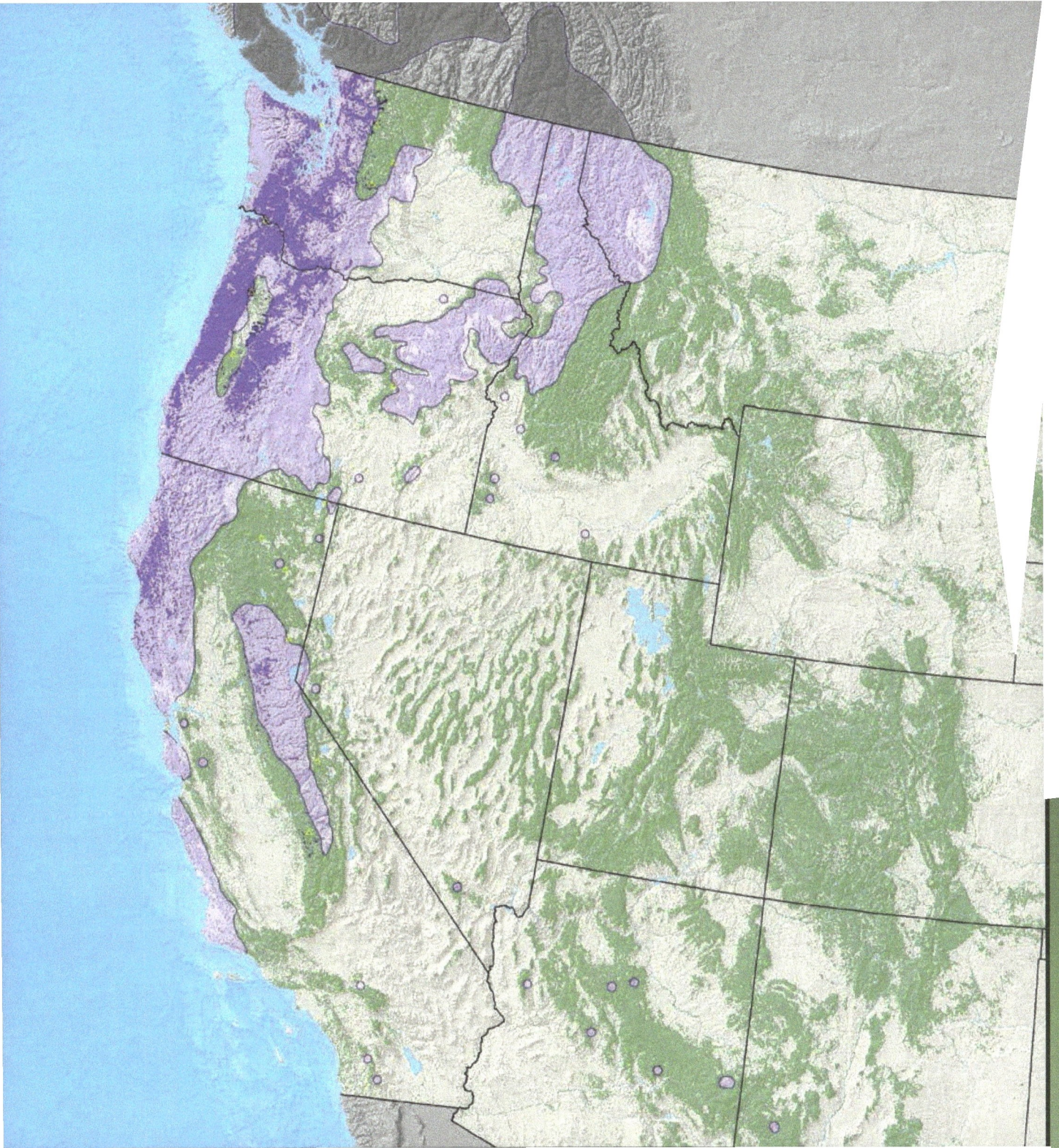
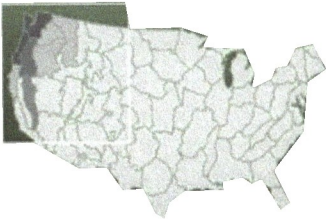

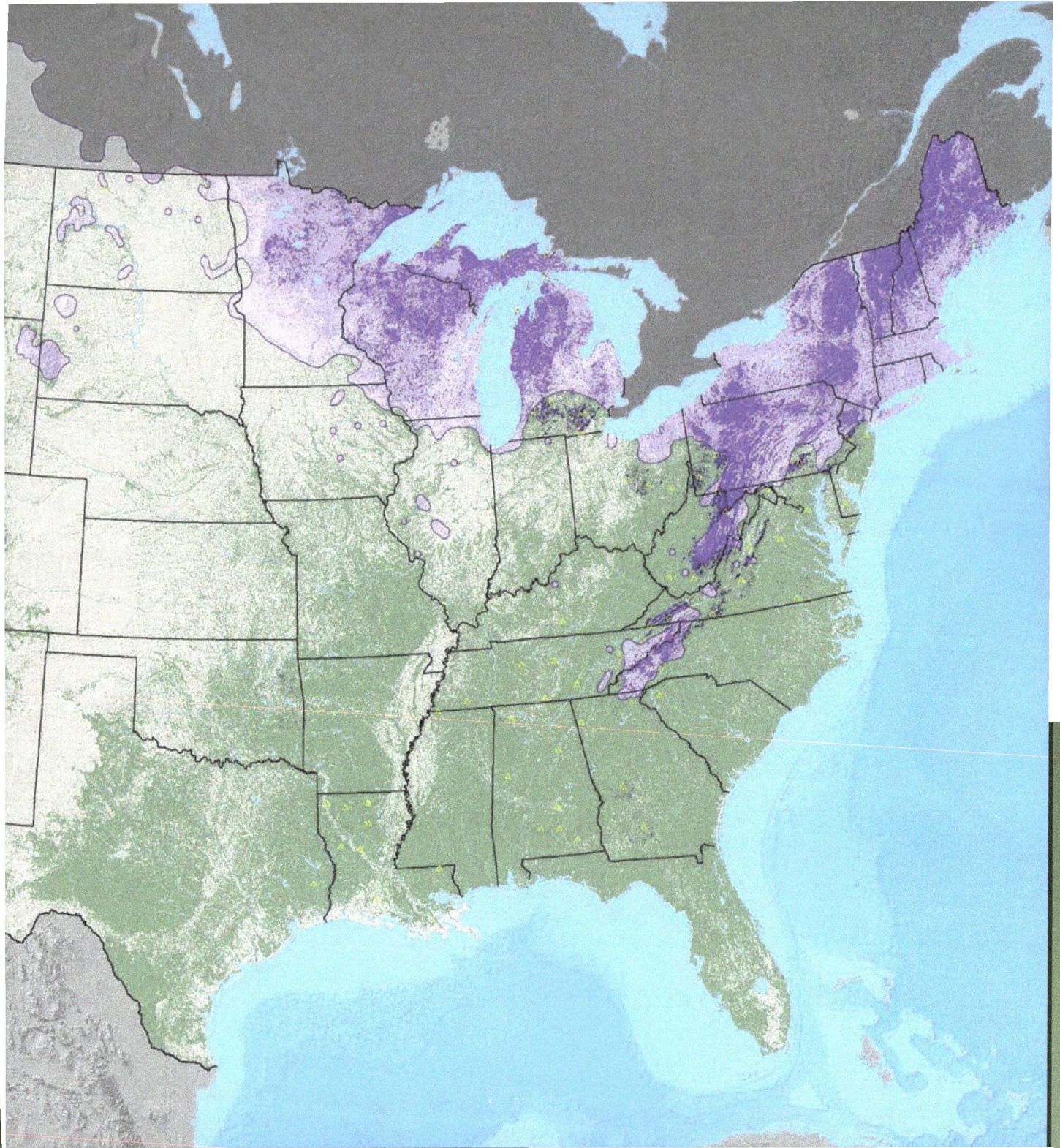

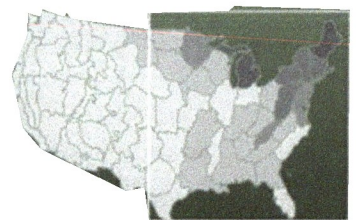

National Individual Tree Species Atlas

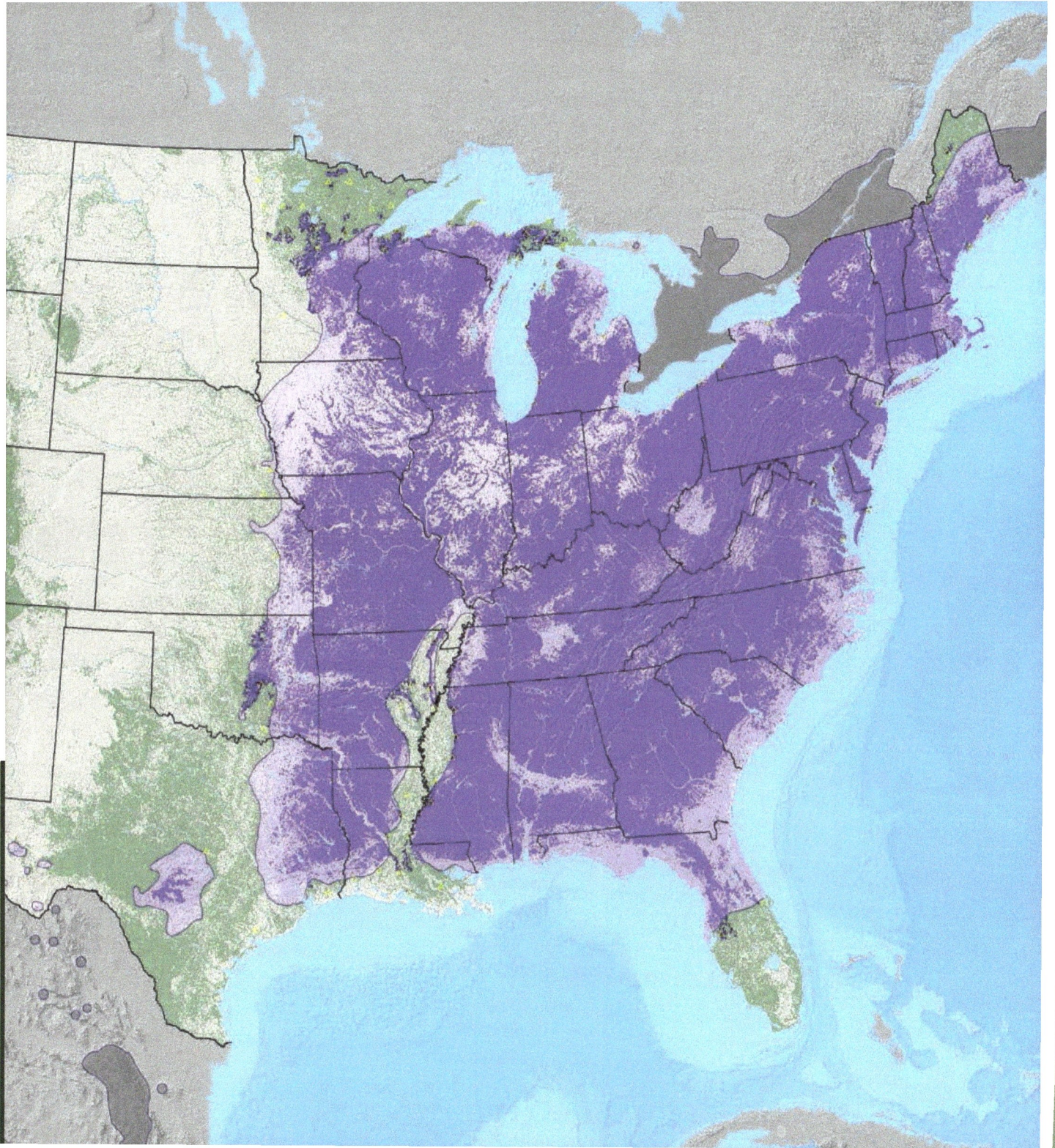

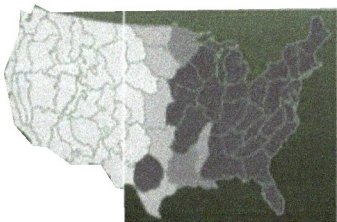

National Individual Tree Species Atlas

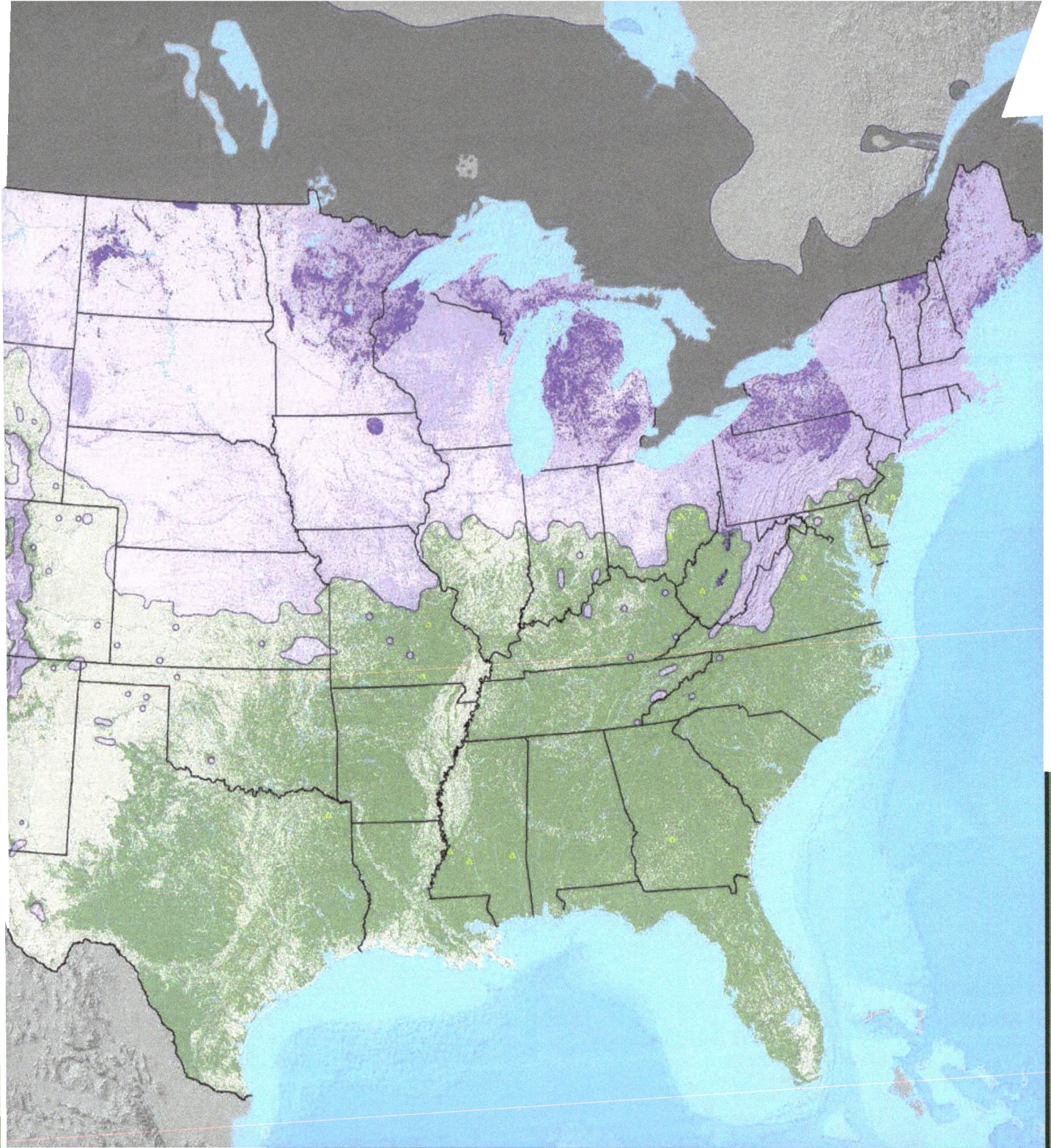

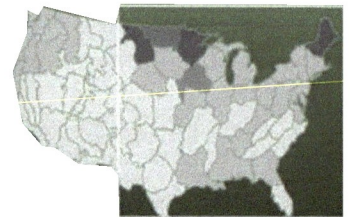

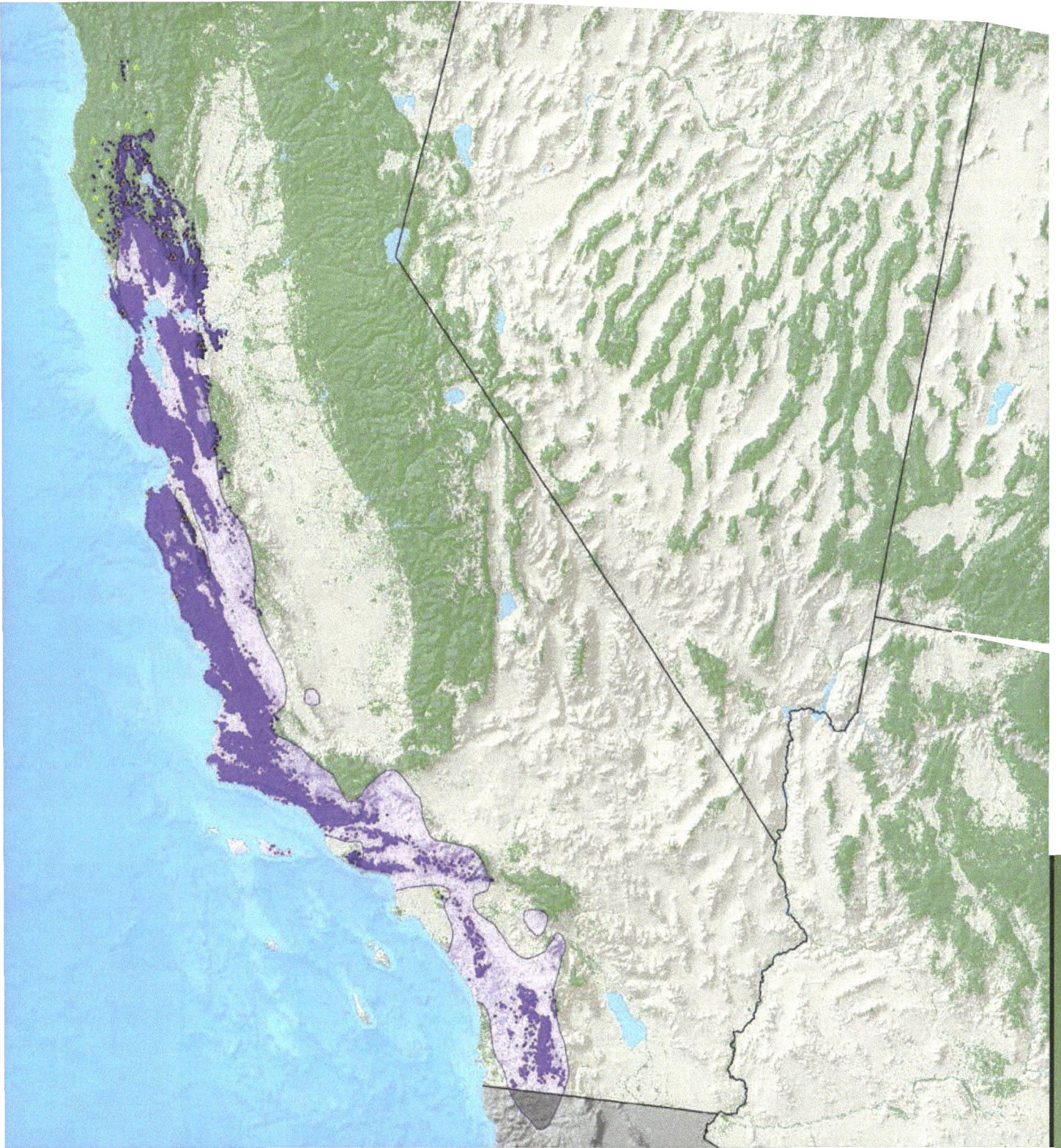
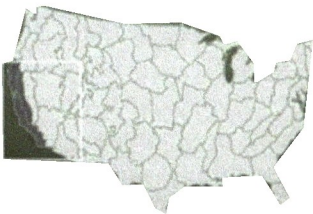

National Individual Tree Species Atlas

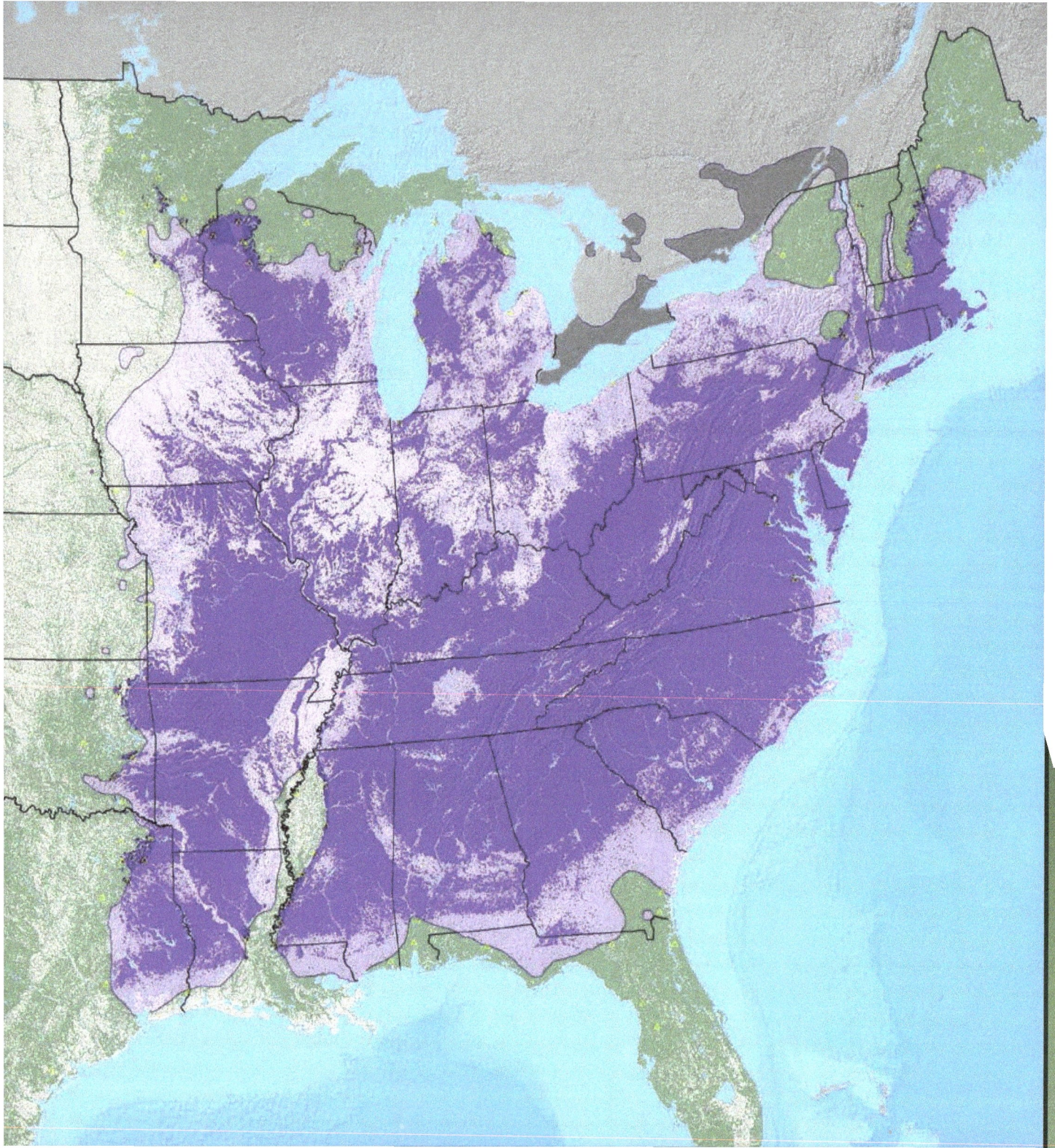

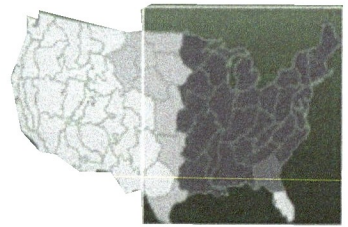

National Individual Tree Species Atlas

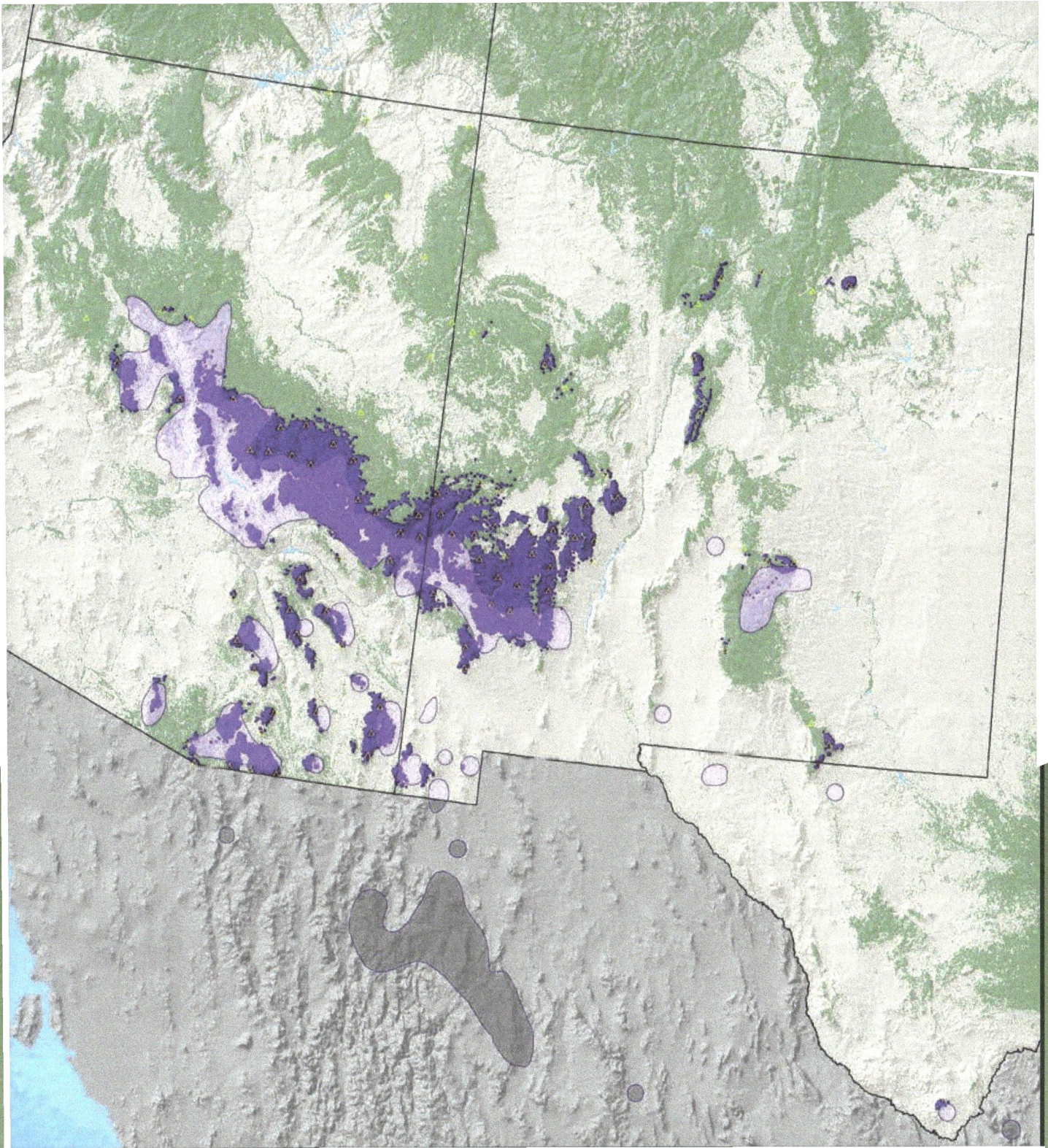

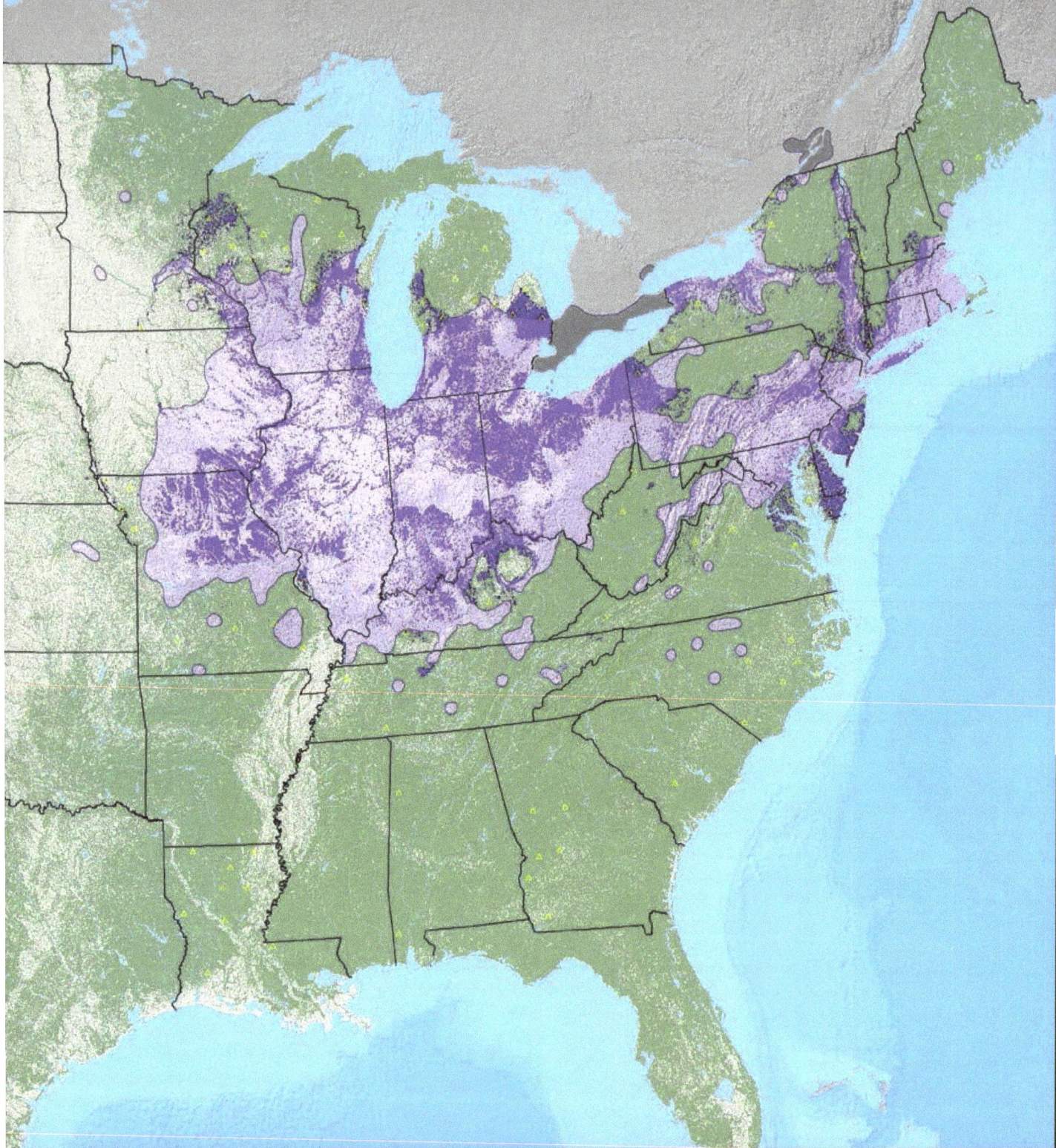

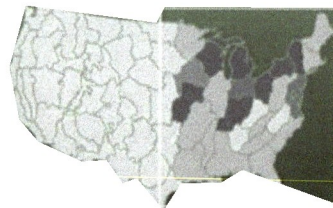

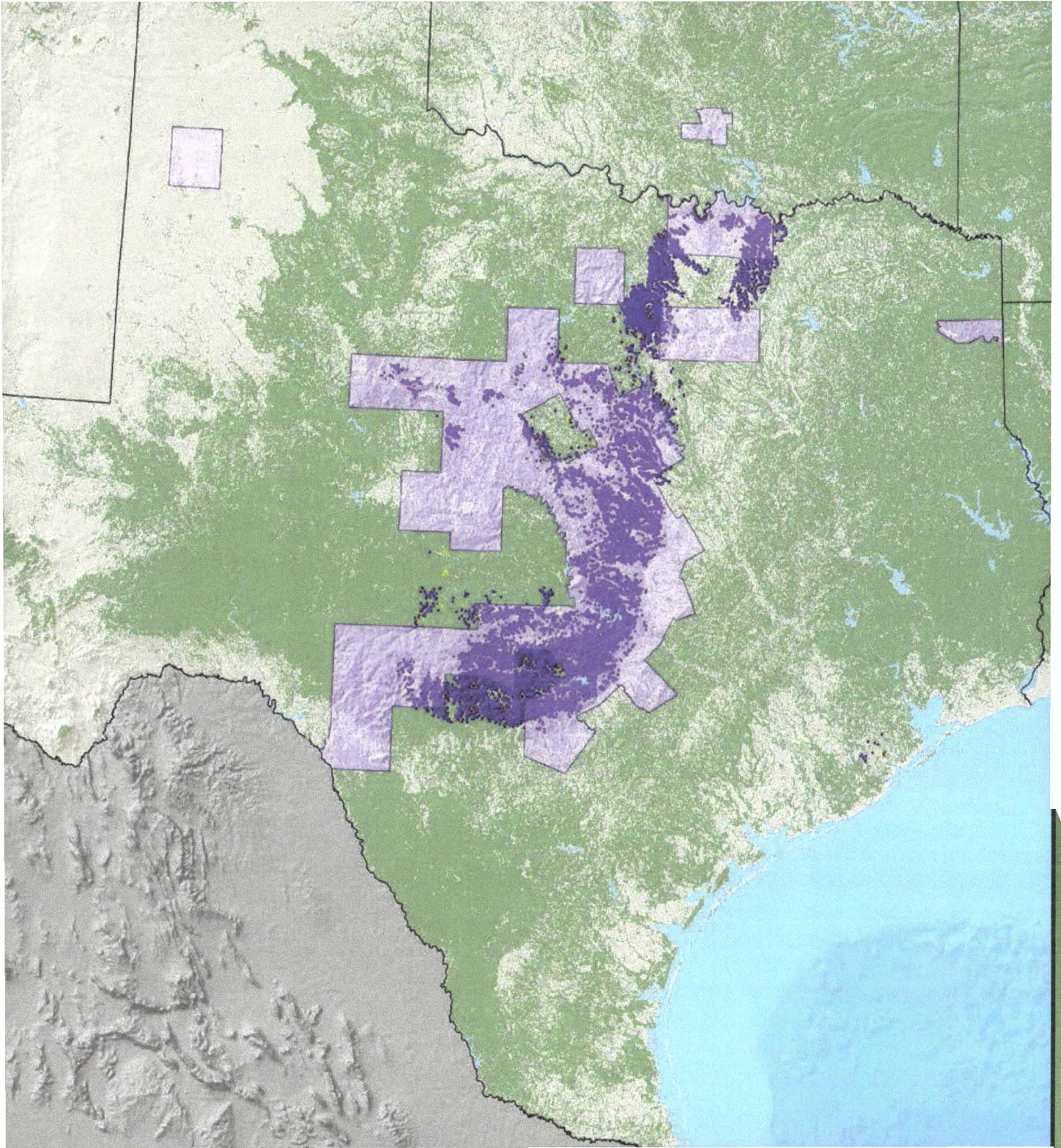

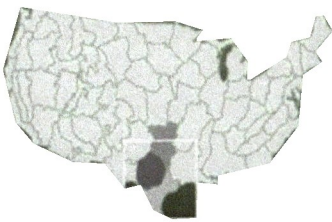

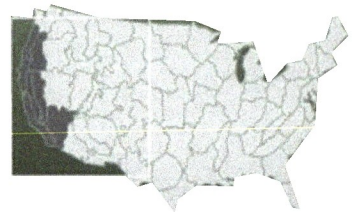

National Individual Tree Species Atlas

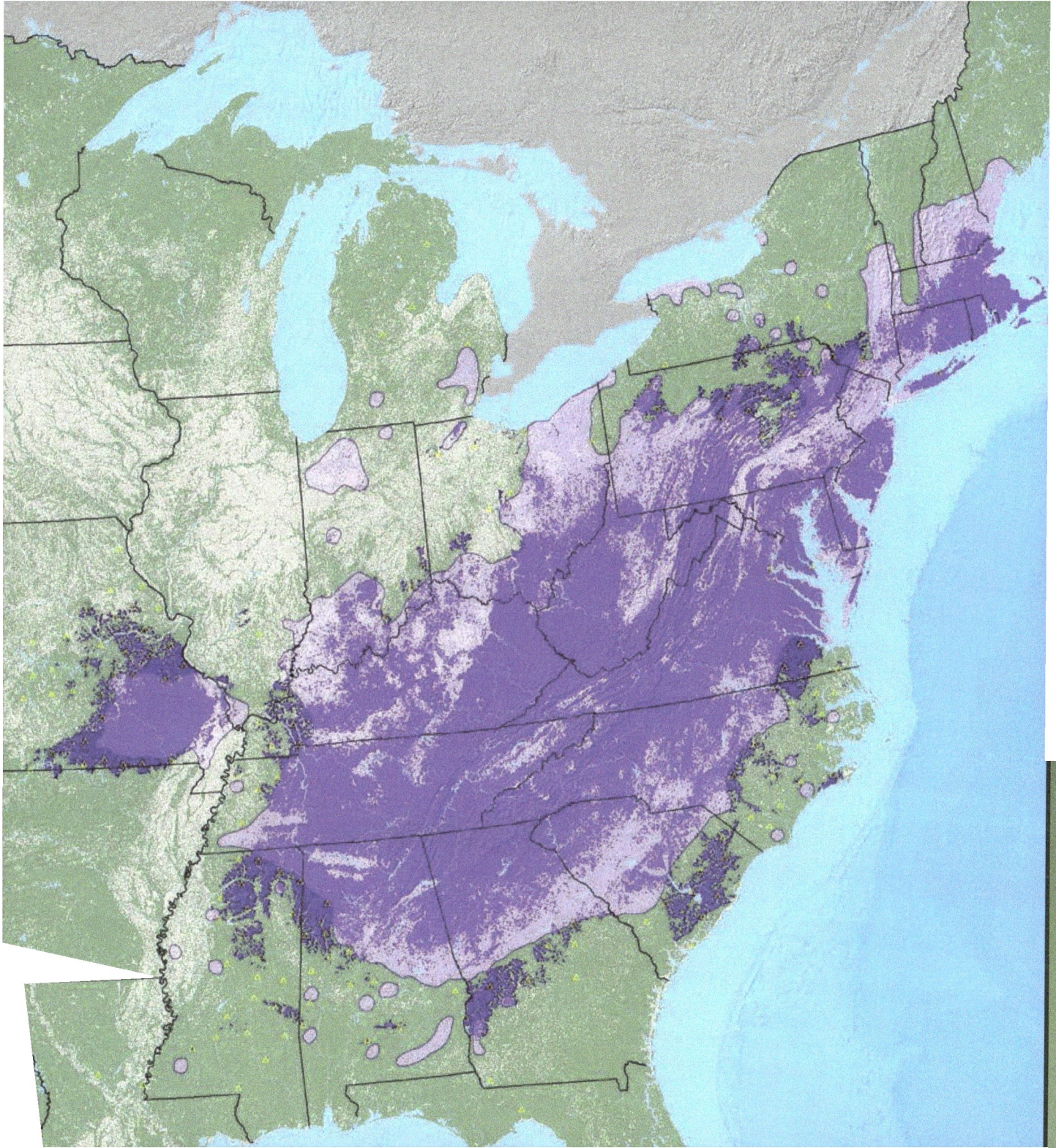

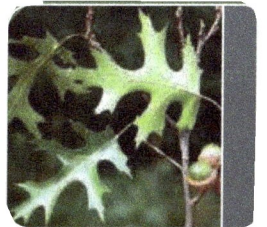

National Individual Tree Species Atlas

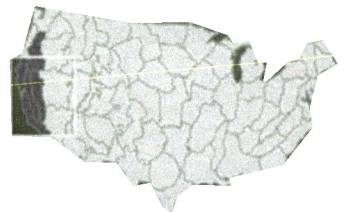

National Individual Tree Species Atlas

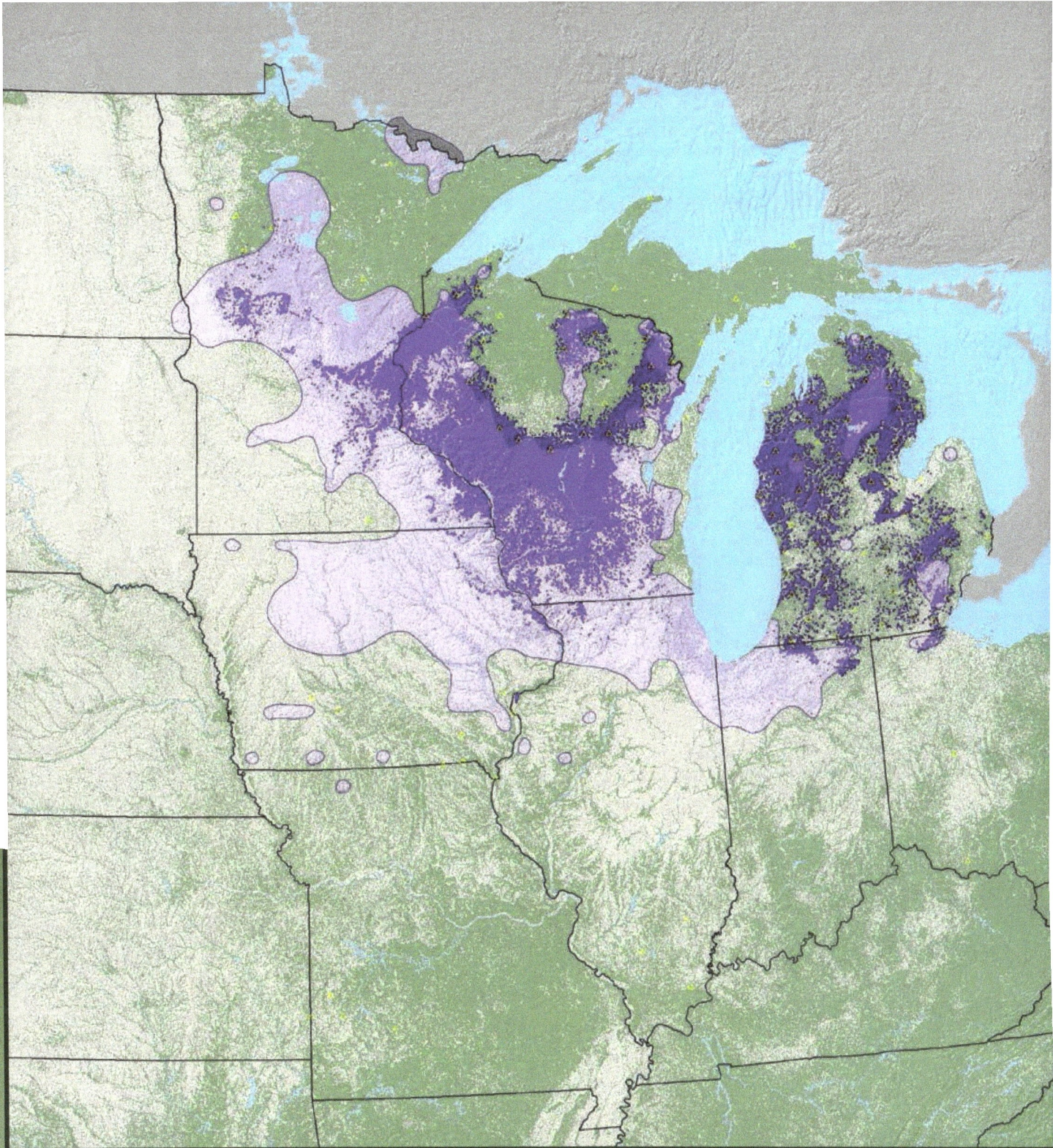

northern pin oak
Quercus ellipsoidalis E.J. Hill
Fagaceae – Beech Family
FIA Code **809** – Plants Symbol **QUEL**

National Individual Tree Species Atlas

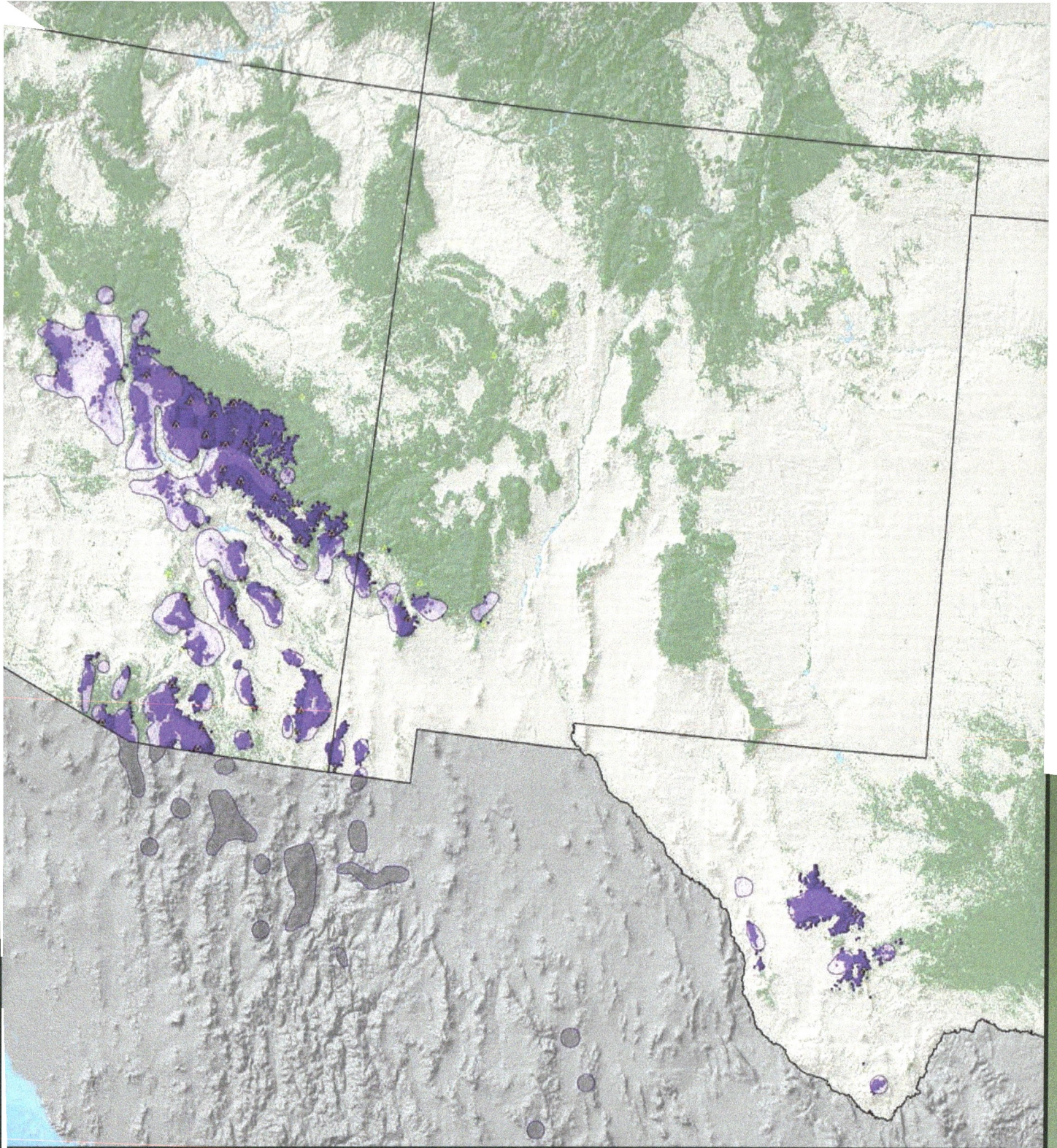

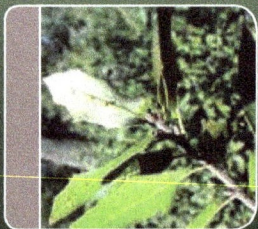

Emory oak
Quercus emoryi Torr.
Fagaceae – Beech Family
FIA Code **810** – Plants Symbol **QUEM**

National Individual Tree Species Atlas

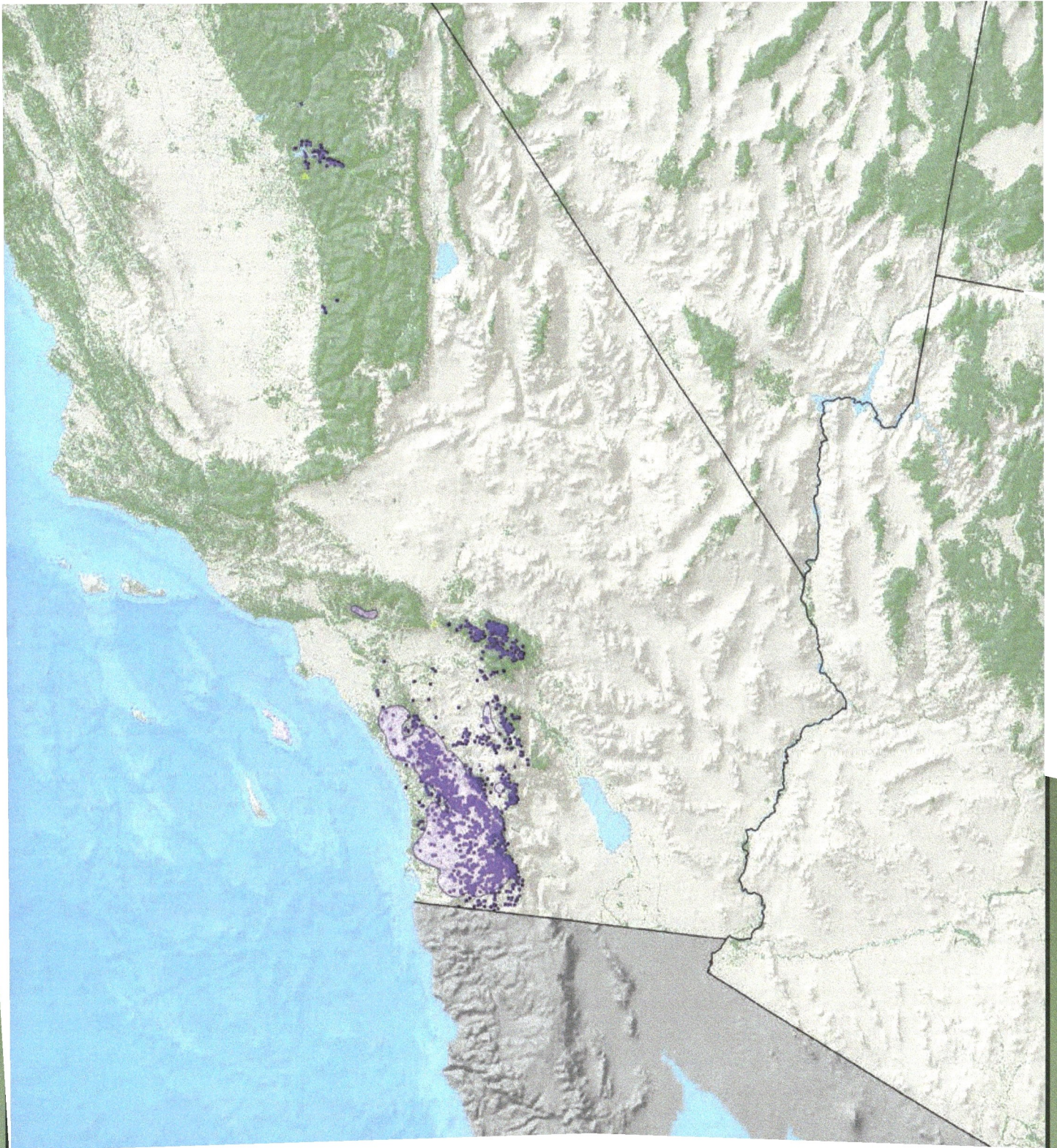

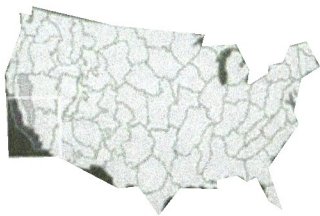

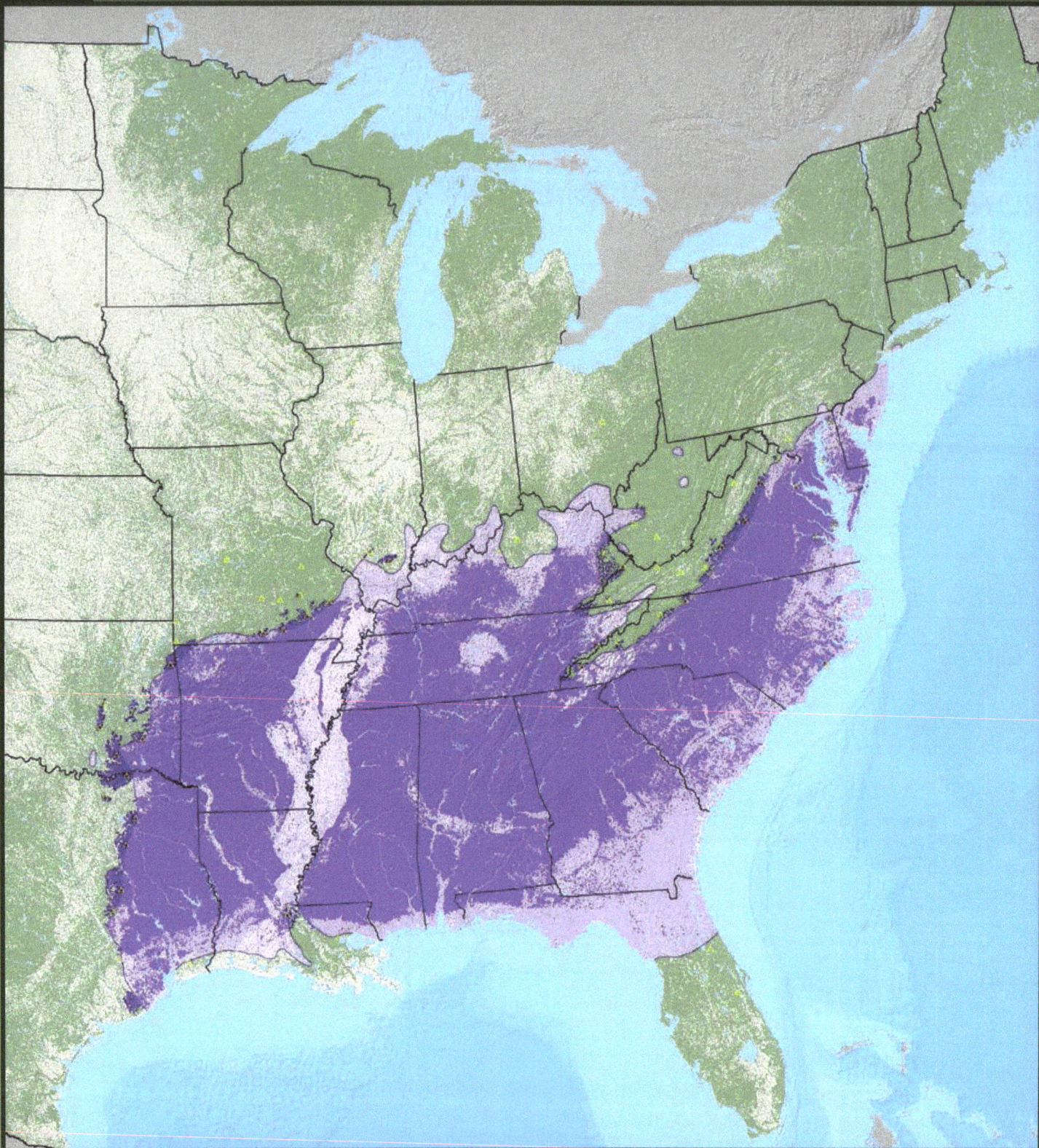

southern red oak

Quercus falcata Michx.

Fagaceae – Beech Family

FIA Code **812** – Plants Symbol **QUFA**

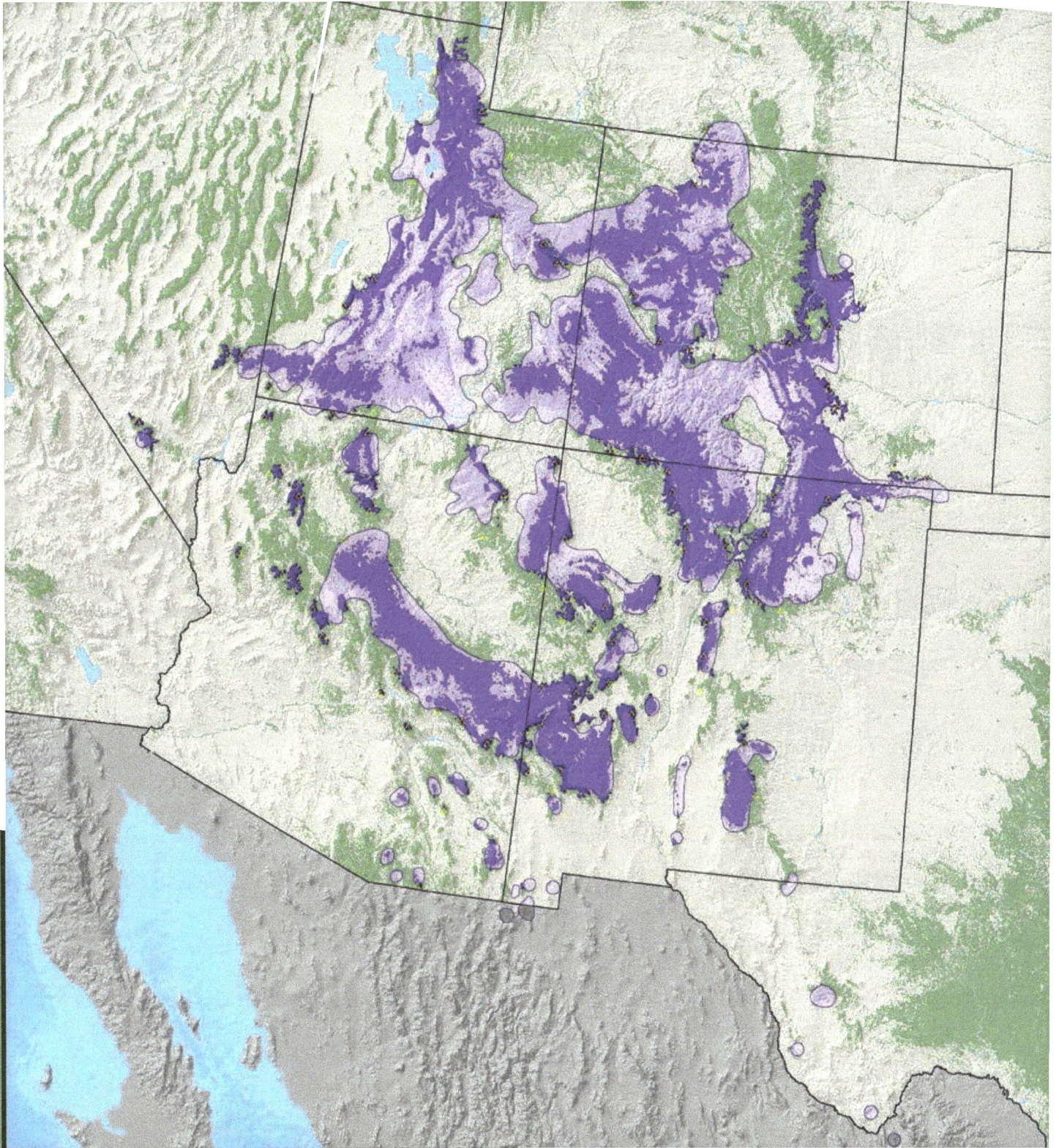

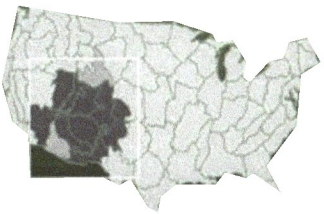

National Individual Tree Species Atlas

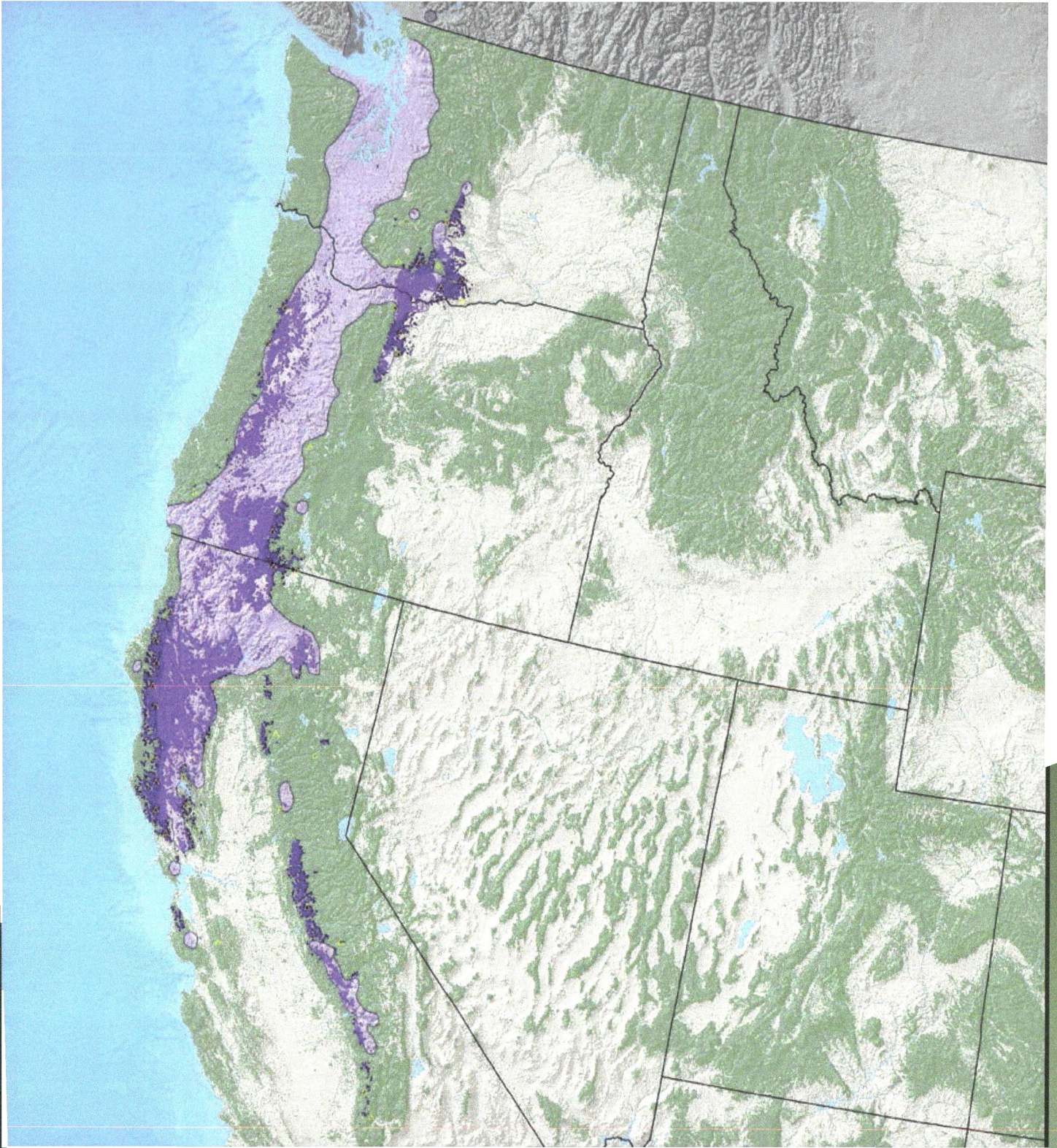

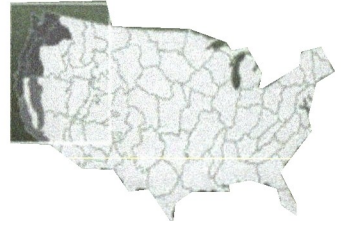

National Individual Tree Species Atlas

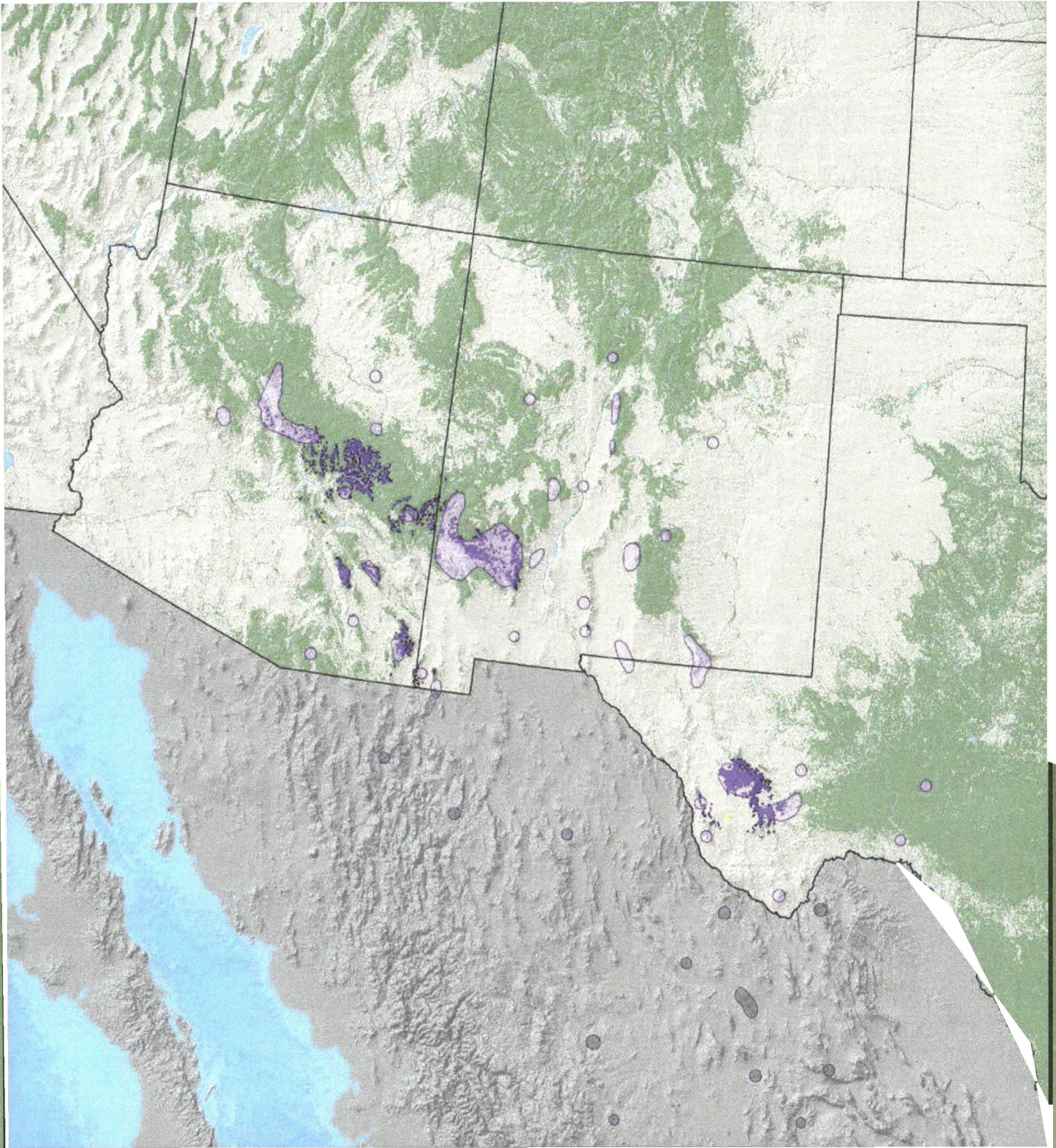

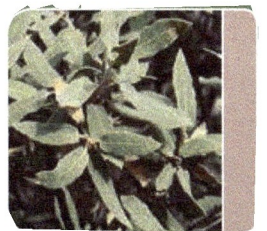

National Individual Tree Species Atlas

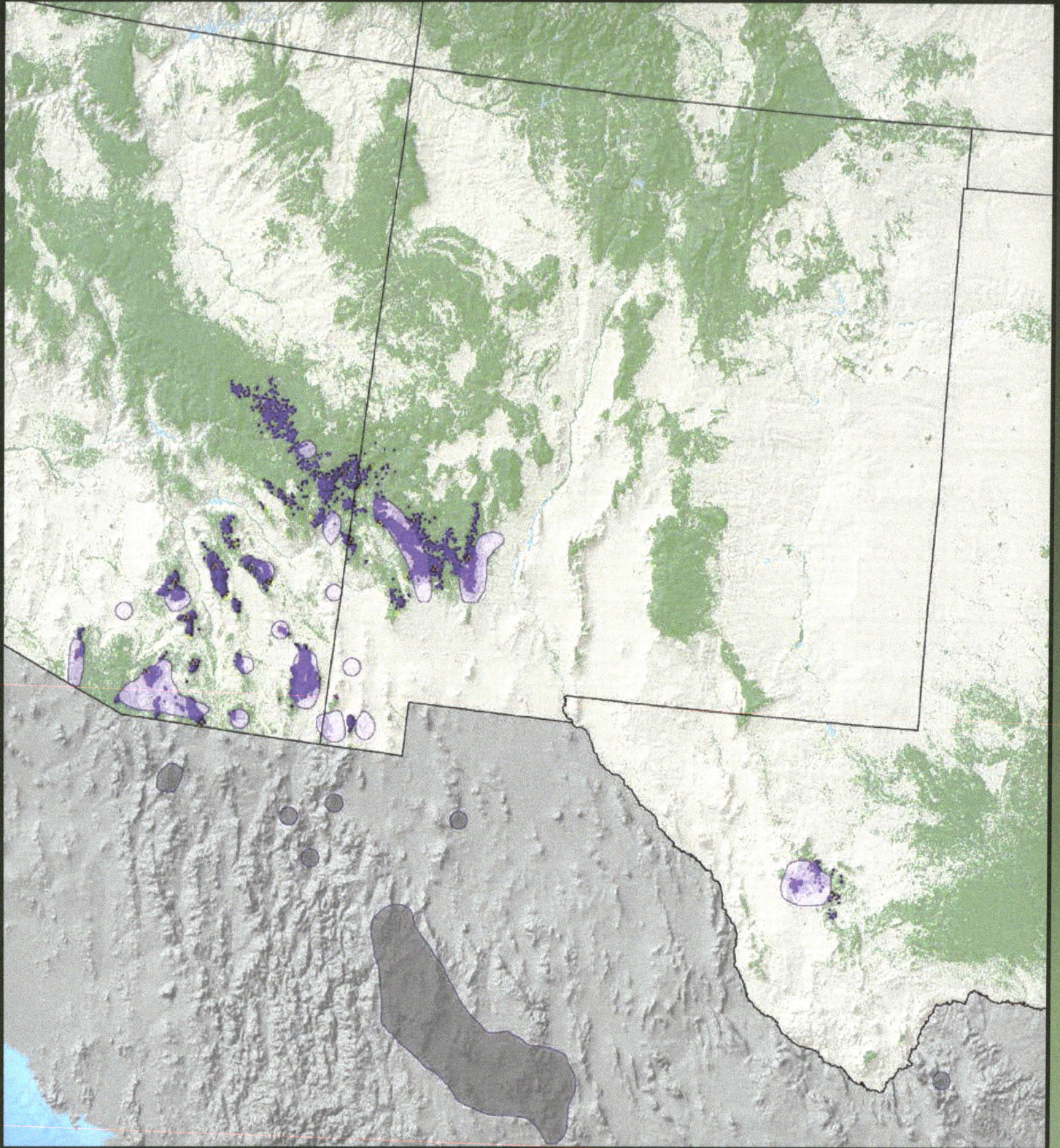

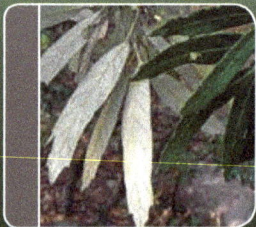

silverleaf oak

silverleaf oak
Quercus hypoleucoides A. Camus
Fagaceae – Beech Family
FIA Code **843** – Plants Symbol **QUHY**

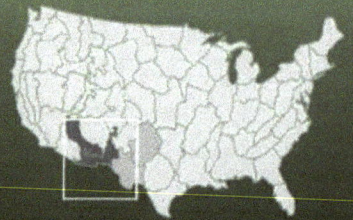

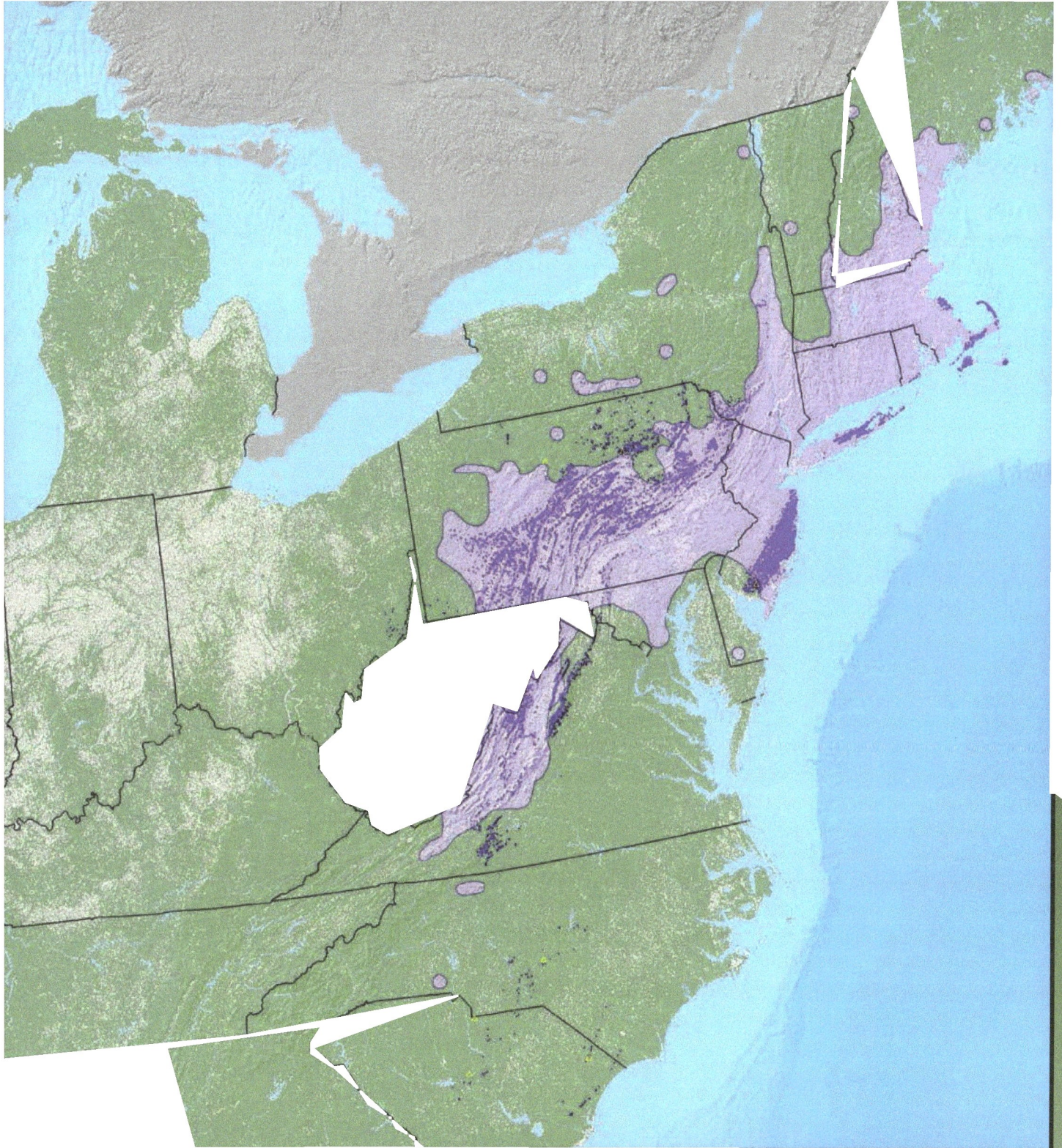
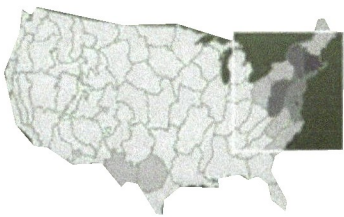
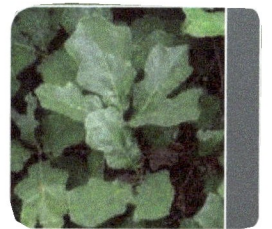

National Individual Tree Species Atlas

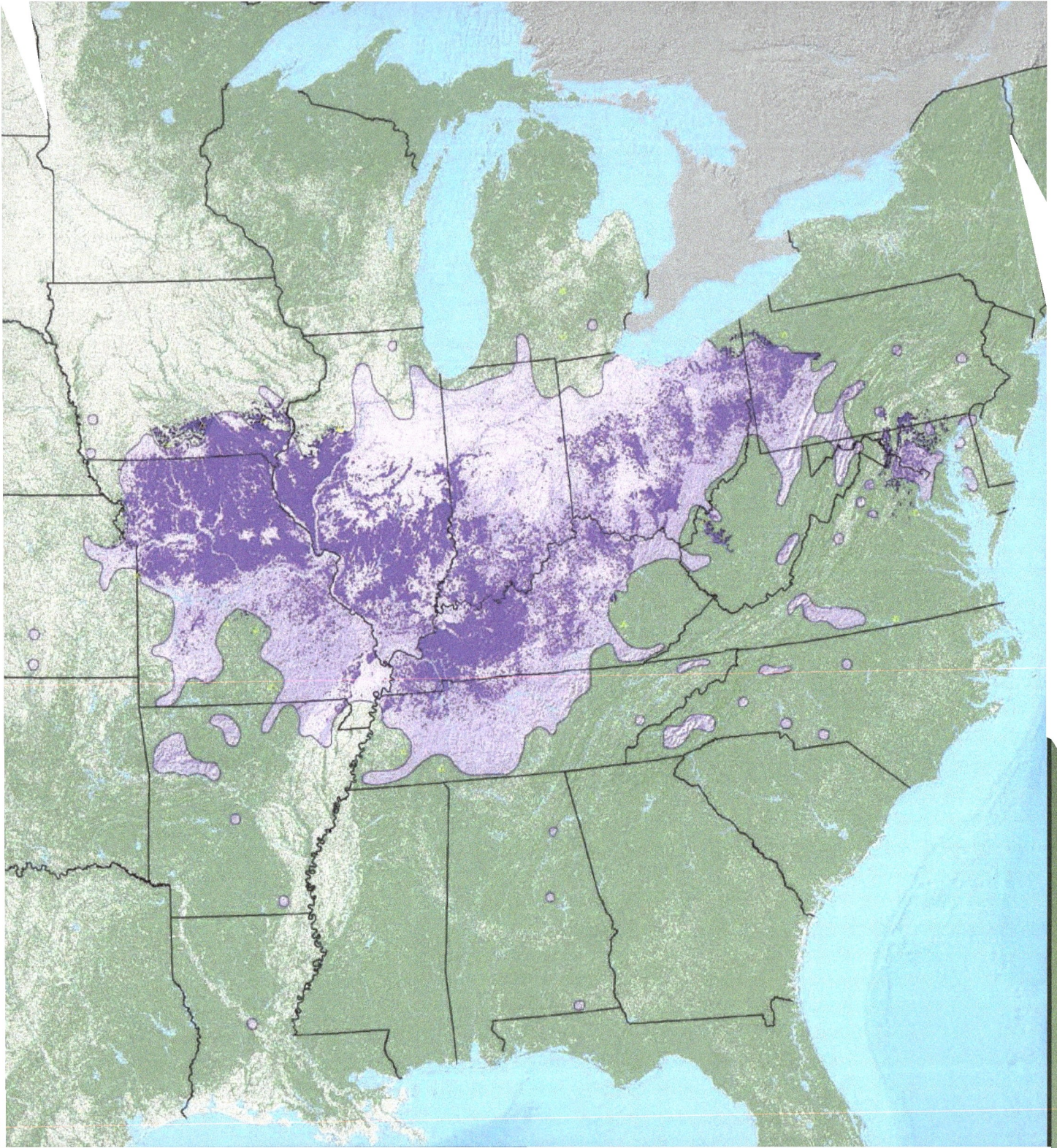

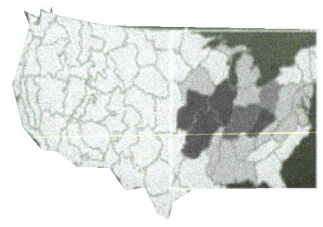

National Individual Tree Species Atlas

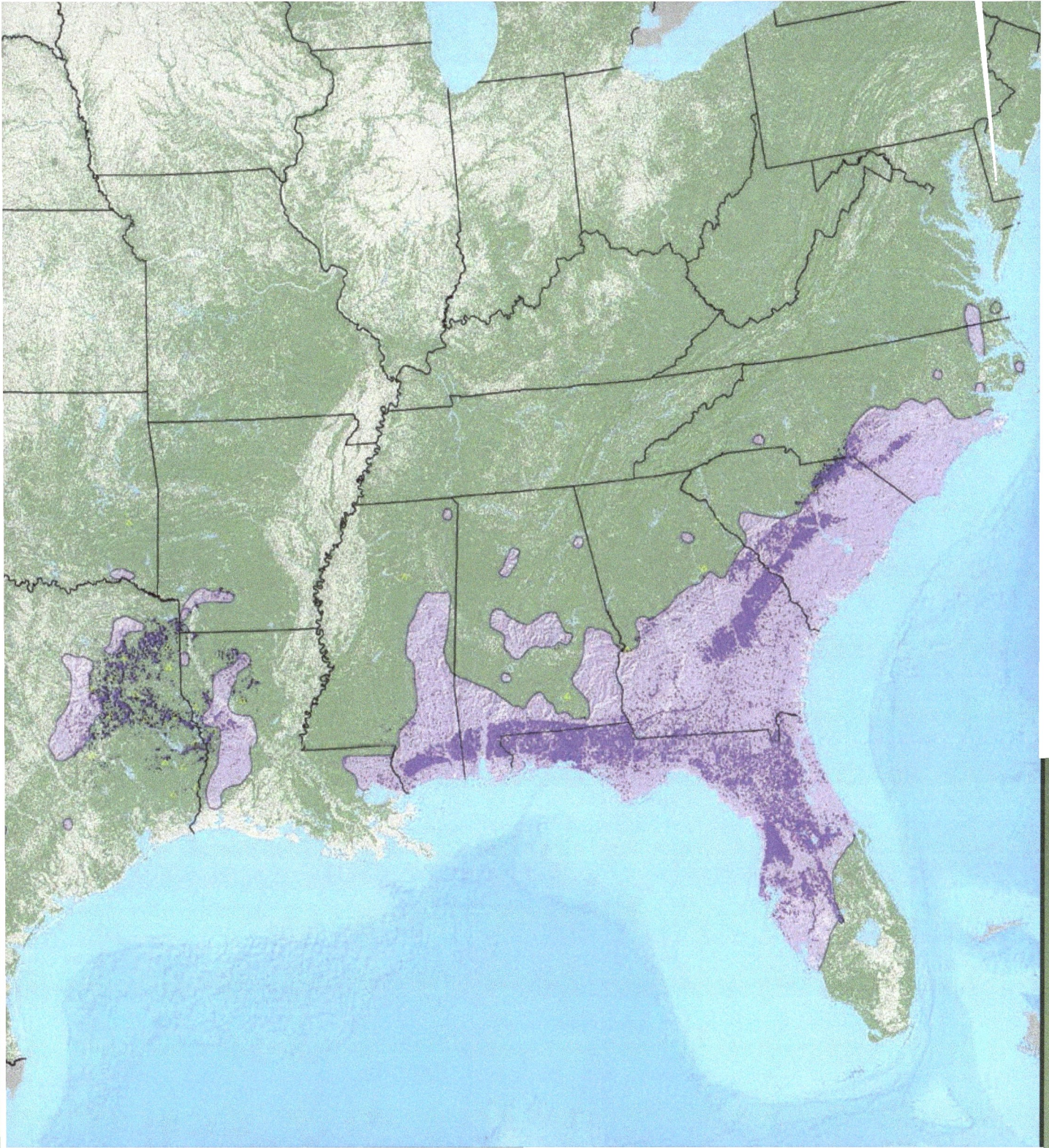

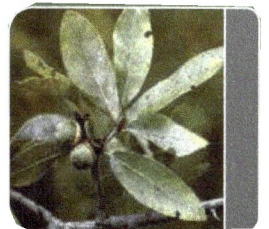

National Individual Tree Species Atlas

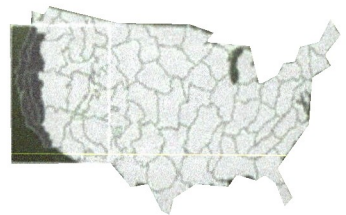

National Individual Tree Species Atlas

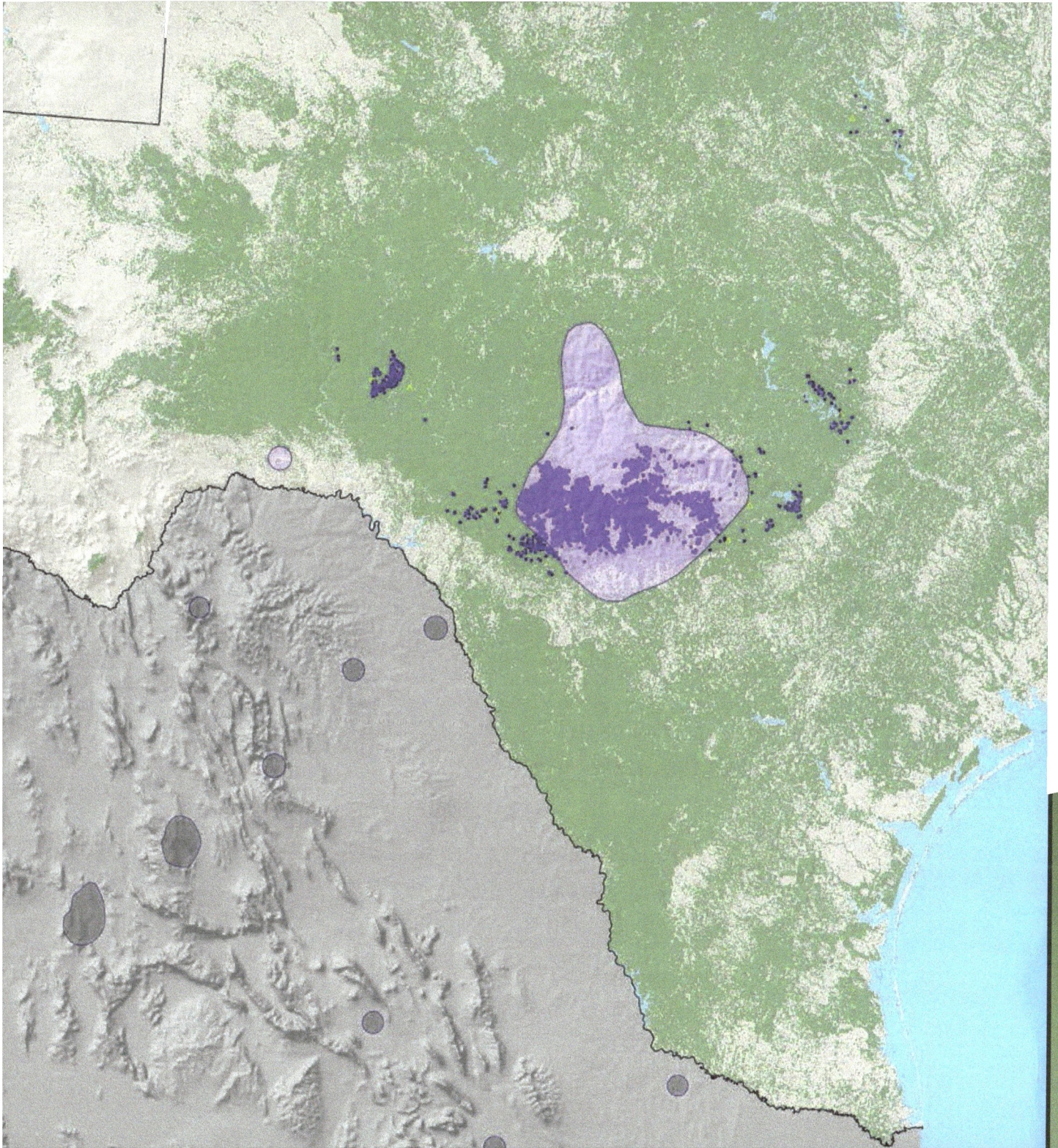

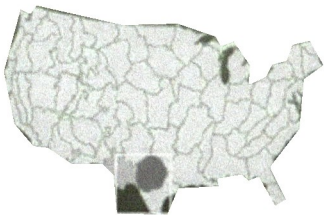

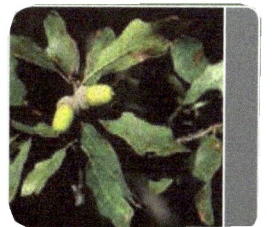

National Individual Tree Species Atlas

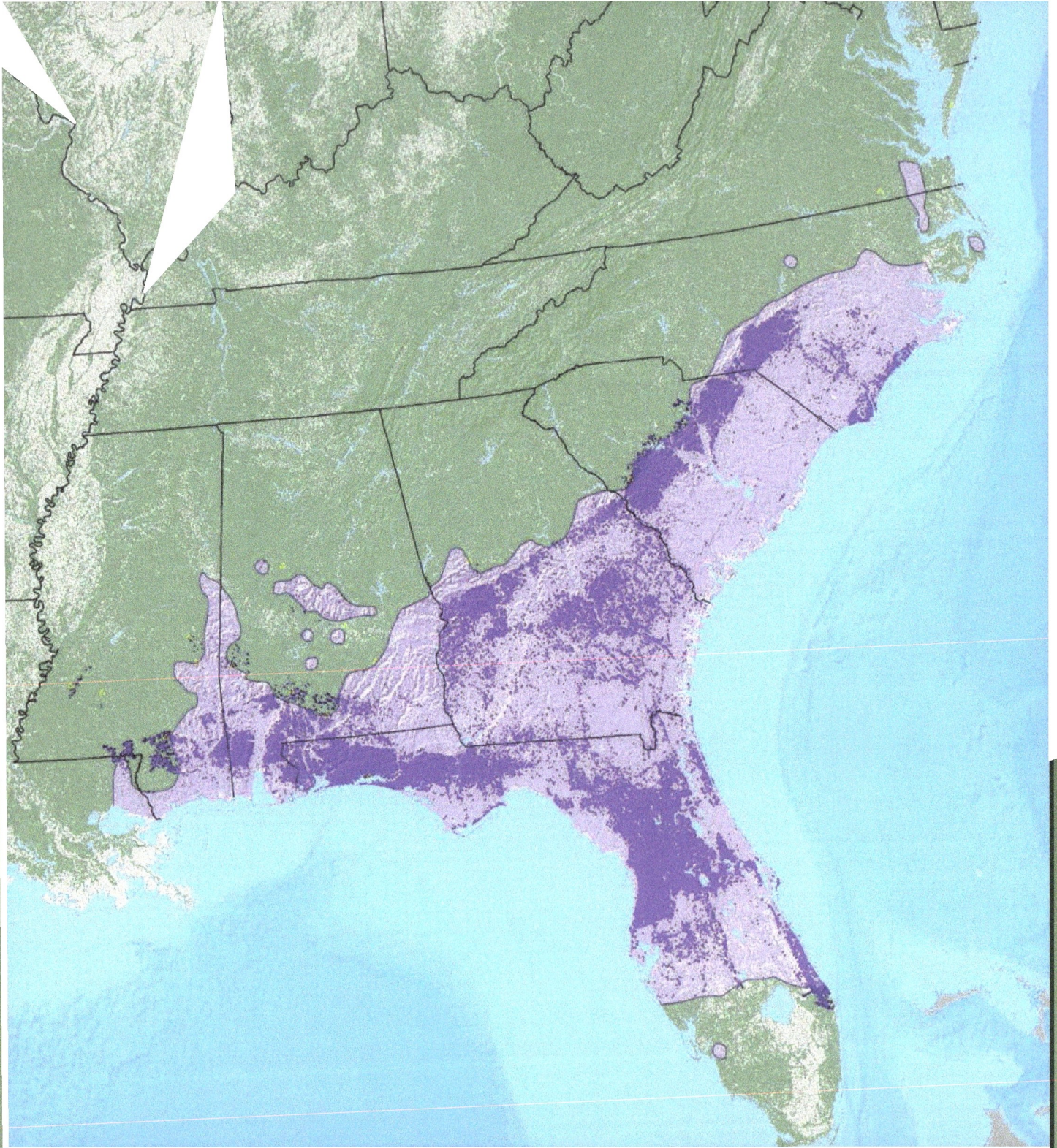

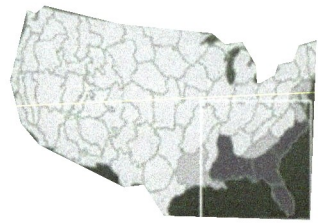

National Individual Tree Species Atlas

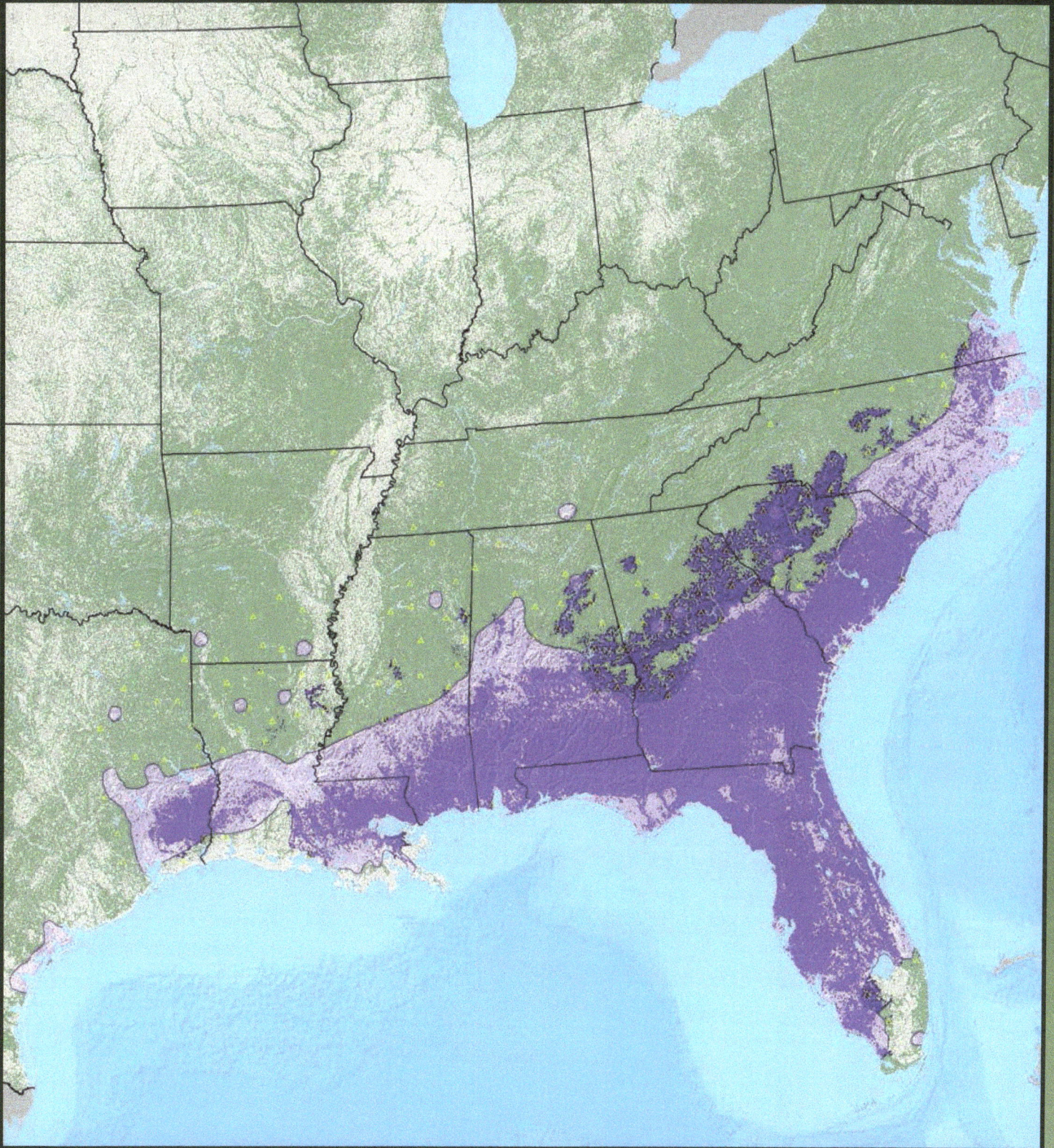

laurel oak

laurel oak
Quercus laurifolia Michx.
Fagaceae – Beech Family
FIA Code **820** – Plants Symbol **QULA3**

National Individual Tree Species Atlas

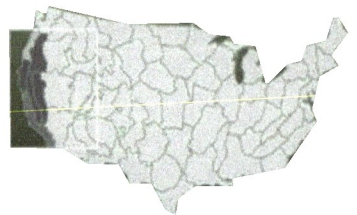

National Individual Tree Species Atlas

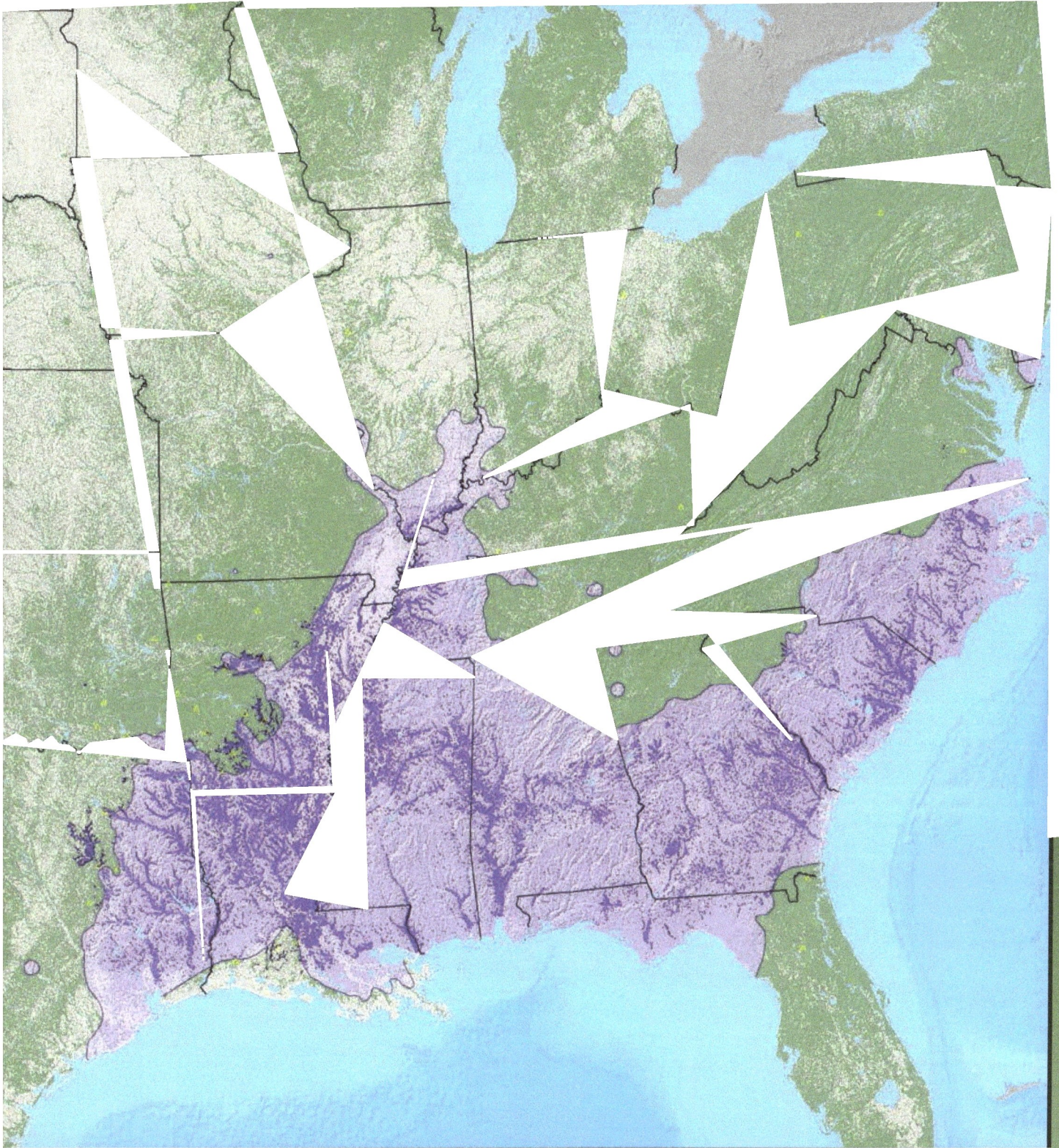

National Individual Tree Species Atlas

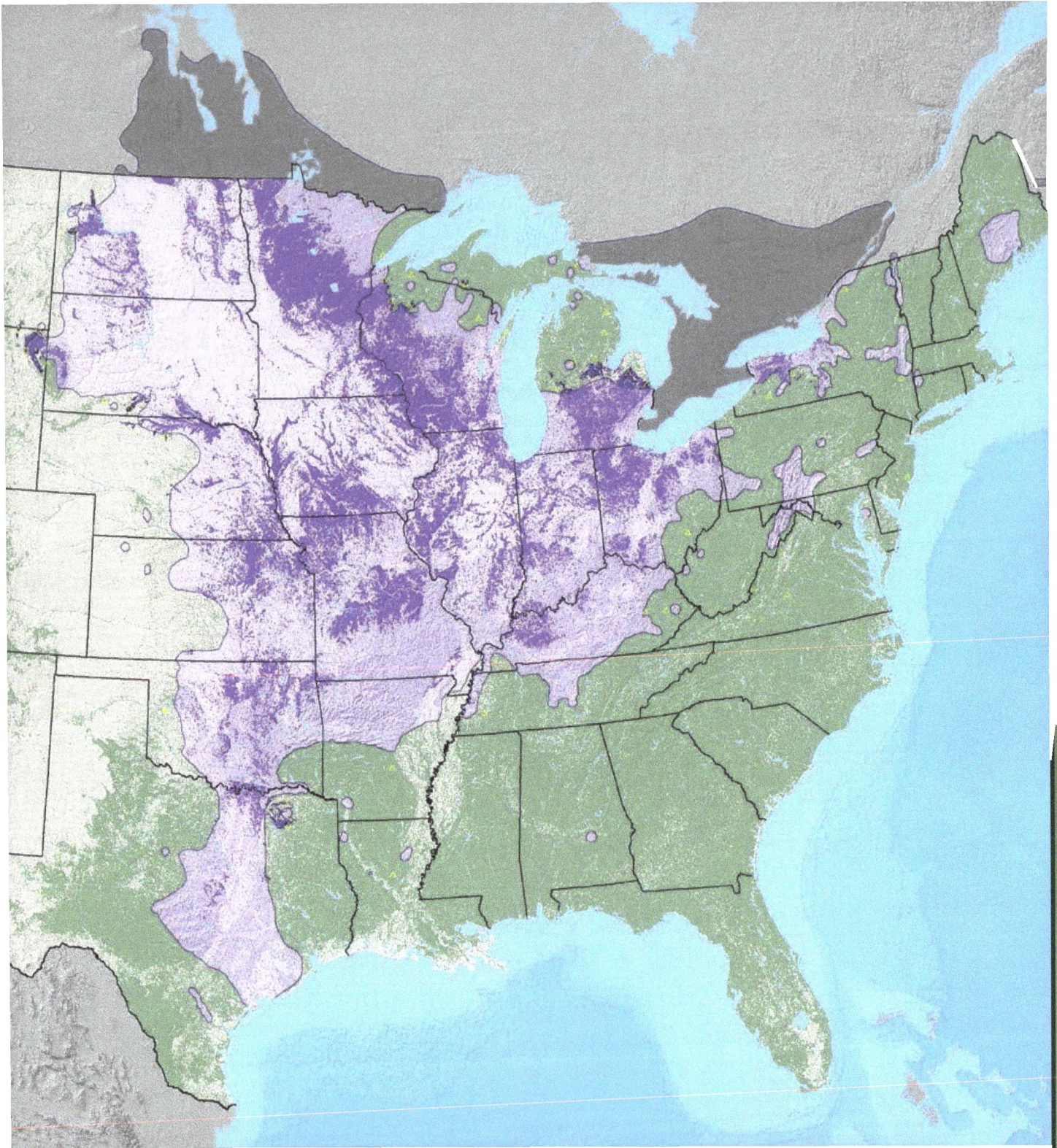

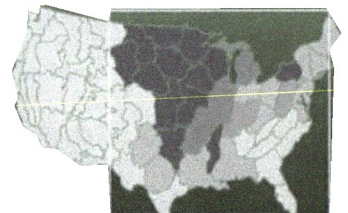

National Individual Tree Species Atlas

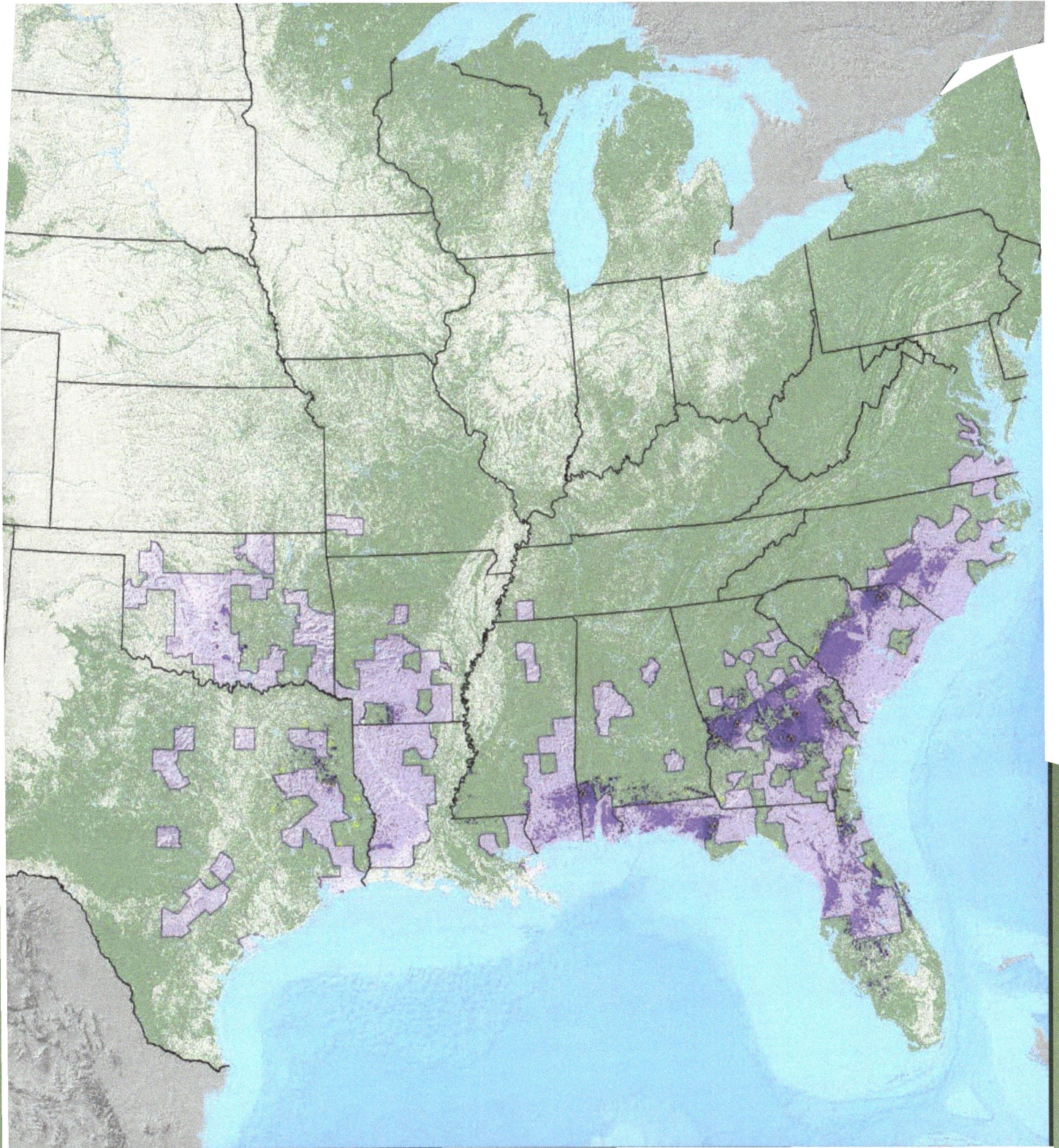

National Individual Tree Species Atlas

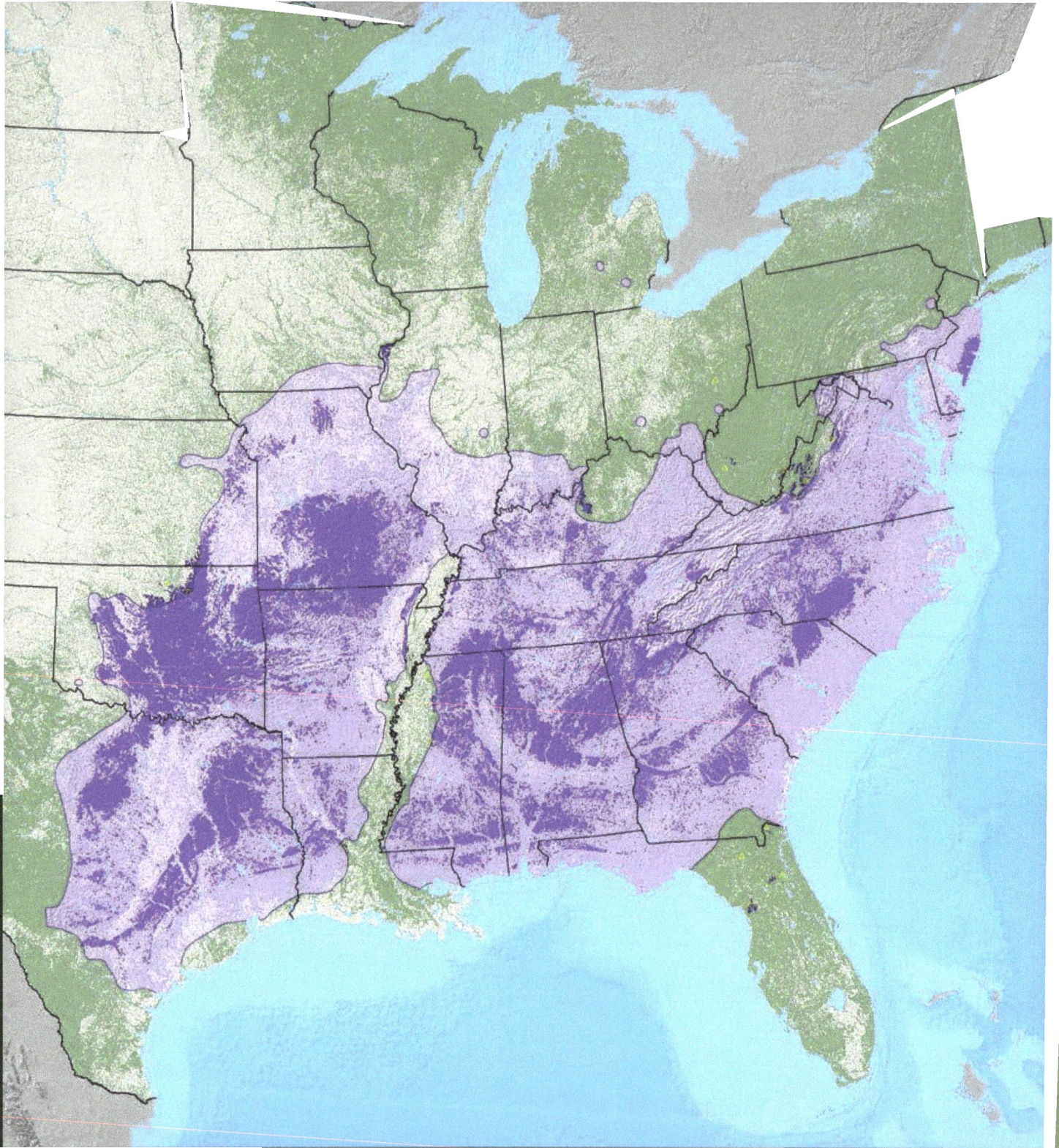

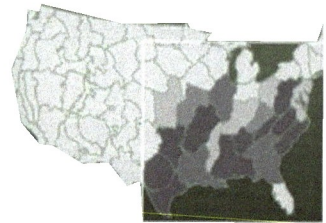

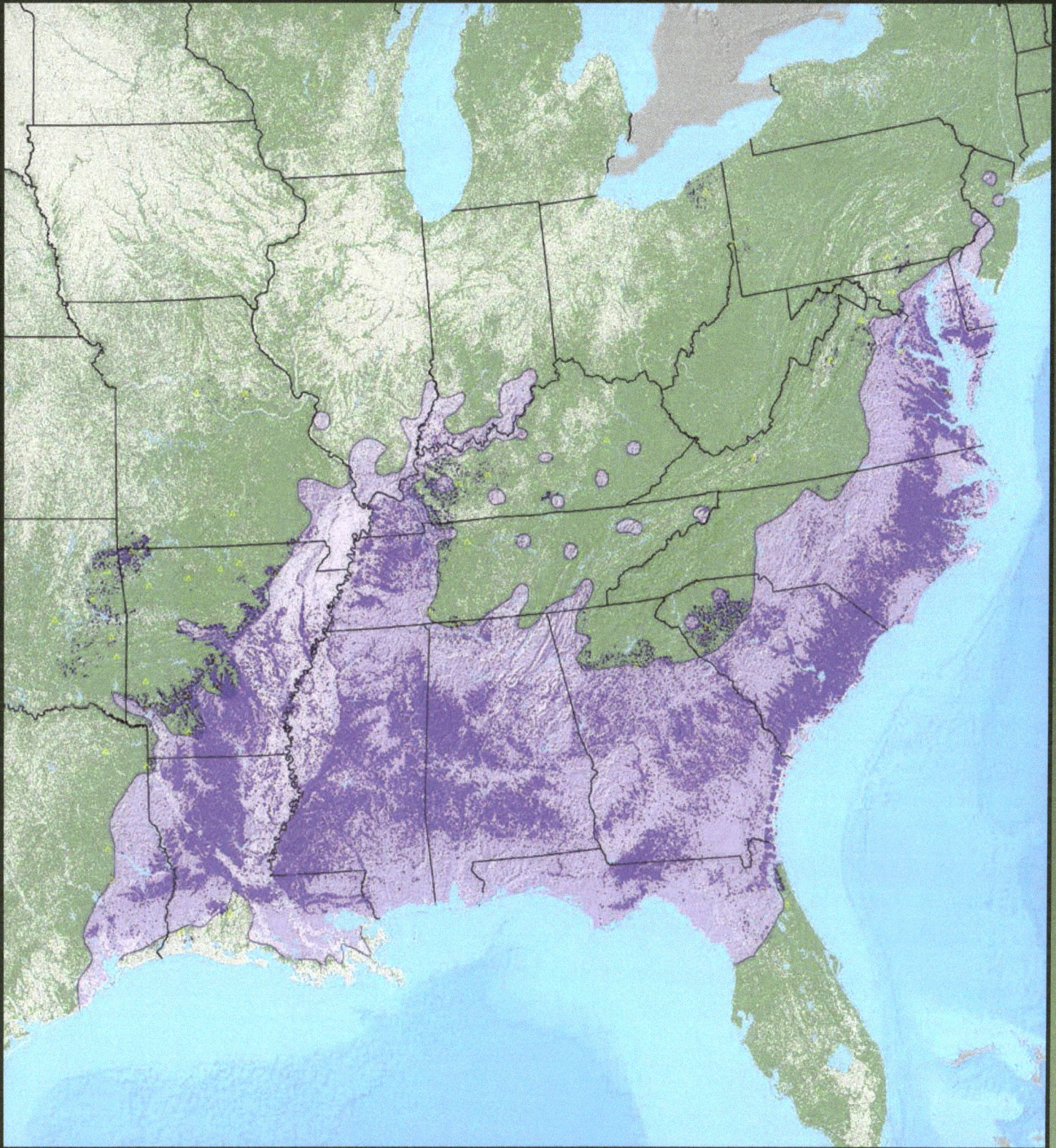

swamp chestnut oak

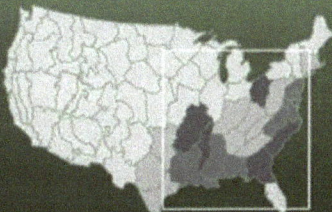

swamp chestnut oak
Quercus michauxii Nutt.
Fagaceae – Beech Family
FIA Code **825** – Plants Symbol **QUMI**

dwarf live oak

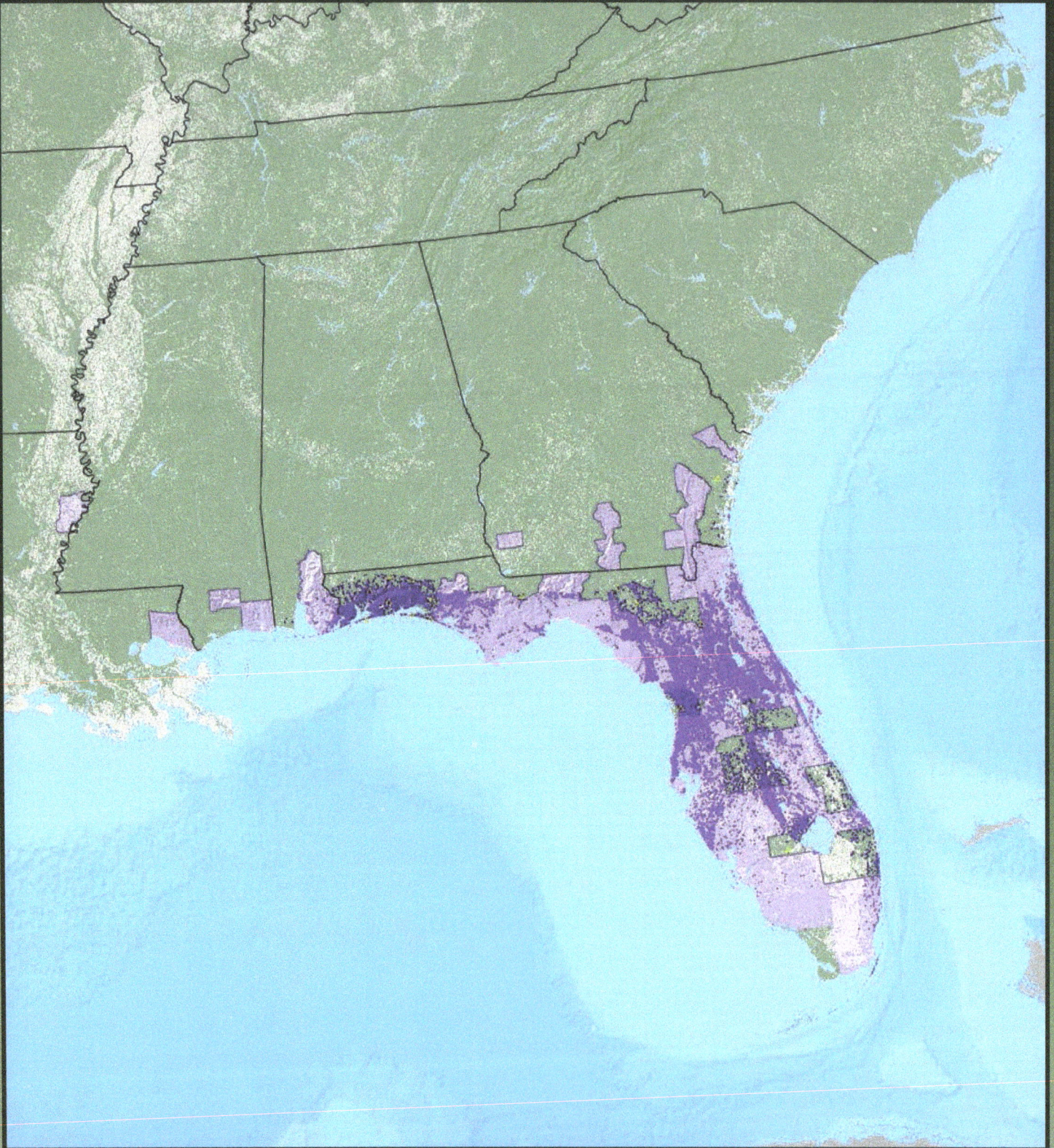

dwarf live oak

Quercus minima (Sarg.) Small

Fagaceae – Beech Family

FIA Code **841** – Plants Symbol **QUMI2**

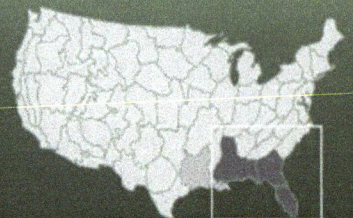

National Individual Tree Species Atlas

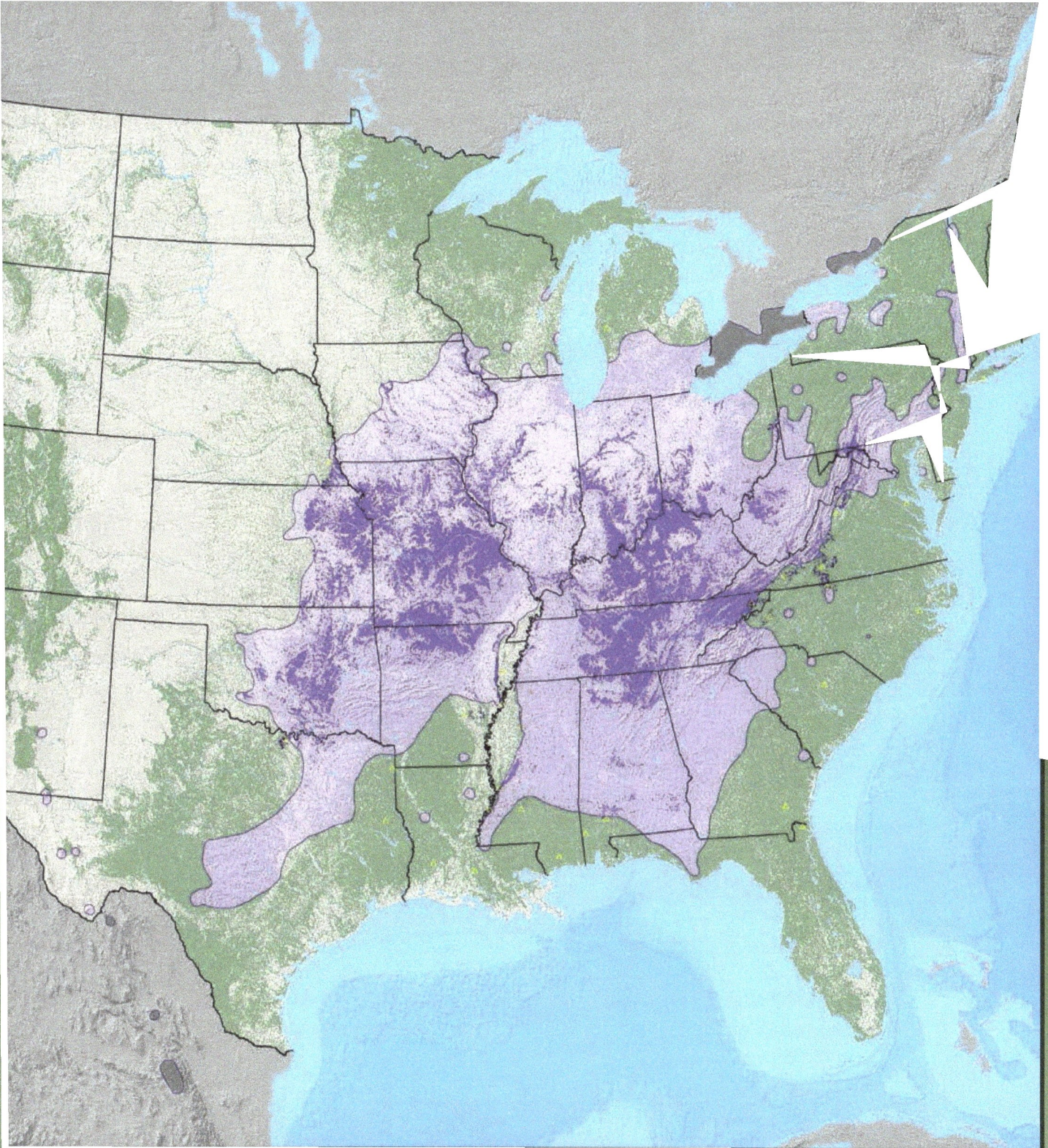
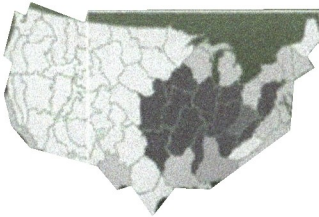
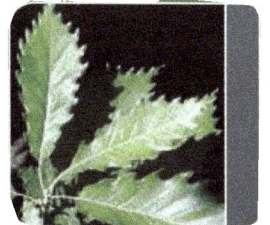

National Individual Tree Species Atlas

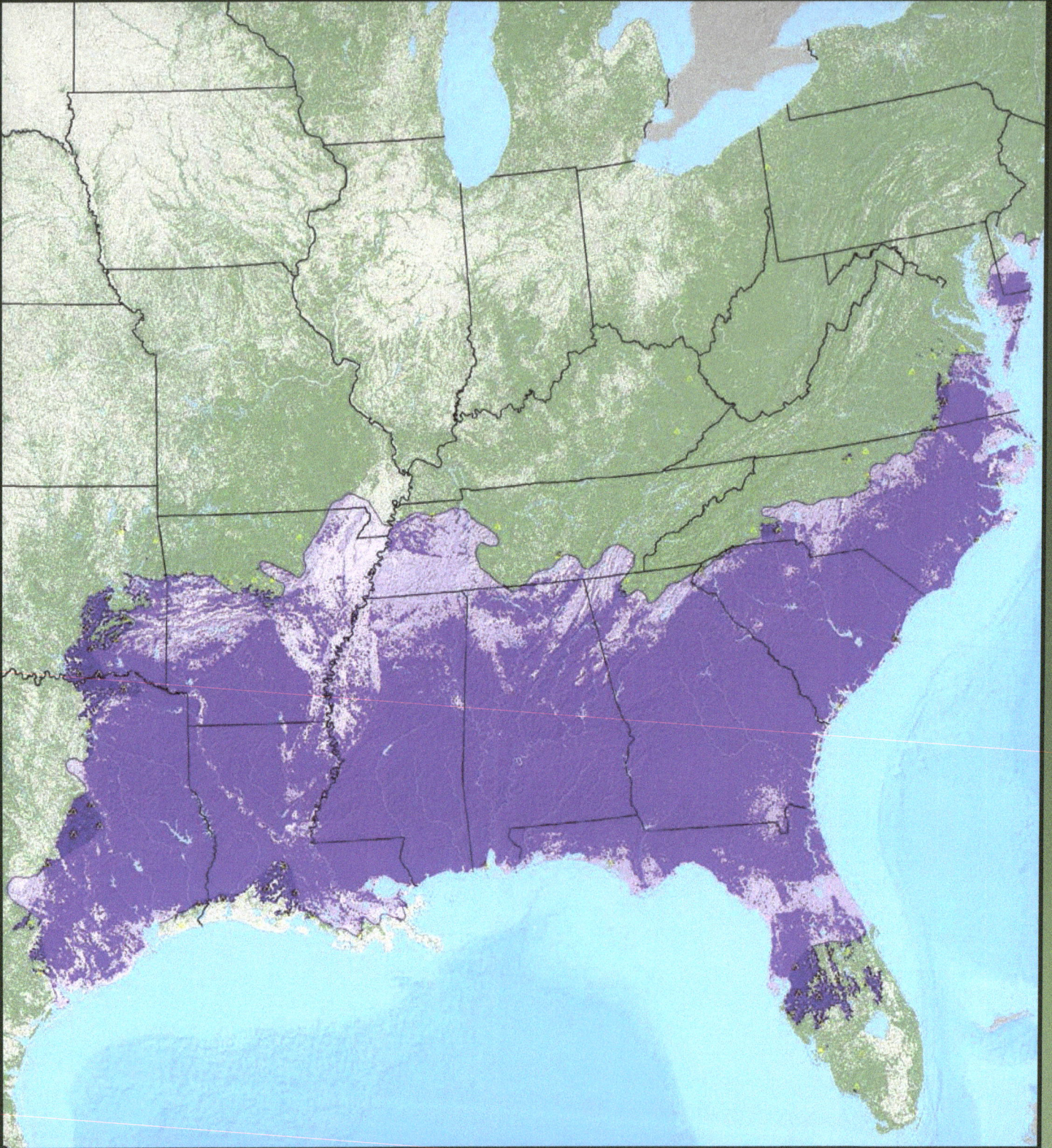

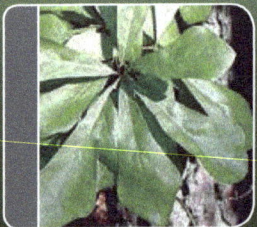

water oak

water oak
Quercus nigra L.
Fagaceae – Beech Family
FIA Code **827** – Plants Symbol **QUNI**

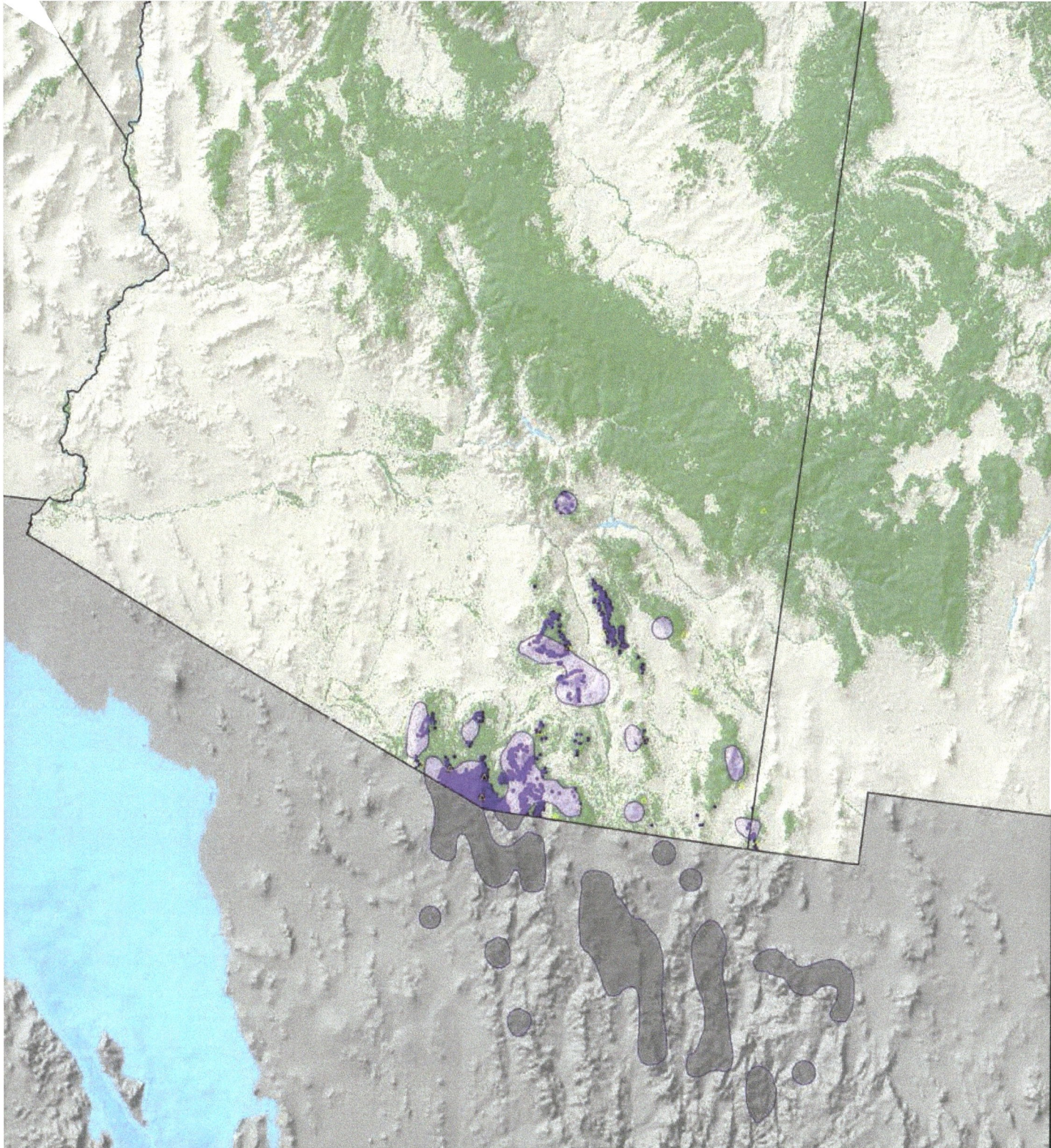

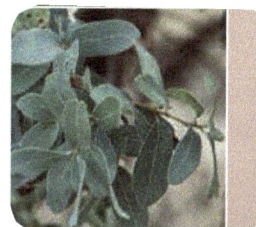

National Individual Tree Species Atlas

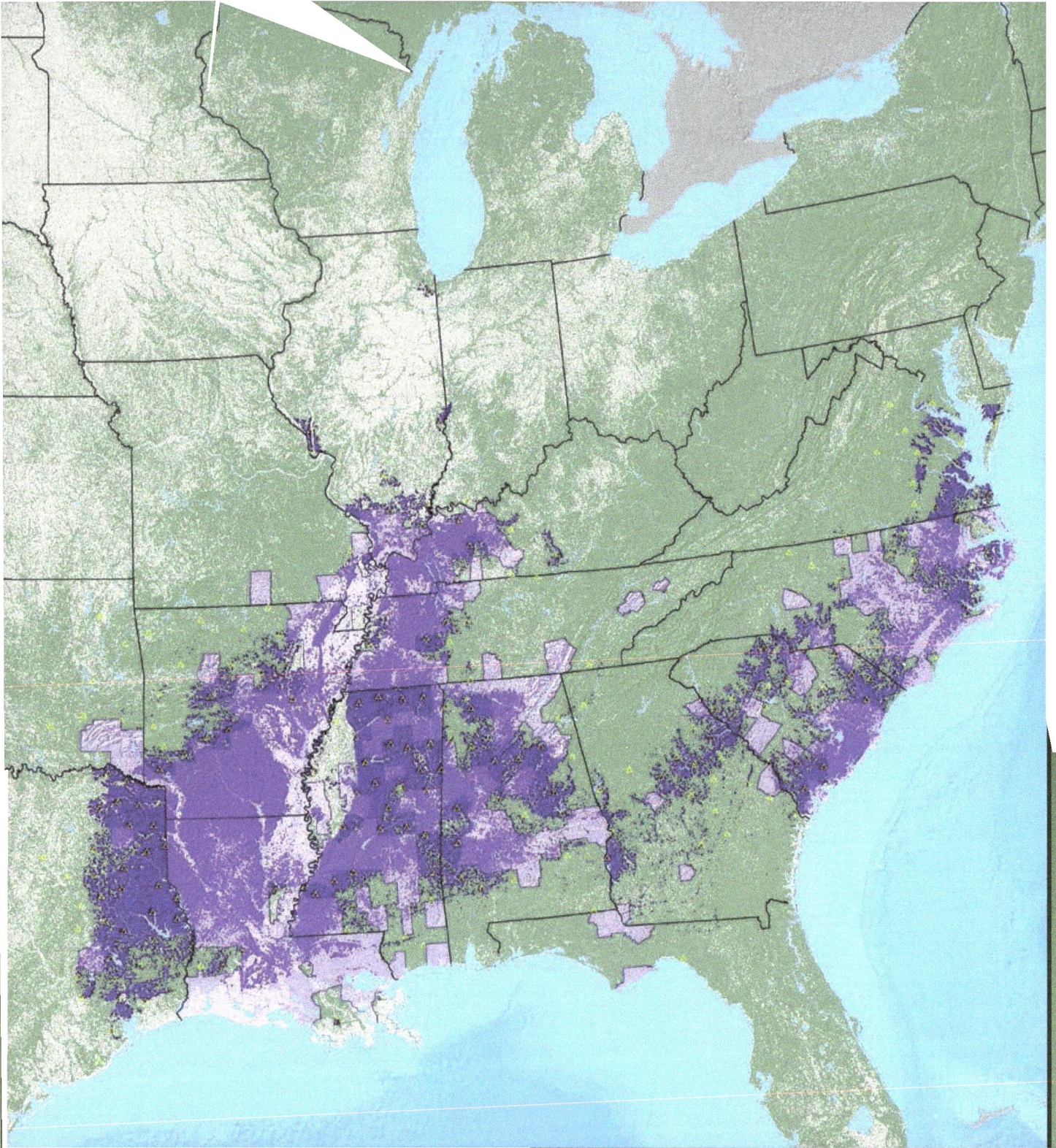

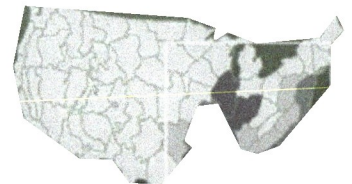

National Individual Tree Species Atlas

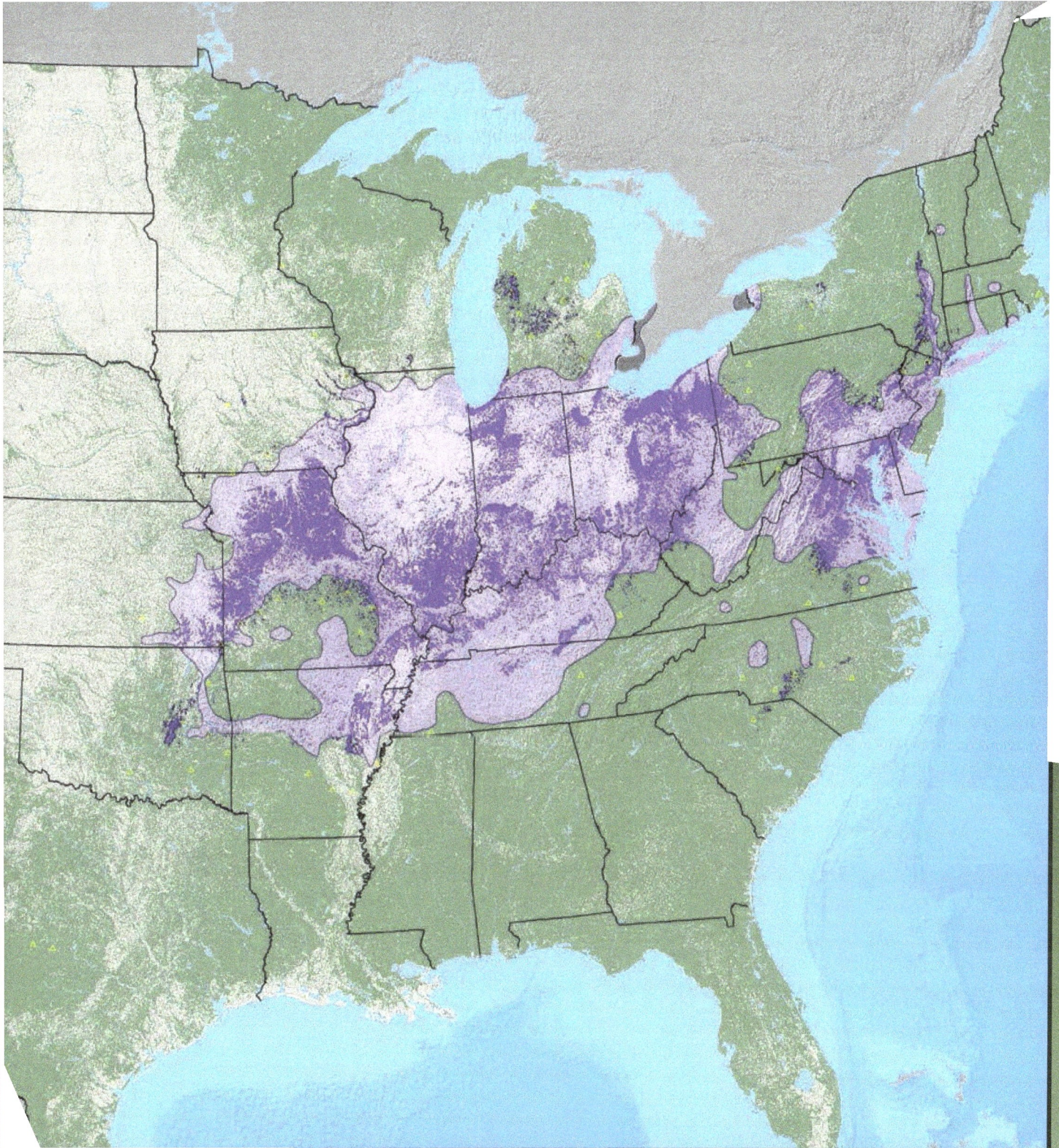

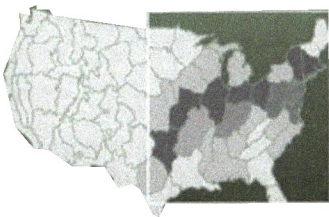

National Individual Tree Species Atlas

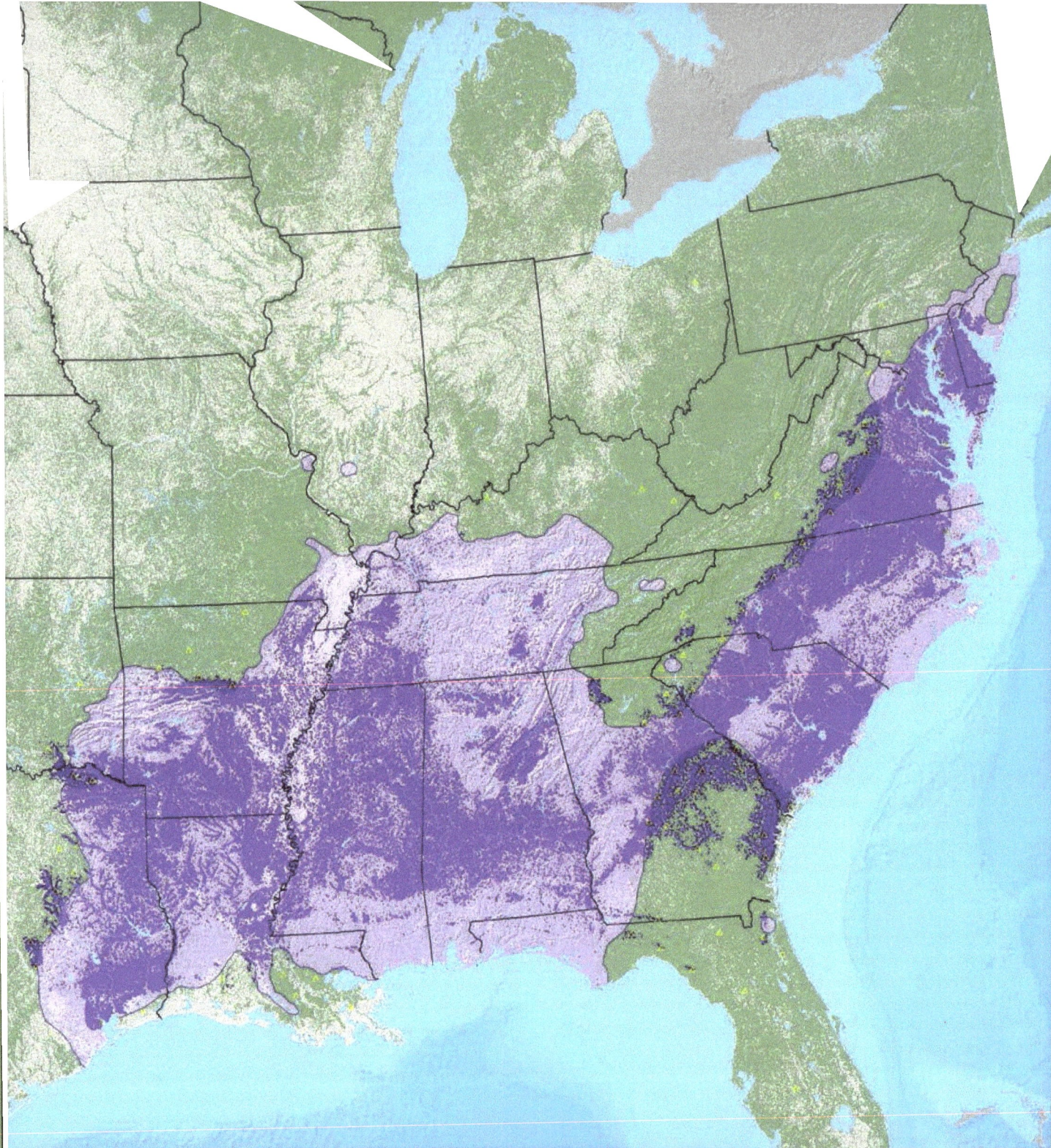

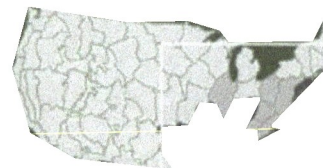

National Individual Tree Species Atlas

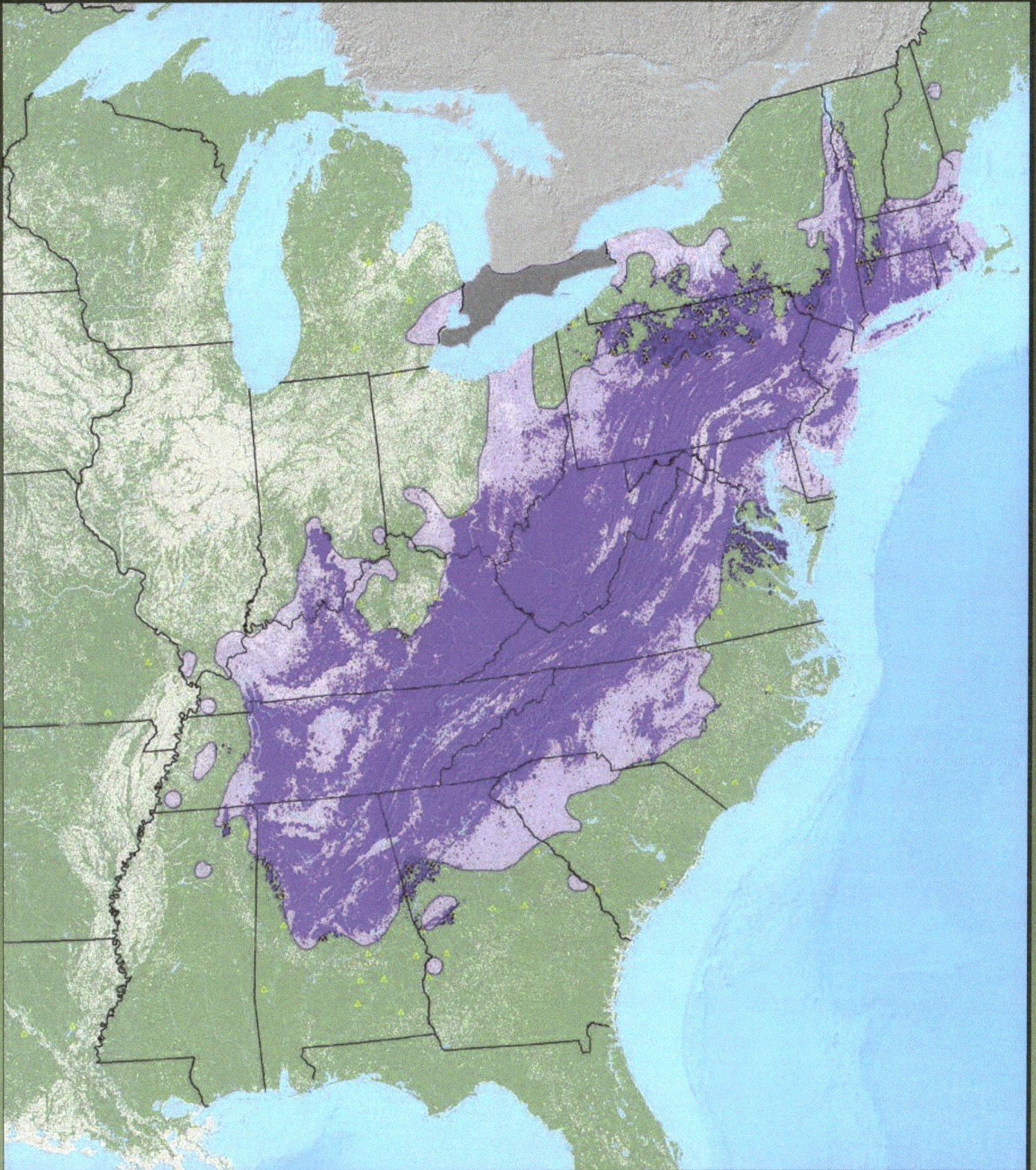

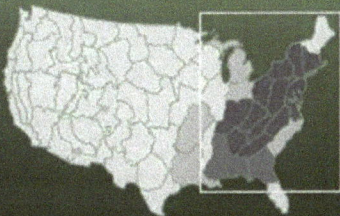

chestnut oak

chestnut oak
Quercus prinus L.
Fagaceae – Beech Family
FIA Code **832** – Plants Symbol **QUPR2**

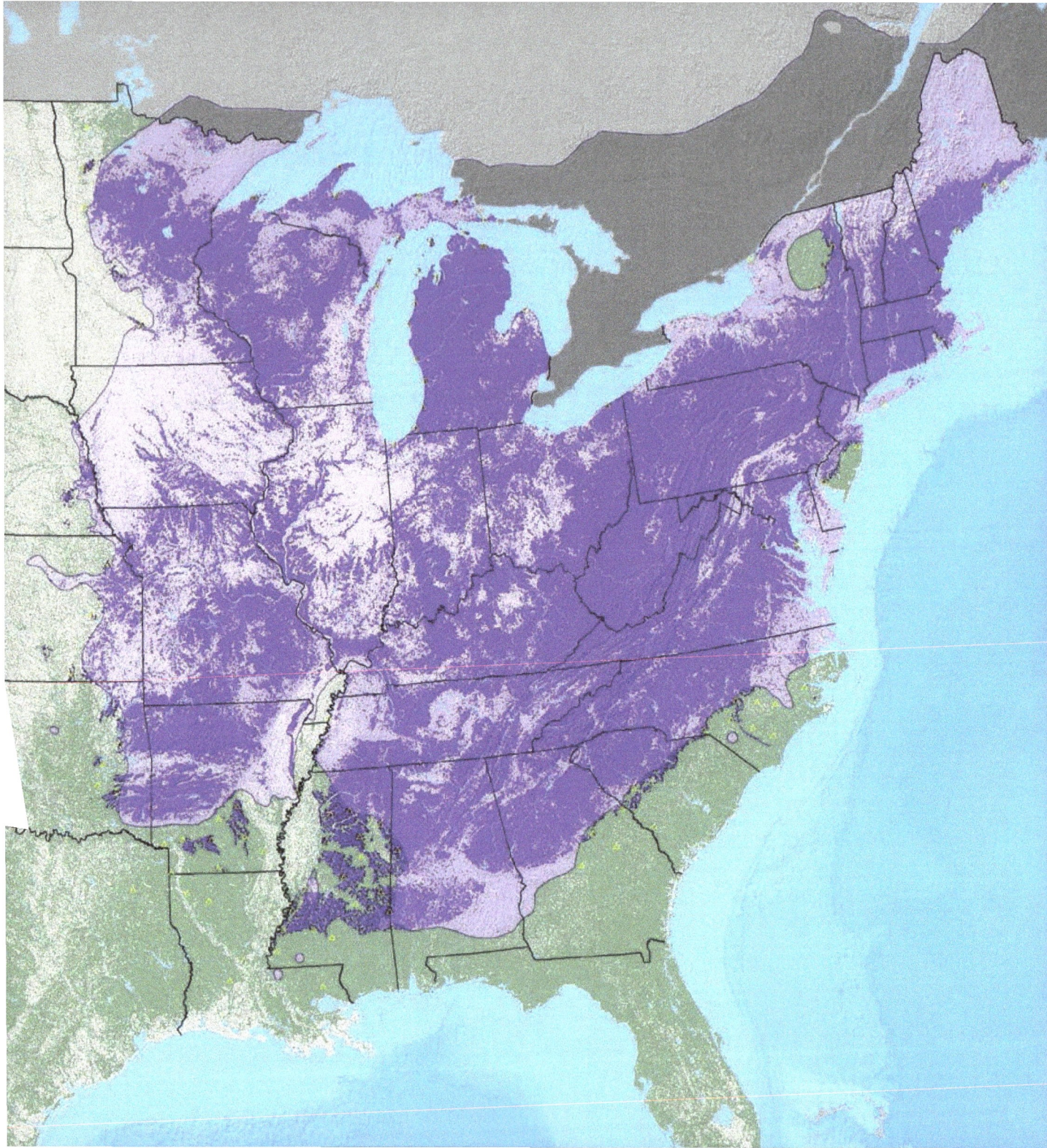

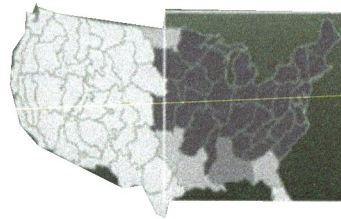

National Individual Tree Species Atlas

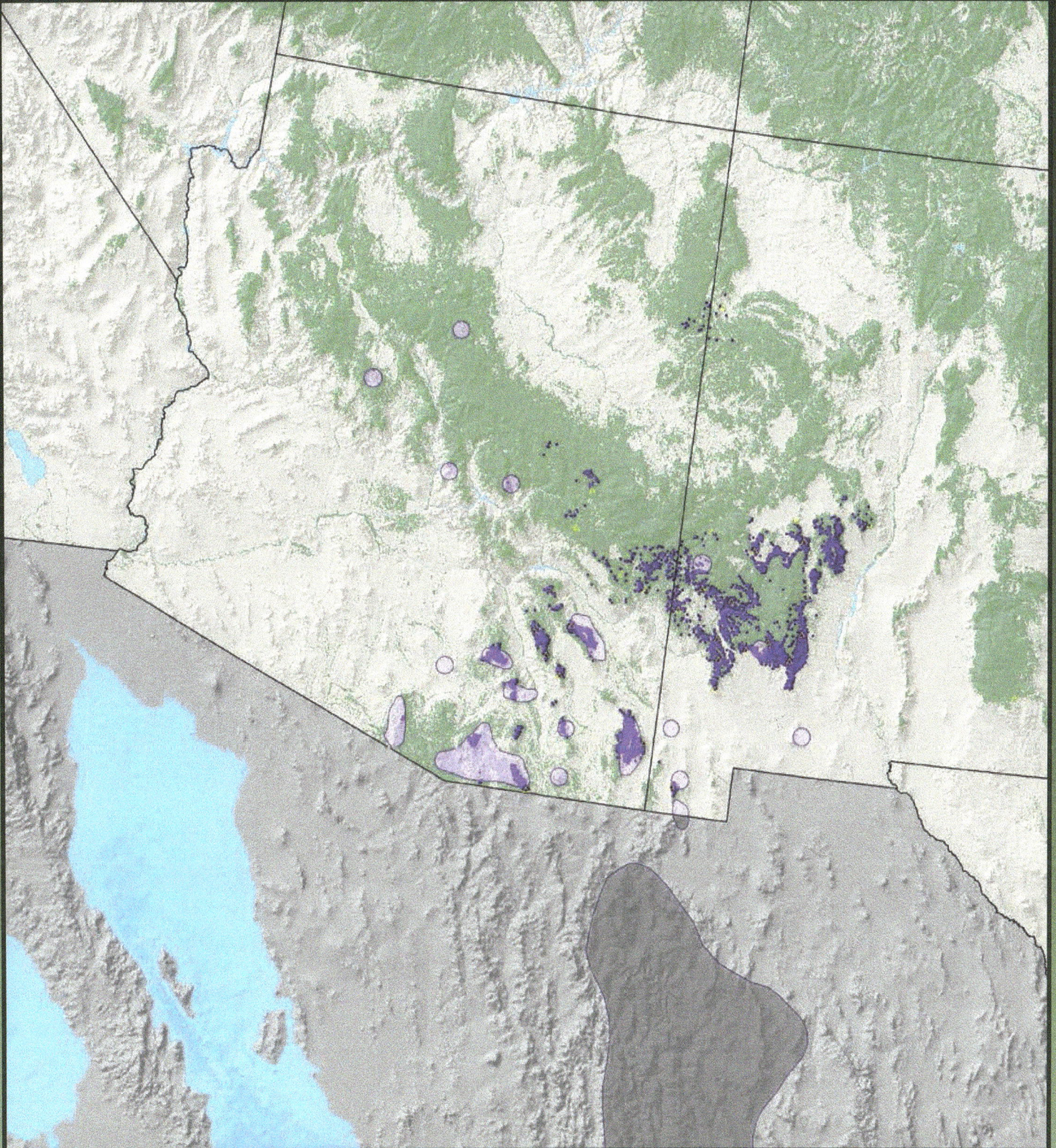

netleaf oak

Quercus rugosa Née

Fagaceae – Beech Family

FIA Code **847** – Plants Symbol **QURU4**

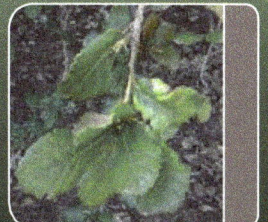

National Individual Tree Species Atlas

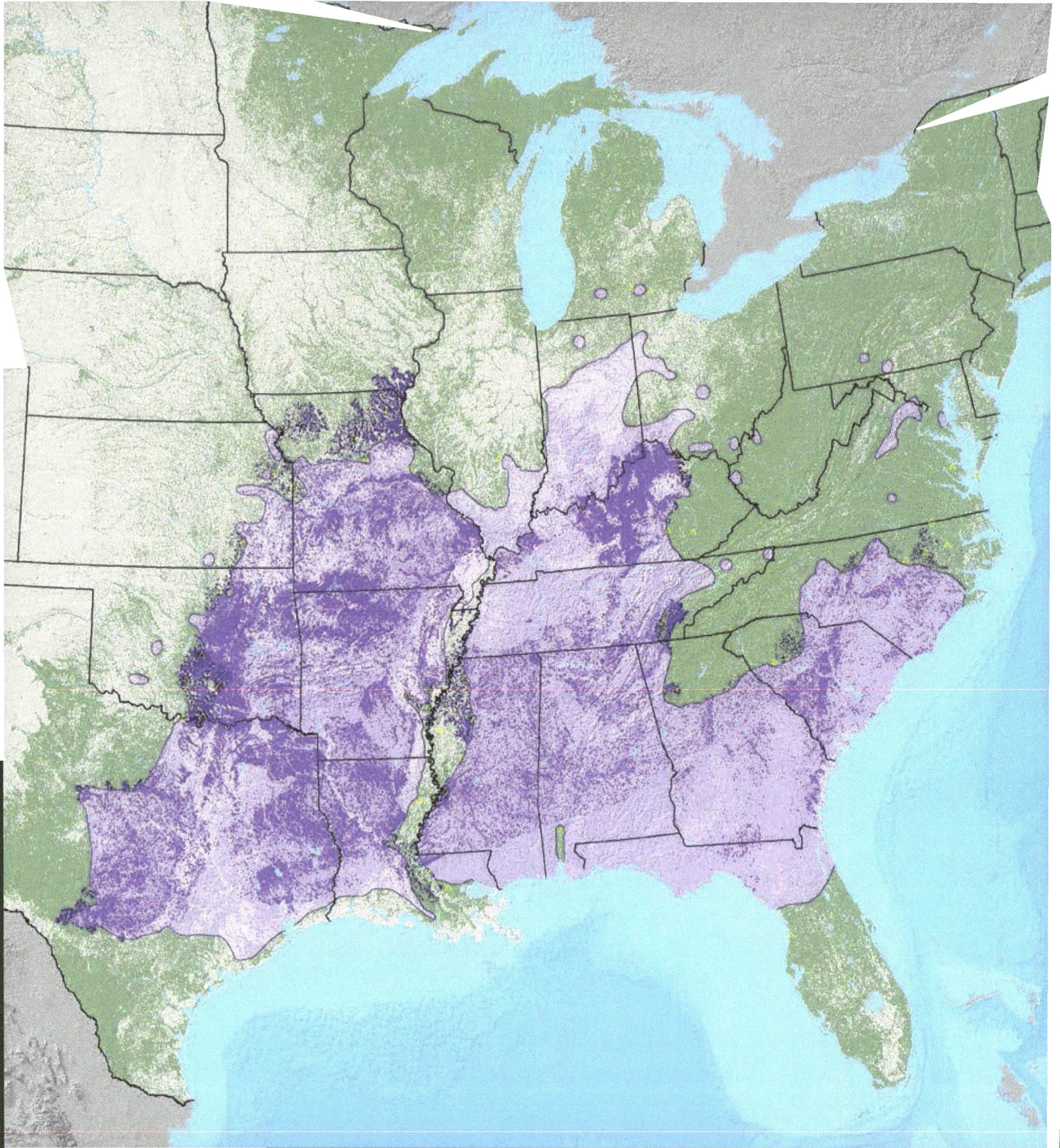

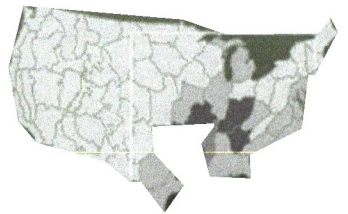

National Individual Tree Species Atlas

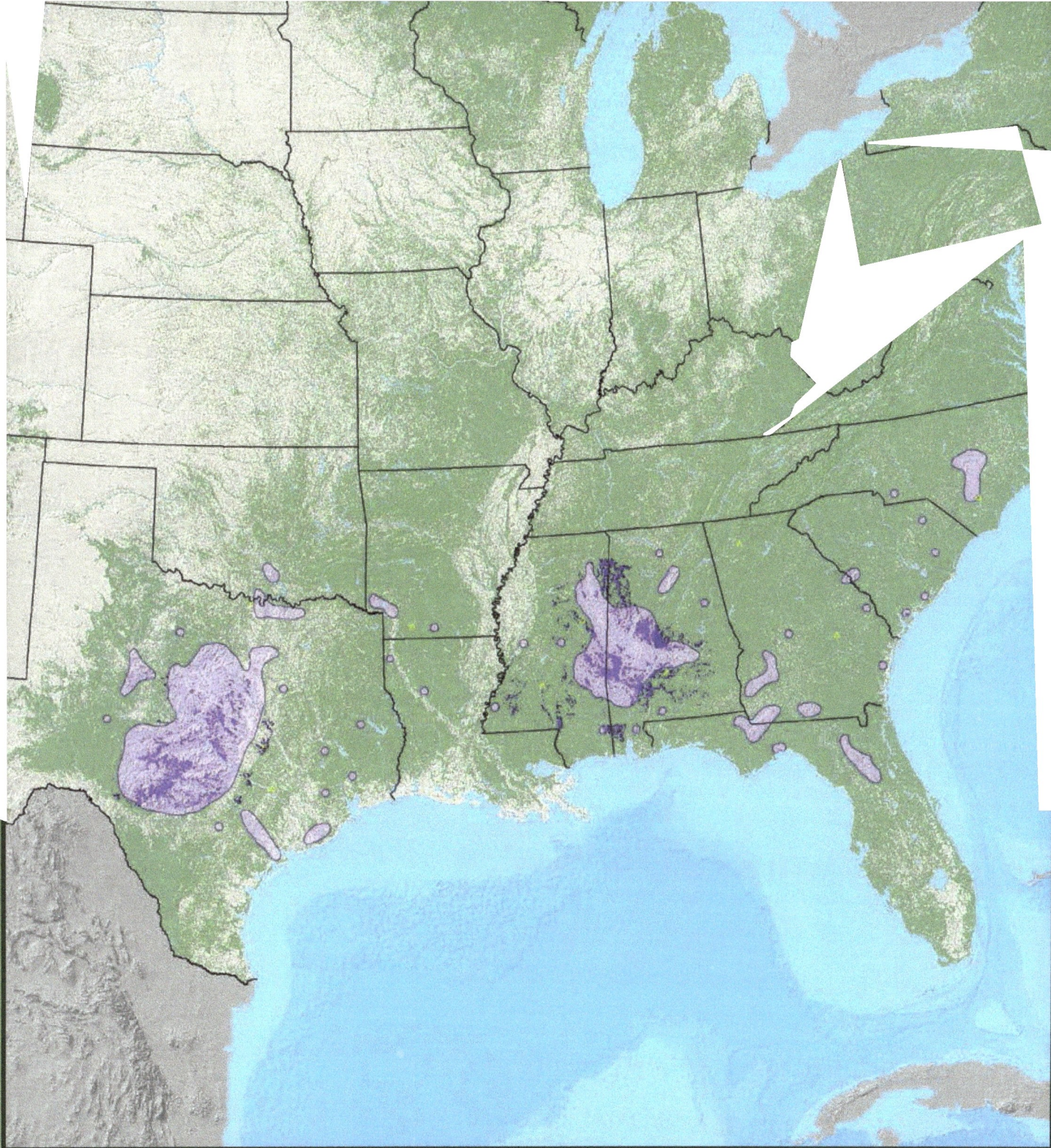

Durand oak
Quercus sinuata Walter var. *sinuata*

Fagaceae – Beech Family

FIA Code **808** – Plants Symbol **QUSIS**

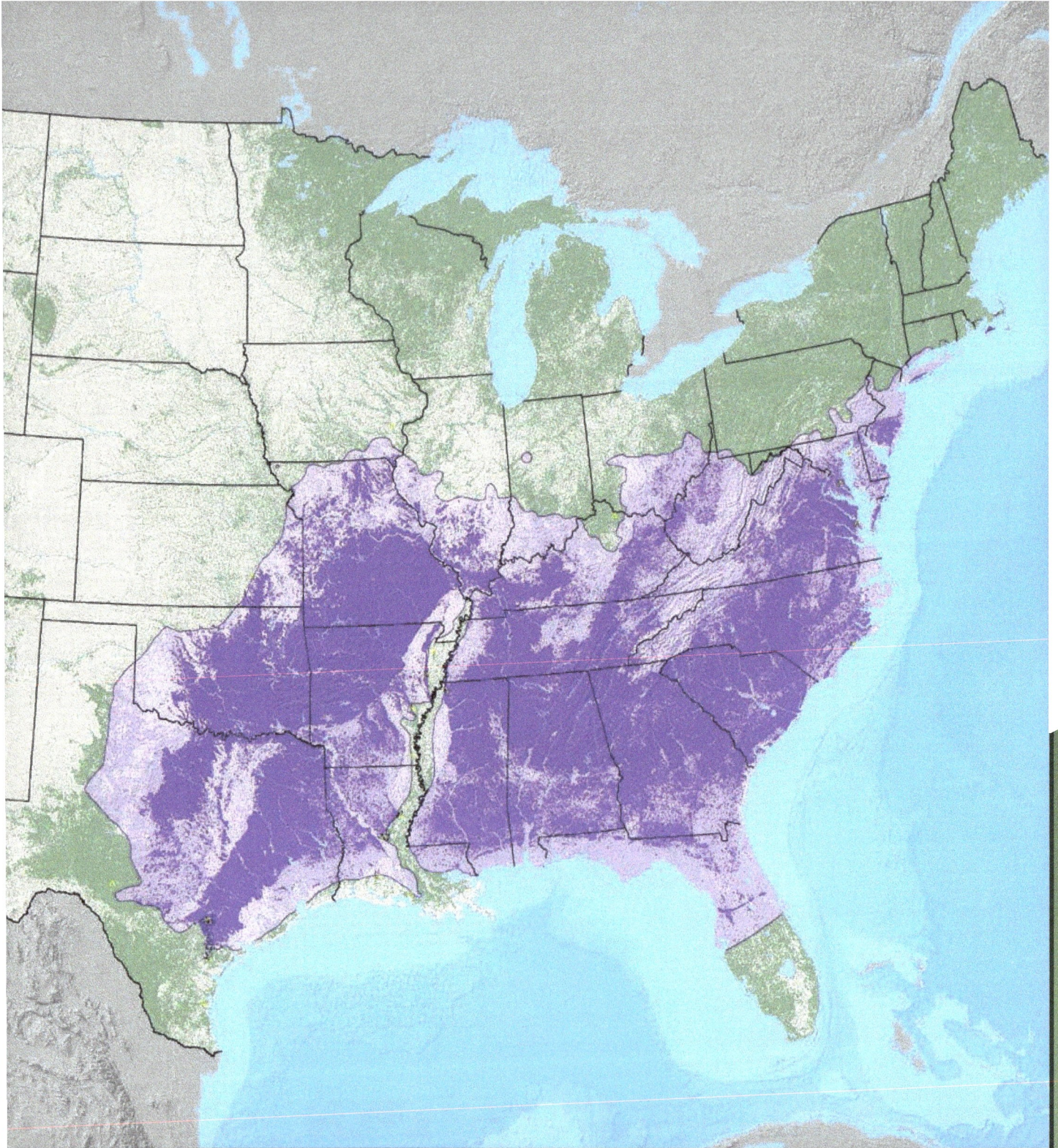

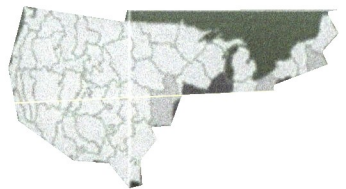

National Individual Tree Species Atlas

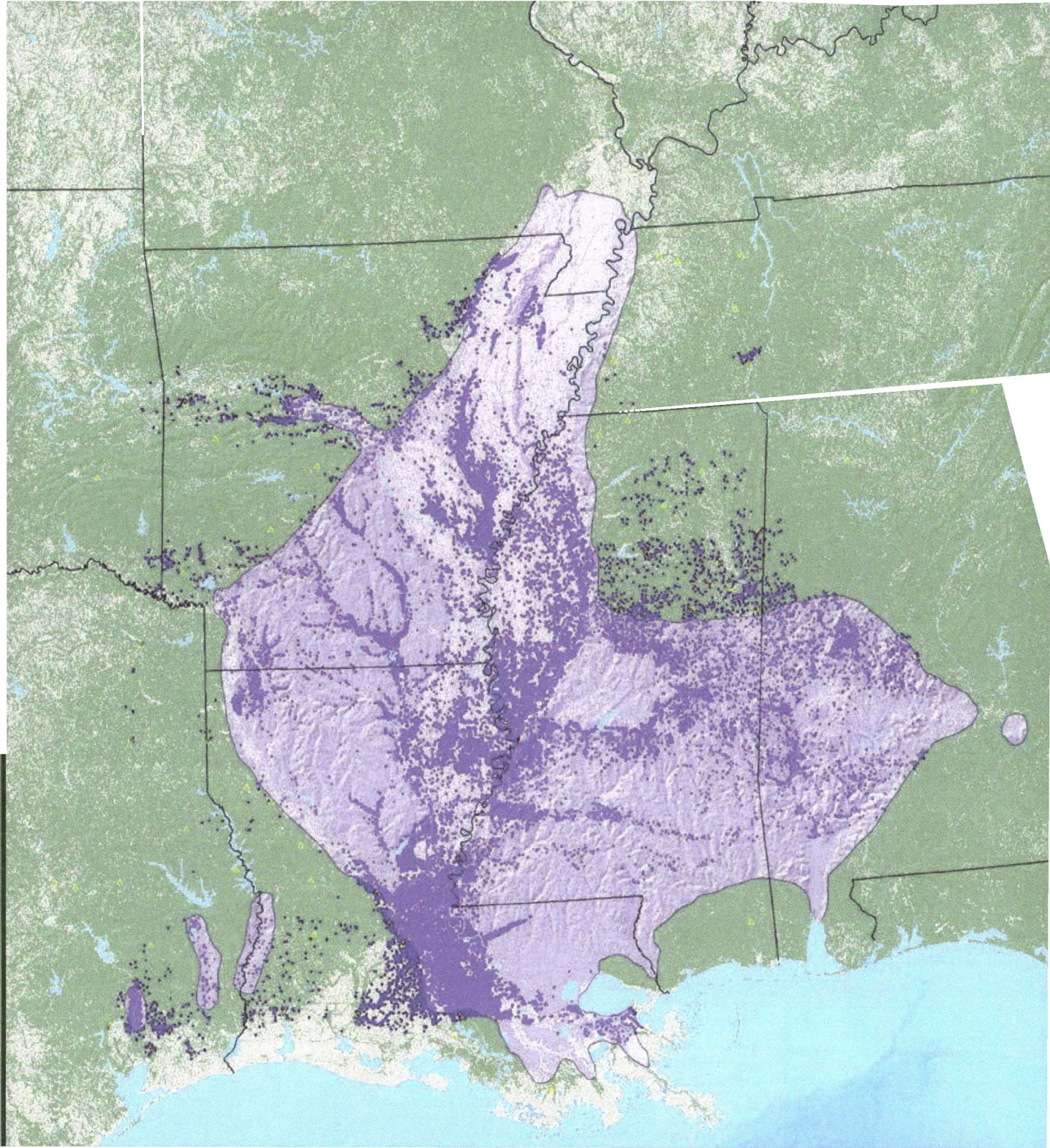

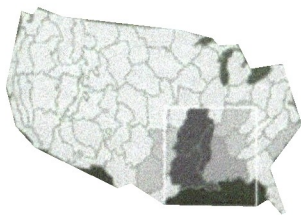

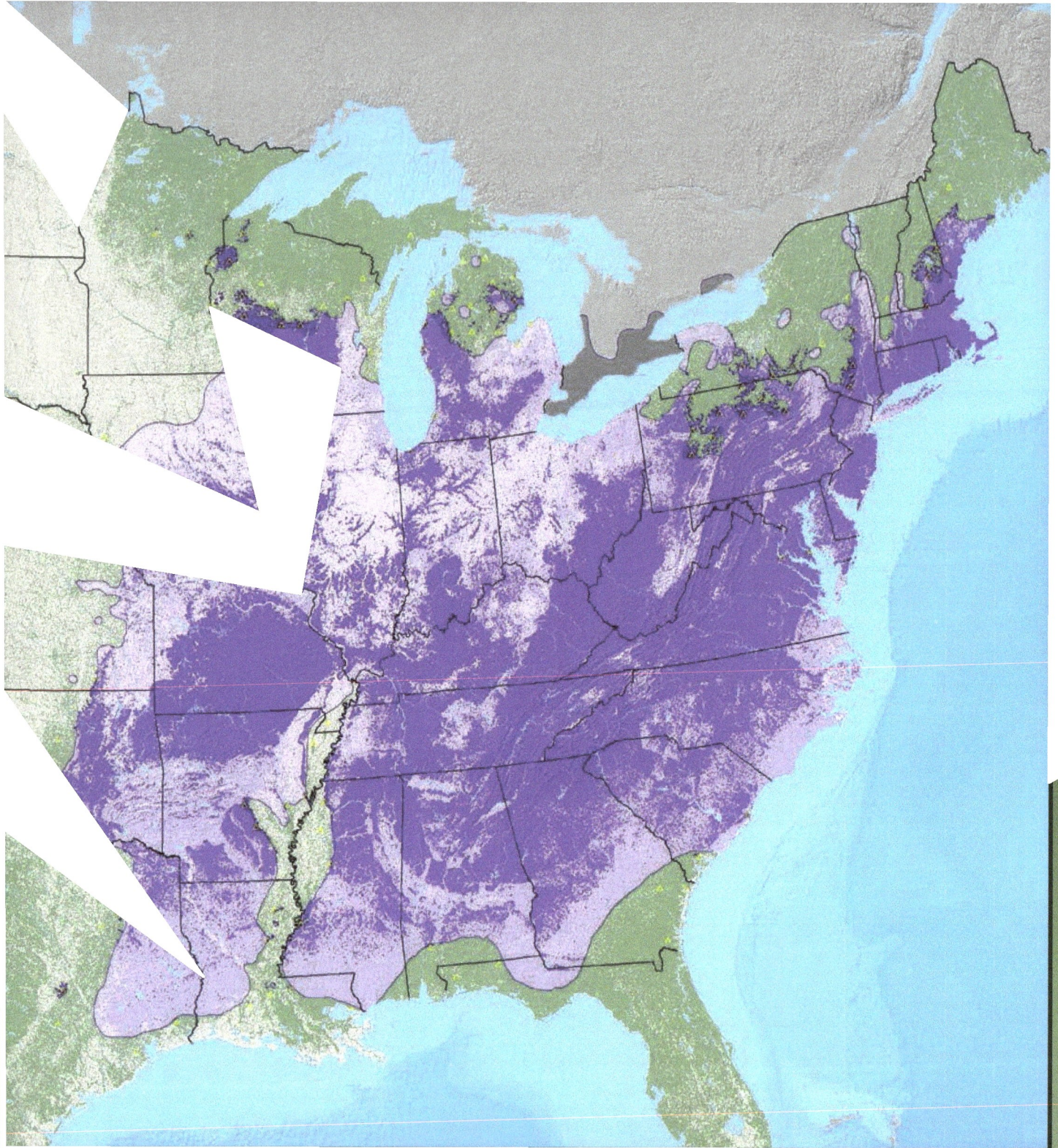

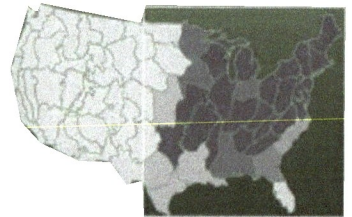

National Individual Tree Species Atlas

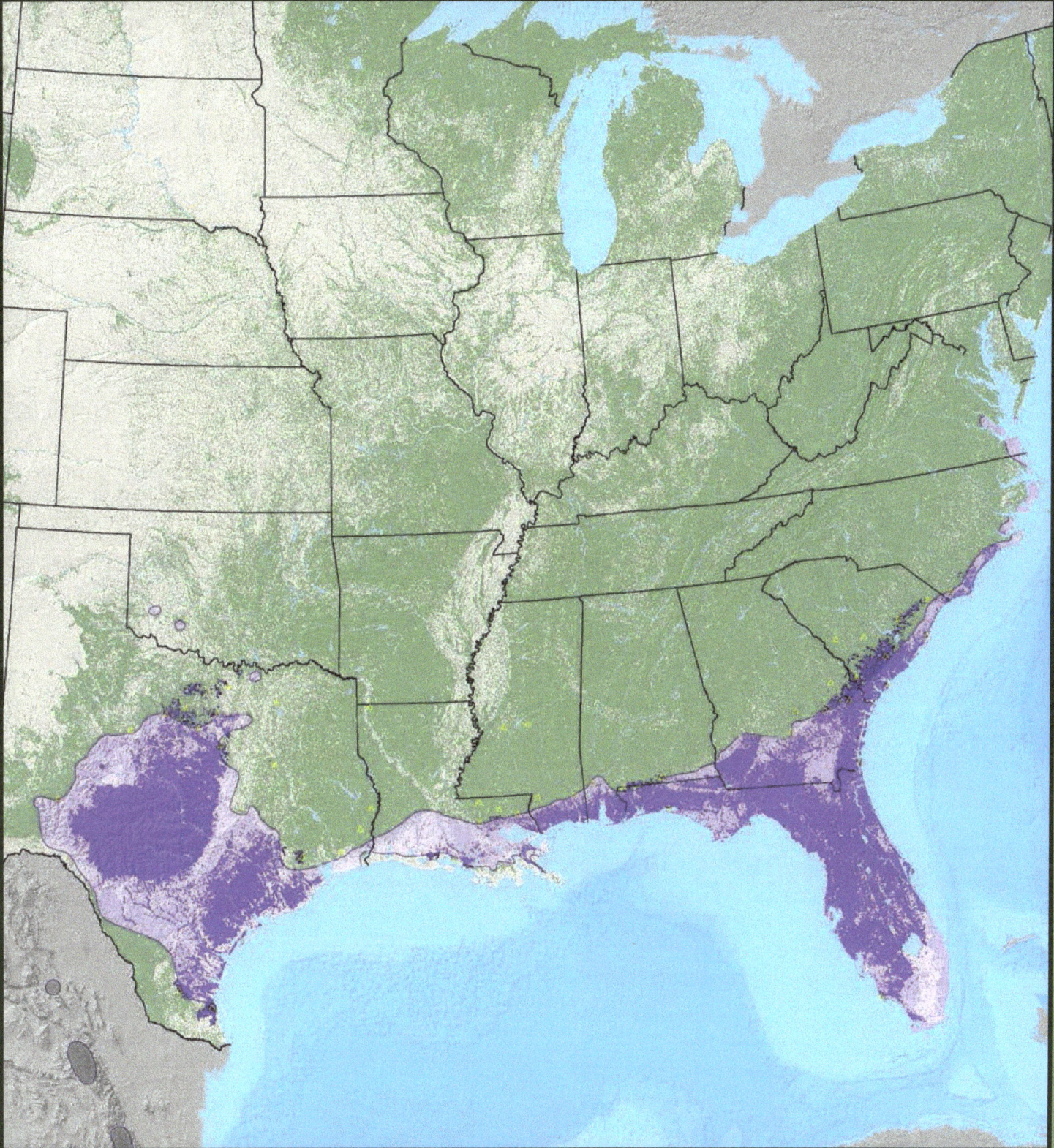

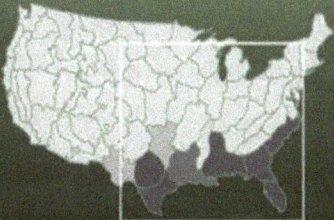

live oak

live oak

Quercus virginiana Mill.

Fagaceae – Beech Family

FIA Code **838** – Plants Symbol **QUVI**

National Individual Tree Species Atlas

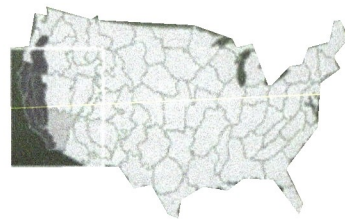

National Individual Tree Species Atlas

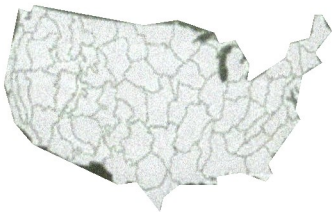
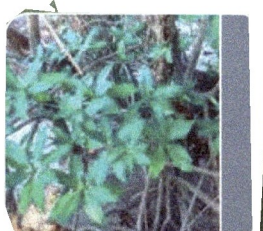

National Individual Tree Species Atlas

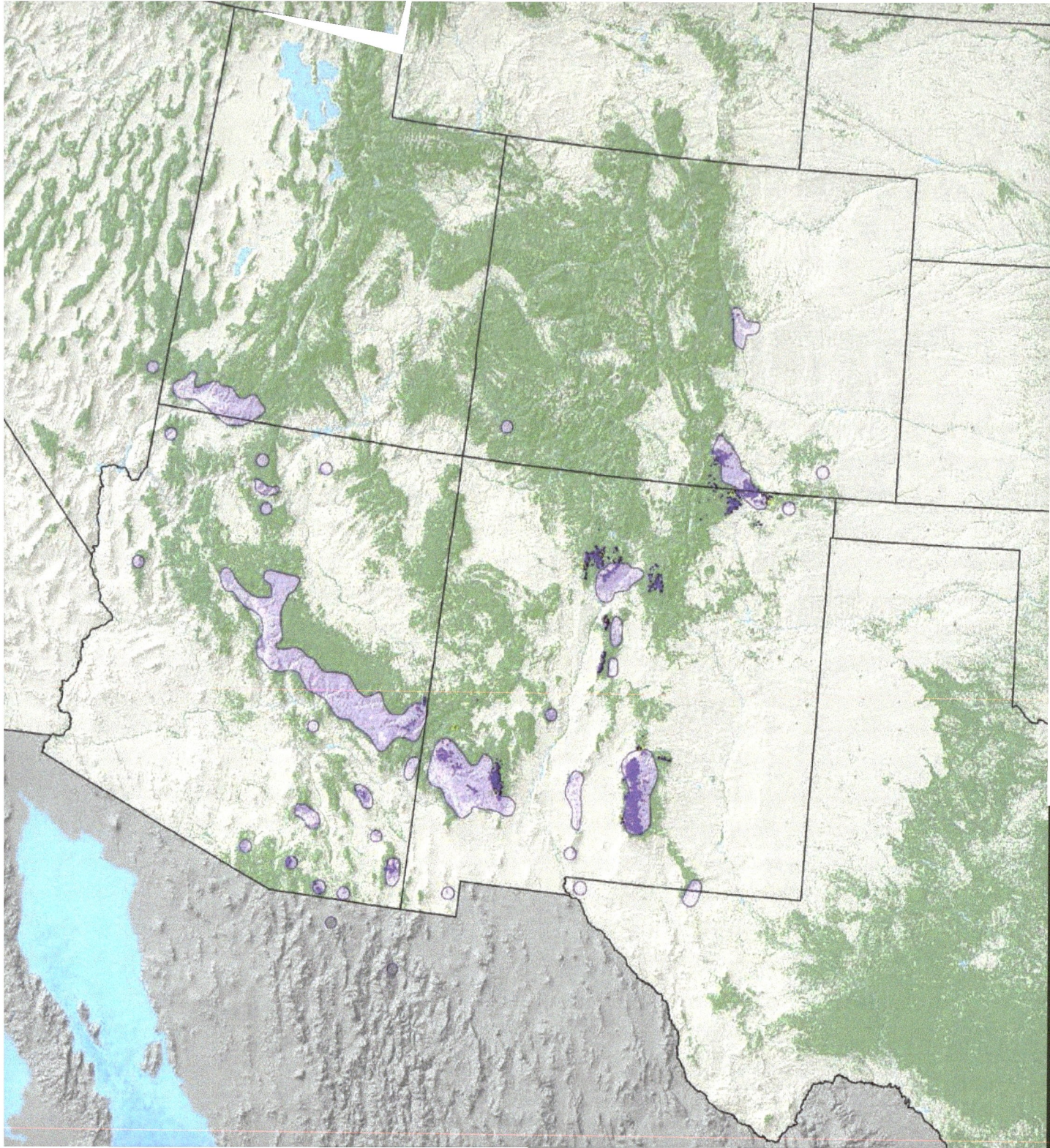

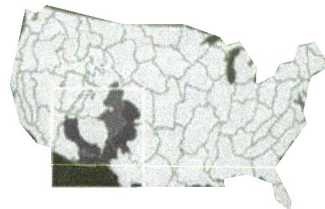

National Individual Tree Species Atlas

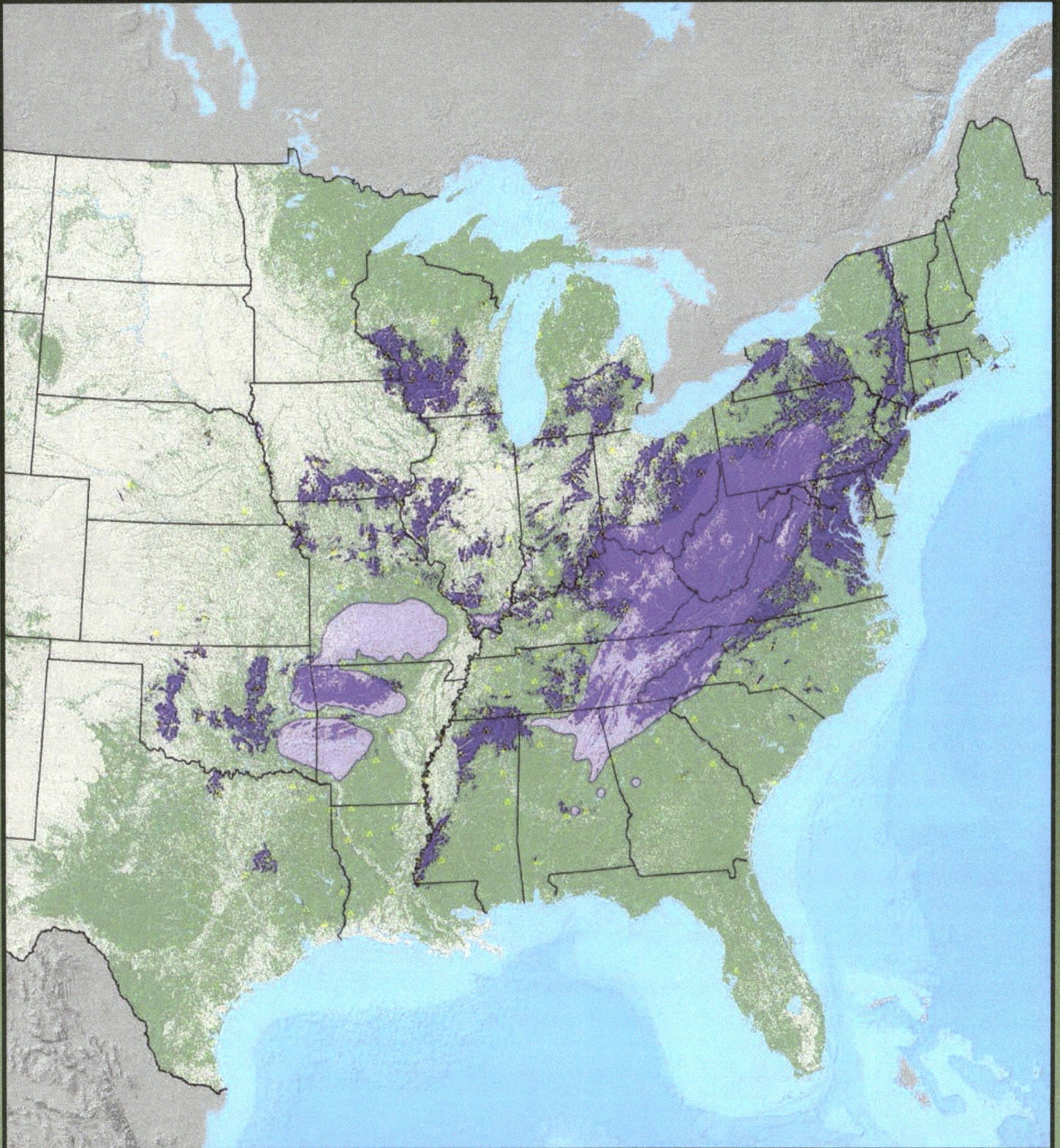

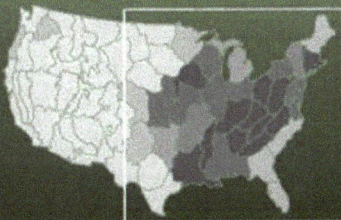

black locust

black locust
Robinia pseudoacacia L.
Fabaceae – Pea Family
FIA Code **901** – Plants Symbol **ROPS**

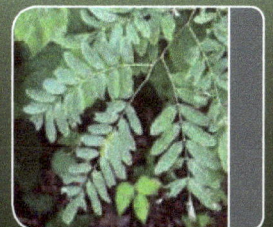

National Individual Tree Species Atlas

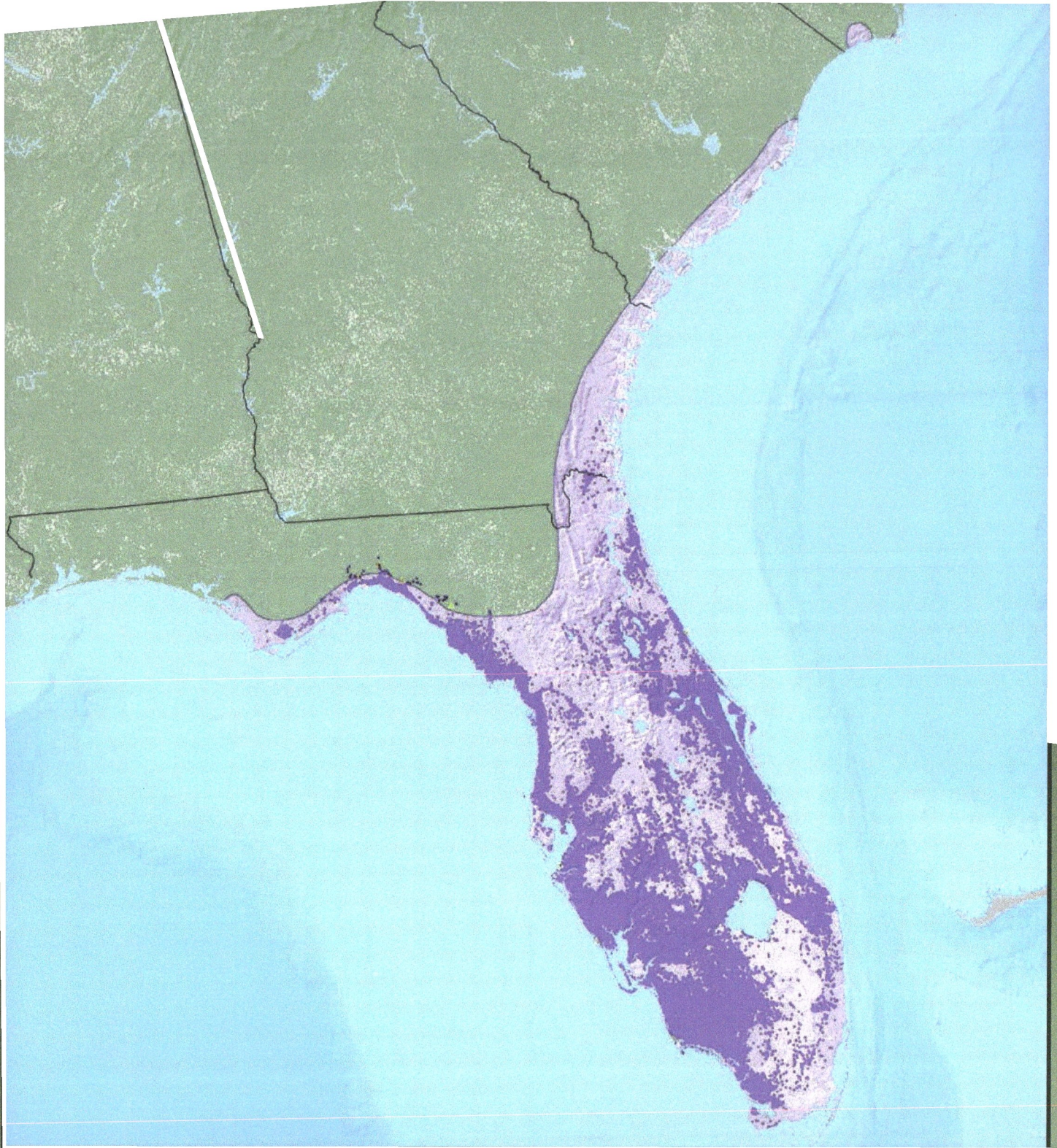

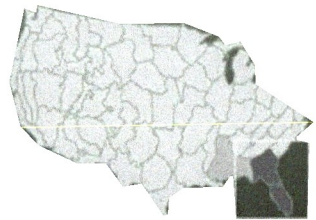

National Individual Tree Species Atlas

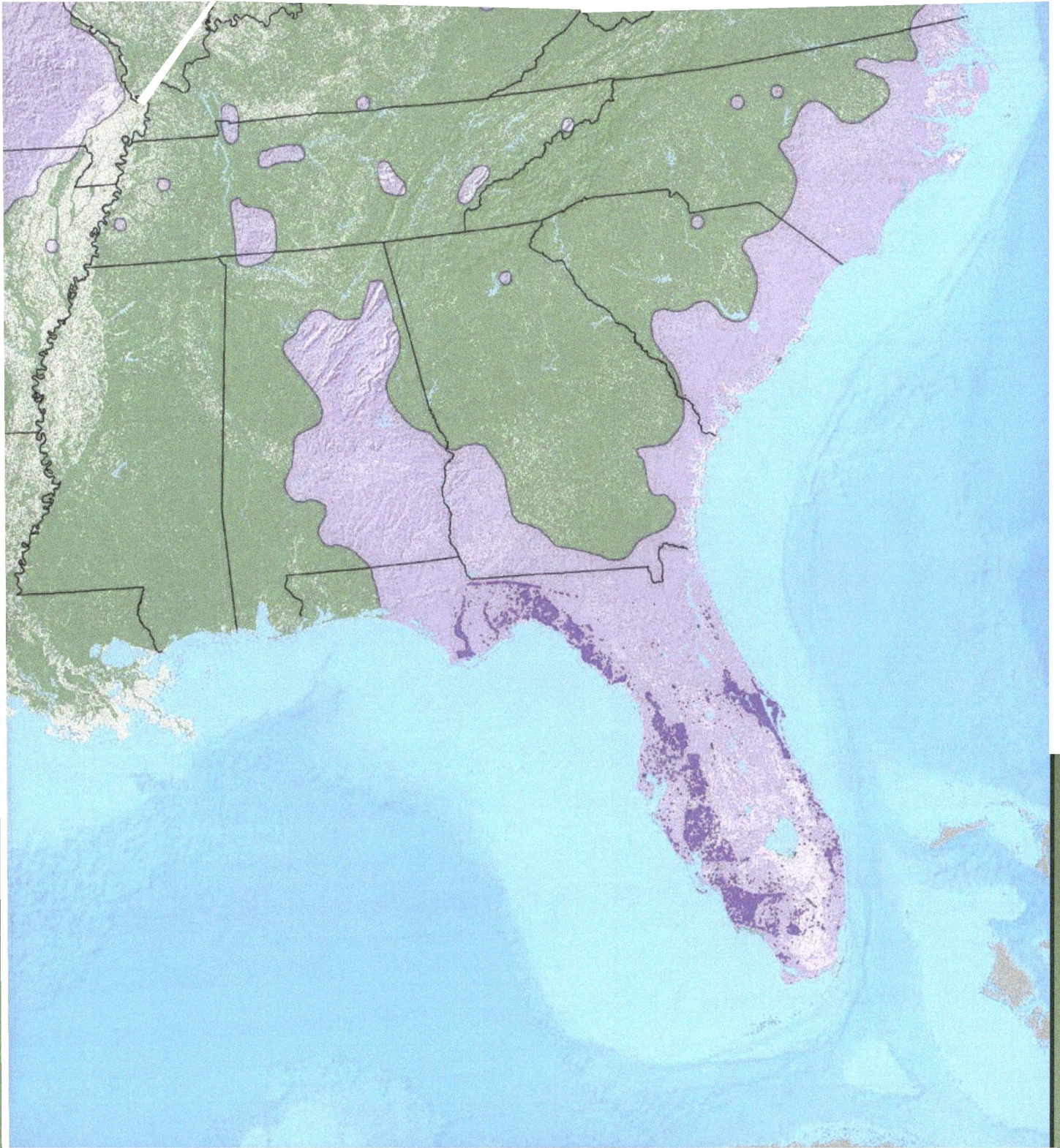
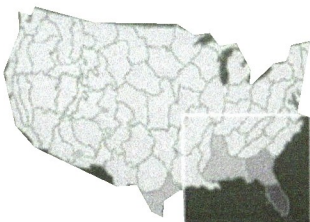
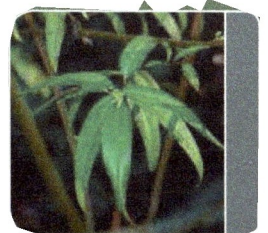

National Individual Tree Species Atlas

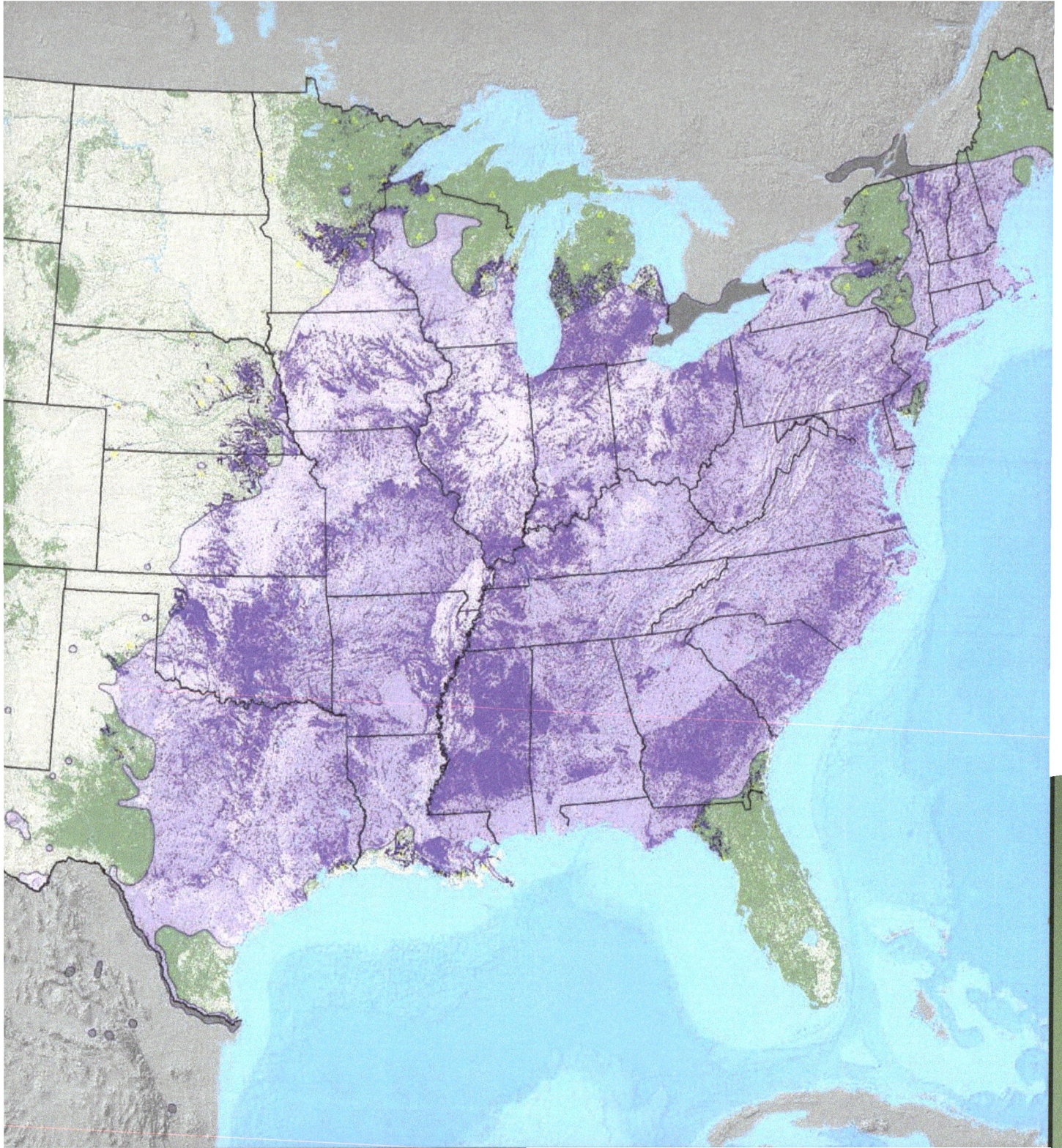

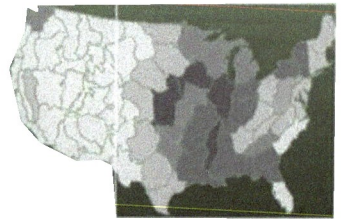

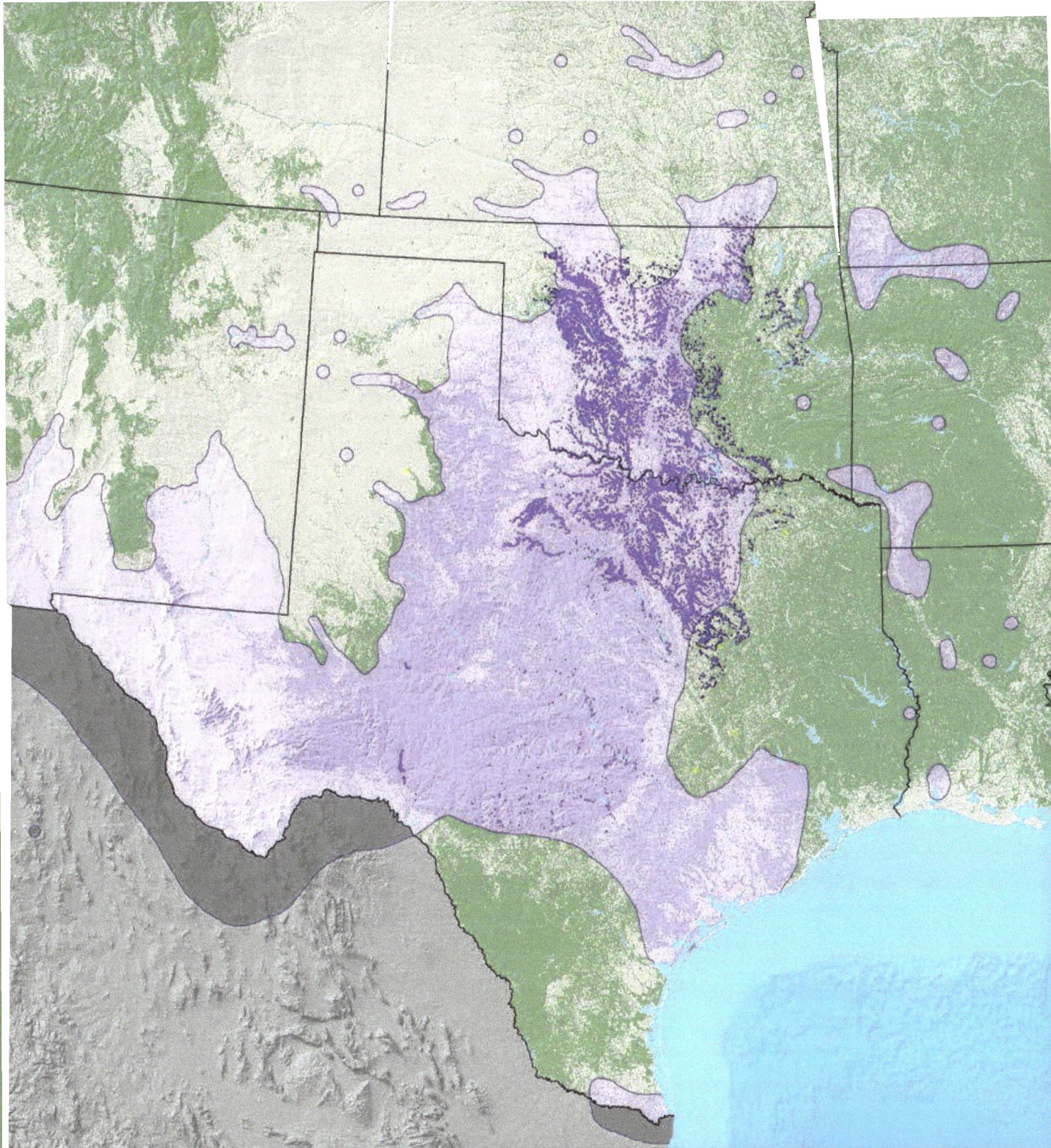
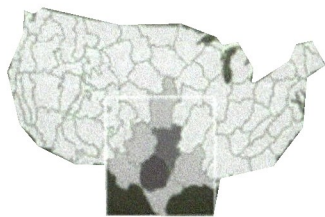
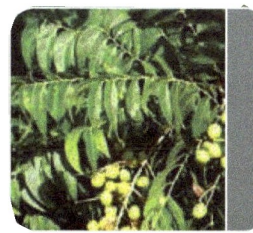

National Individual Tree Species Atlas

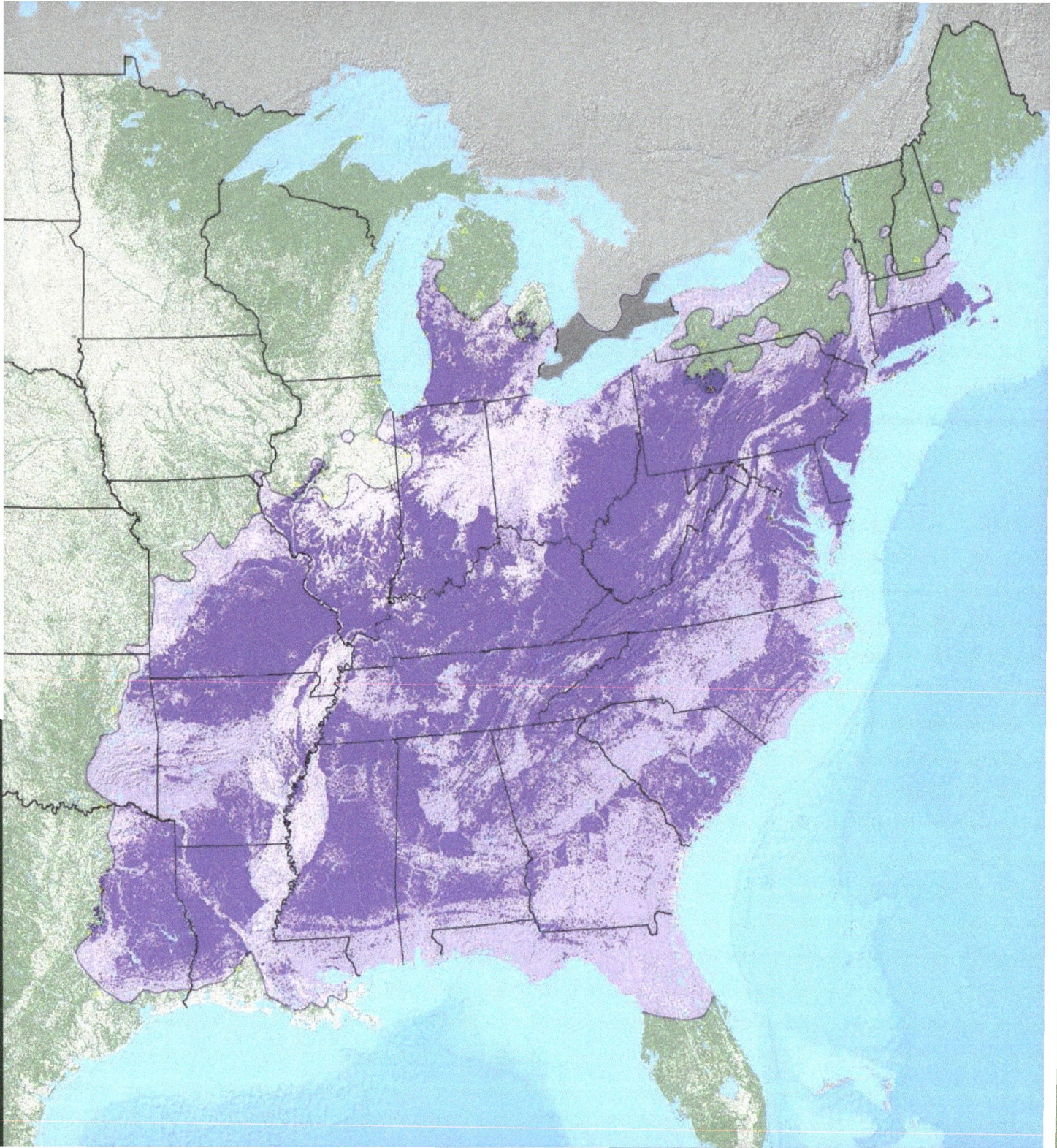

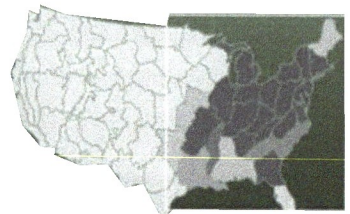

National Individual Tree Species Atlas

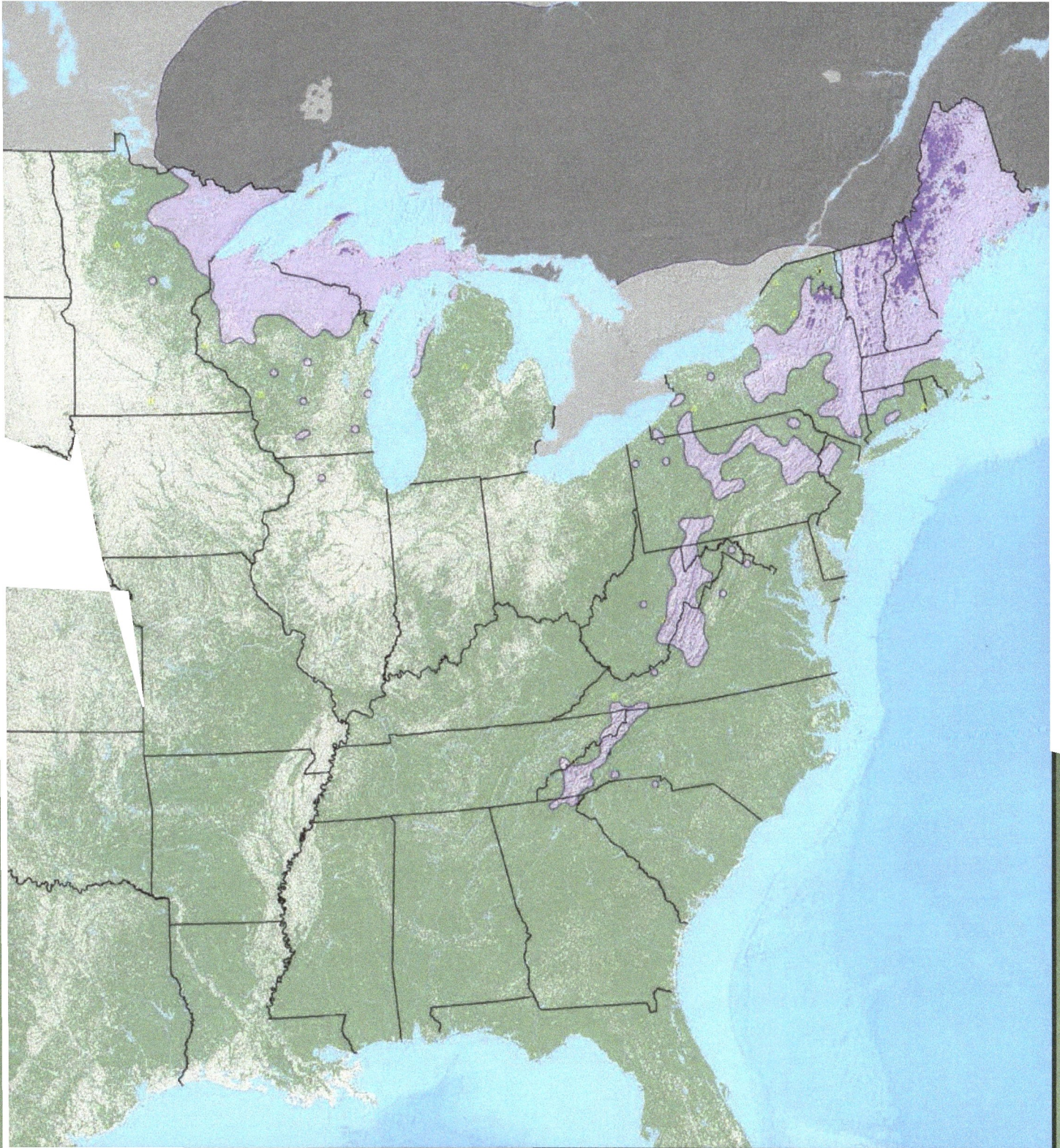

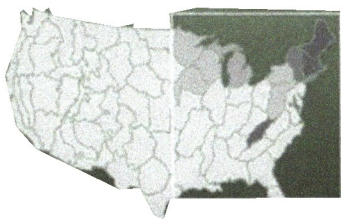

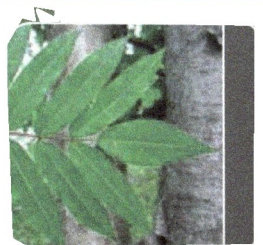

National Individual Tree Species Atlas

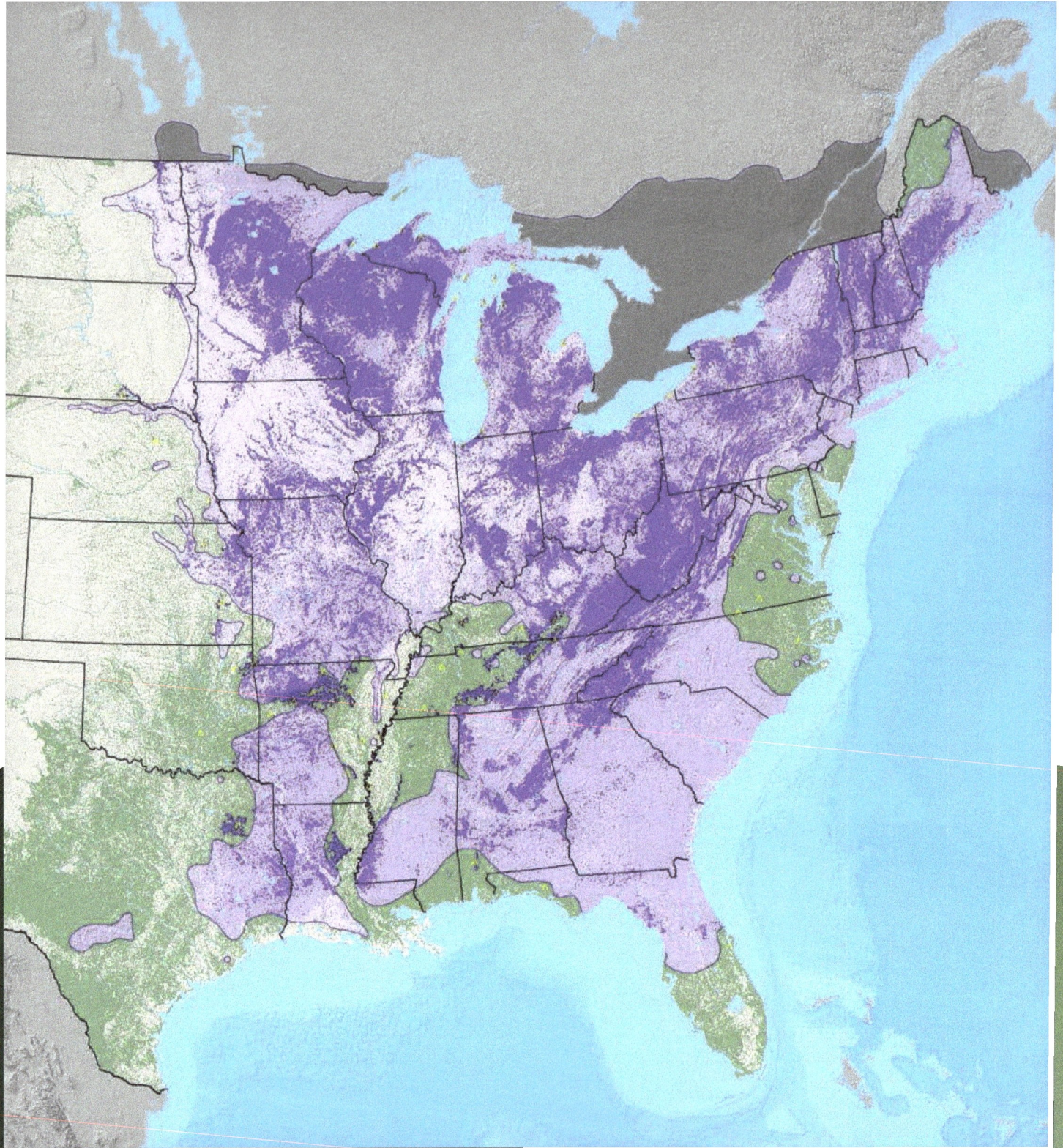

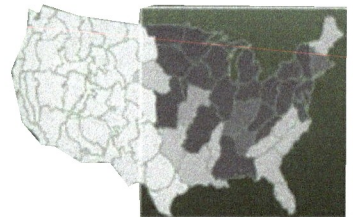

National Individual Tree Species Atlas

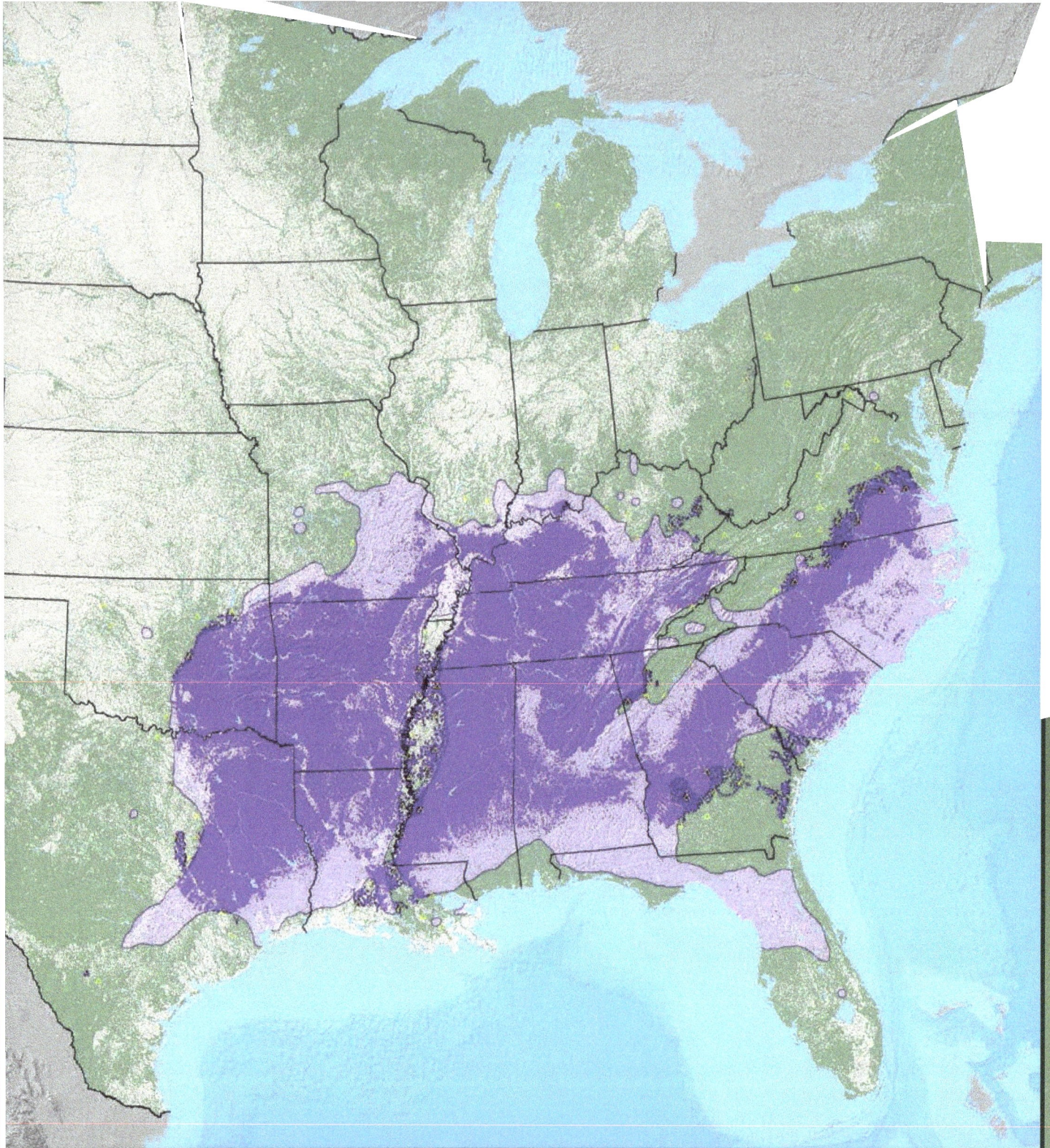

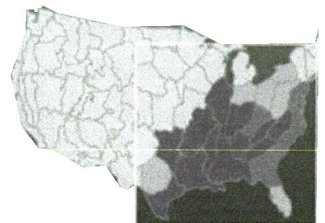

National Individual Tree Species Atlas

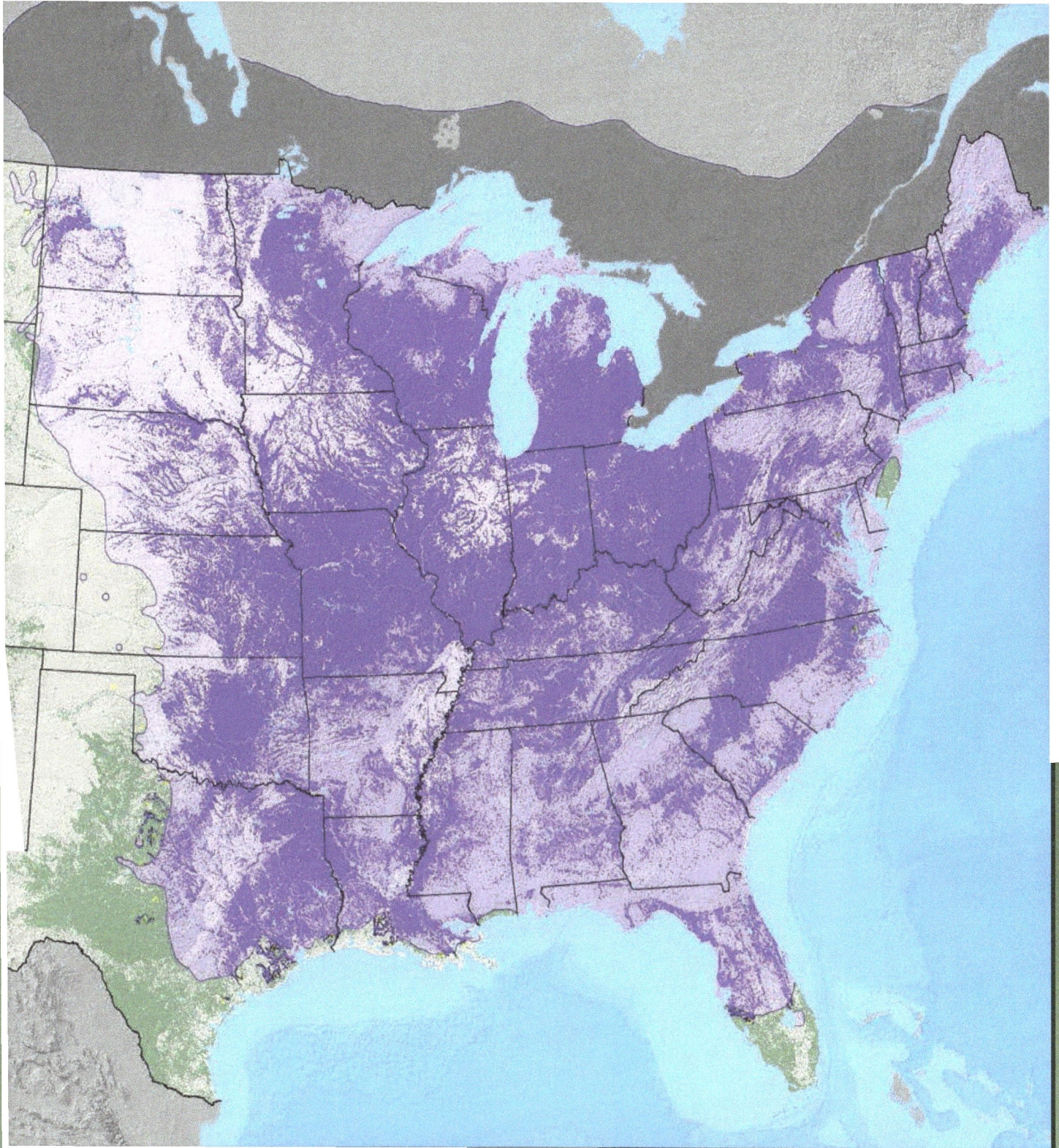
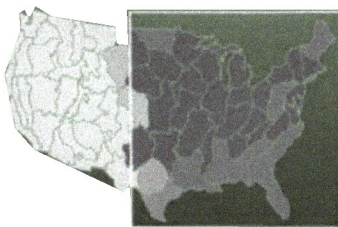

National Individual Tree Species Atlas

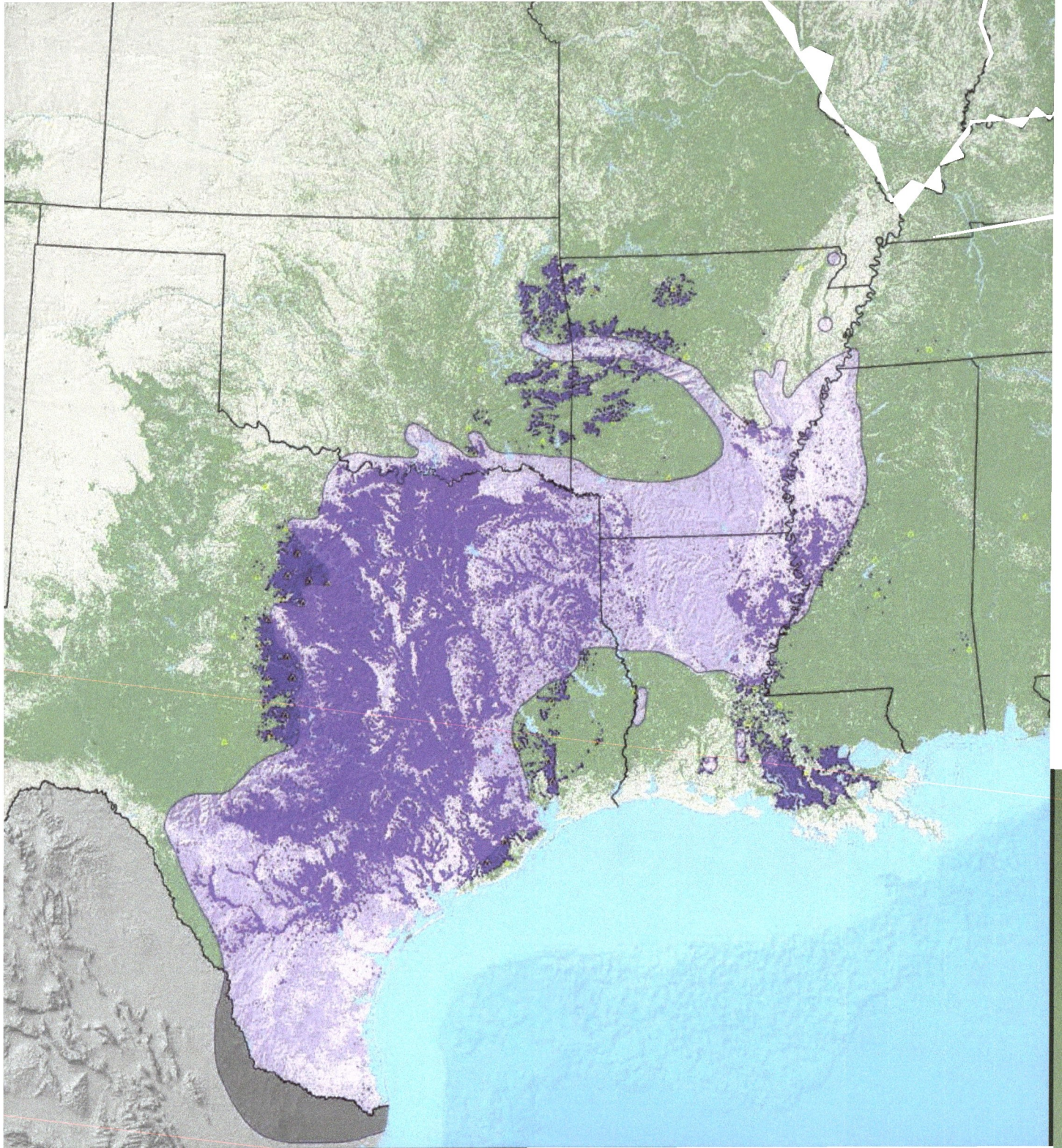

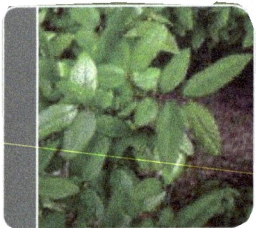

National Individual Tree Species Atlas

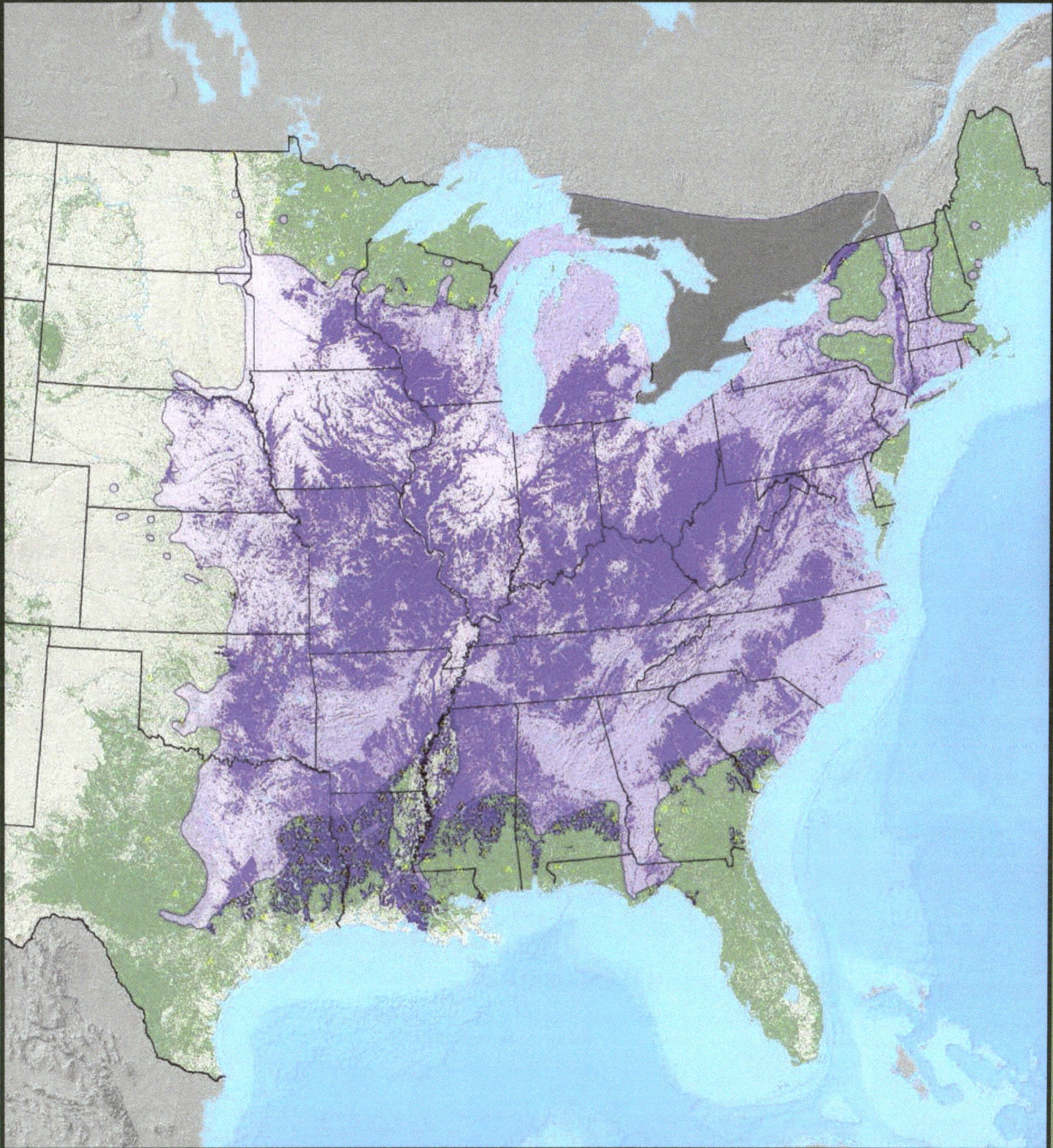

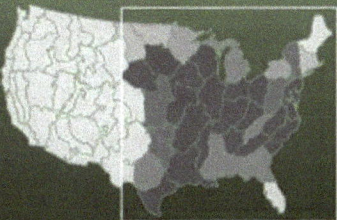

slippery elm

slippery elm
Ulmus rubra Muhl.
Ulmaceae – Elm Family
FIA Code **975** – Plants Symbol **ULRU**

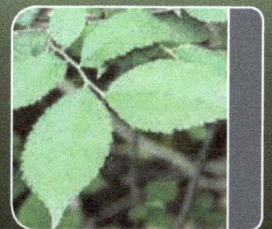

National Individual Tree Species Atlas

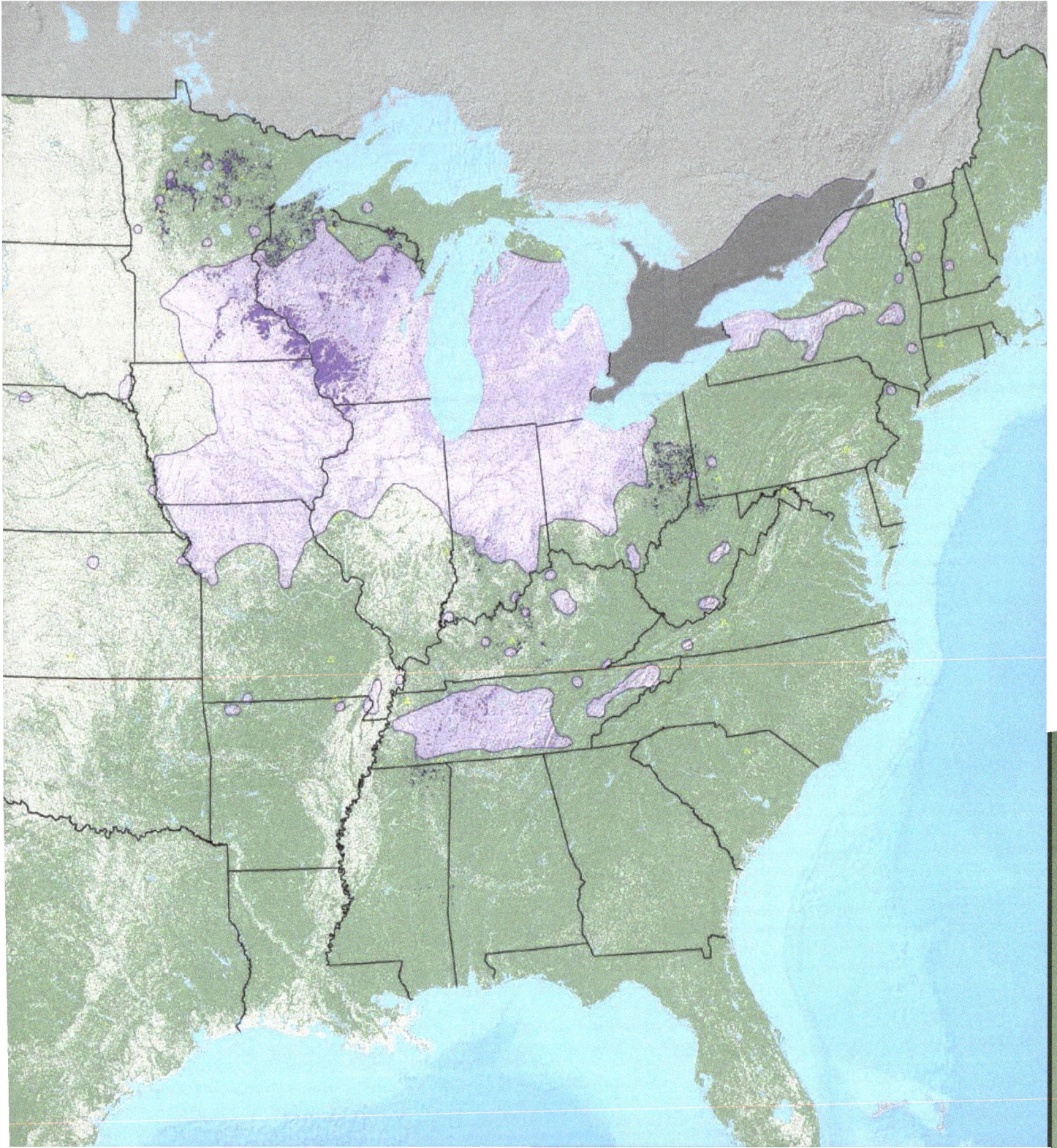

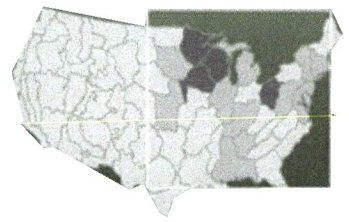

National Individual Tree Species Atlas

Appendix – A (Predictor Layers)

Variable	Group	Source	Definition	e	i	j	k	g	Top	Top5
lofrc1	101	NLCD	Landsat 5/7, NLCD 2002 mosaic, leaf-off reflectance band 1 (visible blue)	1	1	1	1	1	1	5
lofrc2	101	NLCD	Landsat 5/7, NLCD 2002 mosaic, leaf-off reflectance band 2 (visible green)	1	1				0	1
lofrc3	101	NLCD	Landsat 5/7, NLCD 2002 mosaic, leaf-off reflectance band 3 (visible red)	1	1				2	4
lofrc4	101	NLCD	Landsat 5/7, NLCD 2002 mosaic, leaf-off reflectance band 4 (near-infrared)	1	1	1	1	1	0	3
lofrc5	101	NLCD	Landsat 5/7, NLCD 2002 mosaic, leaf-off reflectance band 5 (shortwave infrared)	1	1	1	1	1	0	5
lofrc6	101	NLCD	Landsat 5/7, NLCD 2002 mosaic, leaf-off reflectance band 7 (short-wave infrared)	1	1	1	1	1	0	3
lonrc1	101	NLCD	Landsat 5/7, NLCD 2002 mosaic, leaf-on reflectance band 1 (visible blue)	1	1	1	1	1	1	6
lonrc2	101	NLCD	Landsat 5/7, NLCD 2002 mosaic, leaf-on reflectance band 2 (visible green)	1	1				1	3
lonrc3	101	NLCD	Landsat 5/7, NLCD 2002 mosaic, leaf-on reflectance band 3 (visible red)	1	1				2	14
lonrc4	101	NLCD	Landsat 5/7, NLCD 2002 mosaic, leaf-on reflectance band 4 (near-infrared)	1	1	1	1	1	2	19
lonrc5	101	NLCD	Landsat 5/7, NLCD 2002 mosaic, leaf-on reflectance band 5 (shortwave infrared)	1	1	1	1	1	0	4
lonrc6	101	NLCD	Landsat 5/7, NLCD 2002 mosaic, leaf-on reflectance band 7 (shortwave infrared)	1	1	1	1	1	0	3
sprrc1	101	NLCD	Landsat 5/7, NLCD 2002 mosaic, spring reflectance band 1 (visible blue)	1	1	1	1	1	0	4
sprrc2	101	NLCD	Landsat 5/7, NLCD 2002 mosaic, spring reflectance band 2 (visible green)	1	1				0	4
sprrc3	101	NLCD	Landsat 5/7, NLCD 2002 mosaic, spring reflectance band 3 (visible red)	1	1				2	10
sprrc4	101	NLCD	Landsat 5/7, NLCD 2002 mosaic, spring reflectance band 4 (near-infrared)	1	1	1	1	1	2	15
sprrc5	101	NLCD	Landsat 5/7, NLCD 2002 mosaic, spring reflectance band 5 (shortwave infrared)	1	1	1	1	1	2	13
sprrc6	101	NLCD	Landsat 5/7, NLCD 2002 mosaic, spring reflectance band 7 (shortwave infrared)	1	1	1	1	1	0	10
lofc1	102	NLCD	leaf-off tasseled cap band 1 (brightness)	1					1	4
lofc2	102	NLCD	leaf-off tasseled cap band 2 (greenness)	1	1		1	1	1	11
lofc3	102	NLCD	leaf-off tasseled cap band 3 (wetness)	1					0	4
lonc1	102	NLCD	leaf-on tasseled cap band 1 (brightness)	1					2	10
lonc2	102	NLCD	leaf-on tasseled cap band 2 (greenness)	1	1		1	1	8	38
lonc3	102	NLCD	leaf-on tasseled cap band 3 (wetness)	1					1	3
sprc1	102	NLCD	spring tasseled cap band 1 (brightness)	1					0	5
sprc2	102	NLCD	spring tasseled cap band 2 (greenness)	1	1		1	1	4	33
sprc3	102	NLCD	spring tasseled cap band 3 (wetness)	1					1	9
lofdy	103	NLCD-D	leaf-off imagery Julian day	1					5	34
londy	103	NLCD-D	leaf-on imagery Julian day	1					7	28
sprdy	103	NLCD-D	spring imagery Julian day	1					9	44
loft	104	NLCD-D	leaf-off 3x3 standard deviation normalized index band 5/band2	1	1	1	1	1	0	4
lont	104	NLCD-D	leaf-on 3x3 standard deviation normalized index band 5/band2	1	1	1	1	1	0	1
sprt	104	NLCD-D	spring 3x3 standard deviation normalized index band 5/band2	1	1	1	1	1	0	8
lc	150	NLCD	land cover from NLCD 2002							
cc	151	NLCD	canopy cover from NLCD 2002							
imp	152	NLCD	impervious surface from NLCD 2002	1					0	0
zone	160	USGS	USGS mapping zone	1			1		4	16
ecoreg	161	USFS	Bailey's eco-region section							
ecomap	162	USFS	Ecomap subsection identifier							
sec	163	USFS	Ecomap section							
su	164	USFS	Ecomap subsection							
dem	200	NLCD	digital elevation model - NLCD 2002 corrected	1	1	1	1	1	16	58

Variable	Group	Source	Definition	e	i	j	k	g	Top	Top5
curv	201	USGS-D	Arc/Info curvature – derived from digital elevation model	1	1	1	1	1	0	0
ewslp	201	USGS-D	slope* cos(aspect) – derived from digital elevation model	1	1	1	1	1	0	0
nsslp	201	USGS-D	slope* sin(aspect) – derived from digital elevation model	1	1	1	1	1	1	4
posidx	201	NLCD	slope position index, NLCD 2002 corrected	1	1	1	1	1	1	7
slp	201	USGS-D	slope – 0.5 percent integer scale, derived from digital elevation model	1		1			4	16
srad	202	USGS-D	shortwave radiation – derived from digital elevation model							
shwv	203	USGS-D	direct shortwave radiation – derived from digital elevation model	1	1	1	1	1	4	14
shwvd	203	USGS-D	diffuse shortwave radiation – derived from digital elevation model	1	1	1	1	1	5	23
tscl	203	USGS-D	topographic scale	1	1	1	1	1	1	17
ctii	204	USGS-D	compound topographic index (integer)	1					0	2
landform	204	USGS-D	landform	1					0	0
trmi	204	USGS-D	topographic relative moisture index	1					1	1
trmim	204	USGS-D	topographic relative moisture index – modified	1					3	6
di	300	NRCS-D	soil drainage index – derived from SSURGO/STATSGO/NFS	1					11	49
fpi	300	NRCS-D	fertility index – derived from SSURGO/STATSGO/NFS	1					4	40
cdom	301	NRCS	soil component dominance	1					0	8
cfreq	301	NRCS	soil component frequency	1					0	13
di_src	301	NRCS-D	soil data source – SSURGO/STATSGO/USFS	1					0	5
ppt[l,k,h,r]1	402	AnusplinD	1st principal component monthly precipitation – l48, ak, hi, Rehfeldt		1		1	1		
ppt[l,k,r]2	402	AnusplinD	2nd principal component monthly precipitation – l48, ak, Rehfeldt		1		1	1		
tavg[l,k,h]1	402	AnusplinD	1st principal component average temperature – l48, ak, hi							
tmax[l,k,h]1	402	AnusplinD	1st principal component maximum temperature – l48, ak, hi							
tmin[l,k,h]1	402	AnusplinD	1st principal component minimum temperature – l48, ak, hi							
tmin[l,k]2	402	AnusplinD	2nd principal component minimum temperature – l48, ak							
admi	405	AnusplinD	annual moisture index, the ratio of variable dd5 to variable map	1	1	1	1	1	10	37
d100	405	AnusplinD	Julian day when the sum of degree-days above 5 °C reaches 100							
dd0	405	AnusplinD	total average temperature degree-days below 0 °C	1	1	1	1	1	12	27
dd5	405	AnusplinD	total average temperature degree-days above 5 °C	1					13	29
fday	405	AnusplinD	Julian day of the first below-freezing day of autumn							
ffp	405	AnusplinD	length of the frost-free period – number of days							
gsdd5	405	AnusplinD	total degree-days above 5 °C accumulating within the frost-free period	1					9	23
gsp	405	AnusplinD	growing season precipitation, months between sday and fday, inclusive	1	1	1	1	1	6	33
map	405	AnusplinD	annual precipitation	1					10	49
mat	405	AnusplinD	mean annual temperature	1					11	36
mmax	405	AnusplinD	maximum temperature in the warmest month	1	1	1	1	1	6	40
mmin	405	AnusplinD	minimum temperature in the coldest month	1					5	29
mmindd0	405	AnusplinD	total minimum temperature degree-days below 0 °C	1					2	24
mtcm	405	AnusplinD	average temperature in the coldest month	1					18	57
mtwm	405	AnusplinD	average temperature in the warmest month	1					14	52
pratio	405	AnusplinD	ratio of growing season precipitation to annual precipitation	1					13	53
sday	405	AnusplinD	Julian day of the last below-freezing day of spring							
sdmi	405	AnusplinD	seasonal moisture index, the ratio of variable gsdd5 to gsp	1	1	1	1	1	7	29
fd	407	DAYMET	frost-free days – number of days							
gdd	407	DAYMET	growing degree-days							
swr	407	DAYMET	shortwave radiation							
wvp	407	DAYMET	water vapor pressure							
prod	501	FIA	timber productivity – modeled from FIA site index data [ordered classes 1–7]	1					1	8
		Total		67	33	35	34	33		

NOTE: Tree presence/absence is from the j models; individual tree models are from the i, k, g models. Models e, k, and g are global models with all zones merged together. Top = the top variable from the e models, Top5 = one of the top five variables from the e models.

Appendix – B (Species Metadata)

B1 — Species observed on an FIA plot but not included in the *Modeled Atlas*

Broad	scrub oak spp., other tropical, unknown dead conifer, unknown dead hardwood, unknown hardwood tree
Family	other palms
Genus	apple spp., ash spp., basswood spp., birch spp., buckeye, catalpa spp., cherry and plum spp., cottonwood and poplar spp., citrus spp., cypress, elm spp., eucalyptus spp., fir spp., hackberry spp., hawthorn, hemlock spp., hickory spp., honeylocust spp., larch spp., magnolia spp., maple, mesquite, mountain-ash spp., mulberry spp., oak/deciduous, pine spp., redcedar / juniper, spruce spp., tupelo spp., walnut, willow
Too Few	Allegheny chinkapin, Arizona madrone, Arizona pine, Australian pine, Baker cypress, balsam willow, blackbead ebony, camphor tree, Canada plum, Carolina silverbell, Chinese chestnut, Chisos oak, citrus spp., common serviceberry, drooping juniper, dwarf chinakapin oak, European alder, European mountain-ash, fish poison tree, Florida poisontree, Ginkgo/maidenhair tree, grand eucalyptus, Graves oak, gumbo limbo, Java plum, key thatch palm, Monterey cypress, northern mountain-ash, nutmeg hickory, Oglethorpe oak, Ozark chinkapin, painted buckeye, peach, prairie crabapple, red hickory, red stopper, royal palm spp., saltcedar, Santa Lucia fir, Scouler's willow, September elm, silver poplar, smoketree, Northern California black walnut, southern catalpa, southern crabapple, southern redcedar, southern shagbark hickory, Tasmanian bluegum, Texas buckeye, Texas madrone, Texas mulberry, Texas sophora, Texas walnut, tung oil tree, two-wing silverbell, umbrella magnolia, water birch, weeping willow, white willow, yellowwood
Poor Fit	Bebb willow, chittamwood gum bumelia, cockspur hawthorn, Delta post oak, downy hawthorn, northern catalpa, peachleaf willow, sand hickory, serviceberry, shellbark hickory, silverbell, sweet crabapple, white mulberry
Introduced	Austrian pine, chinaberry, melaleuca, mimosa/silktree, Norway maple, Norway spruce, Russian-olive, Scotch pine, Siberian elm, sweet cherry/domesticated

B2 — Alaska Model Fit Validation Summary

Page	Common name	Scientific name	FIA code	Plots	Global	Validation	Five most significant variables*
16	Pacific silver fir	*Abies amabilis*	11	3	75	66.7	none
22	subalpine fir	*Abies lasiocarpa*	19	5	100	60	mtwm, mmin
30	Alaska yellow-cedar	*Chamaecyparis nootkatensis*	42	383	99.8	99	gsp, lonc3, mmax, mtcm, lonrc4
45	tamarack	*Larix laricina*	71	21	97.6	9.5	gsp, map, lonrc6, lonc2, sprc2
51	white spruce	*Picea glauca*	94	1504	91.9	89.8	map, mtwm, londy, tscl, lonrc5
53	black spruce	*Picea mariana*	95	821	97.1	69.7	mmindd0, sprdy, mmin, dem, mmax
57	Sitka spruce	*Picea sitchensis*	98	805	99.1	96.6	mmindd0, di, map, mmin, londy
66	lodgepole pine	*Pinus contorta*	108	174	99.5	94.3	lonc3, di, di_src, dd0, dem
106	western redcedar	*Thuja plicata*	242	199	98.1	95.5	mat, mmax, mmin, fpi, lonrc5
111	western hemlock	*Tsuga heterophylla*	263	770	99.1	99.6	pratio, zone, lonrc5, sprdy, dem
113	mountain hemlock	*Tsuga mertensiana*	264	634	99.3	96.2	gsdd5, di, mmin, admi, di_src
138	red alder	*Alnus rubra*	351	53	97.6	56.6	dem, posidx, pratio, tscl, gsdd5
147	paper birch	*Betula papyrifera*	375	1090	95.3	73.5	map, gsp, lonc2, shwvd, lonrc5
219	balsam poplar	*Populus balsamifera*	741	291	94.4	22	slp, mmindd0, dd5, dem, fpi
228	quaking aspen	*Populus tremuloides*	746	462	97.8	52.8	mmax, sdmi, lonc2, mmin, mtcm
221	black cottonwood	*Populus balsamifera* ssp. *trichocarpa*	747	127	98.2	62.2	mminddo, mtwm, cfreq, sprc2, dd0

NOTE: The five most significant variables for *Pinus contorta* var. *contorta* includes: lonc3, dd5, tscl, di, mminddo and for *Pinus contorta* var. *latifolia*: lonrc6, mat, map, lonc1, lonrc1. *From the e models.

B3 — Coterminous U.S. Model Fit Validation Summary

Page	Common name	Scientific name	FIA code	Plots	Global	Validation	Five most significant variables*
16-17	Pacific silver fir	*Abies amabilis*	11	1796	99.4	95.4	admi, mtcm, cdom, mmax, zone
18	balsam fir	*Abies balsamea*	12	9909	98.3	96.1	dd5, sprc2, gsp, admi, pratio
19	white fir	*Abies concolor*	15	4395	99	88.4	pratio, zone, dem, admi, sdmi
20	Fraser fir	*Abies fraseri*	16	12	72.2	25	sprdy, mmax
21	grand fir	*Abies grandis*	17	3241	98.7	86.8	mtcm, map, lofdy, pratio, sprdy
22-23	subalpine fir	*Abies lasiocarpa*	19	4685	99.5	87.9	mat, admi, sprrc6, pratio, lofdy
24	corkbark fir	*Abies lasiocarpa* var. *arizonica*	18	174	96.8	83.3	dem, gsp, cfreq, mmax, pratio
25	California red fir	*Abies magnifica*	20	807	98.8	90	dem, admi, pratio, lofdy, mmax
26	noble fir	*Abies procera*	22	657	97.3	79.9	admi, mtcm, dem, mtwm, mat
27	Shasta red fir	*Abies shastensis*	21	287	99	90.2	gsdd5, mtwm, gsp, mmax, londy
28	incense-cedar	*Calocedrus decurrens*	81	2600	99.5	88.1	zone, dd0, gsp, mmax, mtwm

*From the e models.

Page	Common name	Scientific name	FIA code	Plots	Global	Validation	Five most significant variables*
29	Port-Orford-cedar	*Chamaecyparis lawsoniana*	41	185	93.4	73	mmin, londy, map, lofdy, cfreq
30-31	Alaska yellow-cedar	*Chamaecyparis nootkatensis*	42	222	96.6	67.1	admi, mat, tscl, map, dem
32	Atlantic white-cedar	*Chamaecyparis thyoides*	43	104	48.6	46.2	londy, sprrc2, fpi, mtwm, lonc1
33	Arizona cypress	*Cupressus arizonica*	51	13	87.5	76.9	pratio, shwv, trmim, gsp, cfreq
34	Sargent's cypress	*Cupressus sargentii*	55	5	81.8	80	sprdy, fpi
35	Ashe juniper	*Juniperus ashei*	61	886	99.3	82.4	mtcm, sprrc5, dd0, di, mmindd0
36	California juniper	*Juniperus californica*	62	133	97.5	70.7	dd0, lonrc2, mtcm, sprc2, shwvd
37	redberry juniper	*Juniperus coahuilensis*	59	344	99.2	62.8	sprrc5, map, dd0, prod, shwv
38	alligator juniper	*Juniperus deppeana*	63	1049	99.5	83.8	mtcm, gsp, admi, zone, mmax
39	oneseed juniper	*Juniperus monosperma*	69	1684	99.3	79.4	mtwm, mmindd0, map, pratio, mmin
40	western juniper	*Juniperus occidentalis*	64	1650	98.9	79.8	map, lonc2, dem, lofdy, lofc2
41	Utah juniper	*Juniperus osteosperma*	65	4004	99.6	92.3	admi, lofdy, lonc2, sprdy, sdmi
42	Pinchot juniper	*Juniperus pinchotii*	58	333	99.2	67.9	gsp, di, sdmi, londy, mat
43	Rocky Mountain juniper	*Juniperus scopulorum*	66	1873	97.5	75.3	mmin, admi, sprdy, lofdy, sdmi
44	eastern redcedar	*Juniperus virginiana*	68	6564	95.4	65.1	mmax, mtwm, fpi, mtcm, dem
45-46	tamarack	*Larix laricina*	71	2225	94.6	77.6	di, mmindd0, lonc1, lonc2, sprrc6
47	subalpine larch	*Larix lyallii*	72	51	94.6	80.4	mat, mmax, cdom, sprc1, shwv
48	western larch	*Larix occidentalis*	73	2342	99.3	87.1	mtcm, dem, sprdy, zone, mmin
49	Brewer spruce	*Picea breweriana*	92	14	61.5	35.7	lofc1, cfreq
50	Engelmann spruce	*Picea engelmannii*	93	4532	99.5	88.3	mat, dem, mtcm, sdmi, posidx
51-52	white spruce	*Picea glauca*	94	3479	95.6	64.2	dd5, lofrc5, mat, pratio, sprc2
53-54	black spruce	*Picea mariana*	95	3337	97.3	86.2	lonc1, dd5, fpi, mtwm, di
55	blue spruce	*Picea pungens*	96	183	67.8	54.1	lonrc3, pratio, mmax, lonc2, lofc2
56	red spruce	*Picea rubens*	97	3062	99.2	81.5	dd5, sprdy, pratio, map, lonrc5
57-58	Sitka spruce	*Picea sitchensis*	98	357	96.3	70.9	sdmi, sprdy, gsp, admi, mat
59	whitebark pine	*Pinus albicaulis*	101	1113	98.2	81.8	dd5, shwvd, sprdy, gsdd5, dem
60	Rocky Mountain bristlecone pine	*Pinus aristata*	102	95	98	82.1	pratio, cfreq, sdmi, dd5, sprdy
61	knobcone pine	*Pinus attenuata*	103	186	65.3	67.2	lofdy, pratio, londy, shwvd, mmin
62	foxtail pine	*Pinus balfouriana*	104	23	95.5	73.9	dem, di_src, lonc3, slp, mmax
63	jack pine	*Pinus banksiana*	105	1801	96.5	85.3	fpi, lonrc4, map, lonc2, di
64	Mexican piñon pine	*Pinus cembroides*	140	57	98.6	78.9	di, gsp, shwvd, mmin, fpi
65	sand pine	*Pinus clausa*	107	183	95.8	80.9	di, sdmi, prod, sprdy, londy
66-67	lodgepole pine	*Pinus contorta*	108	6406	99.8	91.6	dd5, mtwm, pratio, mmin, posidx
68	Coulter pine	*Pinus coulteri*	109	42	79.7	83.3	mmindd0, lonc2, zone, shwv, cdom
69	border piñon	*Pinus discolor*	134	49	97.6	83.7	dd0, gsp, mmindd0, map, mtcm
70	shortleaf pine	*Pinus echinata*	110	5589	97.5	69.2	dem, mmindd0, lonc1, mtcm, sprc2
71	common piñon	*Pinus edulis*	106	4423	99.5	82.2	dd0, admi, lofdy, mtcm, dem
72	slash pine	*Pinus elliottii*	111	3244	98.3	82.3	mtcm, lofdy, di, mtwm, sprc2
73	Apache pine	*Pinus engelmannii*	112	8	58.3	12.5	No signficant variables
74	limber pine	*Pinus flexilis*	113	900	97.1	70	lofdy, mmindd0, gsp, map, dem
75	spruce pine	*Pinus glabra*	115	288	61.8	37.8	No signficant variables
76	Jeffrey pine	*Pinus jeffreyi*	116	1417	98.9	92.6	dem, cfreq, lonrc3, dd5, map
77	sugar pine	*Pinus lambertiana*	117	2097	99	82	mat, sdmi, map, di, sprdy
78	Chihuahua pine	*Pinus leiophylla*	118	16	88	56.3	trmi, sdmi, sprdy
79	Great Basin bristlecone pine	*Pinus longaeva*	142	22	88.5	77.3	sprdy, mtcm, fpi, lonc2, tscl
80	singleleaf piñon	*Pinus monophylla*	133	1014	98.7	88.4	zone, lofdy, map, dem, sdmi
81	Arizona piñon pine	*Pinus monophylla* var. *fallax*	143	138	97.3	72.5	mtcm, lonc2, pratio, di_src, map
82	western white pine	*Pinus monticola*	119	1353	95.2	72.9	pratio, mmax, dem, mmin, sprrc2

Page	Common name	Scientific name	FIA code	Plots	Global	Validation	Five most significant variables*
83	bishop pine	*Pinus muricata*	120	11	23.5	81.8	mtcm, di, admi
84	longleaf pine	*Pinus palustris*	121	1722	94.6	68.6	sdmi, mtcm, fpi, prod, map
85	ponderosa pine	*Pinus ponderosa*	122	10859	99.5	92.6	admi, zone, mat, sdmi, dem
86	Table Mountain pine	*Pinus pungens*	123	100	74.8	60	lonc2, sprrc3, lofdy, sprrc4, sprdy
87	Monterey pine	*Pinus radiata*	124	5	87.5	80	pratio, dem
88	papershell piñon pine	*Pinus remota*	141	11	67.7	45.5	mtwm, lofdy, gsp, sprrc6, pratio
89	red pine	*Pinus resinosa*	125	2676	94.3	84.5	lonc1, map, di, dd0, lofc2
90	pitch pine	*Pinus rigida*	126	833	71	67.5	dem, mmindd0, mtwm, gsdd5, sprrc1
91	California foothills pine	*Pinus sabiniana*	127	306	96.6	61.4	gsdd5, mtwm, dem, lonrc3, lonc3
92	pond pine	*Pinus serotina*	128	417	83.3	58.5	shwv, di, sprc2, fpi, sprc3
93	southwestern white pine	*Pinus strobiformis*	114	163	96	74.8	sprrc3, sdmi, gsp, sprdy, mmax
94	eastern white pine	*Pinus strobus*	129	7278	96.8	75.1	mtwm, sprc2, lofc2, mmindd0, mat
95	loblolly pine	*Pinus taeda*	131	15578	98.8	91	mtcm, dd0, sprc2, lonrc6, mtwm
96	Virginia pine	*Pinus virginiana*	132	3219	96.8	72.5	shwvd, dd0, sprc2, lofdy, dd5
97	Washoe pine	*Pinus washoensis*	137	5	100	80	No signficant variables
98	bigcone Douglas-fir	*Pseudotsuga macrocarpa*	201	50	97	76	sprc3, slp, sprdy, mmin, pratio
99	Douglas-fir	*Pseudotsuga menziesii*	202	17544	99.3	96.2	pratio, dem, shwvd, gsp, dd5
100	redwood	*Sequoia sempervirens*	211	271	98.6	87.1	pratio, zone, mtcm, fpi, mtwm
101	giant sequoia	*Sequoiadendron giganteum*	212	28	100	92.9	londy, sprc2, sprrc6, lonrc3, lonrc1
102	pondcypress	*Taxodium ascendens*	222	801	94.9	73.4	shwvd, di, sprrc4, mtcm, di_src
103	baldcypress	*Taxodium distichum*	221	967	88.3	63.9	sprrc4, dem, mtwm, shwvd, di
104	Pacific yew	*Taxus brevifolia*	231	680	75.6	68.7	dd0, mtcm, tscl, dem, sprc2
105	northern white-cedar	*Thuja occidentalis*	241	4398	98.2	85.7	sprrc5, di, dem, gsp, admi
106-107	western redcedar	*Thuja plicata*	242	2736	99.4	89.5	mtcm, sdmi, londy, dem, lofdy
108	California torrey	*Torreya californica*	251	39	46.6	43.6	trmim, lonrc1, londy, gsp
109	eastern hemlock	*Tsuga canadensis*	261	5598	96.9	82.6	pratio, sprc2, mat, sprrc1, mtwm
110	Carolina hemlock	*Tsuga caroliniana*	262	21	0	28.6	No signficant variables
111-112	western hemlock	*Tsuga heterophylla*	263	4472	99.6	95	map, dem, lonrc3, lofdy, londy
113-114	mountain hemlock	*Tsuga mertensiana*	264	1260	98.8	89.4	mtwm, mmin, sprdy, mtcm, admi
118	sweet acacia	*Acacia farnesiana*	303	123	93.9	61	dd5, gsp, admi, fpi, sprdy
119	Florida maple	*Acer barbatum*	311	571	26.4	40.6	admi, mmax, dd5, pratio, sprrc6
120	Rocky Mountain maple	*Acer glabrum*	321	275	43.8	66.5	mtwm, lonrc5, lonc2, shwvd, londy
121	bigtooth maple	*Acer grandidentatum*	322	199	98.3	80.4	sprc2, lonc2, nsslp, lofc2, sdmi
122	chalk maple	*Acer leucoderme*	323	35	100	17.1	No signficant variables
123	bigleaf maple	*Acer macrophyllum*	312	1848	97.5	72.8	dem, admi, lonrc3, dd5, mmax
124-125	boxelder	*Acer negundo*	313	2953	83.4	56.4	mtwm, pratio, fpi, shwvd, map
126	black maple	*Acer nigrum*	314	120	19.8	22.5	lonc2, sprdy, dem, posidx, lofc2
127	striped maple	*Acer pensylvanicum*	315	1959	85.5	63.7	mtwm, admi, mat, gsdd5, mmax
128	red maple	*Acer rubrum*	316	35626	90.5	81.3	mmax, lonc2, di, sprc2, map
129	silver maple	*Acer saccharinum*	317	1209	80.2	53.7	mtcm, mmin, gsdd5, dd5, shwvd
130	sugar maple	*Acer saccharum*	318	16263	99.3	85.7	zone, lonc2, gsdd5, pratio, mmax
131	mountain maple	*Acer spicatum*	319	752	53.9	49.1	slp, loft, sprt, sprdy, lonrc4
132	California buckeye	*Aesculus californica*	333	140	97.6	40.7	londy, mtwm, fpi, cfreq, shwv
133	yellow buckeye	*Aesculus flava*	332	514	54.4	50.4	pratio, sprc2, mtcm, map, sprt
134	Ohio buckeye	*Aesculus glabra*	331	273	26.6	48.4	lonc3, admi, tscl, dem, sprrc1
135	tree-of-heaven	*Ailanthus altissima*	341	581	37	43.7	dem, cfreq, admi, sprrc5, mtwm
136	Arizona alder	*Alnus oblongifolia*	353	5	83.3	60	No signficant variables
137	white alder	*Alnus rhombifolia*	352	123	31.2	41.5	tscl, sprc2, slp, lonrc1, sprdy
138-139	red alder	*Alnus rubra*	351	2195	98.3	78.2	gsp, lonc2, sprrc4, dd5, mmindd0
140	pond-apple	*Annona glabra*	853	6	18.2	66.7	lonrc3, loft, cdom

Page	Common name	Scientific name	FIA code	Plots	Global	Validation	Five most significant variables*
141	Pacific madrone	*Arbutus menziesii*	361	1810	98.9	85.3	map, dem, dd0, sprrc1, mtwm
142	pawpaw	*Asimina triloba*	367	337	0	25.8	No signficant variables
143	black-mangrove	*Avicennia germinans*	986	5	100	60	sprdy, trmim
144	yellow birch	*Betula alleghaniensis*	371	6686	97.3	72	mmax, pratio, gsdd5, mat, mmin
145	sweet birch	*Betula lenta*	372	3501	98.5	76.4	sprc2, mmax, lofrc2, mtwm, lofdy
146	river birch	*Betula nigra*	373	788	37.2	42.9	di, pratio, mmax, sprrc5, lonrc3
147-149	paper birch	*Betula papyrifera*	375	9464	98.9	85.2	mmin, pratio, shwvd, mat, dem
150	gray birch	*Betula populifolia*	379	680	77.3	52.1	mmin, dem, admi, sprrc4, lofdy
151	American hornbeam	*Carpinus caroliniana*	391	4424	77.6	52.5	map, lonc2, sprdy, tscl, lonrc4
152	mockernut hickory	*Carya alba*	409	8427	96.5	66	slp, sprrc3, mmindd0, pratio, lonc2
153	water hickory	*Carya aquatica*	401	477	71.2	42.8	dem, sprc2, mtwm, lofc1, fpi
154	bitternut hickory	*Carya cordiformis*	402	3215	75.5	50.8	mtcm, fpi, mat, lonc2, sprrc5
155	pignut hickory	*Carya glabra*	403	7716	94.6	63.5	mtcm, lonrc4, pratio, mat, lonrc2
156	pecan	*Carya illinoinensis*	404	598	46.5	40.6	mtwm, lonc2, sprrc3, sprc2, mmindd0
157	shagbark hickory	*Carya ovata*	407	4928	91.9	61.8	mat, mtcm, fpi, pratio, sprc2
158	black hickory	*Carya texana*	408	2816	96.5	71	lofdy, sprdy, zone, map, lofc3
159	American chestnut	*Castanea dentata*	421	179	0	14.5	No signficant variables
160	sugarberry	*Celtis laevigata*	461	1891	79.9	58.1	mtwm, gsdd5, mmindd0, fpi, lonrc4
161	netleaf hackberry	*Celtis laevigata* var. *reticulata*	463	171	74.4	42.7	map, lonc1, lonc2, sdmi, gsdd5
162	hackberry	*Celtis occidentalis*	462	2839	82.4	58.9	sdmi, mtcm, mtwm, pratio, sprc2
163	eastern redbud	*Cercis canadensis*	471	2073	53.2	32.8	lonc2, fpi, mat, slp, mmindd0
164	curlleaf mountain-mahogany	*Cercocarpus ledifolius*	475	817	94	77.7	dem, lonrc3, sdmi, mtcm, map
165	giant chinkapin	*Chrysolepis chrysophylla* var. *chrysophylla*	431	885	94.8	72.1	sprdy, sdmi, sprc2, gsp, dem
166	bluewood	*Condalia hookeri*	867	154	96.3	53.9	pratio, lonc2, admi, mtwm, mtcm
167	buttonwood-mangrove	*Conocarpus erectus*	987	3	100	66.7	No signficant variables
168	flowering dogwood	*Cornus florida*	491	8029	94.4	62.1	posidx, mat, dd0, shwv, lonrc6
169	Pacific dogwood	*Cornus nuttallii*	492	525	62.9	47.4	mmax, mtwm, pratio, mtcm, sprdy
170	Texas persimmon	*Diospyros texana*	522	232	74.2	65.1	sprrc3, lofrc3, admi, mat, sprrc5
171	common persimmon	*Diospyros virginiana*	521	2873	34.3	30.4	mat, mmindd0, gsp, sprrc6, sprrc2
172	anacua knockaway	*Ehretia anacua*	523	7	12.5	28.6	No signficant variables
173	American beech	*Fagus grandifolia*	531	9034	96.8	72.4	admi, mtwm, lonrc4, dd0, dd5
174	Florida strangler fig	*Ficus aurea*	876	7	50	14.3	No signficant variables
175	white ash	*Fraxinus americana*	541	10635	96.8	59.5	dd0, fpi, map, gsp, dem
176	Carolina ash	*Fraxinus caroliniana*	548	88	68.6	72.7	shwvd, cdom, mmax, lonc2, lonrc3
177	Oregon ash	*Fraxinus latifolia*	542	97	65.3	32	mat, di, cfreq, sdmi, posidx
178	black ash	*Fraxinus nigra*	543	4108	96.1	72.7	map, di, lonc2, pratio, tscl
179	green ash	*Fraxinus pennsylvanica*	544	8331	91.7	56.9	fpi, pratio, slp, di, lonrc4
180	pumpkin ash	*Fraxinus profunda*	545	73	62.5	76.7	dd5, sprrc4, sprdy, fpi, mmin
181	blue ash	*Fraxinus quadrangulata*	546	155	38.9	32.9	fpi, loft, posidx, lofdy, mmax
182	Texas ash	*Fraxinus texensis*	549	42	67.7	54.8	sprdy, lonc2, tscl, sprrc4, lonrc4
183	velvet ash	*Fraxinus velutina*	547	8	100	12.5	mtcm
184	waterlocust	*Gleditsia aquatica*	551	74	11.7	35.1	No signficant variables
185	honeylocust	*Gleditsia triacanthos*	552	1388	62.7	42.9	lonrc2, sprc3, mmin, lofrc6, mtwm
186	loblolly-bay	*Gordonia lasianthus*	555	370	85.5	60.3	sdmi, fpi, admi, sprrc3, di
187	Kentucky coffeetree	*Gymnocladus dioicus*	571	75	0	13.3	No signficant variables
188	American holly	*Ilex opaca*	591	2645	82.6	58.4	dem, shwvd, lofdy, sdmi, lonrc5
189	Southern California black walnut	*Juglans californica*	604	3	60	33.3	gsdd5, fpi
190	butternut	*Juglans cinerea*	601	358	31.3	15.1	No signficant variables

Page	Common name	Scientific name	FIA code	Plots	Global	Validation	Five most significant variables*
191	Arizona walnut	*Juglans major*	606	12	10.5	41.7	No signficant variables
192	black walnut	*Juglans nigra*	602	3773	88.3	56.4	dd5, fpi, pratio, dd0, lofrc1
193	white-mangrove	*Laguncularia racemosa*	988	7	100	71.4	trmim, di
194	sweetgum	*Liquidambar styraciflua*	611	15978	98.1	84.9	mmindd0, mtcm, sprc2, shwv, map
195	yellow-poplar	*Liriodendron tulipifera*	621	10765	98.7	77.2	dd0, mtwm, lofdy, sprdy, tscl
196	tanoak	*Lithocarpus densiflorus*	631	1092	99.4	90	mmin, cfreq, zone, londy, lofdy
197	Osage-orange	*Maclura pomifera*	641	1074	80.9	49.7	shwvd, mtcm, fpi, pratio, mmindd0
198	cucumbertree	*Magnolia acuminata*	651	902	56.5	52	mat, map, cfreq, trmim, sprt
199	mountain magnolia	*Magnolia fraseri*	655	271	66.5	56.1	dem, lofc3, lofrc1, loft, sprrc5
200	southern magnolia	*Magnolia grandiflora*	652	514	39.2	44.7	lofrc3, mtwm, mtcm, sprc3, mat
201	bigleaf magnolia	*Magnolia macrophylla*	654	166	11.3	44.6	londy, shwv, map, gsp, di
202	sweetbay	*Magnolia virginiana*	653	2168	89.8	54.2	lonc2, di, tscl, sprrc5, sprc3
203	Oregon crab apple	*Malus fusca*	661	26	14.7	15.4	mtcm, map, di, sprt, lonc1
204	red mulberry	*Morus rubra*	682	1513	27.7	30.3	dd5, pratio, map, sprdy, sprc1
205	water tupelo	*Nyssa aquatica*	691	506	84.5	69.4	sprrc4, dem, gsp, mtcm, di
206	swamp tupelo	*Nyssa biflora*	694	2632	96.1	70.3	di, mtcm, mmax, lonc2, dem
207	Ogeechee tupelo	*Nyssa ogeche*	692	62	48.5	56.5	sdmi, mtcm, sprc1, lofrc5, sprrc3
208	blackgum	*Nyssa sylvatica*	693	11413	98.4	62.1	mtcm, fpi, dem, londy, di
209	desert ironwood	*Olneya tesota*	990	8	100	25	lofc2
210	eastern hophornbeam	*Ostrya virginiana*	701	6758	89.8	60.6	lonrc4, mmin, slp, mmindd0, map
211	sourwood	*Oxydendrum arboreum*	711	4975	97.8	71.5	dem, shwvd, mtcm, slp, map
212	empress-tree	*Paulownia tomentosa*	712	102	1.8	14.7	No signficant variables
213	redbay	*Persea borbonia*	721	1077	73	54	mtcm, pratio, mat, gsp, fpi
214	water-elm	*Planera aquatica*	722	144	45.1	47.2	lonrc1, sprc2, mmin, tscl, sprdy
215	American sycamore	*Platanus occidentalis*	731	2129	54	34.8	mtcm, dem, tscl, gsdd5, admi
216	California sycamore	*Platanus racemosa*	730	27	97.2	55.6	gsdd5, pratio, ctii, mat, sprt
217	Arizona sycamore	*Platanus wrightii*	732	5	100	100	di
218	narrowleaf cottonwood	*Populus angustifolia*	749	75	72.9	42.7	mmax, sprrc4, sprdy, posidx, cdom
219-220	balsam poplar	*Populus balsamifera*	741	1713	82.3	61.5	di, prod, mmax, map, lonrc4
221-222	black cottonwood	*Populus balsamifera* ssp. *trichocarpa*	747	368	64.6	51.6	lonc2, lofrc6, lofrc4, sprrc5, slp
223	eastern cottonwood	*Populus deltoides*	742	878	64.7	35.1	pratio, mtwm, mmax, fpi, lonc1
224	plains cottonwood	*Populus deltoides* ssp. *monilifera*	745	43	97.8	60.5	slp, map, mmin, dem, fpi
225	Fremont cottonwood	*Populus fremontii*	748	36	64.3	36.1	admi, lonrc3, mtwm, cdom, slp
226	bigtooth aspen	*Populus grandidentata*	743	4129	92.8	68.4	sdmi, sprrc6, mtwm, di, lofc2
227	swamp cottonwood	*Populus heterophylla*	744	43	0	30.2	No signficant variables
228-230	quaking aspen	*Populus tremuloides*	746	12665	98.9	88.1	mtcm, lonc2, dem, map, mmindd0
231	honey mesquite	*Prosopis glandulosa*	756	2736	99.7	85.6	pratio, map, mmax, mtcm, sprdy
232	screwbean mesquite	*Prosopis pubescens*	758	9	40	11.1	map
233	velvet mesquite	*Prosopis velutina*	757	270	99.3	75.6	mat, gsp, admi, di, mtcm
234	American plum	*Prunus americana*	766	155	5.2	0	map, cfreq
235	bitter cherry	*Prunus emarginata*	768	274	44.2	41.2	lonc2, sprrc4, sprrc5, londy, admi
236	pin cherry	*Prunus pensylvanica*	761	974	25.6	38.2	mtwm, gsdd5, sprdy, lonc1, sprrc6
237	black cherry	*Prunus serotina*	762	16346	97.2	63.1	dd0, mmin, gsp, di, sprdy
238	chokecherry	*Prunus virginiana*	763	515	0	10.3	No signficant variables
239	coast live oak	*Quercus agrifolia*	801	303	99.1	72.6	dd0, mtcm, map, mtwm, mmax
240	white oak	*Quercus alba*	802	17761	98.4	77.5	gsdd5, pratio, lofdy, sprc3, mmindd0
241	Arizona white oak	*Quercus arizonica*	803	622	99.3	83.4	gsp, sprdy, mtcm, pratio, mmax
242	swamp white oak	*Quercus bicolor*	804	449	33.4	29.4	lonc2, nsslp, tscl, sprc2, lonrc6
243	Buckley oak	*Quercus buckleyi*	8513	122	94.9	58.2	zone, mmindd0, lofc2, lonc2, map
244	canyon live oak	*Quercus chrysolepis*	805	1765	99.6	83.1	di, slp, mtcm, sdmi, sprdy

Page	Common name	Scientific name	FIA code	Plots	Global	Validation	Five most significant variables*
245	scarlet oak	*Quercus coccinea*	806	4906	96	69.2	fpi, map, mmax, lofc2, dd5
246	blue oak	*Quercus douglasii*	807	477	98.8	79.5	dd5, lonrc3, mtwm, lofdy, mmin
247	northern pin oak	*Quercus ellipsoidalis*	809	1234	91.5	64.8	di, mmax, fpi, pratio, dd0
248	Emory oak	*Quercus emoryi*	810	294	99.3	74.1	map, mtwm, londy, di_src, mmin
249	Engelmann oak	*Quercus engelmannii*	811	5	37.5	80	shwvd, nsslp, di
250	southern red oak	*Quercus falcata*	812	6714	97.5	64.3	di, dd0, shwvd, londy, dem
251	Gambel oak	*Quercus gambelii*	814	2144	99.1	84.3	lonc2, lofc3, dd5, sprrc5, mtcm
252	Oregon white oak	*Quercus garryana*	815	604	95.1	72.2	mtwm, sprc3, dem, mat, map
253	gray oak	*Quercus grisea*	846	28	80	50	lofdy, mtcm, londy, di_src
254	silverleaf oak	*Quercus hypoleucoides*	843	87	95.9	78.2	londy, sprc2, mtcm, mmin, fpi
255	bear oak	*Quercus ilicifolia*	816	76	62.4	42.1	prod, lont, londy, lonc1, sprt
256	shingle oak	*Quercus imbricaria*	817	655	74.3	48.4	gsp, admi, lonrc4, londy, dem
257	bluejack oak	*Quercus incana*	842	198	46.1	11.6	sprdy, prod, lonrc3, lofrc1, sprt
258	California black oak	*Quercus kelloggii*	818	1858	99.6	85.8	mtwm, lonc2, admi, sdmi, mat
259	Lacey oak	*Quercus laceyi*	8514	25	75.6	44	lofrc1, zone, dem, shwvd
260	turkey oak	*Quercus laevis*	819	494	82	70.2	di, lofrc4, lofdy, lonrc3, sprrc4
261	laurel oak	*Quercus laurifolia*	820	2866	91.4	63.5	dd0, mtcm, lonrc4, sprc2, mtwm
262	California white oak	*Quercus lobata*	821	108	76.4	52.8	dd5, prod, sprdy, cdom, sprc3
263	overcup oak	*Quercus lyrata*	822	707	77.9	62.2	sprc2, di, mtwm, sprrc4, ctii
264	bur oak	*Quercus macrocarpa*	823	2354	92.2	59.4	admi, shwvd, pratio, mmax, mmin
265	dwarf post oak	*Quercus margarettiae*	840	183	54.6	59	sprdy, sprrc3, sprc2, lonc2, di
266	blackjack oak	*Quercus marilandica*	824	1894	78.2	50.4	gsdd5, sprrc6, dem, pratio, londy
267	swamp chestnut oak	*Quercus michauxii*	825	710	0.3	28.9	dd5, dem, sprrc5, sprrc4, sprc3
268	dwarf live oak	*Quercus minima*	841	131	88.9	65.6	gsp, di, tscl, shwv, lonrc1
269	chinkapin oak	*Quercus muehlenbergii*	826	1637	83.1	52.5	lofdy, fpi, mat, di, dd5
270	water oak	*Quercus nigra*	827	9292	99	77	sprc2, dd0, lonrc4, slp, mmindd0
271	Mexican blue oak	*Quercus oblongifolia*	829	32	96.8	75	mat, shwvd, fpi, gsp, mmax
272	cherrybark oak	*Quercus pagoda*	813	1825	78	53.4	londy, lofrc5, fpi, mmax, sprc2
273	pin oak	*Quercus palustris*	830	497	60.2	43.1	shwv, di, mtcm, mat, londy
274	willow oak	*Quercus phellos*	831	2018	77.6	51.2	shwv, sprrc4, di, lofc3, tscl
275	chestnut oak	*Quercus prinus*	832	5246	98.4	81.7	slp, mtcm, admi, sprdy, fpi
276	northern red oak	*Quercus rubra*	833	13939	98.3	72.2	mat, dd5, shwv, lonc2, di
277	netleaf oak	*Quercus rugosa*	847	68	85.4	47.1	shwv, sprc2, londy, sprc1, mmin
278	Shumard oak	*Quercus shumardii*	834	502	18.8	22.3	No signficant variables
279	Durand oak	*Quercus sinuata* var. *sinuata*	808	45	0	11.1	No signficant variables
280	post oak	*Quercus stellata*	835	7908	97.1	71.7	dem, prod, mmax, lofdy, dd5
281	Texas red oak	*Quercus texana*	828	303	63.2	59.1	londy, di, mtwm, lonrc4, dem
282	black oak	*Quercus velutina*	837	9996	97.1	67.2	gsdd5, mat, sprc1, lofdy, sprrc3
283	live oak	*Quercus virginiana*	838	1549	94	65	dem, lofdy, di, mat, admi
284	interior live oak	*Quercus wislizeni*	839	390	95.9	72.6	dd0, mtwm, lonc2, zone, map
285	American mangrove	*Rhizophora mangle*	989	10	100	80	trmim, lofrc5, lonrc1
286	New Mexico locust	*Robinia neomexicana*	902	30	40.4	50	lofrc3, gsdd5, dem, di, sprrc2
287	black locust	*Robinia pseudoacacia*	901	2777	82.7	60.6	gsdd5, dd0, mtwm, sprdy, mmin
288	cabbage palmetto	*Sabal palmetto*	912	330	98.5	82.4	dem, shwvd, dd5, gsp, lonrc4
289	coastal plain willow	*Salix caroliniana*	925	39	38.6	35.9	sprdy, lofrc3, lofrc1, cfreq, lofdy
290	black willow	*Salix nigra*	922	1055	42	31.2	dd5, pratio, di, lofrc6, lonrc5
291	western soapberry	*Sapindus saponaria* var. *drum-mondii*	919	57	0	33.3	admi, sprrc4, sprrc3, tscl, sprc2
292	sassafras	*Sassafras albidum*	931	5176	84.2	57	dd0, mmin, tscl, mmindd0, pratio
293	American mountain-ash	*Sorbus americana*	935	180	27.2	31.7	mmax, sprrc6, sprt, lonrc4, lofc1

Page	Common name	Scientific name	FIA code	Plots	Global	Validation	Five most significant variables*
294	American basswood	*Tilia americana*	951	5444	91.3	62.4	map, dd0, lofc2, lonc2, gsdd5
295	Chinese tallowtree	*Triadica sebifera*	994	534	83.9	59.9	sdmi, lofdy, gsdd5, lofrc5, shwvd
296	winged elm	*Ulmus alata*	971	6644	97.3	65.8	gsdd5, mmax, fpi, londy, zone
297	American elm	*Ulmus americana*	972	11388	93.6	65.1	lonrc4, map, fpi, mmin, gsdd5
298	cedar elm	*Ulmus crassifolia*	973	561	89.9	66	mtwm, gsp, lofc1, sdmi, lofrc4
299	slippery elm	*Ulmus rubra*	975	4341	82.8	55.3	mtwm, mtcm, fpi, mmindd0, map
300	rock elm	*Ulmus thomasii*	977	105	12.5	33.3	No signficant variables
301	California-laurel	*Umbellularia californica*	981	503	97	73.4	dd0, slp, pratio, mmax, mtwm

B4 — Changes to Species Classifications

FIA Code	Scientific name	Notes	Distribution
18	*Abies lasiocarpa* var. *arizonica*	Variety of *A. lasiocarpa* mapped by FIA	Plants
21	*Abies shastensis*	Not mapped by Little, recognized by FIA, NRCS Plants recognizes as *A. magnifica* var. *shastensis*.	Plants
42	*Chamaecyparis nootkatensis*	NRCS Plants recongizes genus as *Callitropsis*.	
58	*Juniperus pinchotii*	Changed from Little (*J. erythrocarpa*).	
59	*Juniperus coahuilensis*	Changed from Little (*J. erythrocarpa* var. *coahuilensis*).	
81	*Calocedrus decurrens*	Changed genus from *Libocedrus* (Little).	
134	*Pinus discolor*	Formerly *P. cembroides* var. *discolor*, now mapped by FIA.	Plants
137	*Pinus washoensis*	Recongized by NRCS-Plants as *P. ponderosa* var. *washoensis*.	
141	*Pinus remota*	Formerly *P. cembroides* var. *remota*, now mapped by FIA.	Plants
143	*Pinus monophylla* var. *fallax*	Variety of *P. monophylla*, now mapped by FIA.	
222	*Taxodium ascendens*	Formerly *T. distichum* var. *imbricarium*, now mapped by FIA.	Plants
332	*Aesculus flava*	Changed species from *A. octandra* (Little).	
341	*Ailanthus altissima*	Tree-of-heaven, introduced, not mapped by Little.	Plants
409	*Carya alba*	Changed species from Little (*C. tomentosa*).	
431	*Chrysolepis chrysophylla* var. *chrysophylla*	Not mapped by Little.	Other
463	*Celtis laevigata* var. *reticulata*	Changed species from Little (*C. reticulata*).	
661	*Malus fusca*	Changed species from *M. diversifolia* (Little).	
694	*Nyssa biflora*	Formerly *N. sylvatica* var. *biflora*, now mapped by FIA.	Plants
712	*Paulownia tomentosa*	Empress-tree, introduced, not mapped by Little.	
745	*Populus deltoides* ssp. *monilifera*	Variety of *P. deltoides*, mapped by FIA.	
747	*Populus balsamifera* ssp. *trichocarpa*	Changed species from Little (*P. trichocarpa*).	
756	*Prosopis glandulosa*	Formerly *P. juliflora* var. *glandulosa*, now mapped by FIA.	
757	*Prosopis velutina*	Variety of mesquite, mapped by FIA, no longer inventoried in New Mexico by FIA.	Plants
808	*Quercus sinuata* var. *sinuata*	Changed species from Little (*Q. durandii*).	
813	*Quercus pagoda*	Former var. of *Q. falcata*.	eFloras.org
816	*Quercus ilicifolia*	Texas FIA inventory erroneously codes "scrub oak," which is a common name for five other species of oaks not listed in their official coding lists.	
828	*Quercus texana*	Changed species from Little (*Q. nuttallii*).	
840	*Quercus margarettiae*	Former var. of *Q. stellata*, NRCS-Plants recongizes species to *margarettae*.	Plants
841	*Quercus minima*	Former var. of *Q. virginiana*.	Plants
846	*Quercus grisea*	Not inventoried in New Mexico 2002 (considered a shrub).	
8513	*Quercus buckleyi*	Species not mapped by Little.	Plants
8514	*Quercus laceyi*	Changed species from *Q. glaucoides*.	
902	*Robinia neomexicana*	No longer inventoried in New Mexico (as of 2004 considered a shrub).	
919	*Sapindus saponaria* var. *drummondii*	Species mapped by Little, variety recorded by FIA .	
951	*Tilia americana*	*Tilia americana* var. *caroliniana*; *Tilia americana* var. *heterophylla*; *Tilia americana* var. *americana* combined for *modeled Atlas*.	
994	*Triadica sebifera*	Chinese tallowtree, introduced, not mapped by Little.	

Appendix – C (Indices)

C1 — Maps Listed by Scientific Name

Scientific name	Page
Abies amabilis	16 - 17
Abies balsamea	18
Abies concolor	19
Abies fraseri	20
Abies grandis	21
Abies lasiocarpa	22 - 23
Abies lasiocarpa var. arizonica	24
Abies magnifica	25
Abies procera	26
Abies shastensis	27
Acacia farnesiana	118
Acer barbatum	119
Acer glabrum	120
Acer grandidentatum	121
Acer leucoderme	122
Acer macrophyllum	123
Acer negundo	124 - 125
Acer nigrum	126
Acer pensylvanicum	127
Acer rubrum	128
Acer saccharinum	129
Acer saccharum	130
Acer spicatum	131
Aesculus californica	132
Aesculus flava	133
Aesculus glabra	134
Ailanthus altissima	135
Alnus oblongifolia	136
Alnus rhombifolia	137
Alnus rubra	138 - 139
Annona glabra	140
Arbutus menziesii	141
Asimina triloba	142
Avicennia germinans	143
Betula alleghaniensis	144
Betula lenta	145
Betula nigra	146
Betula papyrifera	147 - 149
Betula populifolia	150
Calocedrus decurrens	28
Carpinus caroliniana	151
Carya alba	152
Carya aquatica	153
Carya cordiformis	154
Carya glabra	155

Scientific name	Page
Carya illinoinensis	156
Carya ovata	157
Carya texana	158
Castanea dentata	159
Celtis laevigata	160
Celtis laevigata var. reticulata	161
Celtis occidentalis	162
Cercis canadensis	163
Cercocarpus ledifolius	164
Chamaecyparis lawsoniana	29
Chamaecyparis nootkatensis	30 - 31
Chamaecyparis thyoides	32
Chrysolepis chrysophylla var. chrysophylla	165
Condalia hookeri	166
Conocarpus erectus	167
Cornus florida	168
Cornus nuttallii	169
Cupressus arizonica	33
Cupressus sargentii	34
Diospyros texana	170
Diospyros virginiana	171
Ehretia anacua	172
Fagus grandifolia	173
Ficus aurea	174
Fraxinus americana	175
Fraxinus caroliniana	176
Fraxinus latifolia	177
Fraxinus nigra	178
Fraxinus pennsylvanica	179
Fraxinus profunda	180
Fraxinus quadrangulata	181
Fraxinus texensis	182
Fraxinus velutina	183
Gleditsia aquatica	184
Gleditsia triacanthos	185
Gordonia lasianthus	186
Gymnocladus dioicus	187
Ilex opaca	188
Juglans californica	189
Juglans cinerea	190
Juglans major	191
Juglans nigra	192
Juniperus ashei	35
Juniperus californica	36

Scientific name	Page
Juniperus coahuilensis	37
Juniperus deppeana	38
Juniperus monosperma	39
Juniperus occidentalis	40
Juniperus osteosperma	41
Juniperus pinchotii	42
Juniperus scopulorum	43
Juniperus virginiana	44
Laguncularia racemosa	193
Larix laricina	45 - 46
Larix lyallii	47
Larix occidentalis	48
Liquidambar styraciflua	194
Liriodendron tulipifera	195
Lithocarpus densiflorus	196
Maclura pomifera	197
Magnolia acuminata	198
Magnolia fraseri	199
Magnolia grandiflora	200
Magnolia macrophylla	201
Magnolia virginiana	202
Malus fusca	203
Morus rubra	204
Nyssa aquatica	205
Nyssa biflora	206
Nyssa ogeche	207
Nyssa sylvatica	208
Olneya tesota	209
Ostrya virginiana	210
Oxydendrum arboreum	211
Paulownia tomentosa	212
Persea borbonia	213
Picea breweriana	49
Picea engelmannii	50
Picea glauca	51 - 52
Picea mariana	53 - 54
Picea pungens	55
Picea rubens	56
Picea sitchensis	57 - 58
Pinus albicaulis	59
Pinus aristata	60
Pinus attenuata	61
Pinus balfouriana	62
Pinus banksiana	63
Pinus cembroides	64

Scientific name	Page
Pinus clausa	65
Pinus contorta	66 - 67
Pinus coulteri	68
Pinus discolor	69
Pinus echinata	70
Pinus edulis	71
Pinus elliottii	72
Pinus engelmannii	73
Pinus flexilis	74
Pinus glabra	75
Pinus jeffreyi	76
Pinus lambertiana	77
Pinus leiophylla	78
Pinus longaeva	79
Pinus monophylla	80
Pinus monophylla var. fallax	81
Pinus monticola	82
Pinus muricata	83
Pinus palustris	84
Pinus ponderosa	85
Pinus pungens	86
Pinus radiata	87
Pinus remota	88
Pinus resinosa	89
Pinus rigida	90
Pinus sabiniana	91
Pinus serotina	92
Pinus strobiformis	93
Pinus strobus	94
Pinus taeda	95
Pinus virginiana	96
Pinus washoensis	97
Planera aquatica	214
Platanus occidentalis	215
Platanus racemosa	216
Platanus wrightii	217
Populus angustifolia	218
Populus balsamifera	219 - 220
Populus balsamifera ssp. trichocarpa	221 - 222
Populus deltoides	223
Populus deltoides ssp. monilifera	224
Populus fremontii	225
Populus grandidentata	226
Populus heterophylla	227
Populus tremuloides	228 -230
Prosopis glandulosa	231
Prosopis pubescens	232
Prosopis velutina	233
Prunus americana	234

Scientific name	Page
Prunus emarginata	235
Prunus pensylvanica	236
Prunus serotina	237
Prunus virginiana	238
Pseudotsuga macrocarpa	98
Pseudotsuga menziesii	99
Quercus agrifolia	239
Quercus alba	240
Quercus arizonica	241
Quercus bicolor	242
Quercus buckleyi	243
Quercus chrysolepis	244
Quercus coccinea	245
Quercus douglasii	246
Quercus ellipsoidalis	247
Quercus emoryi	248
Quercus engelmannii	249
Quercus falcata	250
Quercus gambelii	251
Quercus garryana	252
Quercus grisea	253
Quercus hypoleucoides	254
Quercus ilicifolia	255
Quercus imbricaria	256
Quercus incana	257
Quercus kelloggii	258
Quercus laceyi	259
Quercus laevis	260
Quercus laurifolia	261
Quercus lobata	262
Quercus lyrata	263
Quercus macrocarpa	264
Quercus margarettiae	265
Quercus marilandica	266
Quercus michauxii	267
Quercus minima	268
Quercus muehlenbergii	269
Quercus nigra	270
Quercus oblongifolia	271
Quercus pagoda	272
Quercus palustris	273
Quercus phellos	274
Quercus prinus	275
Quercus rubra	276
Quercus rugosa	277
Quercus shumardii	278
Quercus sinuata var. sinuata	279
Quercus stellata	280
Quercus texana	281

Scientific name	Page
Quercus velutina	282
Quercus virginiana	283
Quercus wislizeni	284
Rhizophora mangle	285
Robinia neomexicana	286
Robinia pseudoacacia	287
Sabal palmetto	288
Salix caroliniana	289
Salix nigra	290
Sapindus saponaria var. drummondii	291
Sassafras albidum	292
Sequoia sempervirens	100
Sequoiadendron giganteum	101
Sorbus americana	293
Taxodium ascendens	102
Taxodium distichum	103
Taxus brevifolia	104
Thuja occidentalis	105
Thuja plicata	106 - 107
Tilia americana	294
Torreya californica	108
Triadica sebifera	295
Tsuga canadensis	109
Tsuga caroliniana	110
Tsuga heterophylla	111 - 112
Tsuga mertensiana	113 - 114
Ulmus alata	296
Ulmus americana	297
Ulmus crassifolia	298
Ulmus rubra	299
Ulmus thomasii	300
Umbellularia californica	301

C2 — Maps Listed by Common Name

Common	Name	Page
acacia	sweet	118
alder	Arizona	136
alder	red	138 - 139
alder	white	137
ash	black	178
ash	blue	181
ash	Carolina	176
ash	green	179
ash	Oregon	177
ash	pumpkin	180
ash	Texas	182
ash	velvet	183
ash	white	175
aspen	bigtooth	226
aspen	quaking	228 - 230
baldcypress		103
basswood	American	294
beech	American	173
birch	gray	150
birch	paper	147 - 149
birch	river	146
birch	sweet	145
birch	yellow	144
blackgum		208
bluewood		166
boxelder		124 - 125
buckeye	California	132
buckeye	Ohio	134
buckeye	yellow	133
butternut		190
cherry	bitter	235
cherry	black	237
cherry	pin	236
chestnut	American	159
chinkapin	giant	165
chokecherry		238
coffeetree	Kentucky	187
cottonwood	black	221 - 222
cottonwood	eastern	223
cottonwood	Fremont	225
cottonwood	narrowleaf	218
cottonwood	plains	224
cottonwood	swamp	227
crab apple	Oregon	203
cucumbertree		198
cypress	Arizona	33
cypress	Sargent's	34
dogwood	flowering	168
dogwood	Pacific	169
Douglas-fir	bigcone	98

Common	Name	Page
Douglas-fir		99
elm	American	297
elm	cedar	298
elm	rock	300
elm	slippery	299
elm	winged	296
empress-tree		212
fir	balsam	18
fir	California red	25
fir	corkbark	24
fir	Fraser	20
fir	grand	21
fir	noble	26
fir	Pacific silver	16 - 17
fir	Shasta red	27
fir	subalpine	22 - 23
fir	white	19
fig	Florida strangler	174
hackberry	netleaf	161
hackberry		162
hemlock	Carolina	110
hemlock	eastern	109
hemlock	mountain	113 - 114
hemlock	western	111 - 112
hickory	bitternut	154
hickory	black	158
hickory	mockernut	152
hickory	pignut	155
hickory	shagbark	157
hickory	water	153
holly	American	188
hophornbeam	eastern	210
hornbeam	American	151
incense-cedar		28
ironwood	desert	209
juniper	alligator	38
juniper	Ashe	35
juniper	California	36
juniper	oneseed	39
juniper	Pinchot	42
juniper	redberry	37
juniper	Rocky Mountain	43
juniper	Utah	41
juniper	western	40
knockaway	anacua	172
larch	subalpine	47
larch	tamarack	45 - 46
larch	western	48
laurel	California	301
loblolly-bay		186

Common	Name	Page
locust	black	287
locust	honeylocust	185
locust	New Mexico	286
locust	waterlocust	184
madrone	Pacific	141
magnolia	bigleaf	201
magnolia	mountain	199
magnolia	southern	200
magnolia	sweetbay	202
mangrove	American	285
mangrove	black-	143
mangrove	buttonwood-	167
mangrove	white-	193
maple	bigleaf	123
maple	bigtooth	121
maple	black	126
maple	chalk	122
maple	Florida	119
maple	mountain	131
maple	red	128
maple	Rocky Mountain	120
maple	silver	129
maple	striped	127
maple	sugar	130
mesquite	honey	231
mesquite	screwbean	232
mesquite	velvet	233
mountain-ash	American	293
mountain-mahogany	curlleaf	164
mulberry	red	204
oak	Arizona white	241
oak	bear	255
oak	black	282
oak	blackjack	266
oak	blue	246
oak	bluejack	257
oak	Buckley	243
oak	bur	264
oak	California black	258
oak	California white	262
oak	canyon live	244
oak	cherrybark	272
oak	chestnut	275
oak	chinkapin	269
oak	coast live	239
oak	Durand	279
oak	dwarf live	268
oak	dwarf post	265
oak	Emory	248

National Individual Tree Species Atlas

Common	Name	Page
oak	Engelmann	249
oak	Gambel	251
oak	gray	253
oak	interior live	284
oak	Lacey	259
oak	laurel	261
oak	live	283
oak	Mexican blue	271
oak	netleaf	277
oak	northern pin	247
oak	northern red	276
oak	Oregon white	252
oak	overcup	263
oak	pin	273
oak	post	280
oak	scarlet	245
oak	shingle	256
oak	Shumard	278
oak	silverleaf	254
oak	southern red	250
oak	swamp chestnut	267
oak	swamp white	242
oak	Texas red	281
oak	turkey	260
oak	water	270
oak	white	240
oak	willow	274
Osage-orange		197
palmetto	cabbage	288
pawpaw		142
pecan		156
persimmon	common	171
persimmon	Texas	170
pine	Apache	73
pine	Arizona piñon	81
pine	bishop	83
pine	border piñon	69
pine	California foothills	91
pine	Chihuahua	78
pine	common piñon	71
pine	Coulter	68
pine	eastern white	94
pine	foxtail	62
pine	Great Basin bristlecone	79
pine	jack	63
pine	Jeffrey	76
pine	knobcone	61
pine	limber	74

Common	Name	Page
pine	loblolly	95
pine	lodgepole	66 - 67
pine	longleaf	84
pine	Mexican piñon	64
pine	Monterey	87
pine	papershell piñon	88
pine	pitch	90
pine	pond	92
pine	ponderosa	85
pine	red	89
pine	Rocky Mountain bristlecone	60
pine	sand	65
pine	shortleaf	70
pine	singleleaf piñon	80
pine	slash	72
pine	southwestern white	93
pine	spruce	75
pine	sugar	77
pine	Table Mountain	86
pine	Virginia	96
pine	Washoe	97
pine	western white	82
pine	whitebark	59
plum	American	234
pond-apple		140
pondcypress		102
poplar	balsam	219 - 220
Port-Orford-cedar		29
redbay		213
redbud	eastern	163
redcedar	eastern	44
redcedar	western	106 - 107
redwood		100
sassafras		292
sequoia	giant	101
soapberry	western	291
sourwood		211
spruce	black	53 - 54
spruce	blue	55
spruce	Brewer	49
spruce	Engelmann	50
spruce	red	56
spruce	Sitka	57 - 58
spruce	white	51 - 52
sugarberry		160
sweetgum		194
sycamore	American	215

Common	Name	Page
sycamore	Arizona	217
sycamore	California	216
tallowtree	Chinese	295
tanoak		196
torrey	California	108
tree-of-heaven		135
tupelo	Ogeechee	207
tupelo	swamp	206
tupelo	water	205
walnut	Arizona	191
walnut	black	192
walnut	Southern California black	189
water-elm		214
white-cedar	Atlantic	32
white-cedar	northern	105
willow	black	290
willow	coastal plain	289
yellow-cedar	Alaska	30 - 31
yellow-poplar		195
yew	Pacific	104

Appendix – D (Photo Credits)

D — List of photographers

Photo Credit	Photo Description
Adam A. Agusto, School of Renewable Natural Resources, Louisiana State University AgCenter[3]	*Fraxinus profunda*
James R. Allison, Georgia Department of Natural Resources, Bugwood.org[6]	*Triadica sebifera*
Anonymous[1]	*Quercus lobata*
Arizona-Sonora Desert Museum[3]	*Prosopis velutina*
Steven J. Baskauf, www.cas.vanderbilt.edu[2]	*Platanus wrightii*
Chuck Benedict, USDA Forest Service - FHP/FHTET[1]	Elbert Little's Atlas of the United States Trees Volume 1 photo (Pg. 3)
Ronald F. Billings, Texas Forest Service, Bugwood.org[6]	*Sapindus saponaria* var. *drummondii*
William D. Boyer, USDA Forest Service, Bugwood.org[6]	*Pinus palustris* (Pg. 15 - top right)
Pat Breen, Oregon State University[2]	*Celtis laevigata* var. *reticulata, Fraxinus americana, Pinus lambertiana*
Dana Bressette, Native Plants PNW[3]	*Abies grandis*
Br. Alfred Brousseau, calphotos.berkley.edu[3]	*Quercus garryana*
Mary PK Burns, flickr.com[4]	*Ulmus crassifolia*
John D. Byrd, Mississippi State University, Bugwood.org[6]	*Albizia julibrissin* (Pg. 5 - top right)
J. Cavender-Bares, quercus.lifedesks.org[6]	*Quercus oblongifolia*
Dendroica cerulea, flickr.com[4]	*Betula populifolia*
George P. Chamuris, Bloomsburg University of Pennsylvania[2]	*Fagus grandifolia*
Michael L. Charters, calflora.net[2]	*Pinus coulteri, Populus balsamifera trichocarpa*
Connecticut Agricultural Experiment Station, Bugwood.org[6]	*Chamaecyparis nootkatensis, Juniperus virginiana*
Bill Cook, Michigan State University, Bugwood.org[6]	*Tsuga canadensis, Pinus resinosa*
cotinis, flickr.com[4]	*Acer barbatum*
Whitney Cranshaw, Colorado State University, Bugwood.org[6]	*Juglans californica, Juglans major, Juniperus deppeana*
Dick Culbert, Gibsons, B.C., Canada[6]	*Pinus discolor*
Rachel Cywinski, Lady Bird Johnson Wildflower Center, University of Texas, Austin[3]	*Ehretia anacua*
Tom DeGomez, University of Arizona, Bugwood.org[6]	*Juniperus monosperma*
Joseph M. DiTomaso, University of California - Davis, Bugwood.org[6]	*Juniperus occidentalis*
C.J. Earle, www.conifers.org[2]	*Pinus balfouriana, Pinus cembroides, Pinus longaeva, Pinus muricata, Pinus strobiformis, Pseudotsuga macrocarpa*
Cindy Ellenwood[2]	Rocky Mountain Naitonal Park, CO (Pg. 10)
Jim Ellenwood, USDA Forest Service - FHP/FHTET[2]	*Abies lasiocarpa*, Misty Fjords National Monument, AK (Pg. iv), Kancamagus Highway, White River National Forest, NH (Pg. x), Grand Teton National Park, WY (Pg. 1 - top left), The Loch, Rocky Mountain National Park, CO (Pg. 1 - second from right), Grand Teton National Park, WY (Pg. 2), Misty Fjords National Monument, AK (Pg. 4), Libby Lake, Medicine Bow-Routt National Forest, WY (Pg 5 - second from right), Rocky Mountain Naitonal Park, CO (Pg. 10), Grand Teton National Park, WY (Pg. 14 & 115), Rocky Gorge, White Mountain National Forest, NH (Pg. 302), Sequoia Grove, Big Trees Trail, Sequoia-Kings Canyon National Park (Pg. 303 - upper left), Tulip Poplar, Hoosier National Forest (Pg. 303 - upper right), Kancamagus Highway, White Mountain National Forest, NH (Pg. 319 - upper left), Banff National Park, Alberta, CA (Pg. 319 - second from left), Kancamagus Highway, White Mountain National Forest, NH (Pg. 319 - second from right)
Kendra Ellenwood[2]	Aspen, Loch Vale, Rocky Mountain National Park, CO (Pg. iii - top right)
Chris Evans, Illinois Wildlife Action Plan, Bugwood.org[7]	*Acer glabrum, Aesculus flava, Carya aquatica, Carya glabra, Celtis laevigata, Celtis occidentalis, Cornus nuttallii, Fraxinus caroliniana, Gleditsia aquatica, Liquidambar styraciflua, Liriodendron tulipifera, Magnolia fraseri, Nyssa aquatica, Nyssa biflora, Nyssa sylvatica, Oxydendrum arboreum, Persea borbonia, Pinus echinata, Pinus glabra, Pinus ponderosa, Pinus pungens, Pinus virginiana, Quercus velutina, Robinia pseudoacacia, Salix caroliniana, Sassafras albidum, Ulmus alata*, Appalachian mixed hardwoods (Pg. iii - second from left), *Aralia spinosa* (Pg. 1 - second from left), Big Creek, NC, United States (Pg. 11 - top left)
Troy Evans, Great Smoky Mountains National Park, Bugwood.org[6]	*Planera aquatica*
Wendy VanDyk Evans, Bugwood.org[6]	*Cornus florida, Quercus virginiana*
M Fletcher, flickr.com[4]	*Magnolia macrophylla*

Photo Credit	Photo Description
Norbert Frank, University of West Hungary, Bugwood.org[6]	Natural regeneration (Pg. iii - top left)
Tony Frates, www.flickr.com[4]	*Juniperus scopulorum*
David B. Gleason, www.flickr.com[2]	*Abies fraseri*
Jean-Pol Grandmont, wikimedia[6]	*Fraxinus latifolia*
Marlene Hahn, quercus.lifedesks.org[6]	*Quercus rugosa, Quercus texana, Quercus wislizeni*
Mary Ellen (Mel) Harte, Bugwood.org[7]	*Abies lasiocarpa* var. *arizonica, Juniperus osteosperma, Populus balsamifera*
Billy Hathorn, Wikimedia.org[6]	*Sabal palmetto*
Eric Hunt, Wikimedia.org[6]	*Cupressus sargentii*
Lesley Ingram, Bugwood.org[7]	*Chamaecyparis lawsoniana*
Sky Jacobs, Southwest Envoronmental Information Network[3]	*Quercus hypoleucoides*
Keith Kanoti, Maine Forest Service, Bugwood.org[6]	*Acer spicatum, Betula alleghaniensis, Betula papyrifera, Fraxinus nigra, Picea mariana, Picea rubens, Sorbus americana*
Steven Katovich, USDA Forest Service, Bugwood.org[6]	*Pinus strobus* (Pg. 5 - top left)
Mary Keim, flickr.com[4]	*Ficus aurea*
Russ Kleinman and Richard Felger, Western New Mexico University Department of Natural Sciences, Dale A. Zimmerman Herbarium[3]	*Juniperus coahuilensis*
Russ Kleinman, Carey Anne Lafferty, Shawn White, Western New Mexico University Department of Natural Sciences, Dale A. Zimmerman Herbarium[2]	*Pinus monophylla* var. *fallax*
Tihomir Kostadinov, Department of Geography and the Environment, University of Richmond[2]	*Quercus coccinea, Quercus falcata, Quercus nigra*
Ron Lance[2]	*Quercus buckleyi, Quercus incana, Quercus laceyi, Quercus laevis, Quercus laurifolia, Quercus lyrata, Quercus margarettiae, Quercus marilandica, Quercus michauxii, Quercus minima, Quercus pagoda, Quercus phellos, Quercus prinus, Quercus shumardii, Quercus sinuata* var. *sinuata*
Matt Lavin[3]	*Olneya tesota*
Gerald Lenhard, Louisiana State University[2]	*Taxodium ascendens*
Max Licher, Consortium of Intermountain Herbaria[6]	*Alnus oblongifolia*
Nancy Loewenstein, Auburn University, Bugwood.org[7]	*Acer leucoderme*
Campbell and Lynn Loughmiller, Lady Bird Johnson Wildflower Center, University of Texas, Austin[1]	*Juniperus pinchotii*
Becca MacDonald, Sault College, Bugwood.org[7]	*Abies balsamea, Thuja occidentalis*
Colin Maher, University of Montana[2]	*Pinus albicaulis*
Brett Marshall, Sault College, Bugwood.org[6]	*Pinus banksiana*
J. Maughn, flickr.com[4]	*Pinus attenuata*
Ulf Mehlig[5]	*Conocarpus erectus*
Leslie J. Mehrhoff, University of Connecticut, Bugwood.org[6]	*Ailanthus altissima*
Menchi[6]	*Tsuga heterophylla*
Connie Millar, USDA Forest Service[6]	*Pinus washoensis*
James H. Miller, USDA Forest Service, Bugwood.org[6]	*Paulownia tomentosa*
June Mirabella[2]	Alley Spring, Missouri (Pg. 303 - second from right), Fall Creek Falls State Park, TN (Pg. 319 - upper right)
David J. Moorhead, University of Georgia, Bugwood.org[7]	*Asimina triloba*
Daniel Mosquin, UBC Botanical Garden[7]	*Cupressus arizonica*
Joe Nicholson[6]	*Acer macrophyllum*
Ronnie Nijboer, Wikimedia Commons[1]	*Ulmus thomasii*
http://nwplants.com[4]	*Alnus rubra*
Joseph O'Brien, USDA Forest Service, Bugwood.org[6]	*Arbutus menziesii, Larix laricina, Lithocarpus densiflorus, Picea glauca, Quercus agrifolia, Quercus kelloggii*, Fall foliage of staghorn sumac (Pg. 1 - top right), *Acer spp.* (Pg. 117 - second from left), *Pseudotsuga menziesii* (Pg. 303 - second from left)
Lukasz Osinski, Wikimedia Commons[6]	*Pinus palustris, Pinus rigida, Torreya californica, Tsuga caroliniana*
Donald Owen, California Department of Forestry and Fire Protection, Bugwood.org[7]	*Abies concolor, Abies magnifica, Calocedrus decurrens, Tsuga mertensiana*
John A. Peterson, Department of Forest Resources and Environmental Conservation Virginia Tech[2]	*Annona glabra, Avicennia germinans, Rhizophora mangle*

Photo Credit	Photo Description
Plant Resources Center, The University of Texas at Austin[3]	*Carya texana, Condalia hookeri, Fraxinus texensis, Juniperus ashei*
Dave Powell, USDA Forest Service, Bugwood.org[6]	*Cercocarpus ledifolius, Pinus edulis, Pinus flexilis, Populus angustifolia, Populus deltoides monilifera, Prunus pensylvanica, Robinia neomexicana, Picea engelmannii* (Pg. 15 -top left), *Abies grandis* (Pg. 3 - second from left), Dense stand (Pg. 3 - top right)
Homer Edward Price[6]	*Laguncularia racemosa, Quercus grisea*
Karan A. Rawlins, University of Georgia, Bugwood.org[7]	*Betula nigra, Chamaecyparis thyoides, Fraxinus pennsylvanica, Gordonia lasianthus, Magnolia grandiflora, Nyssa ogeche, Thuja plicata*
Jennifer Rehm, nwplants.com[6]	*Chrysolepis chrysophylla* var. *chrysophylla*
Jane Shelby Richardson, Duke University, Wikimedia Commons[6]	*Pinus monophylla*
Richtid[6]	*Quercus ilicifolia*
Ewen Roberts[6]	*Pinus jeffreyi*
Rob Routledge, Sault College, Bugwood.org[6]	*Acer pensylvanicum, Betula lenta, Carpinus caroliniana, Carya cordiformis, Cercis canadensis, Fraxinus quadrangulata, Gymnocladus dioicus, Pinus contorta, Platanus occidentalis, Prunus americana, Prunus virginiana, Ulmus rubra*
John Ruter, University of Georgia, Bugwood.org[7]	*Aesculus californica, Alnus rhombifolia, Carya alba, Acer barbatum* (Pg. v - second from right)
Chris Schnepf, University of Idaho, Bugwood.org[6, 2]	*Larix occidentalis*[6], *Pinus monticola*[2], Marshes and swamps (Pg. iii - second from right)[6]
John R. Seiler, Department of Forest Resources and Environmental Conservation Virginia Tech[2]	*Diospyros texana, Fraxinus velutina, Larix lyallii, Pinus engelmannii, Platanus racemosa, Populus fremontii, Prosopis glandulosa, Prosopis pubescens, Quercus arizonica, Quercus chrysolepis, Quercus douglasii, Quercus emoryi, Quercus engelmannii*
Jason Sharman, Vitalitree, Bugwood.org[7]	*Acer saccharum, Acer saccharum* Fall Colors (Pg. 117 - top right)
Walter Siegmund[2]	*Abies amabilis, Abies procera, Juniperus californica*
Cynthia Snyder, USDA Forest Service - Region 5 - FHP[2]	*Pinus sabiniana, Taxus brevifolia*
Terry Spivey, USDA Forest Service, Bugwood.org[7]	*Prunus emarginata*
Forest and Kim Starr, Starr Environmental, Bugwood.org[6]	*Pinus radiata, Sequoia sempervirens, Acacia farnesiana*
James Steakley[6]	*Pinus serotina*
David Stephens, Bugwood.org[7]	*Pinus taeda* (Pg. 15 - second from right)
Roland Tanglao[3]	*Picea sitchensis*
US Fish and Wildlife Service[1]	*Malus fusca*
USDA Forest Service - Ogden Archive[7]	*Pinus jeffreyi* (Pg. 11 - second from left), *Pinus contorta* (Pg. 11 - top right)
USDA Forest Service - Region 2 - Rocky Mountain Region Archive, USDA Forest Service, Bugwood.org[7]	*Picea engelmannii, Pinus aristata, Quercus gambelii*
USDA Forest Service, Region 8 - Southern Archive, Bugwood.org[7]	*Acer rubrum* (Pg. 117 - top left)
USDA Forest Service Southern Research Station Archive, USDA Forest Service, SRS, Bugwood.org[7]	Mixed species stands (Pg. v - second from left), cypress (Pg. v - top right), Mixed species stands (Pg. 11 - second from right)
Carlos Velazco, swbiodiversity.org[3]	*Pinus remota*
Robert Vidéki, Doronicum Kft., Bugwood.org[7]	*Juglans nigra, Pinus strobus, Prunus serotina, Sequoiadendron giganteum, Chamaecyparis lawsoniana* (Pg. 5 - second from left)
Damon E. Waitt, Lady Bird Johnson Wildflower Center, University of Texas, Austin[1]	*Acer grandidentatum*
Rebekah D. Wallace, University of Georgia, Bugwood.org[7]	*Diospyros virginiana, Ilex opaca, Magnolia virginiana, Pinus clausa, Pinus elliottii*
Sally and Andy Wasowski, Lady Bird Johnson Wildflower Center, University of Texas, Austin[1]	*Umbellularia californica*
Richard Webb[6]	*Magnolia acuminata, Pinus taeda*
WhatsAllThisThen, flickr.com[4]	*Castanea dentata*
Wikimedia Commons[6]	*Abies shastensis, Picea breweriana*
Craig Wilcox, Coronado National Forest[2]	*Pinus leiophylla*
Vern Wilkins, Indiana University, Bugwood.org[7]	*Aesculus glabra*
Paul Wray, Iowa State University, Bugwood.org[7]	*Acer negundo, Acer nigrum, Acer rubrum, Acer saccharinum, Carya illinoinensis, Carya ovata, Gleditsia triacanthos, Juglans cinerea, Maclura pomifera, Morus rubra, Ostrya virginiana, Picea pungens, Populus deltoides, Populus grandidentata, Populus heterophylla, Populus tremuloides, Pseudotsuga menziesii, Quercus alba, Quercus bicolor, Quercus ellipsoidalis, Quercus imbricaria, Quercus macrocarpa, Quercus muehlenbergii, Quercus palustris, Quercus rubra, Quercus stellata, Taxodium distichum, Tilia americana, Ulmus americana, Salix nigra, Pseudotsuga menziesii* (Pg. 15 - second from left), *Quercus alba* (Pg. 117 - second from right), *Taxodium distichum* (Pg. 3 - second from right), *Acer negundo* (Pg. 3 - top left), *Acer saccharinum* (Pg. v - top left)

NOTES: [1]Public Domain, [2]Used by permission of the photographer, [3]Used by permission of the agency, institution, and/or Website, [4]Creative Commons 2.0, [5]Creative Commons 2.5, [6]Creative Commons 3.0, [7]Creative Commons Noncommercial 3.0

References

Austin, M. 2007. Species distribution models and ecological theory: a critical assessment and some possible new approaches. *Ecological Modeling*. 200: 1–19. Online at: http://people.eri. ucsb. edu/~fd/misc/595/austin-revu-ecolmod07.pdf

Critchfield, W. B., and E. L. Little, Jr. 1966. Geographic distribution of the pines of the world. U.S. Department of Agriculture Miscellaneous Publication 991, p. 1–97.

Ellenwood, J. 1984. Factors influencing the abundance and growth of white ash (*Fraxinus americana* L.) on Heiberg Forest, Tully, NY. M.S. Thesis. [Book]. Syracuse, New York. SUNY ESF.

Ellenwood, J. R., F. J. Krist, Jr., and F. J. Sapio. In press. Modeling individual tree species parameters on a national scale. Fort Collins, Colorado: U.S. Department of Agriculture, Forest Service, Forest Health Technology Enterprise Team.

Ellenwood, M., L. Dilling, and J. Milford. 2012. Managing United States Public Lands in Response to Climate Change: A View From the Ground Up. *Environmental Management*. 49:954–967.

Elith, J. and C. H. Graham. 2009. Do they? How do they? Why do they differ? On finding reasons for differing performances of species distribution models. *Ecography*. 32: 66 77. Online at: http://www.acera.unimelb.edu.au/materials/publications/Elith_Graham 2009.pdf

Gesch, D. B., G. A. Evans, J. Mauck, J. A. Hutchinson, and W. J. Carswell, Jr. 2009. The national map—elevation: U.S. Geological Survey fact sheet, 2009-3053, 4 p. Online at: http://pubs.usgs.gov/ fs/2009/3053/pdf/fs2009_3053.pdf

Harvey, A. E., J. W. Byler, G. I. McDonald, L. F. Neuenschwander, and R. T. Jonalea. 2008. Death of an ecosystem: perspectives on western white pine ecosystems of North America at the end of the twentieth century. Gen. Tech. Rep. RMRS-GTR-208. Fort Collins, Colorado: U.S. Department of Agriculture, Forest Service, Rocky Mountain Research Station. 10 p.

Homer, C., C. Huang, L. Yang, B. Wylie, and M. Coan. 2004. Development of a 2001 national landcover database for the United States. *Photogrammetric Engineering and Remote Sensing*. 70(7): 829–840. Online at: www.mrlc.gov/pdf/July_PERS.pdf.

Hutchinson, M. F. 1991. Continent wide data assimilation using thin plate smoothing splines. *In*: Jasper, J. D., ed. Data assimilation systems: papers presented at the second BMRC modeling workshop, September 1990. BMRC Research Report 27, Bureau of Meteorology, Melbourne, Australia: 104–113.

Krist, Jr., F. J., J. R. Ellenwood, M. Woods, A. McMahan, J. Cowardin, D. Ryerson, F. Sapio, M. Zweifler, and S. A. Romero. 2014. 2013–2027 National Insect and Disease Forest Risk Assessment. Fort Collins, Colorado: U.S. Department of Agriculture, Forest Service, Forest Health Technology Enterprise Team, FHTET-14-01.

Kumar, L., A. K. Skidmore, and E. Knowles. 1997. Modeling Topographic Variation in Solar Radiation in a GIS Environment. *Geographical Information Science*. 11(5): 475–497.

LANDFIRE: Public LANDFIRE Reference Database–LFRDB –Alaska. U.S. Department of Interior, Geological Survey. [Online]. Online at: http://landfire.cr.usgs.gov/viewer/ [extracted 2011,May]

Lister, A. J., C. T. Scott, and S. Rasmussen. 2012. Inventory methods for trees in nonforest areas in the great plains states. Environmental Monitoring and Assessment. 184: 2465– 2474. Online at: http://link.springer.com/content/pdf/10.1007%2Fs10661-011-2131-6

Little, E. L., Jr. 1971. Atlas of United States trees, volume 1, conifers and important hardwoods: U.S. Department of Agriculture Miscellaneous Publication 1146, 9 p., 200 maps.

Little, E. L., Jr. 1976. Atlas of United States trees, volume 3, minor Western hardwoods: U.S. Department of Agriculture Miscellaneous Publication 1314, 13 p., 290 maps.

Little, E. L., Jr. 1977. Atlas of United States trees, volume 4, minor Eastern hardwoods: U.S. Department of Agriculture Miscellaneous Publication 1342, 17 p., 230 maps.

Little, E. L., Jr. 1978. Atlas of United States trees, volume 5, Florida: U.S. Department of Agriculture Miscellaneous Publication 1361, 262 maps.

Little, E. L., Jr. 1981 Atlas of United States trees, volume 6, Supplement: U.S. Department of Agriculture Miscellaneous Publication 1410, 36 maps.

Morgan, J., D. R. Daugherty, A. Hilchie, and B. Carey. 2003. Sample size and modeling accuracy of decision tree based data mining tools. *Academy of Information and Management Science Journal*. 6(2):71-99.

National Climatic Data Center (NCDC). 2001. U.S. Climate Normals 1971–2000, Products. NOAA's National Climatic Data Center, Asheville, North Carolina.

Quinlan, R. 2012. Rulequest, Inc.: Cubist/See5. Online at: http://www.rulequest.com

Rehfeldt, G. L. 2006. A spline model of climate for the western United States. General Technical Report. RMRS-GTR-165. Fort Collins, Colorado: U.S. Department of Agriculture, Forest Service, Rocky Mountain Research Station. 21 p. Online at: www.fs.fed.us/rm/pubs/rmrs_gtr165.pdf

Ruefenacht, B., M. V. Finco, M. D. Nelson, R. Czaplewski, El H. Helmer, J. Blackard, G. R. Holden, A. J. Lister, D. Salajanu, D. Weyermann, and K. Winterberger. 2008. Conterminous U.S. and Alaska forest type mapping using forest inventory and analysis data. *Photogrammetric Engineering and Remote Sensing*. 74(11): 1379–1388. Online at: www.nrs.fs.fed.us/pubs/jrnl/2008/nrs_2008_ruefenacht_001.pdf

Sargent, C. S. 1884. Report on the forests of North America (exclusive of Mexico). Washington. Government printing office.

Schaetzl, R. J., F. J. Krist, Jr., K. Stanley, and C. M. Hupy. 2009. The natural soil drainage index: an ordinal estimate of longterm soil wetness. *Physical Geography*. 30: 383–409. Online at: http://foresthealth.fs.usda.gov/soils/Content/Downloads/DI_paper.pdf

Schaetzl, R. J., F. J. Krist, Jr., B. A. Miller. 2012. A taxonomically based, ordinal estimate of soil productivity for landscape-scale analyses. *Soil Science*. 177: 288–299. Online at: http://foresthealth.fs.usda.gov/soils/Content/Downloads/PIsoilscience.pdf

Soil Survey Staff. 2011. Natural Resources Conservation Service, United States Department of Agriculture Soil Survey Geographic (SSURGO) Database. 15 June 2011. Online at: http://soildatamart.nrcs.usda.gov

USDA, NRCS. 2014. The PLANTS Database. 12 June 2014. National Plant Data Team, Greensboro, North Carolina 27401-4901 USA. Online at: http://plants.usda.gov

U.S. Geological Survey. 1999. Digital representation of Atlas of United States Trees, Elbert L. Little, Jr.

Viereck, L. A. and E. L. Little, Jr. 1975. Atlas of United States trees, volume 2, Alaska trees and common shrubs: U.S. Department of Agriculture Miscellaneous Publication 1293. 82 maps.

Viereck, L. A. and E. L. Little, Jr. 2007. Alaska trees and shrubs. 2nd ed. Fairbanks. University of Alaska Press. 359 p.

Woudenberg, S. W., B. L. Conkling, B. M. O'Connell, E. B. LaPoint, J. Turner, and K. L. Waddell. 2010. The forest inventory and analysis database: database description and user's manual, version 4.0, for phase 2. General Technical Report. RMRS-GTR-245. Fort Collins, Colorado: U.S. Department of Agriculture, Forest Service, Rocky Mountain Research Station. 336 p. Online at: www. forestthreats.org/publications/rtp/rmrs_gtr245.pdf.

Zimmermann, N.E., T. C. Edwards, G. G. Moisen, T. S. Frescino, and J. A. Blackard. 2007. Remote Sensing-based predictors improve distribution models of rare, early successional and broadleaf tree species in Utah. *Applied Ecology*. 44:1057–1067.

www.ingramcontent.com/pod-product-compliance
Lightning Source LLC
Chambersburg PA
CBHW042337030426
42335CB00030B/3382